Quantum Trajectories

ATOMS, MOLECULES, AND CLUSTERS
Structure, Reactivity, and Dynamics
Series Editor: Pratim Kumar Chattaraj

Aromaticity and Metal Clusters
Pratim Kumar Chattaraj

Quantum Trajectories
Pratim Kumar Chattaraj

Atoms, Molecules, and Clusters
Structure, Reactivity, and Dynamics

Quantum Trajectories

EDITED BY
Pratim Kumar Chattaraj

CRC Press
Taylor & Francis Group
Boca Raton London New York

CRC Press is an imprint of the
Taylor & Francis Group, an **informa** business

CRC Press
Taylor & Francis Group
6000 Broken Sound Parkway NW, Suite 300
Boca Raton, FL 33487-2742

First issued in paperback 2017

ISBN-13: 978-1-4398-2561-7 (hbk)
ISBN-13: 978-1-138-11473-9 (pbk)

Library of Congress Cataloging-in-Publication Data

Quantum trajectories / editor, Pratim Kumar Chattaraj.
    p. cm.
    Summary: "Applying quantum mechanics to many-particle systems is an active area of research. The popularity of using quantum trajectories as a computational tool has exploded over the last decade, finally bringing this methodology to the level of practical application. This book explores this powerful tool to efficiently solve both static and time-dependent systems across a large area of quantum mechanics. Many of the pioneers who have either developed the subject from its inception or have improved on the subject contribute to this volume. They offer their insights on Lagrangian and Eulerian methods as well as various mixed quantum-classical techniques, among other topics"-- Provided by publisher.
    Includes bibliographical references and index.
    ISBN 978-1-4398-2561-7 (hardback)
    1. Quantum trajectories. I. Chattaraj, Pratim Kumar.

QC174.17.Q385Q36 2010
530.12--dc22
                                            2010035467

Visit the Taylor & Francis Web site at
http://www.taylorandfrancis.com

and the CRC Press Web site at
http://www.crcpress.com

Contents

Foreword

It is a characteristic of contemporary culture that the aim of much public discourse in quantum theory is not to educate but to mystify. One encounters a relentless message that "nobody understands quantum mechanics," that "an electron can be at two places at once," and that the world is ruled by "quantum weirdness." In this context the special claim of the scientist is just that she comprehends more deeply than the populace why she does not comprehend. It's a lucrative market and it is rarely mentioned that a rational explanation exists.

A currently popular approach that is often presented as if it is the official quantum worldview is the "many-worlds" interpretation. This interpretation asserts that everything that can happen according to quantum mechanics does happen, somewhere. The suggestion that an individual can have multiple versions may chime with the public but should such a view be afforded special status in science? It seems an extraordinarily inflated (and still to be proven consistent) solution to a comparatively mundane physics problem: achieving consistency between the superposition principle and the individuality of events. That is not to say that the puzzle is not daunting, but there should be proportion between problem and solution. A fashionable argument often adduced in support of the many-worlds option is Occam's razor, roughly the claim that the "less" the premises the "better" the theory. Even if such a complex business as the commensurability of theories could be formulated rigorously through counting concepts, it is not clear why this should be the exclusive criterion, or even a relevant one, in deciding the relative merit of theories, or why "less" means "better" (one can conceive of theories of gravity much simpler than general relativity that are wrong). For example, we might judge a theory's value by examining the range of questions it can answer. The real issue is the *quality of explanation*. A lesser count could, after all, mean that we are operating with an inadequate system of concepts relative to the full set of physical questions it is reasonable to pose.

This is, in fact, the starting point of the rational quantum mechanics initiated by de Broglie and Bohm, in which the origin of the conceptual dilemmas posed by quantum theory is located in the arbitrary rejection of the notion that a material system has a well-defined space–time trajectory. The insistence that the wavefunction provides the most complete physical description that is in principle possible has resulted in nearly a century of confusion. It is as if one removed a key organ from the body and struggled on in ill health insisting all the while that nothing was wrong. The simple step of retaining the trajectory gives the quantum trajectorician a powerful *constructive* tool to counter claims about the alleged incomprehensibility of the quantum world (only one of which is now needed). We shall mention three key benefits of the idea that offer clear advantages over other interpretations, such as many-worlds.

First, the trajectory opens a door—partially closed by Copenhagen—to a discourse of conceptual precision about *physics*. In giving physical meaning to the wavefunction the theory has recourse to a full range of physical concepts including energy, force, momentum, mass, and associated notions of stability, structure, size, and form. This allows a significant improvement in the clarity and consistency with which language is used in quantum theory, where concepts are now no longer just words attached to symbols but are set in correspondence with the actual state of matter. In one way or another, conventional quantum mechanics uses these terms and it is incumbent on a serious interpretation to provide a benchmark against which to measure the efficacy with which it does this. It is surely beneficial that we all agree on what we are talking about, as was unexceptional in pre-quantum physics. Central to this discourse is Bohm's perception that the novelty of quantum theory may be expressed through a new notion of energy. The quantum potential (whose form for two-slit interference is now an iconic figure in the subject) provides a post-mechanical mode of analysis where the interactions between the parts of a system are a function of the whole that contains them, a clear break with classical mechanism where the whole is no more than the sum of parts whose interactions are prescribed uniquely by their intrinsic properties. The relative magnitudes of the quantum potential and force can determine the limits of applicability of the classical description.

In his well-known essays on the foundations of quantum theory, Bell advanced a version of the de Broglie–Bohm theory that truncates its broad conceptual spectrum, the theory comprising just a trajectory determined by a first-order guidance law with an ensemble distributed according to the quantum formula. Certainly, if confronted by a sceptical audience, tactical necessity may well dictate such a pared-down account. Whether Bell regarded this emasculated version as the entire theory is not known but at this primitive stage in the development of trajectory theories such an Occamist stance would surely be misplaced if elevated to a principle. It is, of course, conceivable that, in the quantum context, the deeper conceptual content we have alluded to above may become inapplicable, or at least require modification. Such an analysis has, however, never been given. Conceptual abstinence may even encourage a reactionary mechanistic reading of de Broglie and Bohm that misses how far the old modes of thought have been transcended. For, as indicated above, this is not a theory just about trajectories. Being able to say more—about, say, the meaning of operator eigenvalues, or the physical content of a spinor field, or the energy changes that account for tunneling, or the forces responsible for molecular stability—enhances rather than diminishes our understanding and is hardly redundant knowledge. The particle trajectory realizes its full constructive role only in this wider historical context of explanation.

The original de Broglie–Bohm guidance equation occupies a position somewhat analogous to Einstein's equations in gravitational theory. It was the first, it is battle-tested, but it is only one of many possible laws that are empirically sufficient within the domain of phenomena to which the theory applies. In the gravitational case the contenders have been whittled down by consistency conditions and by enhanced experimental constraints. In contrast, a comparable quantum-theoretical analysis has begun only recently. Indeed, what the trajectory theory most needs is a universal and necessary founding principle, analogous to the principle of equivalence of gravitation and inertia. It is for this reason, perhaps, that this subject still lacks a generally agreed name that captures its essence, a necessary milestone in the full emergence of a scientific theory. And, of course, in the quantum case one cannot appeal to experiment since the predictions of the formalism being interpreted do not depend on the existence of the trajectory. To demonstrate the legitimacy of the idea we must envisage going beyond the conventional formalism. Fortunately, and this is its second benefit, the trajectory potentially provides a way to do this by allowing the precise formulation of questions concerning particle-like concepts such as speed and separation that are ambiguous in a pure-wave context (examples include tunneling times and chaos). Whether these ideas can be made empirical is not yet known. Advances in this direction will presumably entail restrictions on the range of valid trajectory laws.

The third benefit we shall mention, and arguably the most significant recent development in this field, is the observation that the trajectory is a constructive aid in a very practical sense: it can be made the basis of the quantum description in that the evolution of a physical system may be deduced from the dynamics of an ensemble of trajectories. This gives the trajectory theory a status somewhat analogous to Feynman's path integral technique but with a more comprehensible model. This insight has been propelled not by the foundations of the physics community (who are largely unaware of it) but by Robert Wyatt and the quantum chemists. In particular, it highlights the power of a particular analogy between quantum mechanics and hydrodynamics. Not only does the trajectory picture provide an alternative method of solving the wave equation, it may even enable one to arrive at novel facts about quantum dynamics, such as its symmetries, which otherwise would have been difficult to obtain.

The advantages of the trajectory outlook we have mentioned, and others, are strikingly exhibited in the contributions to this book. The breadth of topics attests to the vibrancy of the field. In view of the trajectory's proven value, a reasonable question to ask of conventional quantum treatments is: why has it been dropped?

Peter Holland
Green Templeton College
University of Oxford
Oxford, England, United Kingdom

Preface

The quantum dynamics of many-particle systems is traditionally understood by solving the time-dependent Schrödinger equation (TDSE) which involves the time evolution of the wavefunction, a function of $(3N + 1)$ variables. Attempts have been made since the birth of quantum mechanics to express the dynamics in three-dimensional space using quantities like charge- and current densities. In time-dependent density functional theory (TDDFT) the map between an arbitrary time-dependent (TD) potential and the density is uniquely invertible up to an additive TD function in the potential. This TDDFT allows us to formulate quantum fluid dynamics (QFD) wherein the quantum dynamics is described in terms of the flow of a probability fluid associated with the charge- and current densities by solving an equation of continuity and an Euler-type equation of motion. A quantum fluid density functional theory may also be envisaged in this context.

In order to understand the epistemological significance, one needs to know the quantum theory of motion (QTM) as well, in the sense of the classical interpretation of quantum mechanics as developed by de Broglie and Bohm. In QTM, the complete description of a physical system needs the simultaneous presence of the "wave" and the "particle" as opposed to conventional quantum mechanics involving "wave–particle duality" ("wave" or "particle" picture). The wave motion is governed by the solution of the TDSE, and the motion of a point particle guided by that wave for a given initial position is characterized by a velocity defined as the gradient of the phase of the wavefunction. An assembly of initial positions will constitute an ensemble of particle motions (so-called quantum trajectories or Bohmian trajectories) guided by the same wave, and the probability of having the particle in a given region of space at a given time is provided by the quantum mechanical TD probability density.

A crucial link between QTM and QFD is the quantum potential. In QTM, the particle experiences forces originating from both classical and quantum potentials, and in QFD, the fluid motion takes place under the influence of the external classical potential augmented by the quantum potential.

Apart from providing conceptual insights into many-body quantum dynamics, the quantum trajectory method provides an efficient on-the-fly computational methodology for solving both stationary and time-evolving states and hence encompassing a large area of quantum mechanics. Both Lagrangian and Eulerian techniques may be developed. Owing to the strong classical flavor in the de Broglie–Bohm representation of quantum mechanics, various mixed quantum–classical techniques have been tried. It also provides an easy way to analyze the quantum domain behavior of classically chaotic systems.

Several experts who have either developed the subject from inception or have improved upon the subject to take it to the state-of-the-art level have written chapters for this volume. The Foreword has been written by Professor Peter Holland, an authority in this subject. This book will be useful to graduate students and other research workers in the related fields of chemistry, physics, mathematics, and computer science.

I am grateful to all the authors who cooperated with me so that the book could be published on time. I would like to thank Professor Peter Holland, Lance Wobus and David Fausel, Santanab Giri and Soma Duley for their help in various ways. Finally, I must express my gratitude toward my wife Samhita and my daughter Saparya for their whole-hearted support.

Pratim Kumar Chattaraj
IIT Kharagpur, India

Editor

Pratim Kumar Chattaraj joined the faculty of the Indian Institute of Technology (Kharagpur) after obtaining his BSc and MSc degrees from Burdwan University and his PhD degree from the Indian Institute of Technology (Bombay). He is now a professor and the head of the Department of Chemistry and also the convener of the Center for Theoretical Studies there. He has visited the University of North Carolina (Chapel Hill) as a postdoctoral research associate and several other universities throughout the world as a visiting professor. Apart from teaching, Professor Chattaraj is involved in research on density functional theory, the theory of chemical reactivity, aromaticity in metal clusters, ab initio calculations, quantum trajectories, and nonlinear dynamics. He has been invited to deliver special lectures at several international conferences and to contribute chapters to many edited volumes. Professor Chattaraj is a member of the editorial board of the *Journal of Molecular Structure (Theochem)* and the *Journal of Chemical Science*, among others. He is a council member of the Chemical Research Society of India and a fellow of the Indian Academy of Sciences (Bangalore), the Indian National Science Academy (New Delhi), the National Academy of Sciences, India (Allahabad), and the West Bengal Academy of Science and Technology. He is a J.C. Bose National Fellow.

List of Contributors

J. Alberto Beswick
Laboratoire Collisions Agrégats
    Réactivité
IRSAMC, Université Paul Sabatier
Toulouse, France

Eric R. Bittner
Department of Chemistry
University of Houston
Houston, Texas

F. Borondo
Departamento de Química
Instituto Mixto de Ciencias
    Matemáticas
CSIC–UAM–UC3M–UCM
Universidad Autónoma de Madrid
Cantoblanco, Madrid, Spain

Gary E. Bowman
Department of Physics and Astronomy
Northern Arizona University
Flagstaff, Arizona

Irene Burghardt
Département de Chimie
Ecole Normale Supérieure
Paris, France

Pratim K. Chattaraj
Department of Chemistry and Center
    for Theoretical Studies
Indian Institute of Technology
Kharagpur, India

Chia-Chun Chou
Institute for Theoretical Chemistry and
    Department of Chemistry and
    Biochemistry
The University of Texas at Austin
Austin, Texas

D.-A. Deckert
Mathematisches Institut LMU
München, Germany

Arnaldo Donoso
Laboratorio de Física Estadística de
    Sistemas Desordenados
Centro de Física, Instituto Venezolano
    de Investigaciones Científicas
Caracas, Venezuela

Soma Duley
Department of Chemistry and Center for
    Theoretical Studies
Indian Institute of Technology
Kharagpur, India

D. Dürr
Mathematisches Institut LMU
München, Germany

Alon E. Faraggi
Department of Mathematical Sciences
University of Liverpool
Liverpool, United Kingdom

Edward R. Floyd
Coronado, California

Sophya Garashchuk
Department of Chemistry and
    Biochemistry
University of South Carolina
Columbia, South Carolina

Swapan K. Ghosh
Theoretical Chemistry Section
Bhabha Atomic Research Centre
Mumbai, India

Santanab Giri
Department of Chemistry and Center
  for Theoretical Studies
Indian Institute of Technology
Kharagpur, India

Sheldon Goldstein
Departments of Mathematics, Physics
  and Philosophy
Rutgers University
Piscataway, New Jersey

Gebhard Grübl
Theoretical Physics Institute
Universität Innsbruck
Innsbruck, Austria

Peter Holland
Green Templeton College
University of Oxford
Oxford, United Kingdom

Dipankar Home
Department of Physics
Center for Astroparticle Physics and
  Space Science
Bose Institute
Calcutta, India

Keith H. Hughes
School of Chemistry
Bangor University
Bangor, United Kingdom

Moncy V. John
Department of Physics
St. Thomas College
Kozhencherry, Kerala, India

Brian K. Kendrick
Theoretical Division
Los Alamos National Laboratory
Los Alamos, New Mexico

Munmun Khatua
Department of Chemistry and Center for
  Theoretical Studies
Indian Institute of Technology
Kharagpur, India

Donald J. Kouri
Department of Chemistry
University of Houston
Houston, Texas

Craig C. Martens
Department of Chemistry
University of California, Irvine
Irvine, California

Marco Matone
Department of Physics "G. Galilei"—
  Istituto
Nazionale di Fisica Nucleare
University of Padova
Via Marzolo, Padova, Italy

Christoph Meier
Laboratoire Collisions Agrégats
  Réactivité
IRSAMC, Université Paul Sabatier
Toulouse, France

Salvador Miret-Artés
Instituto de Física Fundamental
Consejo Superior de Investigaciones
  Científicas
Madrid, Spain

Alok Kumar Pan
Department of Physics
Center for Astroparticle Physics and
  Space Science
Bose Institute, Sector-V
Calcutta, India

Gérard Parlant
Institut Charles Gerhardt,
Université Montpellier 2
Montpellier, France

Markus Penz
Theoretical Physics Institute
Universität Innsbruck
Innsbruck, Austria

P. Pickl
Institut für Theoretische Physik
ETH Zürich, Switzerland

Bill Poirier
Department of Chemistry and
  Biochemistry, and Department of
  Physics
Texas Tech University
Lubbock, Texas

Vitaly Rassolov
Department of Chemistry and
  Biochemistry
University of South Carolina
Columbia, South Carolina

Àngel S. Sanz
Instituto de Física Fundamental
Consejo Superior de Investigaciones
  Científicas
Madrid, Spain

Utpal Sarkar
Department of Physics
Assam University
Silchar, India

Dmitrii V. Shalashilin
School of Chemistry
University of Leeds
Leeds, United Kingdom

Roderich Tumulka
Department of Mathematics
Rutgers University
Piscataway, New Jersey

Robert E. Wyatt
Department of Chemistry and
  Biochemistry
Institute for Theoretical Chemistry
The University of Texas at Austin
Austin, Texas

Tarik Yefsah
Laboratoire Kastler Brossel
Département de Physique de l'Ecole
  Normale Supérieure
Paris, France

Nino Zanghì
Dipartimento di Fisica
Università di Genova and INFN sezione
  di Genova
Genova, Italy

Yujun Zheng
School of Physics
Shandong University
Jinan, People's Republic of China

1 Bohmian Trajectories as the Foundation of Quantum Mechanics

Sheldon Goldstein, Roderich Tumulka, and Nino Zanghì

CONTENTS

Bohmian trajectories have been used for various purposes, including the numerical simulation of the time-dependent Schrödinger equation and the visualization of time-dependent wave functions. We review the purpose they were invented for: to serve as the foundation of quantum mechanics, i.e., to explain quantum mechanics in terms of a theory that is free of paradoxes and allows an understanding that is as clear as that of classical mechanics. Indeed, they succeed in serving that purpose in the context of a theory known as Bohmian mechanics, to which this article is an introduction.

1.1 BOHMIAN TRAJECTORIES

Let us consider a wave function $\psi_t(q)$ of non-relativistic quantum mechanics, defined on the configuration space $\mathbb{R}^{3N}$ of N particles, taking values in the set $\mathbb{C}$ of complex numbers, and evolving with time t according to the non-relativistic Schrödinger equation,

$$i\hbar \frac{\partial \psi_t}{\partial t} = -\sum_{k=1}^{N} \frac{\hbar^2}{2m_k} \nabla_k^2 \psi_t + V\psi_t, \qquad (1.1)$$

where m_k is the mass of the k-th particle, $\nabla_k = \nabla_{q_k} = \left(\frac{\partial}{\partial x_k}, \frac{\partial}{\partial y_k}, \frac{\partial}{\partial z_k}\right)$ is the derivative with respect to the coordinates of the k-th particle, and $V : \mathbb{R}^{3N} \to \mathbb{R}$ is a potential function, for example the Coulomb potential

$$V(q_1, \ldots, q_N) = \sum_{1 \leq j < k \leq N} \frac{e_j e_k}{|q_j - q_k|} \qquad (1.2)$$

with e_k the charge of the k-th particle.

With this wave function there is associated a family of trajectories in configuration space $\mathbb{R}^{3N}$, the *Bohmian trajectories*, which are defined to be those trajectories $t \mapsto Q(t) = (Q_1(t), \ldots, Q_N(t))$ satisfying the equation

$$\frac{d Q_k(t)}{dt} = \frac{\hbar}{m_k} \mathrm{Im} \frac{\nabla_k \psi_t}{\psi_t}(Q(t)). \qquad (1.3)$$

Put differently, to every wave function $\psi : \mathbb{R}^{3N} \to \mathbb{C}$ there is associated a vector field v^ψ on configuration space according to

$$v^\psi = (v_1^\psi, \ldots, v_N^\psi), \quad v_k^\psi = \frac{\hbar}{m_k} \mathrm{Im} \frac{\nabla_k \psi}{\psi}, \qquad (1.4)$$

and Equation 1.3 amounts to

$$\frac{d Q(t)}{dt} = v^{\psi_t}(Q(t)). \qquad (1.5)$$

Since Equation 1.5 is an ordinary differential equation (ODE) of first order (or rather, a system of $3N$ coupled ODEs of first order), it has, leaving aside the exceptions, a unique solution for every choice of initial configuration $Q(0)$.

Another way of writing Equation 1.5 is

$$\frac{d Q(t)}{dt} = \frac{j^{\psi_t}}{|\psi_t|^2}(Q(t)), \qquad (1.6)$$

where j^ψ is the vector field on $\mathbb{R}^{3N}$ usually called the *probability current* associated with the wave function ψ,

$$j^\psi = (j_1^\psi, \ldots, j_N^\psi), \quad j_k^\psi = \frac{\hbar}{m_k} \mathrm{Im}(\psi^* \nabla_k \psi). \qquad (1.7)$$

1.2 BOHMIAN MECHANICS

The theory that uses Bohmian trajectories as the foundation of quantum mechanics is known as Bohmian mechanics; it arises if we take a Bohmian trajectory seriously. Namely, Bohmian mechanics claims that in our world, electrons and other elementary particles have precise positions $Q_k(t) \in \mathbb{R}^3$ at every time t that move according to Equation 1.3. That is, for a certain Bohmian trajectory $t \mapsto Q(t)$ in configuration space, it claims that $Q(t) = (Q_1(t), \ldots, Q_N(t))$ is the configuration of particle positions in our world at time t.

This picture is in contrast with the orthodox view of quantum mechanics, according to which quantum particles do not have precise positions, but are regarded as "delocalized" to the extent to which the wave function ψ_t is spread out. It is also in contrast with another picture of the Bohmian trajectories that one often has in mind when using Bohmian trajectories for numerical purposes: the hydrodynamic picture. According to the latter, all the Bohmian trajectories associated with a given wave function (but corresponding to different $Q(0)$) are on an equal footing, none is more real than the others, they are all regarded as flow lines in analogy to the flow lines of a classical fluid. In Bohmian mechanics, however, only one of the Bohmian trajectories corresponds to reality, and all the other ones are no more than mathematical curves, representing possible alternative histories that could have occurred if the initial configuration of our world had been different, but did not occur.

As a consequence, talk of probability makes immediate sense in Bohmian mechanics but not in the hydrodynamic picture: In Bohmian mechanics, with only one trajectory realized, that trajectory may be random. In the hydrodynamic picture, with all trajectories equally real, it is not clear what a probability distribution over the trajectories could be the probability *of*, and what it could mean to say that a trajectory is random.

Bohmian mechanics was first proposed by Louis de Broglie (1892–1987) in the 1920s [5]; it is named after David Bohm (1917–1992), who was the first to realize that this theory provides a foundation for quantum mechanics [4]: The inhabitants of a typical Bohmian world would, as a consequence of the equations of Bohmian mechanics, observe exactly the probabilities predicted by the quantum formalism.

To understand how this comes about requires a rather subtle "quantum equilibrium" analysis [8] that is beyond the scope of this paper. An important element of the analysis is however rather simple. It is the property of *equivariance*, expressing the compatibility between the evolution of the wave function given by Schrödinger's equation and the evolution of the actual configuration given by the *guiding equation* 1.6. This property will be discussed in the next section.

The upshot of the quantum equilibrium analysis is the justification of the *probability postulate* for Bohmian mechanics, that the configuration Q of a system with wave function $\psi = \psi(q)$ is random with probability density $|\psi(q)|^2$. Bohmian mechanics, with the probability postulate, is *empirically equivalent* to standard quantum mechanics. We will return to this point later and explain how this follows from the equations.

1.3   EQUIVARIANCE

Equivariance amounts to the following assertion: If $Q(0)$ is random with probability density given by $|\psi_0|^2$, then $Q(t)$ is also random, with probability density given by $|\psi_t|^2$.

This is easy to see: Let ρ_t be the probability density of $Q(t)$. Then ρ_t evolves according to the continuity equation

$$\frac{\partial \rho_t}{\partial t} = -\mathrm{div}(\rho_t v^{\psi_t}), \tag{1.8}$$

whose right hand side is short for

$$-\sum_{k=1}^{N} \nabla_k \cdot (\rho_t v_k^{\psi_t}),$$

where $\cdot$ denotes the dot product of two vectors in $\mathbb{R}^3$. On the other hand, from the Schrödinger Equation 1.1

$$\frac{\partial}{\partial t}(\psi_t^* \psi_t) = -\mathrm{div}\, j^{\psi_t}, \tag{1.9}$$

which is the same equation as 1.8 with ρ_t replaced by $|\psi_t|^2$. Thus, if $\rho_0 = |\psi_0|^2$ then $\rho_t = |\psi_t|^2$ at any time t.

1.4   THE QUANTUM POTENTIAL

A particle trajectory $Q(t)$ can always be written in the form of Newton's law:

$$m\frac{d^2 Q(t)}{dt^2} = \mathrm{force}(t), \tag{1.10}$$

where m is the mass of the particle; after all, the right hand side can simply be so chosen as to make this equation true. It is sometimes useful to do this for the trajectories of the Bohmian particles; we find, by taking the time derivative of Equation 1.3 and after some calculation:

$$m_k\frac{d^2 Q_k(t)}{dt^2} = -\nabla_k(V + V_{qu}^{\psi_t})(Q(t)), \tag{1.11}$$

where V_{qu}^{ψ} is a function called the *quantum potential*,

$$V_{qu}^{\psi} = -\sum_{j=1}^{N} \frac{\hbar^2}{2m_j} \frac{\nabla_j^2 |\psi|}{|\psi|}. \tag{1.12}$$

For comparison, a classical particle would move according to

$$m_k\frac{d^2 Q_k(t)}{dt^2} = -\nabla_k V(Q(t)). \tag{1.13}$$

That is, a Bohmian trajectory is also a solution to classical mechanics if we add a suitable time-dependent term $V_{qu}^{\psi_t}$ to the potential function V. One particular application of the quantum potential arises in the study of the classical limit of quantum mechanics [1]: As we see from Equations 1.11 and 1.13, the regime in which Bohmian trajectories agree with classical trajectories is characterized by the condition that the gradient of the quantum potential vanishes, or at least, for approximate agreement, is small.

Several things have been, or may be, puzzling about the quantum potential. In David Bohm's 1952 article [4] on Bohmian mechanics (of course, Bohm did not refer to this theory as "Bohmian mechanics"), he presented Equation 1.11 together with Equation 1.12, rather than Equation 1.3, as the basic equation of motion. That created a sense of mystery because it does not appear natural to postulate the existence of an additional potential without specifying which physical object causes this potential, and how and why. Moreover, since the Formula 1.12 is neither obvious nor natural, it created the impression that Bohmian mechanics was a contrived, artificial theory. These unnecessary difficulties vanish when we regard Equation 1.3 as the equation of motion because then we do not just add another term to Equation 1.13 but replace it instead with an equation that is altogether different but equally simple. Furthermore, Equation 1.11 is mathematically not equivalent to Equation 1.3: while every solution of Equation 1.3 is a solution of Equation 1.11, the converse is not true, as Equation 1.11 is a second-order equation that provides a solution for every choice of initial positions and velocities, including choices for which the initial velocities fail to be related to the initial positions in accordance with Equation 1.3. Bohm introduced, in order to exclude these further solutions of Equation 1.11, a constraint condition on the possible velocities—and the condition was Equation 1.3! In fact, it follows from the fact that Equation 1.11 can be obtained from Equation 1.3 that if a solution of the second-order equation 1.11 has initial velocities satisfying Equation 1.3 then also the velocities at any other time will satisfy Equation 1.3. But then Equation 1.3 is satisfied at all times, and instead of calling it a constraint we can simply call it the equation of motion. To call it a constraint just creates another unnecessary mystery: why should nature impose such a constraint? Would it not be simpler to have only Equation 1.11? Not, of course, if the alternative is to have only Equation 1.3.

1.5 CONNECTION WITH NUMERICAL METHODS

In our definition Equation 1.3 of the Bohmian trajectories, we used ψ_t for every t, supposing that the Schrödinger equation has been solved already. It may thus seem surprising that the Bohmian trajectories can be used for solving the Schrödinger equation (up to a global phase factor). But the second-order Equation 1.11 suggests how that can be done, roughly as follows [9].

Suppose we know the initial wave function ψ_0. Choose an ensemble of points (say $Q^{(1)}, \ldots, Q^{(n)}$, with very large n) in configuration space; note the difference between N points in physical space $\mathbb{R}^3$ (which is one configuration) and n points in configuration space $\mathbb{R}^{3N}$ (which correspond to nN points in $\mathbb{R}^3$). Choose the ensemble so that its distribution density in configuration space is, with sufficient accuracy, $\rho_0 = |\psi_0|^2$. For each $Q^{(i)}$, determine $v^{\psi_0}(Q^{(i)})$, the $3N$-vector of velocities, from the velocity law Equation 1.4. For each i, solve the second-order equation of motion (Equation 1.11) with initial positions as in $Q^{(i)}$, initial velocities as in $v^{\psi_0}(Q^{(i)})$, and the quantum potential as defined in Equation 1.12 but with $|\psi_t|$ replaced by $\sqrt{\rho_t}$, where ρ_t is the density in configuration space of the ensemble $Q^{(1)}(t), \ldots, Q^{(n)}(t)$. That is, for every time step $t \to t + \delta t$, determine the quantum potential at time t from the density ρ_t of the ensemble points according to

$$V_{qu}(t) = -\sum_{j=1}^{N} \frac{\hbar^2}{2m_j} \frac{\nabla_j^2 \sqrt{\rho_t}}{\sqrt{\rho_t}}, \qquad (1.14)$$

then use Equation 1.11 to propagate $Q^{(i)}(t)$ and $\frac{dQ^{(i)}}{dt}(t)$ by one-time step and obtain $Q^{(i)}(t + \delta t)$ and $\frac{dQ^{(i)}}{dt}(t + \delta t)$.

Equivariance, together with the fact that Equation 1.11 follows from the Schrödinger Equation 1.1 and the equation of motion Equation 1.3, implies that this method would yield exactly the family of Bohmian trajectories if we used infinitely many sample points with an initial distribution given *exactly* by $|\psi_0|^2$, if the time step δt were infinitesimal, and if no numerical error were involved in solving Equation 1.11. With finite n and finite δt, we may obtain an approximation to the family of Bohmian trajectories.

Once we have the trajectories, we can (more or less) recover the wave function $\psi_t(q)$ up to a time-dependent phase factor as follows. Note that the right hand side of the equation of motion 1.3 is proportional to the q_k-derivative of the phase of the wave function; i.e., if we write

$$\psi(q) = |\psi(q)| e^{iS(q)/\hbar} \qquad (1.15)$$

with a real-valued function S then

$$v_k^\psi(q) = \frac{1}{m_k} \nabla_k S. \qquad (1.16)$$

Now, if we know all Bohmian trajectories then we can read-off the $3N$-velocity $v(q,t) = (v_1(q,t), \ldots, v_N(q,t))$ of the trajectory that passes through $q \in \mathbb{R}^{3N}$ at time t, solve

$$v_k(q,t) = \frac{1}{m_k} \nabla_k S(q,t) \qquad (1.17)$$

for $S(q,t)$, and finally set

$$\psi_t(q) = \sqrt{\rho_t(q)} e^{iS(q,t)/\hbar}. \qquad (1.18)$$

Note that Equation 1.17 determines the function $q \mapsto S(q,t)$ up to a real constant $\theta(t)$; i.e., any $S(q,t) + \theta(t)$ is another solution of Equation 1.17 and, conversely, if $S_1(q,t)$ and $S_2(q,t)$ are two solutions of Equation 1.17 then $S_2(q,t) - S_1(q,t)$ is (real and) independent of q and can thus be called $\theta(t)$. As a consequence, $\psi_t(q)$ has been determined up to a (global, i.e., q-independent) phase factor $e^{i\theta(t)}$.

1.6 THE QUANTUM POTENTIAL AGAIN

The previous section has illustrated how the Second-Order Equation 1.11 and the concept of the quantum potential can be useful even if we regard the First-Order Equation 1.3 as the fundamental equation of motion. But the algorithm outlined there leads to another puzzle about the quantum potential: It involves a picture in which an ensemble of points in configuration space, with density ρ_t, leads to a quantum potential via Equation 1.14, which in turn acts on every trajectory. This may suggest that this ensemble of points in configuration space is the physical *cause* of some kind of real field, the quantum potential, which in turn is the physical *cause* of the shape of the individual trajectory, in particular of its deviation from a classical trajectory. This picture, however, requires that all Bohmian trajectories be physically real, in agreement with the hydrodynamic picture mentioned before, but in conflict with Bohmian mechanics as described before, the theory asserting that only one of the trajectories is real while the others are merely hypothetical. But how could merely hypothetical trajectories push the actual particles around? They cannot. Bohmian mechanics is incompatible with the picture that the density of trajectories causes a quantum potential that pushes in turn every trajectory

1.7 WAVE–PARTICLE DUALITY

So what picture arises instead from Bohmian mechanics? The object that influences the motion of the one actual configuration is the wave function. Note, however, that we should not assume that the configuration would "normally" move along a classical trajectory unless some physical agent (be it other trajectories, the quantum potential, or the wave function) pushed it off to another trajectory; rather, the classical equation of motion 1.13 is replaced by a new equation of motion 1.3, and this new equation does not talk about forces, about pushing, or about causes, but merely defines the trajectory in terms of the wave function. By virtue of the very purpose that it was designed for, the numerical algorithm above avoids referring to the wave function; after all, it is an algorithm for *finding* the wave function. However, for a theory such as Bohmian mechanics (i.e., for a proposal as to how nature might work), it is acceptable to suppose that nature solves the Schrödinger equation independently of trajectories, and then lets the one trajectory depend on the wave function. This is the picture that arises if we insist that only one trajectory is real.

As a consequence, we need to take the wave function seriously as a physical object. Put differently, in a world governed by Bohmian mechanics, there is a wave–particle duality in the literal sense: there is a wave (ψ on $\mathbb{R}^{3N}$), and there are particles (at $Q_1, \ldots, Q_N$). The wave evolves according to the Schrödinger Equation 1.1, and the particles move in a way that depends on the wave, namely according to Equation 1.3. Put differently, the wave guides, or pilots, the particles; that is why this theory has also been called the *pilot-wave theory*.

1.8   THE PHASE FUNCTION $S(q, t)$

Above in Section 1.5, we have made use of writing the wave function in terms of its modulus, often denoted $R(q, t) = |\psi(q, t)|$, and phase function, often denoted $S(q, t)/\hbar$,

$$\psi(q, t) = R(q, t)\, e^{iS(q,t)/\hbar}. \tag{1.19}$$

There are some difficulties with this decomposition, though, and maybe this is a good place to make the reader aware of them.

The first difficulty is that the value of $S(q, t)$ is not uniquely determined by Equation 1.19, but only up to addition of an integer multiple of $2\pi\hbar$. Of course, we can simply choose one of the possible values, but it is not always possible to stick with that choice; more precisely, it is not always possible to choose S as a continuous function, even though ψ is continuous. An example can be found among the eigenstates of the hydrogen atom, which are known to factorize in spherical coordinates into a function of the radial coordinate r, a function of the colatitude θ, and a function of the azimuth φ; what matters here is that the last factor is $e^{im\varphi}$ with integer m, which contributes a summand $m\varphi\hbar$ to the S function. For $m \neq 0$, this S function is discontinuous, as it jumps from $m2\pi\hbar$ to 0 at $\varphi = 2\pi$, while ψ is continuous. (In fact, every choice of the S function will be discontinuous, since for any fixed r and θ such that $\psi(r, \theta, \varphi) \neq 0$,

$$\frac{\partial S}{\partial \varphi} = \hbar\, \mathrm{Im}\Big(\frac{1}{\psi}\frac{\partial \psi}{\partial \varphi}\Big) = m\hbar, \tag{1.20}$$

so S has to grow linearly with φ.)

Note that, at the discontinuity, the S function cannot jump by an arbitrary amount, but only by an integer multiple of $2\pi\hbar$, and that S is not defined where $\psi = 0$. As a consequence, the correspondence between a complex-valued function ψ and the two real-valued functions R and S is a bit complicated. And as a consequence of *that*, the usual pair of real equations for R and S that can be obtained from the Schrödinger equation,

$$\frac{\partial R^2}{\partial t} = -\sum_{k=1}^{N} \nabla_k \cdot \Big(R^2 \frac{\nabla_k S}{m_k}\Big) \tag{1.21}$$

$$\frac{\partial S}{\partial t} = \sum_{k=1}^{N} \Big(\frac{\hbar^2}{2m_k}\frac{\nabla_k^2 R}{R} - \frac{(\nabla_k S)^2}{2m_k}\Big) - V, \tag{1.22}$$

are actually *not* equivalent to Schrödinger's equation. Explicitly, if we started from Equations 1.21 and 1.22 then we would have no reason to allow S to be undefined where $R = 0$; we would have no reason to expect discontinuities in S; and if we allowed discontinuities then we would have no reason to demand that the jump height is an integer multiple of $2\pi\hbar$.

1.9 PREDICTIONS AND THE QUANTUM FORMALISM

Consider a hypothetical world governed by Bohmian mechanics, and let us call this a Bohmian world. We have mentioned already in Section 1.2 that the inhabitants of a Bohmian world would observe exactly the probabilities predicted by the quantum formalism. In this section, we outline why this is so.

Consider an experiment carried out by an observer. In a Bohmian world, of course, also observers and their apparatuses (the detectors, cameras, photographs, display screens, meter pointers, etc.) consist of particles governed by the equations of Bohmian mechanics. For this reason, let us for a moment consider the N-particle system formed by both the object of the experiment and the apparatus. Let us write the configuration of our system as $Q = (X, Y) \in \mathbb{R}^{3N}$ with $X \in \mathbb{R}^{3K}$ the configuration of the object and $Y \in \mathbb{R}^{3L}$ the configuration of the apparatus. The number $N = K + L$ will be huge because L is, in fact usually $L > 10^{23}$, as the apparatus is a macroscopic system; K, in contrast, may be just 1. Correspondingly, we write the wave function Ψ of this N-particle system as

$$\Psi(q) = \Psi(x, y). \tag{1.23}$$

Suppose that, at the time t_0 at which the experiment begins, the wave function factorizes,

$$\Psi_{t_0}(x, y) = \psi(x)\phi(y) \tag{1.24}$$

with ψ the wave function of the object at time t_0 and ϕ the initial state ("ready state") of the apparatus, $\langle \psi | \psi \rangle = 1 = \langle \phi | \phi \rangle$. (Actually, the symmetrization postulate implies that Ψ_{t_0} cannot factorize in this way if both the object and the apparatus contain particles of the same species, say, if both contain electrons. A treatment that takes the symmetrization postulate into account leads to the same conclusions but is harder to follow, and we prefer to simplify the discussion.)

Suppose the experiment is over at time t_1. The wave function at that time is, of course, given by

$$\Psi_{t_1} = e^{-iH(t_1 - t_0)/\hbar} \Psi_{t_0} \tag{1.25}$$

with H the Hamiltonian of the N-particle system. (We are assuming, for simplicity, that the system is isolated during the experiment; this is not a big assumption since we could make the N-particle system as large as we want, even comprising the entire universe.)

Suppose further that for certain wave functions $\psi_\alpha(x)$ that the object might have, the apparatus will yield a predictable result r_α. More precisely, suppose that if $\psi = \psi_\alpha$ in Equation 1.24 then the final wave function Ψ_{t_1} is concentrated on the set S_α of those y-configurations in which the apparatus' pointer points to the value r_α,

$$\int_{\mathbb{R}^{3K}} dx \int_{S_\alpha} dy |\Psi_{t_1}(x, y)|^2 = 1. \tag{1.26}$$

(For example, standard quantum mechanics asserts that this is the case when the experiment is a "quantum measurement of the observable with operator A" and ψ_α is an eigenfunction of A with eigenvalue r_α. In general, an analysis of the experiment shows whether this is the case.)

Let us write $\Psi^{(\alpha)}$ for Ψ_{t_1} arising from $\psi = \psi_\alpha$, i.e.,

$$\Psi^{(\alpha)} = e^{-iH(t_1-t_0)/\hbar}(\psi_\alpha\phi). \tag{1.27}$$

It now follows from the linearity of the Schrödinger equation that if the wave function of the object is a (non-trivial) linear combination of the ψ_α,

$$\psi = \sum_\alpha c_\alpha\psi_\alpha \tag{1.28}$$

then

$$\Psi_{t_1} = \sum_\alpha c_\alpha\Psi^{(\alpha)}, \tag{1.29}$$

which is a (non-trivial) superposition of different wave functions that correspond to different outcomes (and macroscopically different orientations of the pointer). It is known as the *measurement problem of quantum mechanics* that this wave function, the wave function of the object and the apparatus together after the experiment as determined by the Schrödinger equation, does not single out one of the r_α's as the actual outcome of the experiment.

Since Bohmian mechanics assumes the Schrödinger equation, Equation 1.29 is the correct wave function in Bohmian mechanics. Moreover, the configuration $Q(t_1) = (X(t_1), Y(t_1))$ is, by the probability postulate and equivariance, random with probability density $|\Psi_{t_1}|^2$. As a consequence, Y_{t_1} lies in the set S_α with probability

$$\int_{\mathbb{R}^{3K}} dx \int_{S_\alpha} dy |\Psi_{t_1}(x,y)|^2 = \int_{\mathbb{R}^{3K}} dx \int_{S_\alpha} dy |c_\alpha|^2 |\Psi^{(\alpha)}(x,y)|^2 = |c_\alpha|^2 \tag{1.30}$$

because $\Psi^{(\beta)}(x,y) = 0$ for $y \in S_\alpha$ and $\beta \neq \alpha$. But that Y_{t_1} lies in the set S_α means that the pointer is pointing to the value r_α. Thus, in Bohmian mechanics the apparatus (consisting of Bohmian particles) does point to a certain value, and the value is always one of the r_α's (the same values as provided by the quantum formalism), and the value is random, and the probability it is r_α is $|c_\alpha|^2$ (the same probability as provided by the quantum formalism).

1.10 THE GENERALIZED QUANTUM FORMALISM

The above example illustrates why Bohmian mechanics predicts the same probabilities for the results of quantum measurements as the standard quantum formalism. Let us see what we obtain if we drop the assumption that for $\psi = \psi_\alpha$ the experiment yields a predictable result r_α. It is still true, then, that if the configuration Y_{t_1} of the apparatus lies in the set $S_\alpha \subset \mathbb{R}^{3L}$ then the pointer points to the value r_α, the result of the experiment. The probability that that happens is

$$p_\alpha := \int_{\mathbb{R}^{3K}} dx \int_{S_\alpha} dy |\Psi_{t_1}(x, y)|^2. \tag{1.31}$$

A calculation then shows that there is a positive self-adjoint (and uniquely determined) operator E_α such that

$$p_\alpha = \langle \psi_{t_0} | E_\alpha | \psi_{t_0} \rangle. \tag{1.32}$$

Indeed,

$$E_\alpha = \langle \phi | e^{iH(t_1 - t_0)/\hbar}(I \otimes 1_{S_\alpha}) e^{-iH(t_1 - t_0)/\hbar} | \phi \rangle_y, \tag{1.33}$$

where $\langle \cdot | \cdot \rangle_y$ is the partial scalar product taken only over y but not over x, I is the identity operator (in this case, on the x-Hilbert space), and 1_{S_α} is the operator that multiplies by the characteristic function of the set S_α. If the sum of the p_α is 1 for every ψ_{t_0} (which is the case if we can neglect the possibility that the experiment fails to yield any result, and if we have introduced sufficiently many sets S_α so as to cover all possible results r_α) then

$$\sum_\alpha E_\alpha = 1. \tag{1.34}$$

A family of positive operators $\{E_\alpha\}$ obeying Equation 1.34 is called a *positive-operator-valued measure (POVM)*, and the rule that the probability of the result r_α is given by Equation 1.32 is part of the *generalized quantum formalism*. In case each E_α is a projection operator, we obtain back the usual rules of quantum measurement, as the operator

$$A = \sum_\alpha r_\alpha E_\alpha \tag{1.35}$$

is self-adjoint with eigenvalues r_α, and E_α is the projection to the eigenspace of A with eigenvalue r_α.

1.11 WHAT IS UNSATISFACTORY ABOUT STANDARD QUANTUM MECHANICS?

Many physicists, beginning with Einstein and Schrödinger and including the authors, have felt that standard quantum mechanics is not satisfactory as a physical theory. This is because the axioms of standard quantum mechanics concern the results an observer will obtain if he performs a certain experiment. We think that a fundamental physical theory should not be formulated in terms of concepts like "observer" or "experiment," as these concepts are very vague and certainly do not seem fundamental. Is a cat an observer? A computer? Were there any experiments before life existed on Earth?

Instead, a fundamental physical theory should be formulated in terms of rather simple physical objects in space and time like fields, particles, or perhaps strings. Bohmian mechanics is a beautiful example of a theory that is satisfactory as a fundamental physical theory.

It is a frequent misunderstanding that the main problem that the critics of standard quantum mechanics have is that it is different from classical mechanics. A fundamental physical theory can very well be different from classical mechanics, but we should demand that it be as *clear* as classical mechanics. We think that, for it to make clear sense as physics, it must describe matter moving in space. Standard quantum mechanics does not do that, but Bohmian mechanics does, as it describes the motion of point particles. And, indeed, Bohmian mechanics is different from classical mechanics in many crucial respects. Another frequent misunderstanding is that the goal of Bohmian mechanics is to return as much as possible to classical mechanics. The circumstance that it uses point particles and is deterministic does not mean that we are dogmatically committed to point particles or determinism; also indeterministic theories with another ontology may very well be satisfactory. It just so happens that the simplest satisfactory version of quantum mechanics, Bohmian mechanics, involves point particles and determinism.

It is also a misunderstanding to think that the goal of Bohmian mechanics was to *derive* the Schrödinger equation, or to *replace* it with something else. Of course, a physical theory may involve new postulates and introduce new equations, and we see no problem with the Schrödinger equation. Rather, the goal is to replace the *measurement postulate* of standard quantum mechanics with postulates that refer to electrons and nuclei instead of observers, axioms from which the measurement rules can be derived as theorems.

After these more philosophical themes, let us finally turn to two more technical topics: how to incorporate spin and identical particles into Bohmian mechanics.

1.12 SPIN

In order to treat particles with spin, almost no change in the defining equations of Bohmian mechanics is necessary. Recall that the wave function of a spin-$\frac{1}{2}$ particle can be regarded as a function $\psi : \mathbb{R}^3 \to \mathbb{C}^2$, and for N such particles as $\psi : \mathbb{R}^{3N} \to \mathbb{C}^{2^N}$. That is, ψ is now a multi-component (or vector-valued) function. We keep the form (Equation 1.6), i.e., $dQ/dt = j^\psi/|\psi|^2$, of the equation of motion, with the appropriate expressions for j^ψ and $|\psi|^2$ in terms of a multi-component function ψ; viz., we postulate, as the equation of motion [2],

$$\frac{d\boldsymbol{Q}_k(t)}{dt} = \frac{(\hbar/m_k)\mathrm{Im}(\psi_t^\dagger \nabla_k \psi_t)}{\psi_t^\dagger \psi_t}(Q(t)), \qquad (1.36)$$

where

$$\phi^\dagger \psi = \sum_{s=1}^{2^N} \phi_s^* \psi_s \qquad (1.37)$$

is the scalar product in $\mathbb{C}^{2^N}$. Particles with spin other than $\frac{1}{2}$ can be treated in a similar way.

If we introduce an external magnetic field B, the Schrödinger equation needs to be modified appropriately, i.e., replaced (as usual) with the Pauli equation

$$i\hbar\frac{\partial\psi_t}{\partial t} = -\sum_{k=1}^{N}\frac{\hbar^2}{2m_k}\left(\nabla_k - ie_k A(q_k)\right)^2\psi_t + \sum_{k=1}^{N}\mu_k B(q_k)\cdot\sigma_k\psi_t + V\psi_t, \quad (1.38)$$

where A is the vector potential, e_k and μ_k are the charge and the magnetic moment of the k-th particle, $\sigma = (\sigma_x, \sigma_y, \sigma_z)$ is the vector consisting of the three Pauli spin matrices, and σ_k acts on the spin index $s_k \in \{+1, -1\}$ that is associated with the k-th particle when we regard the spin component index s in Equation 1.37 as a multi-index, $s = (s_1, \ldots, s_N)$.

In this theory, since particles are not literally spinning (i.e., not rotating), the word "spin" is an anachronism like the pre-Copernican word *sunrise*. What may be more surprising about the formulation of Bohmian mechanics for particles with spin is that we did not have to introduce further ("hidden") variables, on the same footing as the positions $Q_k(t)$, and that there is no direction of space in which the "spin vector" is "actually" pointing. In particular, a Stern–Gerlach experiment does not *measure* the value of an additional spin variable. How can that be? In a Stern–Gerlach experiment, we arrange an external magnetic field for the particle to pass through, and measure its position afterwards; depending on where the particle was detected, we say that we obtained the result "up" or "down." Using equivariance, which holds for Equation 1.36 just as it does for Equation 1.6, one easily sees that this experiment yields, in Bohmian mechanics, a random result that has the same probability distribution as predicted by standard quantum mechanics. Looking at the experiment in this way, what is happening is completely clear and it is certainly not necessary to postulate that the particle has an actual spin vector, before the experiment or after it.

1.13 THE SYMMETRIZATION POSTULATE

Our description of Bohmian trajectories in Section 1.1 did not include the proper treatment of identical particles. The *symmetrization postulate* of quantum mechanics asserts that if the variables q_i and $q_j \in \mathbb{R}^3$ in the wave function ψ refer to two identical particles (i.e., two particles of the same species, such as, e.g., two electrons) then ψ is either *symmetric* in q_i and q_j,

$$\psi(\ldots q_i \ldots q_j \ldots) = \psi(\ldots q_j \ldots q_i \ldots), \quad (1.39)$$

if the species is *bosonic*, or *anti-symmetric* in q_i and q_j,

$$\psi(\ldots q_i \ldots q_j \ldots) = -\psi(\ldots q_j \ldots q_i \ldots), \quad (1.40)$$

if the species is *fermionic*. In Equations 1.39 and 1.40 it is understood that all other variables remain unchanged, only the variables q_i and q_j get interchanged. The *spin-statistics rule* asserts that every species with integer spin $(0, 1, 2, \ldots)$ is bosonic and every species with half-odd spin $(\frac{1}{2}, \frac{3}{2}, \ldots)$ is fermionic.

In Bohmian mechanics for identical particles, one uses the same type of wave function as in ordinary quantum mechanics, and the same equation of motion (and Schrödinger equation) as in Bohmian mechanics for distinguishable particles. That is, we include the symmetrization postulate among the postulates of Bohmian mechanics.

It is a traditional claim in textbooks on quantum mechanics that the lack of precise trajectories in orthodox quantum mechanics is the reason for the symmetrization postulate. If the particles had trajectories, it is suggested, then they would automatically be distinguishable. From Bohmian mechanics with the symmetrization postulate we see that this suggestion is incorrect.

On the contrary, the Bohmian trajectories actually *enhance* our understanding of the symmetrization postulate. We start from the following observation:

$$\text{If } (\ldots \boldsymbol{Q}_i(0) \ldots \boldsymbol{Q}_j(0) \ldots) \text{ evolves to } (\ldots \boldsymbol{Q}_i(t) \ldots \boldsymbol{Q}_j(t) \ldots)$$
$$\text{then } (\ldots \boldsymbol{Q}_j(0) \ldots \boldsymbol{Q}_i(0) \ldots) \text{ evolves to } (\ldots \boldsymbol{Q}_j(t) \ldots \boldsymbol{Q}_i(t) \ldots). \tag{1.41}$$

In other words, it is unnecessary to specify the *labelling* of the particles. That seems very appropriate as the labelling is unphysical. The fact expressed by Equation 1.41 follows from the symmetry of the velocity vector field,

$$\boldsymbol{v}_i(\ldots \boldsymbol{q}_i \ldots \boldsymbol{q}_j \ldots) = \boldsymbol{v}_j(\ldots \boldsymbol{q}_j \ldots \boldsymbol{q}_i \ldots), \tag{1.42}$$
$$\boldsymbol{v}_k(\ldots \boldsymbol{q}_i \ldots \boldsymbol{q}_j \ldots) = \boldsymbol{v}_k(\ldots \boldsymbol{q}_i \ldots \boldsymbol{q}_j \ldots) \quad \text{for } i \neq k \neq j, \tag{1.43}$$

which can easily be checked from the formula 1.4 for $\boldsymbol{v}_i$ together with Equations 1.39 or 1.40. For another way of putting the fact expressed by Equation 1.41, let us consider a system of N identical particles. The natural configuration space is

$$^N\mathbb{R}^3 = \{Q \subset \mathbb{R}^3 : \#Q = N\}, \tag{1.44}$$

the set of all N-element subsets of $\mathbb{R}^3$. While an element of $\mathbb{R}^{3N}$ is an *ordered* configuration $(\boldsymbol{Q}_1, \ldots, \boldsymbol{Q}_N)$, an element of $^N\mathbb{R}^3$ is an *unordered* configuration $\{\boldsymbol{Q}_1, \ldots, \boldsymbol{Q}_N\}$. Since the labels need not be specified, any point $Q(0) \in {}^N\mathbb{R}^3$ as initial condition will uniquely define a curve $t \mapsto Q(t) \in {}^N\mathbb{R}^3$. So for symmetric or anti-symmetric wave functions, Bohmian mechanics works on the natural configuration space of identical particles. This fact can be regarded as something like an explanation, or derivation, of the symmetrization postulate. For a deeper discussion see [7].

1.14   FURTHER READING

Concerning the extension of Bohmian mechanics to quantum field theory, see [6, 10]. Concerning the extension of Bohmian mechanics to relativistic space-time, see the review article [11] and references therein. Concerning quantum non-locality, see Bell's book [3], which we highly recommend also about foundations of quantum mechanics in general.

BIBLIOGRAPHY

1. V. Allori, D. Dürr, S. Goldstein, and N. Zanghì: Seven Steps Towards the Classical World. *Journal of Optics B* **4**: 482–488 (2002). http://arxiv.org/abs/quant-ph/0112005.
2. J. S. Bell: On the Problem of Hidden Variables in Quantum Mechanics. *Reviews of Modern Physics* **38**: 447–452 (1966). Reprinted as chapter 1 of [3].
3. J. S. Bell: *Speakable and Unspeakable in Quantum Mechanics*. Cambridge: University Press (1987).
4. D. Bohm: A Suggested Interpretation of the Quantum Theory in Terms of "Hidden" Variables, I and II. *Physical Review* **85**: 166–193 (1952).
5. L. de Broglie: in *Electrons et Photons: Rapports et Discussions du Cinquième Conseil de Physique tenu à Bruxelles du 24 au 29 Octobre 1927 sous les Auspices de l'Institut International de Physique Solvay*, Paris: Gauthier-Villars (1928). English translation in G. Bacciagaluppi and A. Valentini: *Quantum Theory at the Crossroads: Reconsidering the 1927 Solvay Conference*, Cambridge: University Press (2009). http://arxiv.org/abs/quant-ph/0609184.
6. D. Dürr, S. Goldstein, R. Tumulka, and N. Zanghì: Bohmian Mechanics and Quantum Field Theory. *Physical Review Letters* **93**: 090402 (2004). http://arxiv.org/abs/quant-ph/0303156.
7. D. Dürr, S. Goldstein, J. Taylor, R. Tumulka, and N. Zanghì: Quantum Mechanics in Multiply-Connected Spaces. *Journal of Physics A: Mathematical and Theoretical* **40**: 2997–3031 (2007). http://arxiv.org/abs/quant-ph/0506173.
8. D. Dürr, S. Goldstein, and N. Zanghì: Quantum Equilibrium and the Origin of Absolute Uncertainty. *Journal of Statistical Physics* **67**: 843–907 (1992). http://arxiv.org/abs/quant-ph/0308039.
9. C. L. Lopreore and R. E. Wyatt: Quantum Wave Packet Dynamics with Trajectories. *Physical Review Letters* **82**: 5190–5193 (1999).
10. W. Struyve: Field beables for quantum field theory. Preprint (2009). http://arxiv.org/abs/0707.3685.
11. R. Tumulka: The "Unromantic Pictures" of Quantum Theory. *Journal of Physics A: Mathematical and Theoretical* **40**: 3245–3273 (2007). http://arxiv.org/abs/quant-ph/0607124.

2 The Equivalence Postulate of Quantum Mechanics: Main Theorems

Alon E. Faraggi and Marco Matone

CONTENTS

2.1   THE HAMILTON–JACOBI EQUATION AND COORDINATE TRANSFORMATIONS

The Hamilton–Jacobi (HJ) equation for a one-dimensional system is obtained by considering the canonical transformation $(q, p) \rightarrow (Q, P)$ so that the old Hamiltonian H maps to a trivialized one, that is $\tilde{H} = 0$. The old and new momenta are expressed in terms of the generating function of such a transformation, the Hamilton's principal function $p = \frac{\partial \mathcal{S}^{cl}}{\partial q}$, $P = \mathrm{cnst} = -\frac{\partial \mathcal{S}^{cl}}{\partial Q}|_{Q=\mathrm{cnst}}$ that satisfies the classical HJ equation

$$H\left(q, p = \frac{\partial \mathcal{S}^{cl}}{\partial q}, t\right) + \frac{\partial \mathcal{S}^{cl}}{\partial t} = 0.$$

In the case of a time-independent potential the time-dependence in the Hamilton's principal function $\mathcal{S}^{cl}$ is linear, that is $\mathcal{S}^{cl}(q, Q, t) = \mathcal{S}_0^{cl}(q, Q) - Et$, with E the energy of the stationary state. It follows that $\mathcal{S}_0^{cl}$, called Hamilton's characteristic function, or reduced action, satisfies the classical stationary HJ equation (CSHJE)

$$H\left(q, p = \frac{\partial \mathcal{S}_0^{cl}}{\partial q}\right) - E = 0,$$

that is $(\mathcal{W}(q) \equiv V(q) - E)$

$$\frac{1}{2m}\left(\frac{\partial \mathcal{S}_0^{cl}}{\partial q}\right)^2 + \mathcal{W} = 0.$$

Note that the canonical transformation $(q, p) \rightarrow (Q, P)$ treats p and q as independent variables. Following [1–6] we now formulate a similar question to that leading to the CSHJE, but considering the transformation on q, with the one on p induced by the relation

$$p = \frac{\partial S_0^{cl}}{\partial q}.$$

More precisely, given a one-dimensional system, with time-independent potential (the higher dimensional time-dependent case is considered in [7]) we look for the coordinate transformation $q \rightarrow q_0$ such that

$$S_0^{cl}(q) \overset{\text{Coord. Transf.}}{\longleftrightarrow} \tilde{S}_0^{cl\,0}(q_0), \tag{2.1}$$

with $\tilde{S}_0^{cl\,0}(q_0)$ denoting the reduced action of the system with vanishing Hamiltonian. Note that in Equation 2.1 we required that this transformation be an invertible one. This is an important point since by compositions of the maps it follows that if for each system there is a coordinate transformation leading to the trivial state, then even two arbitrary systems are equivalent under coordinate transformations. Imposing this apparently harmless analogy immediately leads to rather peculiar properties of classical mechanics (CM). First, it is clear that such an equivalence principle cannot be satisfied in CM, in other words given two arbitrary systems a and b, the condition

$$S_0^{cl\,b}(q_b) = S_0^{cl\,a}(q_a), \tag{2.2}$$

cannot be generally satisfied. In particular, since

$$\tilde{S}_0^{cl\,0}(q_0) = \text{cnst},$$

it is clear that Equation 2.1 is a degenerate transformation. However, in principle, by itself the failure of Equation 2.2 for arbitrary systems would be a possible natural property. Nevertheless, a more careful analysis shows that such a failure is strictly dependent on the choice of the reference frame. This is immediately seen by considering two free particles of mass m_a and m_b moving with relative velocity v. For an observer at rest with respect to particle a the two reduced actions are

$$S_0^{cl\,a}(q_a) = \text{cnst}, \quad S_0^{cl\,b}(q_b) = m_b v q_b.$$

It is clear that there is no way to have an equivalence under coordinate transformations by setting $S_0^{cl\,b}(q_b) = S_0^{cl\,a}(q_a)$. This means that at the level of the reduced action there is no coordinate transformation making the two systems equivalent. However, note that this coordinate transformation exists if we consider the same problem described by an observer in a frame in which both particles have a non-vanishing velocity so that the two particles are described by non-constant reduced actions. Therefore, in CM, it is possible to connect different systems by a coordinate transformation except in the case in which one of the systems is described by a constant reduced action. This means that in CM equivalence under coordinate transformations is frame-dependent. In particular, in the CSHJE description there is a distinguished frame. This seems peculiar as on general grounds what is equivalent under coordinate transformations in all frames should remain so even in the one at rest

2.2 THE EQUIVALENCE POSTULATE

The above investigation already suggests that the concept of point particle itself cannot be consistent with the equivalence under coordinate transformations. In particular, it suggests that the system where a particle is at rest does not exist at all. If this were the case, then the above critical situation would not occur simply because the reduced action is never a constant. This should result in two main features. First the classical concept of point particle should be reconsidered, and secondly the CSHJE should be modified accordingly. A natural suggestion would be to consider particles as a kind of string with a lower bound on the vibrating modes in such a way that there is no way to define a system where the particle is at rest. It should be observed that this kind of string may differ from the standard one, rather its nature may be related to the fact that in general relativity it is impossible to define the concept of relative stability of a system of particles.

In Ref. [8–10] it was suggested that quantum mechanics and gravity are intimately related. In particular, it was argued that the quantum HJ equation of two free particles, which is attractive, may generate the gravitational potential. This is a consequence of the fact that the quantum potential is always non-trivial even in the case of the free particle. It plays the role of intrinsic energy and may in fact be at the origin of fundamental interactions.

The unification of quantum mechanics and general relativity is the central question of theoretical physics. This problem hinges on the viability of the prevailing theories of matter and interactions at the micro-scale, and of the cosmos at the macro-scale. The Galilean paradigm of modern science drives the search for a mathematical formulation of the synthesis of quantum gravity. In such a context string theory provides an attempt at a self-consistent mathematical formulation of quantum gravity.

String theory provides a perturbatively finite S-matrix approach to the calculation of string scattering amplitudes. Due to its unique world-sheet properties, string theory admits a discrete particle spectrum. It accommodates the gauge bosons and fermion matter states that form the bedrock of modern particle physics, as well as a massless spin 2 symmetric state, which is interpreted as the gravitational force mediation field. Consequently, string theory enables the construction of models that admit the structures of the standard particle model and enables the development of a phenomenological approach to quantum gravity. The state of the art in this regard is the construction of minimal heterotic string standard models, which produces in the observable standard model charged sector solely the spectrum of the minimal supersymmetric standard model [11–13]. Progress in the understanding of string theory was obtained by the observation that the five 10-dimensional string theories, as well as 11-dimensional supergravity, can be connected by perturbative and non-perturbative duality tranformations. However, this observation does not provide a rigorous formulation of quantum gravity, akin to the formulations of general relativity and quantum mechanics, which follow from the equivalence principle in the former and the probability interpretation of the wavefunction in the later.

Let us start imposing the equivalence under coordinate transformations. The key point is to consider, like in general relativity, the (analog of the) reduced action as a scalar field under coordinate transformations.

We postulate that for any pair of one-particle states there exists a field $\mathcal{S}_0$ such that

$$\mathcal{S}_0^b(q_b) = \mathcal{S}_0^a(q_a) \tag{2.3}$$

is well defined. We also require that, in a suitable limit, $\mathcal{S}_0$ reduces to $\mathcal{S}_0^{cl}$. Equation 2.3 can be considered as the scalar hypothesis. Since the conjugate momentum is defined by

$$p_i = \frac{\partial}{\partial q^i} \mathcal{S}_0(q),$$

it follows by Equation 2.3 that the conjugate momenta p^a and p^b are related by a coordinate transformation

$$p_i^b = \Lambda_i^j p_j^a, \tag{2.4}$$

where $\Lambda_i^j = \partial q_a^j / \partial q_b^i$. Note that we have the invariant

$$p_i^b dq_b^i = p_i^a dq_a^i. \tag{2.5}$$

Since Equation 2.3 holds for any pair of one-particle states, we have Det $\Lambda(q) \neq 0$, $\forall q$.

The scalar hypothesis Equation 2.3 implies that two one-particle states are always connected by a coordinate transformation; for this reason we may equivalently consider Equation 2.3 as imposing an equivalence postulate (EP). In particular, while in arbitrary dimension the coordinate transformation is given by imposing Equation 2.4, in the one-dimensional case the scalar hypothesis implies

$$q_b = \mathcal{S}_0^{b-1} \circ \mathcal{S}_0^a(q_a).$$

We now consider the consequences of the EP Equation 2.3. Let us denote by $\mathcal{H}$ the space of all possible $W \equiv V - E$. We also call v-transformations those leading from one system to another. Equation 2.3 is equivalent to requiring that

For each pair $\mathcal{W}^a, \mathcal{W}^b \in \mathcal{H}$, there is a v-transformation such that

$$\mathcal{W}^a(q) \longrightarrow \mathcal{W}^{av}(q^v) = \mathcal{W}^b(q^v). \tag{2.6}$$

This implies that there always exists the trivializing coordinate q_0 for which $\mathcal{W}(q) \longrightarrow \mathcal{W}^0(q_0)$, where

$$\mathcal{W}^0(q_0) \equiv 0.$$

In particular, since the inverse transformation should exist as well, it is clear that the trivializing transformation should be locally invertible. We will also see that since classically $\mathcal{W}^0$ is a fixed point, implementation of Equation 2.6 requires that $\mathcal{W}$ states transform inhomogeneously.

The fact that the EP cannot be consistently implemented in CM is true in any dimension. To show this, let us consider the coordinate transformation induced by the identification

$$S_0^{cl\,v}(q^v) = S_0^{cl}(q). \tag{2.7}$$

Then note that the CSHJE

$$\frac{1}{2m}\sum_{k=1}^{D}(\partial_{q_k}S_0^{cl}(q))^2 + \mathcal{W}(q) = 0, \tag{2.8}$$

provides a correspondence between $\mathcal{W}$ and S_0^{cl} that we can use to fix, by consistency, the transformation properties of $\mathcal{W}$ induced by that of S_0^{cl}. In particular, since $S_0^{cl\,v}(q^v)$ must satisfy the CSHJE

$$\frac{1}{2m}\sum_{k=1}^{D}(\partial_{q^{kv}}S_0^{cl\,v}(q^v))^2 + \mathcal{W}^v(q^v) - 0, \tag{2.9}$$

by Equation 2.7 we have

$$\frac{\partial S_0^{cl\,v}(q^v)}{\partial q^{k\,v}} - \Lambda_k^i \frac{\partial S_0^{cl}(q)}{\partial q^i}. \tag{2.10}$$

Let us set $(p^v|p) = p^t \Lambda^t \Lambda p / p^t p$. By Equations 2.8 through 2.10, we have $\mathcal{W}(q) \longrightarrow \mathcal{W}^v(q^v) = (p^v|p)\mathcal{W}(q)$, so that

$$\mathcal{W}^0(q_0) \longrightarrow \mathcal{W}^v(q^v) = (p^v|p^0)\mathcal{W}^0(q_0) = 0.$$

Thus we have [1–6]:

$\mathcal{W}$ states transform as quadratic differentials under classical v-maps. It follows that $\mathcal{W}^0$ is a fixed point in $\mathcal{H}$. Equivalently, in CM the space $\mathcal{H}$ cannot be reduced to a point upon factorization by the classical v-transformations. Hence, the EP Equation 2.6 cannot be consistently implemented in CM. This can be seen as the impossibility of implementing covariance of CM under the coordinate transformation defined by Equation 2.7.

It is therefore clear that in order to implement the EP we have to deform the CSHJE. As we will see, this requirement will determine the equation for S_0.

In Refs [1–6] the function $\mathcal{T}_0(p)$, defined as the Legendre transform of the reduced action, was introduced:

$$\mathcal{T}_0(p) = q^k p_k - S_0(q), \quad S_0(q) = p_k q^k - \mathcal{T}_0(p).$$

While $S_0(q)$ is the momentum generating function, its Legendre dual $\mathcal{T}_0(p)$ is the coordinate generating function

$$p_k = \frac{\partial S_0}{\partial q_k}, \quad q_k = \frac{\partial \mathcal{T}_0}{\partial p_k}.$$

Note that adding a constant to $\mathcal{S}_0$ does not change the dynamics. Then, the most general differential equation $\mathcal{S}_0$ should satisfy has the structure

$$\mathcal{F}(\nabla \mathcal{S}_0, \Delta \mathcal{S}_0, \ldots) = 0. \tag{2.11}$$

Let us write down Equation 2.11 in the general form

$$\frac{1}{2m} \sum_{k=1}^{D} (\partial_{q^k} \mathcal{S}_0(q))^2 + \mathcal{W}(q) + Q(q) = 0.$$

The transformation properties of $\mathcal{W} + Q$ under the v-maps are determined by the transformed equation

$$\frac{1}{2m} \sum_{k=1}^{D} (\partial_{q^{k v}} \mathcal{S}_0^v(q^v))^2 /2m + \mathcal{W}^v(q^v) + Q^v(q^v) = 0, \tag{2.12}$$

so that

$$\mathcal{W}^v(q^v) + Q^v(q^v) = (p^v | p) [\mathcal{W}(q) + Q(q)]. \tag{2.13}$$

A basic guidance in deriving the differential equation for $\mathcal{S}_0$ is that in some limit it should reduce to the CSHJE. In Refs [1–10] it was shown that the parameter which selects the classical phase is the Planck constant. Therefore, in determining the structure of the Q term we have to take into account that in the classical limit

$$\lim_{\hbar \to 0} Q = 0. \tag{2.14}$$

The only possibility to reach any other state $\mathcal{W}^v \neq 0$ starting from $\mathcal{W}^0$ is that it transforms with an inhomogeneous term. Namely as $\mathcal{W}^0 \longrightarrow \mathcal{W}^v(q^v) \neq 0$, it follows that for an arbitrary $\mathcal{W}^a$ state

$$\mathcal{W}^v(q^v) = (p^v | p^a) \mathcal{W}^a(q_a) + (q_a; q^v), \tag{2.15}$$

and by Equation 2.13

$$Q^v(q^v) = (p^v | p^a) Q^a(q_a) - (q_a; q^v). \tag{2.16}$$

Let us stress that the purely quantum origin of the inhomogeneous term $(q_a; q^v)$ is particularly transparent once one consider the compatibility between the classical limit (Equation 2.14) and the transformation properties of Q in Equation 2.16.

The $\mathcal{W}^0$ state plays a special role. Actually, setting $\mathcal{W}^a = \mathcal{W}^0$ in Equation 2.15 yields

$$\mathcal{W}^v(q^v) = (q_0; q^v),$$

so that, according to the EP Equation 2.6, all the states correspond to the inhomogeneous part in the transformation of the $\mathcal{W}^0$ state induced by some v-map.

Let us denote by $a, b, c, \ldots$ different v-transformations. Comparing

$$\mathcal{W}^b(q_b) = (p^b | p^a) \mathcal{W}^a(q_a) + (q_a; q_b) = (q_0; q_b), \tag{2.17}$$

with the same formula with q_a and q_b interchanged we have

$$(q_b; q_a) = -(p^a | p^b)(q_a; q_b),$$ (2.18)

in particular $(q; q) = 0$. More generally, imposing the commutative diagram of maps

$$
\begin{array}{ccc}
 & B & \\
\nearrow & & \searrow \\
A & \longrightarrow & C
\end{array}
$$

that is comparing

$$
\begin{aligned}
\mathcal{W}^b(q_b) &= (p^b | p^c)\mathcal{W}^c(q_c) + (q_c; q_b) \\
&= (p^b | p^a)\mathcal{W}^a(q_a) + (p^b | p^c)(q_a; q_c) + (q_c; q_b),
\end{aligned}
$$

with Equation 2.17, we obtain the basic cocycle condition

$$(q_a; q_c) = (p^c | p^b)[(q_a; q_b) + (q_b; q_c)],$$ (2.19)

which expresses the essence of the EP. In the one-dimensional case we have

$$(q_a; q_c) = \left(\partial_{q_c} q_b\right)^2 (q_a; q_b) + (q_b; q_c).$$ (2.20)

It is well known that this is satisfied by the Schwarzian derivative. However, it turns out that it is essentially the unique solution. More precisely [1–6],

---

Theorem 2.1

Equation 2.20 defines the Schwarzian derivative up to a multiplicative constant and a coboundary term. ∎

Since the differential equation for S_0 should depend only on $\partial_q^k S_0$, $k \geq 1$, it follows that the coboundary term must be zero, so that [1–6]

$$(q_a; q_b) = -\frac{\beta^2}{4m}\{q_a, q_b\},$$

where $\{f(q), q\} = f'''/f' - 3(f''/f')^2/2$ is the Schwarzian derivative and β is a non-vanishing constant that we identify with $\hbar$. As a consequence, S_0 satisfies the quantum stationary Hamilton–Jacobi equation (QSHJE) [1–6]

$$\frac{1}{2m}\left(\frac{\partial S_0(q)}{\partial q}\right)^2 + V(q) - E + \frac{\hbar^2}{4m}\{S_0, q\} = 0.$$ (2.21)

Note that $\psi = S_0'^{-1/2}(Ae^{-\frac{i}{\hbar}S_0} + Be^{\frac{i}{\hbar}S_0})$ solves the Schrödinger equation (SE)

$$\left(-\frac{\hbar^2}{2m}\frac{\partial^2}{\partial q^2} + V\right)\psi = E\psi.$$ (2.22)

The ratio $w = \psi^D/\psi$, where ψ^D and ψ are two real linearly independent solutions of Equation 2.22 is, in deep analogy with uniformization theory, the *trivializing map* transforming any $\mathcal{W}$ to $\mathcal{W}^0 \equiv 0$ [1–6, 14]. This formulation extends to higher dimensions and to the relativistic case as well [1–7].

Let $q_{-/+}$ be the lowest/highest q for which $\mathcal{W}(q)$ changes sign. Then we have Refs [1–6]

Theorem 2.2

If

$$V(q) - E \geq \begin{cases} P_-^2 > 0, & q < q_-, \\ P_+^2 > 0, & q > q_+, \end{cases} \tag{2.23}$$

then w is a local self-homeomorphism of $\hat{\mathbb{R}} = \mathbb{R} \cup \{\infty\}$ if and only if Equation 2.22 has an $L^2(\mathbb{R})$ solution.   ∎

The crucial consequence is that since the QSHJE is defined if and only if w is a local self-homeomorphism of $\hat{\mathbb{R}}$, it follows that the QSHJE by itself implies energy quantization. We stress that this result is obtained without any probabilistic interpretation of the wavefunction.

2.3 PROOF OF THEOREM 1

The main steps in proving Theorem 1 are two lemmas [1–6]. Let us start by observing that if the cocycle condition Equation 2.20 is satisfied by $(f(q); q)$, then this is still satisfied by adding a coboundary term

$$(f(q); q) \longrightarrow (f(q); q) + (\partial_q f)^2 G(f(q)) - G(q). \tag{2.24}$$

Since $(Aq; q)$ evaluated at $q = 0$ is independent of A, we have

$$0 = (q; q) = (q; q)_{|q=0} = (Aq; q)_{|q=0}. \tag{2.25}$$

Therefore, if both $(f(q); q)$ and Equation 2.24 satisfy Equation 2.20, then $G(0) = 0$, which is the unique condition that G should satisfy. We now use Equation 2.11 to fix the ambiguity Equation 2.24. First of all observe that the differential equation we are looking for is

$$(q_0; q) = \mathcal{W}(q). \tag{2.26}$$

Then, recalling that $q_0 = S_0^{0^{-1}} \circ S_0(q)$, we see that a necessary condition to satisfy Equation 2.11 is that $(q_0; q)$ depends only on the first and higher derivatives of q_0. This in turn implies that for any constant B we have $(q_a + B; q_b) = (q_a; q_b)$ that, together with Equation 2.18, gives

$$(q_a + B; q_b) = (q_a; q_b) = (q_a; q_b + B). \tag{2.27}$$

Let A be a non-vanishing constant and set $h(A, q) = (Aq; q)$. By Equation 2.27 we have $h(A, q + B) = h(A, q)$, that is $h(A, q)$ is independent of q. On the other hand, by Equation 2.25 $h(A, 0) = 0$ that, together with Equation 2.18, implies

$$(Aq; q) = 0 = (q; Aq). \tag{2.28}$$

Equation 2.20 implies $(q_a; Aq_b) = A^{-2}((q_a; q_b) - (Aq_b; q_b))$, so that by Equation 2.28

$$(q_a; Aq_b) = A^{-2}(q_a; q_b). \tag{2.29}$$

By Equations 2.18 and 2.29 we have

$$(Aq_a; q_b) = -A^{-2}(\partial_{q_b} q_a)^2 (q_b; Aq_a) = -(\partial_{q_b} q_a)^2 (q_b; q_a) = (q_a; q_b),$$

that is

$$(Aq_a; q_b) = (q_a; q_b). \tag{2.30}$$

Setting $f(q) = q^{-2}(q; q^{-1})$ and noticing that by Equations 2.18 and 2.30 $f(Aq) = -f(q^{-1})$, we obtain

$$(q; q^{-1}) = 0 = (q^{-1}; q) \tag{2.31}$$

Furthermore, since by Equations 2.20 and 2.31 one has $(q_a; q_b^{-1}) = q_b^4(q_a; q_b)$, it follows that

$$(q_a^{-1}; q_b) = -\left(\partial_{q_b} q_a^{-1}\right)^2 (q_b; q_a^{-1}) = -\left(\partial_{q_b} q_a\right)^2 (q_b; q_a) = (q_a; q_b),$$

so that

$$(q_a^{-1}; q_b) = (q_a; q_b) = q_b^{-4}(q_a; q_b^{-1}). \tag{2.32}$$

Since translations, dilatations, and inversion are the generators of the Möbius group, it follows by Equations 2.27, 2.29, 2.30, and 2.32 that

---

Lemma 2.1

Up to a coboundary term, Equation 2.20 implies

$$(\gamma(q_a); q_b) = (q_a; q_b),$$
$$(q_a; \gamma(q_b)) = \left(\partial_{q_b} \gamma(q_b)\right)^{-2} (q_a; q_b),$$

where $\gamma(q)$ is an arbitrary $PSL(2, \mathbb{C})$ transformation. ∎

Now observe that since $(q_a; q_b)$ should depend only on $\partial_{q_b}^k q_a$, $k \geq 1$, we have

$$(q + \epsilon f(q); q) = c_1 \epsilon f^{(k)}(q) + \mathcal{O}(\epsilon^2), \tag{2.33}$$

where $q_a = q + \epsilon f(q)$, $q \equiv q_b$ and $f^{(k)} \equiv \partial_q^k f$, $k \geq 1$. Note that by Lemma 1 and Equation 2.33

$$
\begin{aligned}
(Aq + \epsilon A f(q); Aq) &= (q + \epsilon f(q); Aq) \\
&= A^{-2}(q + \epsilon f(q); q) = A^{-2} c_1 \epsilon f^{(k)}(q) + \mathcal{O}(\epsilon^2); \quad (2.34)
\end{aligned}
$$

on the other hand, setting $F(Aq) = A f(q)$, by Equation 2.33

$$
\begin{aligned}
(Aq + \epsilon A f(q); Aq) &= (Aq + \epsilon F(Aq); Aq) \\
&= c_1 \epsilon \partial_{Aq}^k F(Aq) + \mathcal{O}(\epsilon^2) = A^{1-k} c_1 \epsilon f^{(k)}(q) + \mathcal{O}(\epsilon^2),
\end{aligned}
$$

which, compared with Equation 2.34 gives $k = 3$. The above scaling property generalizes to higher order contributions in ϵ. In particular, to order ϵ^n the quantity $(Aq + \epsilon A f(q); Aq)$ is a sum of terms of the form

$$
c_{i_1 \ldots i_n} \partial_{Aq}^{i_1} \epsilon F(Aq) \cdots \partial_{Aq}^{i_n} \epsilon F(Aq) = c_{i_1 \ldots i_n} \epsilon^n A^{n - \sum i_k} f^{(i_1)}(q) \cdots f^{(i_n)}(q),
$$

and by Equation 2.34 $\sum_{k=1}^n i_k = n + 2$. On the other hand, since $(q_a; q_b)$ depends only on $\partial_{q_b}^k q_a$, $k \geq 1$, we have

$$
i_k \geq 1, \quad k \in [1, n],
$$

so that either

$$
i_k = 3, \quad i_j = 1, \quad j \in [1, n], \quad j \neq k,
$$

or

$$
i_k = i_j = 2, \quad i_l = 1, \quad l \in [1, n], \quad l \neq k, l \neq j.
$$

Hence

$$
(q + \epsilon f(q); q) = \sum_{n=1}^{\infty} \epsilon^n \left(c_n f^{(3)} f^{(1)^{n-1}} + d_n f^{(2)^2} f^{(1)^{n-2}} \right), \quad d_1 = 0. \quad (2.35)
$$

Let us now consider the transformations

$$
q_b = v^{ba}(q_a), \quad q_c = v^{cb}(q_b) = v^{cb} \circ v^{ba}(q_a), \quad q_c = v^{ca}(q_a).
$$

Note that $v^{ab} = v^{ba^{-1}}$, and

$$
v^{ca} = v^{cb} \circ v^{ba}. \quad (2.36)
$$

We can express these transformations in the form

$$
\begin{aligned}
q_b &= q_a + \epsilon^{ba}(q_a), \\
q_c = q_b + \epsilon^{cb}(q_b) &= q_b + \epsilon^{cb}(q_a + \epsilon^{ba}(q_a)), \quad (2.37) \\
q_c &= q_a + \epsilon^{ca}(q_a).
\end{aligned}
$$

Since $q_b = q_a - \epsilon^{ab}(q_b)$, we have $q_b = q_a - \epsilon^{ab}(q_a + \epsilon^{ba}(q_a))$, which, compared with $q_b = q_a + \epsilon^{ba}(q_a)$ yields

$$\epsilon^{ba} + \epsilon^{ab} \circ (1 + \epsilon^{ba}) = 0,$$

where 1 denotes the identity map. More generally, Equation 2.37 gives

$$\epsilon^{ca}(q_a) = \epsilon^{cb}(q_b) + \epsilon^{ba}(q_a) = \epsilon^{cb}(q_b) - \epsilon^{ab}(q_b),$$

so that we obtain Equation 2.36 with $v^{yx} = 1 + \epsilon^{yx}$

$$\epsilon^{ca} = \epsilon^{cb} \circ (1 + \epsilon^{ba}) + \epsilon^{ba} = (1 + \epsilon^{cb}) \circ (1 + \epsilon^{ba}) - 1. \tag{2.38}$$

Let us consider the case in which $\epsilon^{yx}(q_x) = \epsilon f_{yx}(q_x)$, with ϵ infinitesimal. To first order in ϵ (Equation 2.38) reads

$$\epsilon^{ca} = \epsilon^{cb} + \epsilon^{ba}, \tag{2.39}$$

in particular, $\epsilon^{ab} = -\epsilon^{ba}$. Since $(q_a; q_b) = c_1 \epsilon^{ab'''}(q_b) + \mathcal{O}^{ab}(\epsilon^2)$, where $'$ denotes the derivative with respect to the argument, we can use the cocycle condition (Equation 2.20) to get

$$c_1 \epsilon^{ac'''}(q_c) + \mathcal{O}^{ac}(\epsilon^2) = (1 + \epsilon^{bc'}(q_c))^2 (c_1 \epsilon^{ab'''}(q_b) + \mathcal{O}^{ab}(\epsilon^2)$$
$$- c_1 \epsilon^{cb'''}(q_b) - \mathcal{O}^{cb}(\epsilon^2)), \tag{2.40}$$

which, to firstorder in ϵ corresponds to Equation 2.39. We see that $c_1 \neq 0$. For, if $c_1 = 0$, then by Equation 2.40, to second order in ϵ one would have

$$\mathcal{O}^{ac}(\epsilon^2) = \mathcal{O}^{ab}(\epsilon^2) - \mathcal{O}^{cb}(\epsilon^2), \tag{2.41}$$

which contradicts Equation 2.39. In fact, by Equation 2.35 we have

$$\mathcal{O}^{ab}(\epsilon^2) = c_2 \epsilon^{ab'''}(q_b)\epsilon^{ab'}(q_b) + d_2 \epsilon^{ab''^2}(q_b) + \mathcal{O}^{ab}(\epsilon^3),$$

that together with Equation 2.41 provides a relation which cannot be consistent with $\epsilon^{ac}(q_c) = \epsilon^{ab}(q_b) - \epsilon^{cb}(q_b)$. A possibility is that $(q_a; q_b) = 0$. However, this is ruled out by the EP, so that

$$c_1 \neq 0.$$

Higher-order contributions due to a non-vanishing c_1 are obtained by using

$$q_c = q_b + \epsilon^{cb}(q_b), \quad \epsilon^{ac}(q_c) = \epsilon^{ab}(q_b) - \epsilon^{cb}(q_b),$$

and $\epsilon^{bc}(q_c) = -\epsilon^{cb}(q_b)$ in $c_1 \partial^3_{q_c} \epsilon^{ac}(q_c)$ and in

$$c_1 (2\partial_{q_c} \epsilon^{bc}(q_c) + \partial_{q_c} \epsilon^{bc}(q_c)^2)\partial^3_{q_b}(\epsilon^{ab}(q_b) - \epsilon^{cb}(q_b)).$$

Note that one can also consider the case in which both the first- and second-order contributions to $(q_a; q_b)$ are vanishing. However, this possibility is ruled out by a similar analysis. In general, one has that if the first non-vanishing contribution to $(q_a; q_b)$ is of order ϵ^n, $n \geq 2$, then, unless $(q_a; q_b) = 0$, the cocycle condition (Equation 2.20) cannot be consistent with the linearity of Equation 2.39. Observe that we proved that $c_1 \neq 0$ is a necessary condition for the existence of solutions $(q_a; q_b)$ of the cocycle condition Equation 2.20, depending only on the first and higher derivatives of q_a. Existence of solutions follows from the fact that the Schwarzian derivative $\{q_a, q_b\}$ solves Equation 2.20 and depends only on the first and higher derivatives of q_a.

The fact that $c_1 = 0$ implies $(q_a; q_b) = 0$ can also be seen by explicitly evaluating the coefficients c_n and d_n. These can be obtained using the same procedure considered above to prove that $c_1 \neq 0$. Namely, inserting the expansion Equation 2.35 in Equation 2.20 and using $q_c = q_b + \epsilon^{cb}(q_b)$, $\epsilon^{ac}(q_c) = \epsilon^{ab}(q_b) - \epsilon^{cb}(q_b)$ and $\epsilon^{bc}(q_c) = -\epsilon^{cb}(q_b)$, we obtain

$$c_n = (-1)^{n-1} c_1, \quad d_n = \frac{3}{2}(-1)^{n-1}(n-1)c_1, \tag{2.42}$$

which in fact are the coefficients one obtains expanding $c_1\{q + \epsilon f(q), q\}$. However, we now use only the fact that $c_1 \neq 0$, as the relation $(q + \epsilon f(q); q) = c_1\{q + \epsilon f(q), q\}$ can be proved without making the calculations leading to Equation 2.42. Summarizing, we have

---

Lemma 2.2

If

$$q_a = q_b + \epsilon^{ab}(q_b),$$

the unique solution of Equation 2.20, depending only on the first and higher derivatives of q_a, is

$$(q_a; q_b) = c_1 \epsilon^{ab'''}(q_b) + \mathcal{O}^{ab}(\epsilon^2), \quad c_1 \neq 0. \qquad \blacksquare$$

It is now easy to prove that, up to a multiplicative constant and a coboundary term, the Schwarzian derivative is the unique solution of the cocycle condition Equation 2.20. Let us first note that

$$[q_a; q_b] = (q_a; q_b) - c_1\{q_a; q_b\},$$

satisfies the cocycle condition

$$[q_a; q_c] = \left(\partial_{q_c} q_b\right)^2 ([q_a; q_b] - [q_c; q_b]) .$$

In particular, since both $(q_a; q_b)$ and $\{q_a; q_b\}$ depend only on the first and higher derivatives of q_a, we have, as in the case of $(q + \epsilon f(q); q)$, that

$$[q + \epsilon f(q); q] = \tilde{c}_1 \epsilon f^{(3)}(q) + \mathcal{O}(\epsilon^2),$$

where either $\tilde{c}_1 \neq 0$ or $[q + \epsilon f(q); q] = 0$. However, since $\{q + \epsilon f(q); q\} = \epsilon f^{(3)}(q) + \mathcal{O}(\epsilon^2)$ and $(q + \epsilon f(q); q) = \epsilon f^{(3)}(q) + \mathcal{O}(\epsilon^2)$, we have $\tilde{c}_1 = 0$ and the lemma yields $[q + \epsilon f(q); q] = 0$. Therefore, we have that the EP univocally implies that

$$(q_a; q_b) = -\frac{\beta^2}{4m}\{q_a, q_b\},$$

where for convenience we replaced c_1 by $-\beta^2/4m$. This concludes the proof of Theorem 1.

We observe that despite some claims [15], we have not been able to find in the literature a complete and closed proof of the above theorem (see also Ref. [16]). We thank D.B. Fuchs for a bibliographic comment concerning the above theorem.

In deriving the equivalence of states we considered the case of one-particle states with identical masses. The generalization to the case with different masses is straightforward. In particular, the right hand side of Equation 2.20 gets multiplied by m_b/m_a, so that the cocycle condition becomes

$$m_a(q_a; q_c) = m_a \left(\partial_{q_c} q_b\right)^2 (q_a; q_b) + m_b(q_b; q_c),$$

explicitly showing that the mass appears in the denominator and that it refers to the label in the first entry of $(\cdot \, ; \, \cdot)$, that is

$$(q_a; q_b) = \frac{\hbar^2}{4m_a}\{q_a; q_b\}. \tag{2.43}$$

The QSHJE (Equation 2.21) follows almost immediately by Equation 2.43 [1–6].

The above investigation may be applied to Conformal Field Theory (CFT). Let us consider a local conformal transformation of the stress tensor in a 2D CFT. The infinitesimal variation of T is given by

$$\delta_\epsilon T(w) = -\frac{1}{12}c\partial_w^3\epsilon(w) - 2T(w)\partial_w\epsilon(w) - \epsilon(w)\partial_w T(w), \tag{2.44}$$

where c is the central charge. The finite version of such a transformation is

$$\tilde{T}(w) = (\partial_w z)^2 T(z) + \frac{c}{12}\{w, z\}. \tag{2.45}$$

While it is immediate to see that Equation 2.45 implies Equation 2.44, the converse is not evident. A possible way to prove Equation 2.45 is just to set

$$\tilde{T}(w) = (\partial_w z)^2 T(z) + k(w; z), \tag{2.46}$$

and then to impose the cocycle condition which will show that $(w; z)$ is proportional to $\{w, z\}$. Comparison with the infinitesimal transformation Equation 2.44 fixes the constant k.

In Ref. [7] it has been shown that the cocycle condition fixes the higher dimensional version of the Schwarzian derivative. In this respect we observe that its definition seems an open question in the mathematical literature. While in the one-dimensional case the QSHJE reduces to a unique differential equation, this is not immediate in the higher dimensional case. However, it turns out that such a reduction exists upon introducing an antisymmetric tensor [7] (in this respect it is worth noticing that some authors introduce a connection to define the higher dimensional Schwarzian derivative).

A basic feature of the cocycle condition is that it implies, as it should, the higher dimensional Möbious invariance with respect to q_a in $(q_a; q_b)$ (with similar properties with respect to q_b). In particular, in Ref. [7] it has been shown that

$$(q^a; q^b) = -\frac{\hbar^2}{2m}\left[(p^b|p^a)\frac{\Delta^a R^a}{R^a} - \frac{\Delta^b R^b}{R^b}\right]. \tag{2.47}$$

It would be interesting to consider such a definition in the context of the transformation properties of the stress tensor in higher dimensional CFTs.

2.4   PROOF OF THEOREM 2

The QSHJE is equivalent to

$$\{w, q\} = -\frac{4m}{\hbar^2}\mathcal{W}(q), \tag{2.48}$$

where $w = \psi^D/\psi$ with ψ^D and ψ two real linearly independent solutions of the Schrödinger equation. The existence of this equation requires some conditions on the continuity properties of w and its derivatives. Since the QSHJE is a consequence of the EP, we can say that the EP imposes some constraints on $w = \psi^D/\psi$. These constraints are nothing but the existence of the QSHJE (Equation 2.21) or, equivalently, of Equation 2.48. That is, implementation of the EP imposes that $\{w, q\}$ exists, so that

$$w \neq \text{cnst}, \quad w \in C^2(\mathbb{R}) \quad \text{and} \quad \partial_q^2 w \text{ differentiable on } \mathbb{R}. \tag{2.49}$$

These conditions are not complete. The reason is that, as we have seen, the implementation of the EP requires that the properties of the Schwarzian derivative be satisfied. Actually, its very properties, derived from the EP, led to the identification $(q_a; q_b) = -\hbar^2\{q_a, q_b\}/4m$. Therefore, in order to implement the EP, the transformation properties of the Schwarzian derivative and its symmetries must be satisfied. In deriving the transformation properties of $(q_a; q_b)$ we noticed how, besides dilatations and translations, there is a highly non-trivial symmetry under inversion. Therefore, we have that Equation 2.48 must be equivalent to

$$\{w^{-1}, q\} = -\frac{4m}{\hbar^2}\mathcal{W}(q).$$

A property of the Schwarzian derivative is duality between its entries

$$\{w, q\} = -\left(\frac{\partial w}{\partial q}\right)^2 \{q, w\}. \tag{2.50}$$

This shows that the invariance under inversion of w results in the invariance, up to a Jacobian factor, under inversion of q. That is $\{w, q^{-1}\} = q^4\{w, q\}$, so that the QSHJE Equation 2.48 can be written in the equivalent form

$$\{w, q^{-1}\} = -\frac{4m}{\hbar^2}q^4\mathcal{W}(q).$$ (2.51)

In other words, starting from the EP one can arrive at either Equation 2.48 or Equation 2.51. The consequence of this fact is that since under

$$q \to \frac{1}{q},$$

$0^{\pm}$ maps to $\pm\infty$, we have to extend Equation 2.49 to the point at infinity. In other words, Equation 2.49 should hold on the extended real line $\hat{\mathbb{R}} = \mathbb{R} \cup \{\infty\}$. This aspect is related to the fact that the Möbius transformations, under which the Schwarzian derivative transforms as a quadratic differential, map circles to circles. We stress that we are considering the systems defined on $\mathbb{R}$ and not $\hat{\mathbb{R}}$. What happens is that the existence of the QSHJE forces us to impose smoothly joining conditions even at $\pm\infty$, that is Equation 2.49 must be extended to

$$w \neq \text{cnst}, \ w \in C^2(\hat{\mathbb{R}}) \quad and \quad \partial_q^2 w \ differentiable \ on \ \hat{\mathbb{R}}.$$ (2.52)

One may easily check that w is a Möbius transformation of the trivializing map [1–6]. Therefore, Equation 2.50, which is defined if and only if $w(q)$ can be inverted, that is if $\partial_q w \neq 0$, $\forall q \in \mathbb{R}$, is a consequence of the cocycle condition Equation 2.19. By Equation 2.51 we see that also local univalence should be extended to $\hat{\mathbb{R}}$. This implies the following joining condition at spatial infinity

$$w(-\infty) = \begin{cases} w(+\infty), & \text{for } w(-\infty) \neq \pm\infty, \\ -w(+\infty), & \text{for } w(-\infty) = \pm\infty. \end{cases}$$ (2.53)

As illustrated by the non-univalent function $w = q^2$, the apparently natural choice $w(-\infty) = w(+\infty)$, one would consider also in the $w(-\infty) = \pm\infty$ case, does not satisfy local univalence.

We saw that the EP implied the QSHJE Equation 2.21. However, although this equation implies the SE, we saw that there are aspects concerning the canonical variables which arise in considering the QSHJE rather than the SE. In this respect a natural question is whether the basic facts of Quantum Mechanics (QM) also arise in our formulation. A basic point concerns a property of many physical systems such as energy quantization. This is a matter of fact beyond any interpretational aspect of QM. Then, as we used the EP to get the QSHJE, it is important to understand how energy quantization arises in our approach. According to the EP, the QSHJE contains all the possible information on a given system. Then, the QSHJE itself should be sufficient to recover the energy quantization including its structure. In the usual approach the quantization of the spectrum arises from the basic condition that in the case in which $\lim_{q\to\pm\infty}\mathcal{W} > 0$, the wavefunction should vanish at infinity. Once the possible solutions are selected, one also imposes the continuity conditions whose role in determining the possible spectrum is particularly transparent in the case of discontinuous potentials. For example, in the case of the potential well, besides the restriction on the spectrum due to the $L^2(\mathbb{R})$ condition for the wavefunction (a consequence of the probabilistic interpretation of the wavefunction), the spectrum is further restricted by the smoothly joining conditions. Since the SE contains the term $\partial_q^2 \psi$, the continuity conditions correspond to an existence condition for this equation. On the other hand, also in this case, the physical reason underlying this request is the interpretation of the wavefunction in terms of probability amplitude. Actually, strictly speaking, the continuity conditions come from the continuity of the probability density $\rho = |\psi|^2$. This density should also satisfy the continuity equation $\partial_t \rho + \partial_q j = 0$, where $j = i\hbar(\psi \partial_q \bar{\psi} - \bar{\psi} \partial_q \psi)/2m$. Since for stationary states $\partial_t \rho = 0$, it follows that in this case $j = $ cnst. Therefore, in the usual formulation, it is just the interpretation of the wavefunction in terms of probability amplitude, with the consequent meaning of ρ and j, which provides the physical motivation for imposing the continuity of the wavefunction and of its first derivative.

Now observe that in our formulation the continuity conditions arise from the QSHJE. In fact, Equation 2.52 implies continuity of ψ^D, ψ, with $\partial_q \psi^D$ and $\partial_q \psi$ differentiable, that is

$$EP \; \to \; (\psi^D, \psi) \; continuous \; and \; (\psi^{D'}, \psi') \; differentiable. \qquad (2.54)$$

In the following we will see that if $V(q) > E, \forall q \in \mathbb{R}$, then there are no solutions such that the ratio of two real linearly independent solutions of the SE corresponds to a local self-homeomorphism of $\hat{\mathbb{R}}$. The fact that this is an unphysical situation can also be seen from the fact that the case $V > E, \forall q \in \mathbb{R}$, has no classical limit. Therefore, if $V > E$ both at $-\infty$ and $+\infty$, a physical situation requires that there are at least two points where $V - E = 0$. More generally, if the potential is not continuous, $V(q) - E$ should have at least two turning points. Let us denote by q_- (q_+) the lowest (highest) turning point. Note that by Equation 2.23 we have

$$\int_{q_-}^{-\infty} dx \kappa(x) = -\infty, \qquad \int_{q_+}^{+\infty} dx \kappa(x) = +\infty,$$

where $\kappa = \sqrt{2m(V - E)}/\hbar$. Before going further, let us stress that what we actually need to prove is that, in the case Equation 2.23, the joining condition (Equation 2.53) requires that the corresponding SE has an $L^2(\mathbb{R})$ solution. Observe that while Equation 2.52, which however follows from the EP, can be recognized as the standard condition Equation 2.54, the other condition Equation 2.53, which still follows from the existence of the QSHJE, and therefore from the EP, is not directly recognized in the standard formulation. Since this leads to energy quantization, while in the usual approach one needs one more assumption, we see that there is quite a fundamental difference between the QSHJE and the SE. We stress that Equations 2.52 and 2.53 guarantee that w is a local self-homeomorphism of $\hat{\mathbb{R}}$.

Let us first show that the request that the corresponding SE has an $L^2(\mathbb{R})$ solution is a sufficient condition for w to satisfy Equation 2.53. Let $\psi \in L^2(\mathbb{R})$ and denote by ψ^D a linearly independent solution. As we will see, the fact that $\psi^D \not\propto \psi$ implies that if $\psi \in L^2(\mathbb{R})$, then $\psi^D \notin L^2(\mathbb{R})$. In particular, ψ^D is divergent both at $q = -\infty$ and $q = +\infty$. Let us consider the real ratio

$$w = \frac{A\psi^D + B\psi}{C\psi^D + D\psi},$$

where $AD - BC \neq 0$. Since $\psi \in L^2(\mathbb{R})$, we have

$$\lim_{q \to \pm\infty} w = \lim_{q \to \pm\infty} \frac{A\psi^D + B\psi}{C\psi^D + D\psi} = \frac{A}{C}, \tag{2.55}$$

that is $w(-\infty) = w(+\infty)$. In the case in which $C = 0$ we have

$$\lim_{q \to \pm\infty} w = \lim_{q \to \pm\infty} \frac{A\psi^D}{D\psi} = \pm\epsilon \cdot \infty,$$

where $\epsilon = \pm 1$. The fact that $A\psi^D/D\psi$ diverges for $q \to \pm\infty$ follows from the mentioned properties of ψ^D and ψ. It remains to check that if $\lim_{q \to -\infty} A\psi^D/D\psi = -\infty$, then $\lim_{q \to +\infty} A\psi^D/D\psi = +\infty$, and vice versa. This can be seen by observing that

$$\psi^D(q) = c\psi(q) \int_{q_0}^{q} dx \psi^{-2}(x) + d\psi(q),$$

$c \in \mathbb{R}\backslash\{0\}, d \in \mathbb{R}$. Since $\psi \in L^2(\mathbb{R})$ we have $\psi^{-1} \notin L^2(\mathbb{R})$ and $\int_{q_0}^{+\infty} dx \psi^{-2} = +\infty$, $\int_{q_0}^{-\infty} dx \psi^{-2} = -\infty$, implying that $\psi^D(-\infty)/\psi(-\infty) = -\epsilon \cdot \infty = -\psi^D(+\infty)/\psi(+\infty)$, where $\epsilon = \text{sgn } c$.

We now show that the existence of an $L^2(\mathbb{R})$ solution of the SE is a necessary condition to satisfy the joining condition Equation 2.53. We give two different proofs of this, one is based on the WKB approximation while the other one uses Wronskian arguments. In the WKB approximation, we have

$$\psi = \frac{A_-}{\sqrt{\kappa}} \exp\left(-\int_{q_-}^{q} dx\kappa\right) + \frac{B_-}{\sqrt{\kappa}} \exp\left(\int_{q_-}^{q} dx\kappa\right), \quad q \ll q_-, \tag{2.56}$$

and

$$\psi = \frac{A_+}{\sqrt{\kappa}} \exp\left(-\int_{q_+}^{q} dx\kappa\right) + \frac{B_+}{\sqrt{\kappa}} \exp\left(\int_{q_+}^{q} dx\kappa\right), \quad q \gg q_+. \tag{2.57}$$

In the same approximation, a linearly independent solution has the form

$$\psi^D = \frac{A_-^D}{\sqrt{\kappa}} \exp\left(-\int_{q_-}^{q} dx\kappa\right) + \frac{B_-^D}{\kappa} \exp\left(\int_{q_-}^{q} dx\kappa\right), \quad q \ll q_-.$$

Similarly, in the $q \gg q_+$ region we have

$$\psi^D = \frac{A_+^D}{\sqrt{\kappa}} \exp\left(-\int_{q_+}^{q} dx\kappa\right) + \frac{B_+^D}{\sqrt{\kappa}} \exp\left(\int_{q_+}^{q} dx\kappa\right), \quad q \gg q_+.$$

Note that Equations 2.56 and 2.57 are derived by solving the differential equations corresponding to the WKB approximation for $q \ll q_-$ and $q \gg q_+$, so that the coefficients of $\kappa^{-1/2} \exp \pm \int_{q_-}^{q} dx\kappa$, e.g., A_- and B_- in Equation 2.56, cannot be simultaneously vanishing. In particular, the fact that $\psi^D \not\propto \psi$ yields

$$A_- B_-^D - A_-^D B_- \neq 0, \quad A_+ B_+^D - A_+^D B_+ \neq 0. \tag{2.58}$$

Let us now consider the case in which, for a given E satisfying Equation 2.23, any solution of the corresponding SE diverges at least at one of the two spatial infinities, that is

$$\lim_{q \to +\infty} (|\psi(-q)| + |\psi(q)|) = +\infty. \tag{2.59}$$

This implies that there is a solution diverging both at $q = -\infty$ and $q = +\infty$. In fact, if two solutions ψ_1 and ψ_2 satisfy $\psi_1(-\infty) = \pm\infty$, $\psi_1(+\infty) \neq \pm\infty$ and $\psi_2(-\infty) \neq \pm\infty$, $\psi_2(+\infty) = \pm\infty$, then $\psi_1 + \psi_2$ diverges at $\pm\infty$. On the other hand, Equation 2.58 rules out the case in which all the solutions in their WKB approximation are divergent only at one of the two spatial infinities, say $-\infty$. Since, in the case Equation 2.23, a solution which diverges in the WKB approximation is itself divergent (and vice versa), we have that in the case Equation 2.23, the fact that all the solutions of the SE diverge only at one of the two spatial infinities cannot occur.

Let us denote by ψ a solution which is divergent both at $-\infty$ and $+\infty$. In the WKB approximation this means that both A_- and B_+ are non-vanishing, so that

$$\psi \underset{q \to -\infty}{\sim} \frac{A_-}{\sqrt{\kappa}} \exp\left(-\int_{q_-}^{q} dx\kappa\right), \quad \psi \underset{q \to +\infty}{\sim} \frac{B_+}{\sqrt{\kappa}} \exp\left(\int_{q_+}^{q} dx\kappa\right).$$

The asymptotic behavior of the ratio ψ^D/ψ is given by

$$\lim_{q \to -\infty} \frac{\psi^D}{\psi} = \frac{A_-^D}{A_-}, \quad \lim_{q \to +\infty} \frac{\psi^D}{\psi} = \frac{B_+^D}{B_+}.$$

Note that since in the case at hand any divergent solution also diverges in the WKB approximation, we have that Equation 2.59 rules out the case $A_-^D = B_+^D = 0$. Let us then suppose that either $A_-^D = 0$ or $B_+^D = 0$. If $A_-^D = 0$, then $w(-\infty) = 0 \neq w(+\infty)$. Similarly, if $B_+^D = 0$, then $w(+\infty) = 0 \neq w(-\infty)$. Hence, in this case w, and therefore the trivializing map, cannot satisfy Equation 2.53. On the other hand, also in the case in which both A_-^D and B_+^D are non-vanishing, w cannot satisfy Equation 2.53. For, if $A_-^D/A_- = B_+^D/B_+$, then

$$\phi = \psi - \frac{A_-}{A_-^D} \psi^D = \psi - \frac{B_+}{B_+^D} \psi^D,$$

would be a solution of the SE whose WKB approximation has the form

$$\phi = \frac{B_-}{\sqrt{\kappa}} \exp\left(\int_{q_-}^{q} dx\kappa\right), \quad q \ll q_-,$$

and

$$\phi = \frac{A_+}{\sqrt{\kappa}} \exp\left(-\int_{q_+}^{q} dx\kappa\right), \quad q \gg q_+.$$

Hence, if $A_-^D/A_- = B_+^D/B_+$, then there is a solution whose WKB approximation vanishes both at $-\infty$ and $+\infty$. On the other hand, we are considering the values of E satisfying Equation 2.23 and for which any solution of the SE has the property Equation 2.59. This implies that no solutions can vanish both at $-\infty$ and $+\infty$ in the WKB approximation. Hence

$$\frac{A_-^D}{A_-} \neq \frac{B_+^D}{B_+},$$

so that $w(-\infty) \neq w(+\infty)$. We also note that not even the case $w(-\infty) = \pm\infty = -w(+\infty)$ can occur, as this would imply that $A_- = B_+ = 0$, which in turn would imply, against the hypothesis, that there are solutions vanishing at $q = \pm\infty$. Hence, if for a given E satisfying Equation 2.23, any solution of the corresponding SE diverges at least at one of the two spatial infinities, we have that the trivializing map has a discontinuity at $q = \pm\infty$. As a consequence, the EP cannot be implemented in this case so that this value E cannot belong to the physical spectrum.

Therefore, the physical values of E satisfying Equation 2.23 are those for which there are solutions which are divergent neither at $-\infty$ nor at $+\infty$. On the other hand, from the WKB approximation and Equation 2.23, it follows that the non-divergent solutions must vanish both at $-\infty$ and $+\infty$. It follows that the only energy levels satisfying the property Equation 2.23, which are compatible with the EP, are those for which there exists the solution vanishing both at $\pm\infty$. On the other hand, solutions vanishing as $\kappa^{-1/2} \exp \int_{q_-}^{q} dx\kappa$ at $-\infty$ and $\kappa^{-1/2} \exp - \int_{q_+}^{q} dx\kappa$ at $+\infty$, with $P_\pm^2 > 0$, cannot contribute with an infinite value to $\int_{-\infty}^{+\infty} dx\psi^2$. The reason is that existence of the QSHJE requires that $\{e^{\frac{2i}{\hbar}S_0}, q\}$ be defined and this, in turn, implies that any solution of the SE must be continuous. On the other hand, since ψ is continuous, and therefore finite also at finite values of q, we have $\int_{q_a}^{q_b} dx\psi^2 < +\infty$ for all finite q_a and q_b. In other words, the only possibility for a continuous function to have a divergent value of $\int_{-\infty}^{+\infty} dx\psi^2$ comes from its behavior at $\pm\infty$. Therefore, since the implementation of the EP in the case Equation 2.23 requires that the corresponding E should admit a solution with the behavior

$$\psi \underset{q\to-\infty}{\sim} \frac{A_-}{\sqrt{\kappa}} \exp\left(\int_{q_-}^{q} dx\kappa\right), \quad \psi \underset{q\to+\infty}{\sim} \frac{B_+}{\sqrt{\kappa}} \exp\left(-\int_{q_+}^{q} dx\kappa\right),$$

we have the following basic fact

The values of E satisfying

$$V(q) - E \geq \begin{cases} P_-^2 > 0, & q < q_-, \\ P_+^2 > 0, & q > q_+, \end{cases} \tag{2.60}$$

are physically admissible if and only if the corresponding SE has an $L^2(\mathbb{R})$ solution.

We now give another proof of the fact that if W is of the type Equation 2.60, then the corresponding SE must have an $L^2(\mathbb{R})$ solution in order to satisfy Equation 2.53. In particular, we will show that this is a necessary condition. That this is sufficient has already been proved above.

Wronskian arguments, which can be found in Messiah's book [17], imply that if $V(q) - E \geq P_+^2 > 0, q > q_+$, then as $q \to +\infty$, we have $(P_+ > 0)$

  – There is a solution of the SE that vanishes at least as e^{-P_+q}.
  – Any other linearly independent solution diverges at least as e^{P_+q}.

Similarly, if $V(q) - E \geq P_-^2 > 0, q < q_-$, then as $q \to -\infty$, we have $(P_- > 0)$

  – There is a solution of the SE that vanishes at least as e^{P_-q}.
  – Any other linearly independent solution diverges at least as e^{-P_-q}.

These properties imply that if there is a solution of the SE in $L^2(\mathbb{R})$, then any solution is either in $L^2(\mathbb{R})$ or diverges both at $-\infty$ and $+\infty$. Let us show that the possibility that a solution vanishes only at one of the two spatial infinities is ruled out. Suppose that, besides the $L^2(\mathbb{R})$ solution, which we denote by ψ_1, there is a solution ψ_2 which is divergent only at $+\infty$. On the other hand, the above properties show that there exists also a solution ψ_3 which is divergent at $-\infty$. Since the number of linearly independent solutions of the SE is two, we have $\psi_3 = A\psi_1 + B\psi_2$. However, since ψ_1 vanishes both at $-\infty$ and $+\infty$, we see that $\psi_3 = A\psi_1 + B\psi_2$ can be satisfied only if ψ_2 and ψ_3 are divergent both at $-\infty$ and $+\infty$. This fact and the above properties imply that

If the SE has an $L^2(\mathbb{R})$ solution, then any solution has two possible asymptotics:

  – *Vanishes both at $-\infty$ and $+\infty$ at least as e^{P_-q} and e^{-P_+q} respectively.*
  – *Diverges both at $-\infty$ and $+\infty$ at least as e^{-P_-q} and e^{P_+q} respectively.*

Similarly, we have

If the SE does not admit an $L^2(\mathbb{R})$ solution, then any solution has three possible asymptotics:

  – *Diverges both at $-\infty$ and $+\infty$ at least as e^{-P_-q} and e^{P_+q} respectively.*
  – *Diverges at $-\infty$ at least as e^{-P_-q} and vanishes at $+\infty$ at least as e^{-P_+q}.*
  – *Vanishes at $-\infty$ at least as e^{P_-q} and diverges at $+\infty$ at least as e^{P_+q}.*

Let us consider the ratio $w = \psi^D/\psi$ in the latter case. Since any different choice of linearly independent solutions of the SE corresponds to a Möbius transformation of w, we can choose

$$\psi^D \underset{q \to -\infty}{\sim} a_- e^{P_-q}, \qquad \psi^D \underset{q \to +\infty}{\sim} a_+ e^{P_+q},$$

and

$$\psi \underset{q \to -\infty}{\sim} b_- e^{-P_-q}, \qquad \psi \underset{q \to +\infty}{\sim} b_+ e^{-P_+q},$$

were by $\sim$ we mean that ψ^D and ψ either diverge or vanish "at least as." Their ratio has the asymptotic

$$\frac{\psi^D}{\psi} \underset{q \to -\infty}{\sim} c_- e^{2P_- q} \to 0, \qquad \frac{\psi^D}{\psi} \underset{q \to +\infty}{\sim} c_+ e^{2P_+ q} \to \pm\infty,$$

so that w cannot satisfy Equation 2.53. This concludes the alternative proof of the fact that, in the case Equation 2.60, the existence of the $L^2(\mathbb{R})$ solution is a necessary condition in order for Equation 2.53 to be satisfied. The fact that this is a sufficient condition has been proved previously in deriving Equation 2.55.

The above results imply that the usual quantized spectrum arises as a consequence of the EP.

Let us note that we are considering real solutions of the SE. Thus, apparently, in requiring the existence of an $L^2(\mathbb{R})$ solution, one should specify the existence of a real $L^2(\mathbb{R})$ solution. However, if there is an $L^2(\mathbb{R})$ solution ψ, this is unique up to a constant, and since also $\bar{\psi} \in L^2(\mathbb{R})$ solves the SE, we have that an $L^2(\mathbb{R})$ solution of the SE is real up to a phase.

In this chapter we have discussed two main theorems that underlie the formulation of quantum mechanics from an EP. One should regard this postulate as providing a novel starting point for formulating quantum mechanics, and ultimately for formulating quantum gravity. It also provides an arena to reexamine many of the tenants of the conventional approaches, and in this regard we note the related work of several other authors [18–92], and, in particular, that of Floyd [18–29].

BIBLIOGRAPHY

1. A.E. Faraggi and M. Matone, *Phys. Lett. B* **450**, 34 (1999).
2. A.E. Faraggi and M. Matone, *Phys. Lett. B* **437**, 369 (1998).
3. A.E. Faraggi and M. Matone, *Phys. Lett. A* **249**, 180 (1998).
4. A.E. Faraggi and M. Matone, *Phys. Lett. B* **445**, 77 (1998).
5. A.E. Faraggi and M. Matone, *Phys. Lett. B* **445**, 357 (1999).
6. A.E. Faraggi and M. Matone, *Int. J. Mod. Phys. A* **15**, 1869 (2000).
7. G. Bertoldi, A.E. Faraggi and M. Matone, *Class. Quant. Grav.* **17**, 3965 (2000).
8. M. Matone, *Found. Phys. Lett.* **15**, 311 (2002).
9. M. Matone, hep-th/0212260.
10. M. Matone, *Braz. J. Phys.* **35**, 316 (2005).
11. A.E. Faraggi, D.V. Nanopoulos and K.J. Yuan, *Nucl. Phys. B* **335**, 347 (1990).
12. G.B. Cleaver, A.E. Faraggi and D.V. Nanopoulos, *Phys. Lett. B* **455**, 135 (1999).
13. B. Assel, K. Christodoulides, A.E. Faraggi, C. Kounnas and J. Rizos, arXiv:0910.3697 [hep-th].
14. M. Matone, *Int. J. Mod. Phys. A* **10**, 289 (1995).
15. V. Ovsienko, *Lagrange Schwarzian Derivative and Symplectic Sturm Theory*, CPT-93-P-2890.
16. P. Di Francesco, P. Mathieu and D. Sénéchal, *Conformal Field Theory*, Springer, Berlin (1996).
17. A. Messiah, *Quantum Mechanics*, Vol. 1, North-Holland, Amsterdam (1961).
18. E.R. Floyd, *Phys. Rev. D* **25**, 1547 (1982).

19. E.R. Floyd, *Phys. Rev. D* **26**, 1339 (1982).
20. E.R. Floyd, *Phys. Rev. D* **29**, 1842 (1984).
21. E.R. Floyd, *Phys. Rev. D* **34**, 3246 (1986).
22. E.R. Floyd, *Int. J. Theor. Phys.* **27**, 273 (1988).
23. E.R. Floyd, *Phys. Lett. A* **214**, 259 (1996).
24. E.R. Floyd, *Found. Phys. Lett.* **9**, 489 (1996).
25. E.R. Floyd, *Found. Phys. Lett.* **13**, 235 (2000).
26. E.R. Floyd, *Int. J. Mod. Phys. A* **14**, 1111 (1999).
27. E.R. Floyd, *Int. J. Mod. Phys. A* **15**, 1363 (2000).
28. E.R. Floyd, quant-ph/0302128.
29. E.R. Floyd, quant-ph/0307090.
30. G. Reinisch, *Physica A* **206**, 229 (1994).
31. G. Reinisch, *Phys. Rev. A* **56**, 3409 (1997).
32. A.E. Faraggi, hep-th/0411118.
33. R. Carroll, *Can. J. Phys.* **77**, 319 (1999).
34. R. Carroll, quant-ph/0309023.
35. R. Carroll, quant-ph/0401082.
36. R. Carroll, gr-qc/0406004.
37. R. Carroll, quant-ph/0403156.
38. R. Carroll, gr-qc/0501045.
39. A. Bouda, *Found. Phys. Lett.* **14**, 17 (2001).
40. A. Bouda, *Int. J. Mod. Phys. A* **18**, 3347 (2003).
41. A. Bouda and T. Djama, *Phys. Lett. A* **285**, 27 (2001).
42. A. Bouda and T. Djama, *Phys. Scripta* **66**, 97 (2002).
43. A. Bouda and F. Hammad, *Acta Phys. Slov.* **52**, 101 (2002).
44. M.R. Brown, quant-ph/0102102.
45. T. Djama, quant-ph/0111121.
46. T. Djama, quant-ph/0111142.
47. T. Djama, quant-ph/0201003.
48. T. Djama, quant-ph/0404098.
49. A.E. Faraggi, hep-th/9910042.
50. R. Carroll, *Quantum Theory, Deformation and Integrability*, Elsevier, North-Holland, Amsterdan (2000).
51. J.M. Delhotel, quant-ph/0401063.
52. R.E. Wyatt, *Quantum Dynamics with Trajectories*, Springer, New York (2005).
53. M. Matone, *Phys. Lett. B* **357**, 342 (1995).
54. M. Matone, *Phys. Rev. D* **53**, 7354 (1996).
55. G. Bonelli and M. Matone, *Phys. Rev. Lett.* **76**, 4107 (1996).
56. G. Bonelli and M. Matone, *Phys. Rev. Lett.* **77**, 4712 (1996).
57. G. Bonelli, M. Matone and M. Tonin, *Phys. Rev. D* **55**, 6466 (1997).
58. M. Matone, *Phys. Rev. Lett.* **78**, 1412 (1997).
59. A.E. Faraggi and M. Matone, *Phys. Rev. Lett.* **78**, 163 (1997).
60. R.W. Carroll, hep-th/9607219.
61. R.W. Carroll, hep-th/9610216.
62. R.W. Carroll, hep-th/9702138.
63. R.W. Carroll, *Nucl. Phys. B* **502**, 561 (1997).
64. R.W. Carroll, *Lect. Notes Phys.* **502**, 33 (1998).
65. I.V. Vancea, *Phys. Lett. B* **480**, 331 (2000).

66. I.V. Vancea, *Phys. Lett. A* **321**, 155 (2004).
67. M.A. De Andrade and I.V. Vancea, *Phys. Lett. B* **474**, 46 (2000).
68. M.C.B. Abdalla, A.L. Gadelha and I.V. Vancea, *Phys. Lett. B* **484**, 362 (2000).
69. J. Butterfield, quant-ph/0210140.
70. J.M. Isidro, *Int. J. Mod. Phys. A* **16**, 3853 (2001).
71. J.M. Isidro, quant-ph/0105012.
72. J.M. Isidro, *J. Phys. A* **35**, 3305 (2002).
73. J.M. Isidro, quant-ph/0112032.
74. J.M. Isidro, *J. Geom. Phys.* **41**, 275 (2002).
75. J.M. Isidro, *Phys. Lett. A* **301**, 210 (2002).
76. J.M. Isidro, hep-th/0204178.
77. J.M. Isidro, *Mod. Phys. Lett. A* **18**, 1975 (2003).
78. J.M. Isidro, hep-th/0304175.
79. J.M. Isidro, *Mod. Phys. Lett. A* **19**, 349 (2004).
80. J.M. Isidro, *Phys. Lett. A* **317**, 343 (2003).
81. J.M. Isidro, quant-ph/0310092.
82. J.M. Isidro, *Mod. Phys. Lett. A* **19**, 1733 (2004).
83. J.M. Isidro, hep-th/0407161.
84. J.M. Isidro, *Mod. Phys. Lett. A* **19**, 2339 (2004).
85. J.M. Isidro, quant-ph/0411166.
86. D.M. Appleby, *Found. Phys.* **29**, 1863 (1999).
87. D.M. Appleby, *Phys. Rev. A* **65**, 022105 (2002).
88. D.M. Appleby, quant-ph/0308114.
89. B. Poirier, *J. Chem. Phys.* **121**, 4501 (2004).
90. F. Girelli, E.R. Livine and D. Oriti, *Nucl. Phys. B* **708**, 411 (2005).
91. M.V. John, *Found. Phys. Lett.* **15**, 329 (2002).
92. M.V. John, quant-ph/0102087.

3 Quantum Trajectories and Entanglement

Edward R. Floyd

CONTENTS

3.1   INTRODUCTION

Entanglement phenomena offer fine, didactic examples that demonstrate the application of quantum trajectories. Quantum entanglement phenomena are tractable to the quantum trajectory representation. Quantum trajectories render insight into entanglement phenomena of quantum systems. Entanglement is one of the prime features that distinguishes quantum mechanics from classical mechanics. Entanglement implies nonlocality as confirmed by the Aspect experiments [1–3]. Entangled quantum systems have nonclassical correlations that manifest nonlocal behavior for the system. The quantum trajectory representation can describe how entanglement induces nonlocality [4]. Within the foundations of quantum mechanics, a quantum trajectory representation of entangled systems is the key to resolving the nonintuitive behavior of wave packet spreading [4], the which-way paradox of the double-slit experiment [5], and the Einstein–Podolsky–Rosen (EPR) paradox [6].

The quantum trajectory representation of quantum mechanics is a nonlocal, phenomenological theory that is deterministic. Herein, "deterministic" means that if without disturbing a system one can predict with certainty the value of a physical quantity (the quantum trajectory), then there exists an element of physical reality that corresponds to such a physical quantity (the quantum trajectory).

The quantum trajectory representation used in this chapter differs from the representation of Bohmian mechanics: the two representations have different equations of motion despite having a common quantum reduced action (Hamilton's characteristic function) [7–14]. The quantum trajectory representation with its attendant bipolar *Ansatz* [9, 12–14], which may be extended to a multipolar *Ansatz*, is well suited for application to entangled systems. While a bipolar *Ansatz* is used herein for convenience, any *Ansatz* is unnecessary for our quantum trajectory representation. Faraggi and Matone derived a quantum equivalence principle, where the underlying quantum Hamilton–Jacobi equation for various systems may be reversibly mapped into each other by a simple coordinate transform, to render the quantum trajectory representation without any of the Copenhagen postulates [13].

The goal herein is didactic: to make quantum trajectories more accessible to uninitiated workers. Accessibility is provided by incorporating knowledge of the wave function without sacrificing the profoundness and insight of quantum trajectories. The techniques developed herein are not limited to entanglement phenomena.

In this chapter, we present a recipe for describing the quantum trajectory of an entangled system. We consider entangled systems of two components, but the extension to multicomponent systems is straightforward. We synthesize an "entanglement molecule" from its entangled components. The entanglement molecule is the entity that represents the entire entangled system under consideration. This system has entangled components which may be latent quantum anyons or just components (spectral or otherwise) of an anyon. In the former case, entangled anyons remain latent until reified by decoherence. In the latter case, the entanglement molecule is an anyon self-entangled by its own multispectral components. The entanglement molecule represents the entangled system as a single entity. We deliberately investigate the behavior of the entanglement molecule rather than the individual behaviors of its two components. We develop the entanglement molecule's wave function in polar form and its reduced action, both of which manifest entanglement. The reduced action of the entanglement molecule is its generator of quantum motion and contains all the information of the entanglement system. We next apply Jacobi's theorem to the reduced action to generate the equation of quantum motion for the entanglement molecule to render its quantum trajectory. The resultant quantum trajectory manifests entanglement and has retrograde segments interspersed between segments of forward motion. This alternating of forward and retrograde segments generates nonlocality and, within the entanglement molecule, there exists entanglement at a distance. We briefly apply the quantum trajectory representation to three fundamental, entanglement problems of quantum mechanics: wave packet spreading, the double-slit experiment, and EPR. In particular, decomposition of the equation of quantum motion for the EPR entanglement molecule, while rendering the latent classical behavior of the two scattered particles, also reveals an emergent "entanglon" that maintains the entanglement between the two scattered latent particles as they become progressively displaced farther from each other.

3.2   RECIPE

Let us consider two components, P_1 and P_2, with spatial wave functions , $\psi_1(x)$ and $\psi_2(x)$, that interact for mathematical simplicity through an instantaneous common repulsive impulse at time $t = 0$. Subsequently, the components are entangled for $t > 0$. The positions (x_1, x_2) of the two components are co-located at the time of impulse interaction, $t = 0$ at $x_1, x_2 = 0$. The masses of P_1 and P_2 are respectively given by m and $\alpha^2 m$ where $\alpha > 0$. The system mass is $(1 + \alpha^2)m$. The two components, which may be latent while entangled, subsequently recoil from each other in opposite directions after impulse at $t = 0$ with their spatial wave functions given by

$$\psi_1(x) = \exp(ikx), \quad \psi_2(x) = \alpha \exp(-ikx + i\beta); \quad t > 0 \tag{3.1}$$

where $k = [2(1 + \alpha^2)mE]^{1/2}/\hbar$, E is the energy of the entanglement system, and $-\pi < \beta \le \pi$. The center-of-mass inertial reference frame is chosen so that Equation 3.1 is true. The term β represents a phase shift between the two components. In our chosen inertial reference frame, ψ_1 and ψ_2 form a set of independent solutions of the Schrödinger equation for energy E. For the quantum trajectory representation, the Wronskian of the set (ψ_1, ψ_2) is normalized rather than normalization in accordance with the Born probability postulate [11]. However, ψ_1 and ψ_2, as will be shown later, appear in the generator of motion as a Möbius transformation $(a\psi_1 - b\psi_2)/(c\psi_1 - d\psi_2)$, $ad - bc \ne 0$ where a, b, c, and d are coefficients [13]. Hence, the relative normalization of ψ_1 and ψ_2 manifested by the factor α suffices.

The choice of inertial reference system, for which ψ_1 and ψ_2 have the wave numbers k and $-k$, induces the relationship $x_1 = -x_2/\alpha^2$ for the positions of the two components for $t > 0$. This is an extension of Fine's conservation of relative position [15]. Conservation of relative position induces loss of parameter independence and outcome independence [16].

We assume herein that no matter how far apart the two components may become, they will remain entangled consistent with the Aspect experiments [1–3]. In other words, the initially entangled system remains "inseparable" even for $t \gg 0$ and $x_1 - x_2 \gg 1$.

For entanglement in the wave function representation of quantum mechanics, we may synthesize an entanglement system as a simple polar wave function, ψ_{ent}, from the pair (bipolar wave function) of entangled components, ψ_1 and ψ_2, by Bohm's method [7]. Bohm's recipe for expressing complex ψ in polar format is given by

$$\psi_{ent} = [(\Re\psi_{ent})^2 + (\Im\psi_{ent})^2]^{1/2} \exp\left[\frac{i}{\hbar} \tan^{-1}\left(\frac{\Im\psi_{ent}}{\Re\psi_{ent}}\right)\right] \tag{3.2}$$

where $\Re\psi_{ent} = \Re\psi_1 + \Re\psi_2$ and $\Im\psi_{ent} = \Im\psi_1 + \Im\psi_2$. By Bohm, the wave function for the entangled system, $\psi_{ent}(x)$, whose components are specified by Equation 3.1, is given by Refs [4, 8]

$$\overbrace{\psi_{ent}(x) = \psi_1(x) + \psi_2(x) = \exp(ikx) + \alpha \exp(-ikx + i\beta)}^{\text{bipolar wave function}}$$

$$\underbrace{= [1 + \alpha^2 + 2\alpha \cos(2kx + \beta)]^{1/2} \exp\left[i \arctan\left(\frac{\sin(kx) - \alpha \sin(kx + \beta)}{\cos(kx) + \alpha \cos(kx + \beta)}\right)\right]}_{\text{polar wave function is still an eigenfuncion for } E = \hbar^2 k^2/[2m(1 + \alpha^2)]}$$

$$\tag{3.3}$$

where we have dropped, for entanglement, the subscript for a component's (latent) position x by the extension of conservation of relative position. The above construction is just the superpositional principle at work. It converts a bipolar *Ansatz* to a polar *Ansatz*.

The wave function for the entangled molecule, ψ_{ent} by Equation 3.3, does not uniquely specify its entangled components. For example, the entanglement of a standing wave function, $2\alpha \exp(-i\beta/2)\cos(kx + \beta/2)$, and a running wave function, $(1 - \alpha)\exp(ikx)$, would also render the same ψ_{ent}. By the superpositional theorem, ψ_{ent} remains valid for any combination of particles whose collective sum of their spectral components is consistent with the right side of the upper line of Equation 3.3.

In the wave function representation, ψ_{ent} of Equation 3.3 is inherently nonlocal, for ψ_{ent} is not factorable, that is $\psi_{ent} \neq K\psi_1\psi_2$ where K is a constant. Any measurement upon ψ_{ent} concurrently measures ψ_1 and ψ_2. Likewise, in the quantum trajectory representation, entanglement implies that the reduced action for the entangled system, W_{ent} is inseparable into its components, that is $W_{ent} \neq W_{\text{component 1}} + W_{\text{component 2}}$.

A generator of the motion for the entanglement molecule is its reduced action, W_{ent}. For one dimension, x, the reduced action for a quantum anyon, say Z, is specified by the quantum stationary Hamilton–Jacobi equation (QSHJE) given by

$$\underbrace{\frac{(\partial W_Z/\partial x)^2}{2m_Z}}_{\text{classical HJE for } Z} + V - E_Z = \underbrace{\frac{\hbar^2}{4m_Z}\overbrace{\left[\frac{\partial^3 W_Z/\partial x^3}{\partial W_Z/\partial x} - \frac{3}{2}\left(\frac{\partial^2 W_Z/\partial x^2}{\partial W_Z/\partial x}\right)^2\right]}^{\text{Schwarzian derivative}}}_{\text{negative of Bohm's quantum potential}} = 0. \quad (3.4)$$

Rather than solving the QSHJE, Equation 3.4, for the reduced action, W_Z, we offer an alternative method that is applicable to entanglement phenomena with complex wave functions. This alternative method cross pollinates the quantum trajectory representation of quantum mechanics with the wave function representation. The polar *Ansatz* for Z may be given by

$$\psi_Z = \frac{\exp(iW_Z/\hbar)}{(\partial W_Z/\partial x)^{1/2}}. \quad (3.5)$$

Substituting Equation 3.5 into the Schrödinger equation yields the QSHJE, Equation 3.4. This gives us a short cut for deriving the reduced action for the entanglement molecule. Its reduced action may be simply extracted from the entangled wave function ψ_{ent} given in Equations 3.3 and 3.5. The reduced action for the entanglement molecule is given by Refs [4, 8]

$$W_{ent} = \hbar \arctan\left(\frac{\sin(kx) - \alpha\sin(kx + \beta)}{\cos(kx) + \alpha\cos(kx + \beta)}\right). \quad (3.6)$$

Whereas we have extracted the reduced action, W_{ent}, from the Schrödinger wave function herein for convenience, the reduced action still may be derived from Faraggi and Matone's quantum equivalence principle independent of the Schrödinger formulation of quantum mechanics [13]. The reduced action, W_{ent}, still remains the solution [4] of the QSHJE, Equation 3.4, for $E = \hbar^2 k^2/[2m(1 + \alpha^2)]$. As such, the quantum

trajectory representation is not tainted with Copenhagen axioms by our cross polli-
nating procedure that extracts W_{ent} from ψ_{ent}. Equation 3.6 posits a deterministic
W_{ent} in Euclidean space while Copenhagen posits ψ_{ent} with its probability amplitude
in Hilbert space. The absolute value of W_{ent} increases monotonically with x as the
arc tangent function in W_{ent} jumps to the next Riemann sheet wherever the argument
of the arc tangent function becomes singular.

The conjugate momentum for an entanglement molecule is given by

$$\partial W_{ent}/\partial x = \frac{\hbar k}{[1 + \alpha^2 + 2\alpha \cos(2kx + \beta)]}. \tag{3.7}$$

The conjugate momentum manifests entanglement by the cosine term in the denominator on the right side of Equation 3.7. By Equation 3.7, the conjugate momentum
is not the mechanical momentum, i.e., $\partial W_{ent}/\partial x \neq m(1 - \alpha^2)\dot{x}$ as so assumed in
Bohmian mechanics.

In the quantum trajectory representation of quantum mechanics as also in the
classical stationary Hamilton–Jacobi theory, time parametrization is given by Jacobi's
theorem. Hence, the equation of quantum motion for the entanglement molecule is
generated from W_{ent} by Jacobi's theorem as

$$\underbrace{t_{ent} - \tau = \frac{\partial W_{ent}}{\partial E}}_{\text{Jacobi's theorem}} = \frac{mx(1 - \alpha^2)}{\hbar k[1 + \alpha^2 + 2\alpha \cos(2kx + \beta)]} \tag{3.8}$$

where t is time and τ specifies the epoch. Note that our equation of quantum motion,
Equation 3.8, differs with that of Bohmian mechanics.

For completeness, it is noted that a cross pollination strategy between the wave
and quantum trajectory representations for real wave functions (e.g., bound states) is
possible. The reduced action may be represented as a function of the set of independent
solutions (ψ, ψ^D) to the Schrödinger equation by Refs [11, 13, 19]

$$W = \hbar \arctan\left(\frac{a\psi + b\psi^D}{c\psi + d\psi^D}\right), \quad ad - bc \neq 0$$

where the coefficients a, b, c, and d are determined by the initial conditions.

The quantum trajectory for the entangled molecule is deterministic and is in
Euclidean space with time as the fourth dimension. Determinism begets realism.
The quantum trajectories manifest realism for the position of the entangled molecule
as a function of time that can be predicted with certainty without disturbing the system [17]. In the forgoing, "certainty" is appropriate for two reasons. First, in the
Copenhagen interpretation, the Heisenberg uncertainty principle uses an insufficient
subset of initial values of the necessary and sufficient set of initial values that specify
unique quantum motion [10, 13, 18]. Lest we forget, the quantum trajectory representation contains more information than the Schrödinger wave function representation
and renders a unique, deterministic quantum trajectory [10, 13, 18, 19]. And second,
the quantum trajectories exist in Euclidean space with time while the Schrödinger
wave function representation is formulated in Hilbert space [20–22]. Realism follows

for the entangled molecule maintains a precise, theoretical quantum trajectory independent of it being measured. Nevertheless, nothing herein implies that measuring an entanglement molecule does not physically disturb the entanglement molecule in compliance with Bohr's complementarity principle. The measurement process may trigger decoherence.

Generating the equation of quantum motion, Equation 3.8, with Jacobi's theorem is consistent with Peres's quantum clocks [23] where $t - \tau = \hbar(\partial \varphi / \partial E)$ where $\hbar \varphi$ is analogous to the reduced action herein and in turn is the phase of the complex wave function of the particle under consideration. Equation 3.8 is a generalization of Peres's quantum clocks for it applies to situations where the wave function is real [9, 13, 19].

3.3 DIDACTIC EXAMPLES

We present three applications of the quantum trajectory representation to quantum mechanics that give insight into the weirdness of quantum entanglement. These applications are didactic examples. The three examples are wave packet spreading [4], the quantum Young's diffraction experiment (a simplified double-slit diffraction) [5], and EPR phenomena [6]. The purpose herein is didactic rather than investigative. Reporting physical findings are consequently limited herein. For a more complete physical understanding, the interested reader should consult the afore-noted references.

3.3.1 WAVE PACKET SPREADING

Let us consider a dichromatic wave function $\psi_d(x)$ consisting of two spectral components which are given by $\psi_{d1} = \exp(ikx)$ and $\psi_{d2} = (1/2)\exp(-ikx)$. For simplicity, we have set $m = 1$ and $\hbar = 1$. As these are spectral components, we describe ψ_d as self-entangled, which generalizes our entanglement molecule to be a self-entangled particle. We avoid solving the QSHJE for the dichromatic reduced action by calculating W_d from the two spectral components, ψ_{d1} and ψ_{d2}. By Equation 3.6, the dichromatic reduced action is given by

$$W_d = \hbar \arctan \left(\frac{\sin(kx) - (1/2)\sin(kx + \beta)}{\cos(kx) + (1/2)\cos(kx + \beta)} \right). \tag{3.9}$$

The subsequent equation of quantum motion is generated from the reduced action by applying Jacobi's theorem. By Equation 3.8, the equation of quantum motion for the dichromatic particle is given by

$$t_d - \tau_d = \frac{\partial W_d}{\partial E} = \frac{(3/4)mx}{\hbar k[(5/4) + \cos(2kx + \beta)]}. \tag{3.10}$$

We note that the equation of quantum motion could have been solved by directly substituting ψ_{d1} and ψ_{d2} into the right side of Equation 3.8.

The motion for the dichromatic particle is exhibited in Figure 3.1 for $\tau_d = 0$, $k = 2/\pi$ and $\beta = 0, \pi$. The physical insight is limited herein to, firstly, the existence

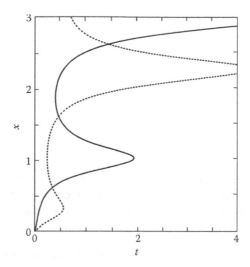

FIGURE 3.1 Quantum motion, $x(t)$, of the dichromatic particle for $\tau = 0$, $\alpha = 0.5$, $k = \pi/2$ and $\beta = 0$ as a solid line and for $\beta = \pi$ as a dashed line. (Reprinted with permission from Floyd, E.R., *Found. Phys.*, 37, 1386–402, 2007.)

of quantum trajectory segments of retrograde motion in time interspersed between segments of forward motion. This induces nonlocality as the dichromatic particle may be at two different locations simultaneously. Secondly, the quantum trajectories are bound within a wedge in Figure 3.1. As the dichromatic particle advances in x within the wedge, the quantum trajectory segments, either forward or retrograde, increase in duration. This manifests wave packet spreading. And thirdly, the turning points at local temporal minima manifest destructive interference between ψ_{d1} and ψ_{d2} while turning points at local temporal maxima manifest reinforcement. Further physical insight has been presented in Ref. [4].

3.3.2 SIMPLIFIED DOUBLE-SLIT EXPERIMENT

We now examine a quantum Young's diffraction experiment for a single quantum particle in the near region. While this near region is well within the domain of Fresnel diffraction, Fresnel approximations are not needed as the results may be determined exactly in closed form. The quantum Young diffraction experiment represents a simplified double-slit experiment with two point sources separated vertically by a distance a. As a double source problem, a prolate spheroidal coordinate system (ξ, η, ϕ) is natural for this quantum Young diffraction experiment. The prolate spheroidal coordinates have scale factors (metrical coefficients) modified by Morse and Feshbach [24] where ξ is the ellipsoidal coordinate, η is the hyperboloidal coordinate, and ϕ is the azimuthal coordinate. The two sources each emit a spherical wave function that are entangled with each other and synthesize a dispherical wave function ψ_Y for the quantum Young diffraction experiment. The two sources are $(\xi, \eta, \phi) = (1, \pm 1, \phi)$

or at the focal points of prolate spheroidal coordinates. The lower source emits the component,

$$\psi_{Y\ell} = \frac{\exp(i k_\ell \cdot r_\ell)}{r_\ell} = \frac{\exp(ikr_\ell)}{r_\ell}, \quad r_\ell = (\xi + \eta)\frac{a}{2}$$

while the upper source emits

$$\psi_{Yu} = \frac{\exp(i k_u \cdot r_u)}{r_u} = \frac{\exp(ikr_u)}{r_u}, \quad r_u = (\xi - \eta)\frac{a}{2}.$$

The self-entangled dispherical wave function, ψ_Y, of the diffracted quantum particle can be synthesized from entanglement of its components, $\psi_{Y\ell}$ and ψ_{Yu}, by using the general Bohm's procedure, Equation 3.2. The simpler Equation 3.3 is not applicable here. This yields [5]

$$\psi_Y = \psi_{Y\ell} + \psi_{Yu} = [r_\ell^{-2} + r_u^{-2} + 2r_\ell^{-1}r_u^{-1}\cos(k_\ell \cdot r_\ell - k_u \cdot r_u)]^{1/2}$$

$$\cdot \exp\left[i \arctan\underbrace{\left(\frac{r_u\sin(k_\ell \cdot r_\ell) + r_\ell\sin(k_u \cdot r_u)}{r_u\cos(k_\ell \cdot r_\ell) + r_\ell\cos(k_u \cdot r_u)}\right)}_{W_Y/\hbar}\right]$$

where W_Y is the the reduced action that accounts for the self-entanglement within the dispherical particle.

The reduced action, W_Y, is given in prolate spheroidal coordinates as

$$W_Y = \hbar \arctan\left(\frac{(\xi - \eta)\sin[k(\xi + \eta)a/2] + (\xi + \eta)\sin[k(\xi - \eta)a/2]}{(\xi - \eta)\cos[k(\xi + \eta)a/2] + (\xi + \eta)\cos[k(\xi - \eta)a/2]}\right).$$

From the above equation, W_Y is azimuthally invariant (independent of ϕ). We also observe that W_Y is not separable in ξ and η (the customary treatment of the double-slit experiment is rendered separable by assuming Fraunhofer approximations for the far region). We also note that neither ξ nor η are cyclic coordinates of W_Y.

We develop the quantum trajectories for the self-entangled quantum particle by Jacobi's theorem $\beta_\eta = \partial W_d/\partial \eta_a$ where η_a is the constant of motion using the value of the η-asymptote, η_a, of the quantum trajectory and β_η is a constant coordinate of Hamilton–Jacobi theory. For a quantum trajectory to egress from either source, we set $\beta_\eta = 0$. The consequent quantum trajectory equation in prolate spheroidal coordinates is given by

$$(\xi - \eta)^2\{\eta_a[(\xi^2 - 1)(1 - \eta^2)]^{1/2} - (1 - \eta_a^2)^{1/2}(\xi\eta + 1)\}$$
$$+ (\xi + \eta)^2\{\eta_a[(\xi^2 - 1)(1 - \eta^2)]^{1/2} - (1 - \eta_a^2)^{1/2}(\xi\eta - 1)\}$$
$$+ 2(\xi^2 - \eta^2)\cos(ka\eta)\{\eta_a[(\xi^2 - 1)(1 - \eta^2)]^{1/2} - (1 - \eta_a^2)^{1/2}\xi\eta\} = 0. \quad (3.11)$$

The quantum trajectory equation is also given for didactic reasons in the more familiar cylindrical coordinates as

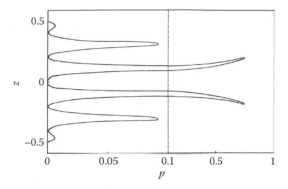

FIGURE 3.2 Quantum trajectory for the quantum disherical particle originating from the upper source with constant of the motion $\eta_a = -\sin(\pi/18)$. Note the change of scale in ρ by a factor of 10 at $\rho = 0.1$, denoted by the *dashed vertical line*, to facilitate exposition. (Reprinted with permission from Floyd, E.R., *Found. Phys.*, 37, 1403–20, 2007.)

$$\underbrace{r_u^2 \left[(z + a/2) - \frac{k_z}{k_\rho} \rho \right]}_{\text{lower source alone}} + \underbrace{r_\ell^2 \left[(z - a/2) - \frac{k_z}{k_\rho} \rho \right]}_{\text{upper source alone}} + \underbrace{2 r_\ell r_u \cos[k(r_u - r_\ell)] \left(z - \frac{k_z}{k_\rho} \rho \right)}_{\text{interference effects}} = 0$$

(3.12)

where $r_\ell = (\rho^2 + z^2)^{1/2}$ and $r_u = (\rho^2 - z^2)^{1/2}$. Both Equations 3.11 and 3.12 render an implicit function for the quantum trajectory in either coordinate system.

The segment of the quantum trajectory between the upper and lower sources for the dichromatic particle originating from the upper source is exhibited in cylindrical coordinates in Figure 3.2 for $k = 15.2$ and constant of the motion $\eta_a = -\sin(\pi/18)$. At the origin, $(\rho, z, \phi) = (0, 0, \phi)$, the azimuthal coordinate jumps [5] by π. The quantum trajectory in the ρ, z-plane has mirror symmetry about the plane $z = 0$ or, for prolate spheroidal coordinates, $\eta = 0$. The quantum trajectory also has forward and retrograde segments interspersed with each other. This induces the nonlocal effect that the quantum trajectory transits both the upper and lower sources simultaneously. Furthermore, the transit time from any point on the quantum trajectory to either the upper or lower source is the same by mirror symmetry. The resolution of the which-way paradox of the double-slit experiment is "both ways, simultaneously." Further physical insight is presented in Ref. [5].

3.3.3 QUANTUM TRAJECTORY FOR EPR-MOLECULE

Our final example applies quantum trajectories to the EPR paradox. This EPR example has much in common with the example of wave packet spreading by a dichromatic particle discussed in Section 3.3.1. While the dichromatic particle is self-entangled by its spectral components, the EPR-molecule is synthesized from two entangled particles whose wave functions are given by $\psi_{EPR1} = \exp(ikx)$ and $\psi_{EPR2} = (1/2)\exp(-ikx)$ where again $m = 1$ and $\hbar = 1$. Then, the EPR-molecule's reduced action, W_{EPR}, and its quantum trajectory follow respectively from Equations 3.9 and 3.10 for the dichromatic particle where now the subscript "*EPR*" is substituted for the subscript "*d*" wherever they appear.

However, the analysis of the equation of quantum motion for the EPR-molecule differs because it is entangled rather than self-entangled. In other words, the entangled components manifest latent particles and not merely spectral components. We may dissect the equation of quantum motion for the EPR-molecule rendering

$$
t_{EPR} - \tau_{EPR} = \partial W_{EPR}/\partial E = \frac{mx(1 - \alpha^2)}{\hbar k[1 + \alpha^2 + 2\alpha \cos(2kx + \beta)]}
$$

$$
= \underbrace{\frac{mx}{\hbar k} \frac{1}{1 + \alpha^2}}_{\text{latent particle 1}} - \underbrace{\frac{mx}{\hbar k} \frac{2\alpha \frac{1-\alpha^2}{1+\alpha^2} \cos(2kx + \beta)}{1 + \alpha^2 + 2\alpha \cos(2kx + \beta)}}_{\text{entanglon}} - \underbrace{\frac{mx}{\hbar k} \frac{\alpha^2}{1 + \alpha^2}}_{\text{latent particle 2}}
$$

$$(3.13)$$

where the latent, expected linear motions for particles 1 and 2 are exhibited. There remains a third term in Equation 3.13 attributed to an emergent "entanglon" that maintains entanglement between the two entangled latent particles. The entanglon does not represent a "force" between the two entangled latent particles for each latent particle maintains its latent linear motion as shown by Equation 3.13. In the limit $\alpha \to 1$ (the historical EPR paradox), then the entanglon maintains spooky entanglement at a distance instantaneously. Note that an attribution of "entanglon" to the analogous "interference effects" term in Equation 3.12 is inappropriate because the diffracted particle of the quantum Young's diffracted experiment is a self-entangled entity that is indivisible by precept. Also note that the quantum trajectory representation needs only one constant of the motion to specify the trajectory of the EPR-molecule, which contains the physics of the EPR paradox. On the other hand, pursuing the EPR paradox as the interference between two particles would generally require a separate constant of the motion for each particle trajectory. Finally, if just a particular latent particle were measured rather than measuring the entanglement molecule, then the measurement process would reify both theretofore latent particles, that is $\psi_{EPR1}(x) \to \psi_1(x_1)$ and $\psi_{EPR_2}(x) \to \psi_2(x_2)$, where under decoherence the reified particles become functions of x_1 and x_2 explicitly [6]. Further physical insight is presented in Reference [6].

BIBLIOGRAPHY

1. Aspect, A., P. Grangier and G. Roger, 1981, Experimental tests of realistic local theories via Bell's theorem. *Phys. Rev. Lett.* **47**: 460–3.
2. Aspect, A., P. Grangier and G. Roger, 1982, Experimental realization of Einstein–Podolsky–Rosen–Bohm gedanken experiment: a new violation of Bell's inequalities. *Phys. Rev. Lett.* **49**: 91–4.
3. Aspect, A., J. Dalibard and G. Roger, 1982, Experimental tests of Bell's inequalities using a time-varying analyser. *Phys. Rev. Lett.* **49**: 1804–7.
4. Floyd, E. R., 2007, Interference, reduced action and trajectories. *Found. Phys.* **37**: 1386–402, arXiv:quant-ph/0605120v3.

5. Floyd, E. R., 2007, *Welcher Weg?* A trajectory representation of a quantum Young's experiment. *Found. Phys.* **37**: 1403–20, arXiv:quant-ph/0605121v3.
6. Floyd, E. R., EPR-Bohr and quantum trajectories: entanglement and nonlocality. http://lanl.arxiv.org/abs/1001.4575.
7. Bohm, D., 1953, A suggested interpretation of quantum mechanics in terms of "hidden" variables. *Phys. Rev.* **85**: 166–79.
8. Holland, P. R., 1993, *The Quantum Theory of Motion*, Cambridge: Cambridge University Press, pp. 86–7, 141–6.
9. Floyd, E. R., 1982, Modified potential and Bohm's quantum potential. *Phys. Rev.* **D 26**: 1339–47.
10. Floyd, E. R., 1984, Arbitrary initial conditions of hidden variables. *Phys. Rev.* **D 29**: 1842–4.
11. Floyd, E. R., 1986, Closed-form solutions for the modified potential. *Phys. Rev.* **D 34**: 3246–9.
12. Floyd, E. R., 2002, The philosophy of the trajectory representation of quantum mechanics. In *Gravitation and Cosmology: From the Hubble Radius to the Planck Scale; Proceedings of Symposium in Honour of the 80th Birthday of Jean-Pierre Vigier*, ed. R. L. Amoroso, G. Hunter, M. Kafatos, and J.-P. Vigier, 401–8. Dordrecht: Kluwer Academic Press. Extended version arXiv:quant-ph/0009070v1.
13. Faraggi, A. E. and M. Matone, 2000, The equivalence principle of quantum mechanics. *Int. J. Mod. Phys.* **A 15**: 1869–2017, arXiv:hep-th/9809127v2.
14. Wyatt, R. E., 2005, *Quantum Dynamics with Trajectories: Introduction to Quantum Hydrodynamics.* New York: Springer.
15. Fine, A. E., 2009, The Einstein–Podolsky–Rosen argument in quantum theory, in *The Stanford Encyclopedia of Philosophy* (Fall 2009 Edition), ed. Edward N. Zalta, URL = <http://plato.stanford.edu/archives/fall2009/entries/qt-epr/> (accessed November 1, 2009).
16. Home, D. and A. Whitaker, 2007, *Einstein's Struggles with Quantum Theory*, New York: Springer, pp. 223–5.
17. Einstein, A., B. Podolski and N. Rosen, 1935, Can quantum mechanical description of physical reality be considered complete? *Phys. Rev.* **47**: 777–80.
18. Floyd, E. R., 2000, Classical limit of the trajectory representation of quantum mechanics, loss of information and residual indeterminacy. *Int. J. Mod. Phys.* **A 15**: 1363–78, arXiv:quant-ph/9907092v3.
19. Floyd, E. R., 1996, Where and why the generalized Hamilton–Jacobi representation describes microstates of the Schrödinger wave function. *Found. Phys. Lett.* **9**: 489–97, arXiv:quant-ph/9707051v1.
20. Carroll, R., 1999, Some remarks on uncertainty and spin. *Can. J. Phys.* **77**: 319–25, quant-ph/9904081.
21. Carroll, R., 2000, *Quantum Theory, Deformation and Integrability*, Amsterdam: Elsevier, pp. 50–56.
22. Carroll, R., Some remarks on time, uncertainty, and spin. http://lanl.arxiv.org/abs/quant-ph/9903081v1.
23. Peres, A., 1980, Measurement of time by quantum clocks. *Amer. J. Phys.* **48**: 552–7.
24. Morse, P. M. and H. Feshbach, 1953, *Methods of Theoretical Physics*, Part II, New York: McGraw-Hill, p. 1284.

4 Quantum Dynamics and Supersymmetric Quantum Mechanics

Eric R. Bittner and Donald J. Kouri

CONTENTS

4.1   A FIRST DATE

Supersymmetry (SUSY) postulates that for every fermion there is boson of equal mass (i.e. energy). In quantum mechanics this comes about because for every quantum Hamiltonian there is a partner Hamiltonian that has the same energy spectrum above the ground state of the original system. In particle physics, one "sector" is populated by bosons and the other sector by fermions and SUSY predicts that the lowest lying fermion state is energetically degenerate with the first excited boson state. Evidence for SUSY has proven to be elusive and it is now believed that SUSY is a broken symmetry. In January 2009 at a conference dedicated to Bob Wyatt, we began to look at SUSY as a way to develop new computational methods and approaches. Up until now, SUSY has been more of a mathematical technique that has been used more or less as a way to obtain stationary solutions to the Schrödinger equation for a variety of

one-dimensional potential systems. In this chapter, we will discuss some of the work
we have been doing in developing "SUSY" inspired methods for performing quantum
many-body calculations and quantum scattering calculations. We begin with a brief
overview of the SUSY theory and some of its elementary results. We then discuss
how we have used the approach to develop both analytical and numerical solutions
of the stationary Schrödinger equation. Finally, we conclude by discussing our recent
extension of SUSY to higher dimensions and for scattering theory.

4.2 MATHEMATICAL CONSIDERATIONS

Before discussing some of our recent results, it is important to introduce briefly the
mathematical formulation of SUSY quantum mechanics.

4.2.1 HAMILTONIAN FORMULATION OF SUSY

In quantum theory, there is a fundamental connection between a bound state and its
potential. This is simple to demonstrate by writing the Schrödinger equation for the
stationary states as

$$V_1(x) - E_n = -\frac{\hbar^2}{2m}\frac{1}{\psi_n}\partial_x^2\psi_n = Q[\psi_n] \qquad (4.1)$$

where we recognize the right-hand side as the Bohm quantum potential which will
certainly be discussed repeatedly in this volume. One of the remarkable consequences
of this equation is that every stationary state of a given potential has the same functional form for its quantum potential Q. Thus, knowing any bound state allows a
global reconstruction of the potential, $V(x)$, up to a constant energy shift.

SUSY is obtained by factoring the Schrödinger equation into the form [1–3]

$$H\psi = A^+A\psi_o^{(1)} = 0 \qquad (4.2)$$

using the operators

$$A = \frac{\hbar}{\sqrt{2m}}\partial_x + W \quad \text{and} \quad A^+ = -\frac{\hbar}{\sqrt{2m}}\partial_x + W. \qquad (4.3)$$

Since we can impose $A\psi_o^{(1)} = 0$, we can immediately write that

$$W(x) = -\frac{\hbar}{\sqrt{2m}}\partial_x \ln \psi_o. \qquad (4.4)$$

$W(x)$ is the *superpotential* which is related to the physical potential by a Riccati
equation.

$$V(x) = W^2(x) - \frac{\hbar}{\sqrt{2m}}W'(x). \qquad (4.5)$$

The SUSY factorization of the Schrödinger equation can always be applied in one dimension.

From this point on we label the original Hamiltonian operator and its associated potential, states, and energies as H_1, V_1, $\psi_n^{(1)}$ and $E_n^{(1)}$. One can also define a partner Hamiltonian, $H_2 = AA^+$ with a corresponding potential

$$V_2 = W^2 + \frac{\hbar}{\sqrt{2m}} W'(x). \tag{4.6}$$

All of this seems rather circular and pointless until one recognizes that V_1 and its partner potential, V_2, give rise to a common set of energy eigenvalues. This principal result of SUSY can be seen by first considering an arbitrary stationary solution of H_1,

$$H_1 \psi_n^{(1)} = A^+ A \psi_n = E_n^{(1)} \psi_n^{(1)}. \tag{4.7}$$

This implies that $(A\psi_n^{(1)})$ is an eigenstate of H_2 with energy $E_n^{(1)}$ since

$$H_2(A\psi_n^{(1)}) = AA^+ A\psi_n^{(1)} = E_n^{(1)}(A\psi_n^{(1)}). \tag{4.8}$$

Likewise, the Schrödinger equation involving the partner potential $H_2\psi_n^{(2)} = E_n^{(2)}\psi_n^{(2)}$ implies that

$$A^+ AA^+ \psi_n^{(2)} - H_1(A^+ \psi_n^{(2)}) = E_n^{(?)}(A^+ \psi_n^{(?)}). \tag{4.9}$$

This (along with $E_o^{(1)} = 0$) allows one to conclude that the eigenenergies and eigenfunctions of H_1 and H_2 are related in the following way: $E_{n+1}^{(1)} = E_n^{(2)}$,

$$\psi_n^{(2)} = \frac{1}{\sqrt{E_{n+1}^{(1)}}} A\psi_{n+1}^{(1)} \quad \text{and} \quad \psi_{n+1}^{(1)} = \frac{1}{\sqrt{E_n^{(2)}}} A^+ \psi_n^{(2)} \tag{4.10}$$

for $n > 0$.[*] Thus, the *ground state of H_2 has the same energy as the first excited state of H_1*. If this state $\psi_o^{(2)}$ is assumed to be nodeless, then $\psi_1^{(1)} \propto A^+\psi_o^{(2)}$ will have a single node. We can repeat this analysis and show that H_2 is partnered with another Hamiltonian, H_3, whose ground state is isoenergetic with the first excited state of H_2 and thus isoenergetic with the second excited state of the original H_1. This hierarchy of partners persists until all of the bound states of H_1 are exhausted.

4.2.2 SUSY Algebra

We can connect the two partner Hamiltonians by constructing a matrix super-Hamiltonian operator

$$\mathbf{H} = \begin{pmatrix} H_1 & 0 \\ 0 & H_2 \end{pmatrix} \tag{4.11}$$

---

[*] Our notation from here on is that $\psi_n^{(m)}$ denotes the nth state associated with the mth partner Hamiltonian with similar notion for related quantities such as energies and superpotentials.

and two matrix "supercharge" operators

$$\mathbf{Q} = \begin{pmatrix} 0 & 0 \\ A & 0 \end{pmatrix} = A\sigma_- \tag{4.12}$$

and

$$\mathbf{Q}^+ = \begin{pmatrix} 0 & A^+ \\ 0 & 0 \end{pmatrix} = A^+\sigma_+ \tag{4.13}$$

where $\sigma_\pm$ are 2×2 Pauli spin matrices. Using these we can rewrite the SUSY Hamiltonian as

$$\mathbf{H} = \left(-\frac{\hbar^2}{2m}\frac{d^2}{dx^2} + W^2\right)\sigma_o + W'\sigma_z. \tag{4.14}$$

The operators $\{\mathbf{H}, \mathbf{Q}, \mathbf{Q}^+\}$ form a closed algebra (termed the Witten superalgebra) with

$$[\mathbf{H}, \mathbf{Q}] = [\mathbf{H}, \mathbf{Q}^+] = 0 \tag{4.15}$$

$$\{\mathbf{Q}, \mathbf{Q}\} = \{\mathbf{Q}^+, \mathbf{Q}^+\} = 0 \tag{4.16}$$

$$\{\mathbf{Q}, \mathbf{Q}^+\} = \mathbf{H}. \tag{4.17}$$

The first algebraic relation is responsible for the degeneracy of the spectra of H_1 and H_2 and the supercharges transform an eigenstate of one sector into an eigenstate of the other sector.

As an example and perhaps a better connection to the physics implied by this structure, consider the case of a one-dimensional particle with an internal spin degree of freedom and with $[x, p] = i$ denoting the position and momentum of the particle. Conserved SUSY would imply that all non-diagonal coupling terms between the bosonic (coordinate) and fermionic (spin) degrees of freedom are exactly zero. This of course is equivalent to making the Born–Oppenheimer approximation for a two-state system coupled to a continuous field $x(t)$. In this case, SUSY is preserved so long as $d_t \psi(x(t), t) = \partial_t \psi(x(t), t)$. SUSY is broken when $\dot{x}(t)\partial_x \psi(x(t), t) \neq 0$ which would lift the degeneracy between the states of H_1 and H_2.

4.2.3 SCATTERING IN ONE DIMENSION

The SUSY approach is not limited to bound-state problems. For a one-dimensional scattering system, it is straightforward to apply the SUSY theory to determine a relation between the transmission and reflection coefficients of the supersymmetric partners. Asymptotically, we can assume that $W(x) \to W_\pm$ as $x \to \pm\infty$. In the same limit, the partner potentials become $V_{1,2} \to W_\pm^2$. For a plane wave incident from the left with energy E scattering from $V_{1,2}$, we require the following asymptotic forms:

$$\lim_{x \to -\infty} \psi^{(1,2)}(k, x) \sim e^{ikx} + R^{(1,2)}e^{-ikx} \tag{4.18}$$

$$\lim_{x \to +\infty} \psi^{(1,2)}(k', x) \sim T^{(1,2)}e^{ik'x}. \tag{4.19}$$

We can derive a relation between the two scattering states by using the relation $\psi^{(1)}(k,x) = NA^+\psi_2(k',x)$. For the left-hand components ($x \to -\infty$).

$$e^{ikx} + R^{(1)}e^{-ikx} = N\left[(-ik + \tilde{W}_-)e^{ikx} + (ik + \tilde{W}_-)e^{-ikx}\right] \quad (4.20)$$

where in the last line we have incorporated the $\hbar/\sqrt{2m}$ in to the normalization and wrote $\tilde{W}_\pm = W_\pm\sqrt{2m}/\hbar$. Likewise for the transmitted coefficients ($x \to +\infty$).

$$T^{(1)}e^{ik'x} = N(-ik' + \tilde{W}_+)T^{(2)}e^{ik'x}. \quad (4.21)$$

Eliminating the common normalization factor and using the fact that $k = \sqrt{2m(E - W_-)}/\hbar$ and $k' = \sqrt{2m(E - W_+)}/\hbar$ from the Schrödinger equation we can arrive at

$$R^{(1)}(k) = \frac{W_- + ik}{W_- - ik}R^{(2)}(k) \quad (4.22)$$

$$T^{(1)}(k) = \frac{W_+ - ik'}{W_- - ik}T^{(2)}(k). \quad (4.23)$$

Consequently, knowledge of the scattering states of V_1 allows one to easily construct scattering states for the partner potential.

4.2.4 Non-Stationary States

Finally, one can use the SUSY approach in a time-dependent context by writing

$$i\hbar\partial_t\psi^{(1)} = H_1\psi^{(1)} = A^+A\psi^{(1)}$$

where $\psi^{(1)}$ is a non-stationary state in the first sector. If V_1 is independent of time, then the superpotential must also be independent of time and so we can write

$$i\hbar A\partial_t\psi^{(1)} = i\hbar\partial_t(A\psi^{(1)}) = AA^+(A\psi^{(1)}).$$

In other words, we have the time-dependent Schrödinger equation for the partner potential

$$i\hbar\partial_t\psi^{(2)} = H_2\psi^{(2)}.$$

The two non-stationary states are partnered, $\psi^{(2)} \propto A^+\psi^{(1)}$. We also note that these states satisfy

$$\psi^{(1)}(t) = e^{-iA^+At/\hbar}\psi(0)$$

and

$$\psi^{(2)}(t) = e^{-iAA^+t/\hbar}\psi(0)$$

for some initial state $\psi(0)$. Using the charge operators we can show that

$$A\psi^{(1)}(t) = e^{-iAA^+t/\hbar}(A\psi(0)).$$

As above in the scattering example, one can use the dynamics of one sector to determine the dynamics in the other sector.

The partnering scheme presents a powerful prescription for developing novel approaches for solving a wide variety of quantum mechanical problems. This allows one to use analytical or numerical solutions of one problem to determine either approximate or exact solutions to some new problem. In the sections that follow, I present some of our attempt to use SUSY in a numerical context. At the moment our numerical results are limited to one spatial dimension. As I shall discuss, extending SUSY to multiple dimensions has proven to be problematic. However, in Section 4.5 we present our extension using the vector-SUSY approach we are developing.

4.3   USING SUSY TO OBTAIN EXCITATION ENERGIES AND EXCITED STATES

The SUSY hierarchy also provides a useful prescription for determining the excited states of H_1 (which may represent the physical problem of interest). The first excited state of H_1 is isoenergetic with the ground state of H_2. Since this state is nodeless, one can use either Ritz variational approaches or Monte Carlo approaches to determine this state to very high accuracy.

Two basic tools used in computational chemistry are the Quantum Monte Carlo (QMC) and the Rayleigh–Ritz variational approaches. Both approaches yield their best and most accurate results for ground-state energies and wave functions. Although the variational method also gives bounds for the excited state energies as well as the ground state (the Hylleraas–Undheim theorem [4]), it is well known that their accuracy is significantly lower than that of the ground state. Even more serious, the wave functions are known to converge much more slowly than the energies.

In the case of the QMC [5–10], there are additional difficulties associated with the presence of nodes in the excited state wave functions [11]. While some progress has been made in dealing with this issue (e.g., the "fixed node" or "guide wave" techniques) [8–12] the computational effort required is greater and the accuracy is lower and in fact, no general solution to the difficulty has been found for reducing the computational effort and increasing the accuracy for excited state calculations in QMC to the same level as is attained for the ground state. In fact, it is very likely the presence and effects of nodes in the excited states that is largely responsible for the lower accuracy and slower convergence of excited state results in the variational method. The precise determination of nodal surfaces is expected to play a crucial role since they reflect changes in the relative phase of the wave function. Because of the ubiquitous importance of both the variational and QMC methods, solving the so-called "node problem" will have enormous impact on computational chemistry.

4.3.1   USING SUSY TO IMPROVE QUALITY OF VARIATIONAL CALCULATIONS

We now turn to the proof of principle for this approach as a computational scheme to obtain improved excited state energies and wave functions in the Rayleigh–Ritz variational method. We should note that these results can be generalized to any system where a hierarchy of Hamiltonians can be generated because of the nature of the Rayleigh–Ritz scheme. In the standard approach one calculates the energies and wave functions variationally, relying on the Hylleraas–Undheim theorem for convergence [4]. This, however, is unattractive for higher-energy states because they require a much larger basis to converge to the same error. We stress that this is true regardless of the specific basis set used. Of course, some bases will be more efficient than others but it is generally true that for a given basis, the Rayleigh–Ritz result is less accurate for excited states. We address this situation by solving for ground states in the variational part of the problem.

To demonstrate our computational scheme, we investigate the first example system from the previous section. For the potential

$$V_1(x) = x^6 + 4x^4 + x^2 - 2 \tag{4.24}$$

exact solutions are known for all states of H_1. We use the exact results to assess the accuracy of the variational calculations. Here we employed an n-point discrete variable representation (DVR) based upon the Tchebechev polynomials to compute the eigenspectra of the first and second sectors [13,14]. In Figure 4.1 we show the numerical error in the first excitation energy by comparing $E_1^1(n)$ and $E_0^2(n)$ from an n-point DVR to the numerically "exact" value corresponding to a 100-point DVR,

$$\epsilon_1^1(n) = \log_{10}|E_1^1(n) - E_1^1(\text{exact})|.$$

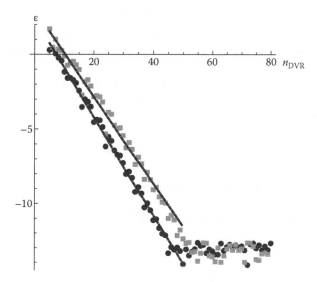

FIGURE 4.1 Convergence of the first excitation energy E_1^1 for the model potential $V_1 = x^6 + 4x^4 + x^2 - 2$ using an n-point discrete variable representation (DVR). Gray squares: $\epsilon = \log_{10}|E_1^1(n) - E_1^1(\text{exact})|$; black squares: $\epsilon = \log_{10}|E_0^2(n) - E_1^1(\text{exact})|$. Dashed lines are linear fits. (Reproduced with permission from Kouri, D. J., Markovich, T., Maxwell, N., and Bittner, E. R., *J. Phys. Chem. A* 2009, http://dx.doi.org/10.1021/jp905798m.)

Likewise,

$$\epsilon_0^2(n) = \log_{10} |E_0^1(n) - E_1^1(\text{exact})|.$$

For any given basis size, $\epsilon_0^2 < \epsilon_1^1$. Moreover, over a range of $15 < n < 40$ points, the excitation energy computed using the second sector's ground state is between 10 and 100 times more accurate than $E_1^1(n)$. This effectively reiterates our point that by using the SUSY hierarchy, one can systematically improve upon the accuracy of a given variational calculation.

4.3.2 MONTE CARLO SUSY

Having defined the basic terms of SUSY quantum mechanics, let us presume that one can determine an accurate approximation to the ground-state density $\rho_o^{(1)}(x)$ of Hamiltonian H_1. One can then use this to determine the superpotential using the Riccati transform

$$W_o^{(1)} = -\frac{1}{2} \frac{\hbar}{\sqrt{2m}} \frac{\partial \ln \rho_o^{(1)}}{\partial x} \tag{4.25}$$

and the partner potential

$$V_2 = V_1 - \frac{\hbar^2}{2m} \frac{\partial^2 \ln \rho_o^{(1)}}{\partial x^2}. \tag{4.26}$$

Certainly, our ability to compute the energy of the ground state of the partner potential V_2 depends on having first obtained an accurate estimate of the ground-state density associated with the original V_1.

For this we turn to an adaptive Monte Carlo-like approach developed by Maddox and Bittner [16]. Here, we assume we can write the trial density as a sum over N Gaussian approximate functions

$$\rho_T(x) = \sum_n G_n(x, \mathbf{c}_n) \tag{4.27}$$

parameterized by their amplitude, center, and width

$$G_n(x, \{\mathbf{c}_n\}) = c_{no} e^{-c_{n2}(x - c_{n3})^2}. \tag{4.28}$$

This trial density is then used to compute the energy

$$E[\rho_T] = \langle V_1 \rangle + \langle Q[\rho_T] \rangle \tag{4.29}$$

where $Q[\rho_T]$ is the Bohm quantum potential,

$$Q[\rho_T] = -\frac{\hbar^2}{2m} \frac{1}{\sqrt{\rho_T}} \frac{\partial^2}{\partial x^2} \sqrt{\rho_T}. \tag{4.30}$$

The energy average is computed by sampling $\rho_T(x)$ over a set of trial points $\{x_i\}$ and then moving the trial points along the conjugate gradient of

$$E(x) = V_1(x) + Q[\rho_T](x). \tag{4.31}$$

After each conjugate gradient step, a new set of coefficients is determined according to an expectation maximization criterion such that the new trial density provides the best N-Gaussian approximation to the actual probability distribution function sampled by the new set of trial points. The procedure is repeated until $\delta\langle E \rangle = 0$. In doing so, we simultaneously minimize the energy and optimize the trial function. Since the ground state is assumed to be nodeless, we will not encounter the singularities and numerical instabilities associated with other Bohmian equations of motion based approaches [16–21]. Moreover, the approach has been extended to very high dimensions and to finite temperature by Derrickson and Bittner in their studies of the structure and thermodynamics of rare gas clusters with up to 130 atoms [22, 23].

4.4 TEST CASE: TUNNELING IN A DOUBLE WELL POTENTIAL

As a non-trivial test case, consider the tunneling of a particle between two minima of a symmetric double potential well. One can estimate the tunneling splitting using semi-classical techniques by assuming that the ground and excited states are given by the approximate form

$$\psi_\pm = \frac{1}{\sqrt{2}}(\phi_o(x) \pm \phi_o(-x)) \tag{4.32}$$

where ϕ_o is the lowest energy state in the right-hand well in the limit when the wells are infinitely far apart. From this, one can easily estimate the splitting as [24]

$$\delta = 4\frac{\hbar^2}{m}\phi_o(0)\phi_o'(0). \tag{4.33}$$

If we assume the localized states (ϕ_o) to be Gaussian, then

$$\psi_\pm \propto \frac{1}{\sqrt{2}}(e^{-\beta(x-x_o)^2} \pm e^{-\beta(x+x_o)^2}) \tag{4.34}$$

and we can write the superpotential as

$$W = \sqrt{\frac{2}{m}}\hbar\beta\,(x - x_o \tanh(2xx_o\beta))\,. \tag{4.35}$$

From this, one can easily determine both the original potential and the partner potential as

$$V_{1,2} = W^2 \pm \frac{\hbar}{\sqrt{2m}}W' \tag{4.36}$$

$$= \frac{\beta^2\hbar^2}{m}\left(2(x - x_o \tanh(2xx_o\beta))^2 \right.$$
$$\left. \pm (2x_o^2 \operatorname{sech}^2(2xx_o\beta) - 1)\right). \tag{4.37}$$

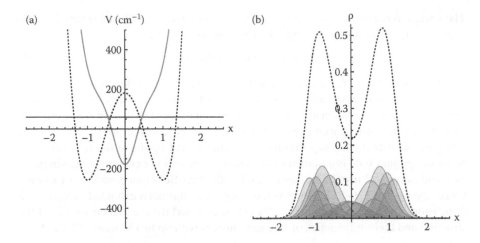

FIGURE 4.2 (a) Model double well potential (dotted line) and partner potential (gray). The energies of the tunneling doublets are indicated by the horizontal lines at $V = 0\,\mathrm{cm}^{-1}$ and $V = 59.32\,\mathrm{cm}^{-1}$ indicating the positions of the sub-barrier tunneling doublet. (b) Final ground-state density (dotted line) superimposed over the Gaussians used in its expansion (gray). (Reproduced with permission from Bittner, E. R., Maddox, J. B., and Kouri, D. J., *J. Phys. Chem. A* 2009, http://dx.doi.org/10.1021/jp9058017.)

While the V_1 potential has the characteristic double minima giving rise to a tunneling doublet, the SUSY partner potential V_2 has a central dimple which in the limit of $x_o \to \infty$ becomes a δ-function which produces an unpaired and nodeless ground state [3]. Using Equation 4.9, one obtains $\psi_1^{(1)} = \psi_- \propto A^\dagger \psi_o^{(2)}$ which now has a single node at $x = 0$.

For a computational example, we take the double well potential to be of the form

$$V_1(x) = ax^4 + bx^2 + E_o \tag{4.38}$$

with $a = 438.9\ \mathrm{cm}^{-1}/(\mathrm{bohr}^2)$, $b = 877.8\ \mathrm{cm}^{-1}/(\mathrm{bohr})^4$, and $E_o = -181.1\ \mathrm{cm}^{-1}$ which (for $m = m_H$) gives rise to exactly two states at below the barrier separating the two minima with a tunneling splitting of $59.32\ \mathrm{cm}^{-1}$ as computed using a DVR approach [26]. The MC/SUSY approach required 1000 sample points and 15 Gaussians in the expansion of the trial density to converge the ground state energy to $1{:}10^{-8}$. This is certainly a bit of an overkill in the number of points and number of Gaussians since far fewer DVR points were required to achieve comparable accuracy (and a manifold of excited states). The numerical results, however, are encouraging since the accuracy of generic Monte Carlo evaluation would be $1/\sqrt{n_p} \approx 3\%$ in terms of the energy.[*] Plots of V_1 and the converged ground state are shown in Figure 4.2.

---

[*] In our implementation, the sampling points are only used to evaluate the requisite integrals and they themselves are adjusted along a conjugate gradient rather than by resampling. One could in principle forego this step entirely and optimize the parameters describing the Gaussians directly.

The partner potential $V_2 = W^2 + \hbar W'/\sqrt{2m}$ can be constructed once we know the superpotential, $W(x)$. Here, we require an accurate evaluation of the ground-state density and its first two log-derivatives. The advantage of our computational scheme is that one can evaluate these analytically for a given set of coefficients. In Figure 4.2a we show the partner potential derived from the ground-state density. Whereas the original V_1 potential exhibits the double well structure with minima near $x_o = \pm 1$, the V_2 partner potential has a pronounced dip about $x = 0$. Consequently, its ground state should have a simple "Gaussian"-like form peaked about the origin.

Once we have determined an accurate representation of the partner potential, it is now a trivial matter to re-introduce the partner potential into the optimization routines. The ground state converges easily and is shown in Figure 4.3a along with its Gaussians. After 1000 CG steps, the converged energy is within 0.1% of the exact tunneling splitting for this model system. Again, this is an order of magnitude better than the $1/\sqrt{n_p}$ error associated with a simple Monte Carlo sampling. Furthermore, Figure 4.3b shows $\psi_1^{(1)} \propto A^\dagger \psi_0^{(2)}$ computed using the converged $\rho_0^{(2)}$ density. As anticipated, it shows the proper symmetry and nodal position.

By symmetry, one expects the node to lie precisely at the origin. However, since we have not imposed any symmetry restriction or bias on our numerical method, the position of the node provides a sensitive test of the convergence of the trial density for $\rho_0^{(2)}$. In the example shown in Figure 4.4, the location of the node oscillates about the origin and appears to converge exponentially with the number of CG steps. This is remarkably good considering that this is ultimately determined by the quality of the third and fourth derivatives of $\rho_o^{(1)}$ that appear when computing the conjugate

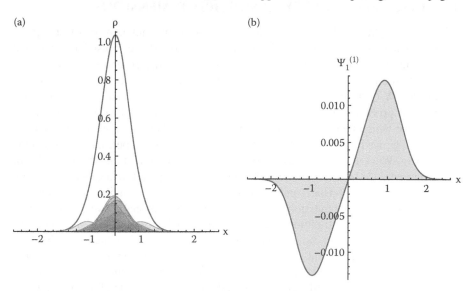

FIGURE 4.3 (a) Ground-state density of the partner Hamiltonian H_2 (blue) superimposed over its individual Gaussian components. (b) Excited state $\psi_1^{(1)}$ derived from the ground state of the partner potential, $\psi_o^{(2)}$. (Reproduced with permission from Bittner, E. R., Maddox, J. B., and Kouri, D. J., *J. Phys. Chem. A* 2009, http://dx.doi.org/10.1021/jp9058017.)

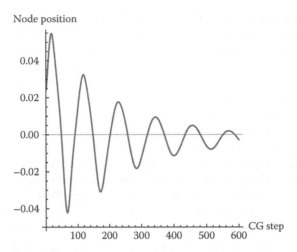

FIGURE 4.4 Location of excited state node for the last 600 CG steps. (Reproduced with permission from Bittner, E. R., Maddox, J. B., and Kouri, D. J., *J. Phys. Chem. A*, 2009, http://dx.doi.org/10.1021/jp9058017.)

gradient of V_2. We have tested this approach on a number of other one-dimensional bound-state problems with similar success.

4.5 EXTENSION OF SUSY TO MULTIPLE DIMENSIONS

While SUSY-QM has also been explored for one-dimensional, non-relativistic quantum mechanical problems [3,27–31], thus far these studies have focused on the formal aspects and on obtaining exact, analytical solutions for the ground state for specific classes of problems. In several recent papers [15, 25, 32, 33], we have begun exploring the SUSY-QM approach as the basis of a general computational scheme for bound-state problems. Our initial studies have been restricted to one-dimensional systems (for which there are, obviously, many powerful computational methods). In our first paper, we found that SUSY-QM (combined with a new periodic version of the Heisenberg–Weyl algebra) yields a robust, natural way to treat an infinite family of hindered rotors [33]. Next we showed that the SUSY-QM leads to a general treatment of an infinite family of an harmonic oscillators (HO), such that highly accurate excited state energies and wave functions could be obtained variationally using significantly smaller basis sets than a traditional variational approach requires [15]. Most recently, we have considered a one-dimensional double well potential in which we solved for the ground-state energy and wave function using a VQMC approach. Then using SUSY-QM, we (numerically) generated an auxiliary Hamiltonian whose *nodeless* ground state is isospectral (degenerate) with the first excited state of the original system Hamiltonian. This ground state was also easily determined by VQMC, yielding excellent accuracy for the first excited state energy [25]. Even more significant, by using the charge operators naturally generated in the SUSY-QM approach, we also obtained excellent accuracy for the first excited state wave function. Furthermore, at no point did we impose a fixed node or symmetry on the excited state wave function and our calculation only involved working with a nodeless ground state.

Of course, all this begs the question: Can this approach be generalized to higher numbers of dimensions and to more than a single particle? There has been substantial effort in the past to do just this [3, 28–30, 34–45]. However, to date, no such generalization has been found that is able to generate all the excited states and energies even for so simple a system as a pair of separable, one-dimensional HO or equivalently, for a separable two-dimensional single HO. In our most recent, unpublished work [32], we have succeeded in obtaining such a generalization and showed that it does, in fact, yield the correct analytical results for separable and non-separable problems. In the next section, we present a succinct summary of our approach. The major question now is whether this formalism provides a basis for a robust, computational method for determining excited state energies and wave functions for large, strongly correlated systems using either QMC or variational algorithms applied solely to nodeless ground-state problems.

4.5.1 Difficulties in Extending Beyond One Dimension

To move beyond one-dimensional SUSY, Ioffe and coworkers have explored the use of higher-order charge operators [42, 43, 46, 47], and Kravchenko has explored the use of Clifford algebras [48]. Unfortunately, this is difficult to do in general, the reason being that the Riccati factorization of the one-dimensional Schrödinger equation does not extend easily to higher dimensions. One remedy is to write the charge operators as vectors $\vec{A} = (+\vec{\partial} + \vec{W})$ and with $\vec{A}^+ = (-\vec{\partial} + \vec{W})^\dagger$ as the adjoint charge operator. The original Schrödinger operator is then constructed as an inner product

$$H_1 = \vec{A}^+ \cdot \vec{A}. \tag{4.39}$$

Working through the vector product produces the Schrödinger equation

$$H_1 \phi = (-\nabla^2 + W^2 - (\vec{\nabla} \cdot \vec{W}))\phi = 0 \tag{4.40}$$

and a Riccati equation of the form

$$U(x) = W^2 - \vec{\nabla} \cdot \vec{W}. \tag{4.41}$$

For a two-dimensional HO, we would obtain a vector superpotential of the form

$$\vec{W} = -\frac{1}{\psi_0^{(1)}} \vec{\nabla} \psi_0^{(1)} = (x, y) = (W_x, W_y). \tag{4.42}$$

Let us look more closely at the $\vec{\nabla} \cdot \vec{W}$ part. If we use the form that $\vec{W} = -\vec{\nabla} \ln \psi$, then $-\vec{\nabla} \cdot \vec{\nabla} \ln \psi = -\nabla^2 \ln \psi$ which for the two-dimensional oscillator results in $\vec{\nabla} \cdot \vec{W} = 2$. Thus,

$$W^2 - \vec{\nabla} \cdot \vec{W} = (x^2 + y^2) - 2 \tag{4.43}$$

which agrees with the original symmetric harmonic potential. Now, we write the scaled partner potential as

$$U_2 = W^2 + \vec{\nabla} \cdot \vec{W} = (x^2 + y^2) + 2. \tag{4.44}$$

This is equivalent to the original potential shifted by a constant amount

$$U_2 = U_1 + 4. \tag{4.45}$$

The ground state in this potential would have the same energy as the states of the original potential with quantum numbers $n + m = 2$. Consequently, even with this naïve factorization, one can in principle obtain excitation energies for higher dimensional systems, but there is no assurance that one can reproduce the entire spectrum of states.

The problem lies in the fact that neither Hamiltonian H_2 nor its associated potential U_2 is given correctly by the form implied by Equations 4.40 and 4.44. Rather, the correct approach is to write the H_2 Hamiltonian as a *tensor* by taking the outer product of the charges $\overline{H}_2 = \vec{A}\vec{A}^+$ rather than as a scalar $\vec{A} \cdot \vec{A}^+$. At first this seems unwieldy and unlikely to lead anywhere since the wave function solutions of

$$\overline{H}_2\vec{\psi} = E\vec{\psi} \tag{4.46}$$

are now vectors rather than scalars. However, rather than adding undue complexity to the problem, it actually simplifies matters considerably. As we shall demonstrate in a forthcoming paper, this tensor factorization preserves the SUSY algebraic structure and produces excitation energies for any n-dimensional SUSY system. Moreover, this produces a scalar $\mapsto$ tensor $\mapsto$ scalar hierarchy as one moves to higher excitations [32].

4.5.2 VECTOR SUSY

We now give a brief summary of our new generalization of SUSY-QM to treat higher dimensionality and more than one particle. Previous attempts generally involved introducing additional, "spin-like" degrees of freedom [29, 30, 37–40, 42, 43, 46, 47, 49]. In our approach, we make use of a vectorial technique that can deal simultaneously with either higher dimensions or more than one particle. In fact, the two problems are dealt with in exactly the same manner. Therefore, for simplicity, we consider a general n-dimensional distinguishable particle system with orthogonal coordinates $\{x_\mu\}$. The Hamiltonian is given by[*]

$$H = -\nabla^2 + V_0(x_1, \ldots, x_n) \tag{4.47}$$

and the nodeless ground state satisfies the Schrödinger equation,

$$H\psi_0^{(1)} = E_0^{(1)}\psi_0^{(1)}. \tag{4.48}$$

---

[*] Our units are such that $\hbar^2/2m = 1$.

We now define a "vector superpotential," $\vec{W}_1$, with components

$$W_{1\mu} = -\frac{\partial}{\partial x_\mu} \ln \psi_0^{(1)}. \tag{4.49}$$

Then it is easily seen that the original Hamiltonian can be recast as

$$H_1 = (-\nabla + \vec{W}_1) \cdot (\nabla + \vec{W}_1) = \vec{A}_1^+ \cdot \vec{A}_1 \tag{4.50}$$

where the $\vec{A}_1$ and $\vec{A}_1^+$ are multi-dimensional generalizations of the SUSY charge operators from Equation 4.39. This defines our "sector-1" (or "boson") Hamiltonian and Equation 4.48 can be written as[*]

$$H_1\psi_0^{(1)} = E_0^{(1)}\psi_0^{(1)}. \tag{4.51}$$

One can show that the vector superpotential is related to the original (scalar) potential via:

$$V_0 = \vec{W}_1 \cdot \vec{W}_1 - \nabla \cdot \vec{W}_1. \tag{4.52}$$

The various components of the charge operators, $\vec{A}_1$ and $\vec{A}_1^+$, are defined by

$$A_{1\mu} = \frac{\partial}{\partial x_\mu} + W_{1\mu} \quad \text{and} \quad A_{1\mu}^+ = -\frac{\partial}{\partial x_\mu} + W_{1\mu}. \tag{4.53}$$

Note that since these are associated with orthogonal degrees of freedom, the charge operators can be applied either by individual components or in vector form.

Next, consider the Schrödinger equation for the first excited state of H. We can write this using the charge operators as

$$H_1\psi_1^{(1)} = E_1^{(1)}\psi_1^{(1)} = (\vec{A}_1^+ \cdot \vec{A}_1 + E_0^{(1)})\psi_1^{(1)}. \tag{4.54}$$

We apply $\vec{A}_1$ to Equation 4.9:

$$(\vec{A}_1\vec{A}_1^+) \cdot \vec{A}_1\psi_1^{(1)} = (E_1^{(1)} - E_1^{(0)})\vec{A}_1\psi_1^{(1)}. \tag{4.55}$$

Here we identify $(\vec{A}_1\vec{A}_1^\dagger)$ as a new, auxiliary Hamiltonian. It is important to note that this is constructed from the outer or tensor product of the charge operators rather than from inner or dot product as used in constructing H_1. Its eigenvector, $\vec{A}_1\psi_1^{(1)}$, is isospectral with the excited state, $\psi_1^{(1)}$ of H_1 (since $E_0^{(1)}$ is known, determining $(E_1^{(1)} - E_1^{(0)})$ yields $E_1^{(1)}$). We therefore define the **tensor** Hamiltonian for the second sector as

$$\overleftrightarrow{H}_2 = \vec{A}_1\vec{A}_1^\dagger \tag{4.56}$$

---

[*] In analogy with the original descriptions of SUSY, we refer to the partner pairs as "boson" and "fermion" sectors or less poetically as "sector-1," "sector-2," and so forth.

and **vector** state function as

$$\vec{\psi}_0^{(2)} = \frac{1}{(E_1^{(1)} - E_1^{(0)})} \vec{A}_1 \psi_1^{(1)}. \tag{4.57}$$

It is easy to show that the ground-state energy of $\overleftrightarrow{H}_2$ is related to the first excitation energy of the original Hamiltonianm

$$E_0^{(2)} = E_1^{(1)} - E_1^{(0)}. \tag{4.58}$$

Furthermore, the ground state of $\overleftrightarrow{H}_2$ is also **nodeless**. This has been explicitly shown to be true for the separable two-particle HOs considered earlier [32]. Therefore, we propose to apply both the VQMC and the standard variational methods to determine $E_0^{(2)}$ and $\vec{\psi}_0^{(2)}$. Of course, knowing the second sector ground-state energy also gives us the first excited state energy of the original Hamiltonian (Equation 4.58). Furthermore, we form the scalar product of

$$\overleftrightarrow{H}_2 \cdot \vec{\psi}_0^{(2)} = E_0^{(2)} \vec{\psi}_0^{(2)} \tag{4.59}$$

with $\vec{A}_1^+$ obtaining

$$(\vec{A}_1^+ \cdot \vec{A}_1)\vec{A}_1^+ \vec{\psi}_0^{(2)} = E_0^{(2)} \vec{A}_1^+ \cdot \vec{\psi}_0^{(2)}. \tag{4.60}$$

Clearly, this is exactly

$$H_1(\vec{A}_1^+ \cdot \vec{\psi}_0^{(2)}) = E_1^{(1)}(\vec{A}_1^+ \cdot \vec{\psi}_0^{(2)}) \tag{4.61}$$

so we can conclude that

$$\psi_1^{(1)} = \frac{1}{\sqrt{E_0^{(2)}}}(\vec{A}_1^+ \cdot \vec{\psi}_0^{(2)}). \tag{4.62}$$

Thus we also obtain the excited state wave function without any significant additional computational effort. This is because applying the charge operator is much simpler than solving an eigenvalue problem (it is a strictly linear operation). Evidence from our one-dimensional studies indicates that the accuracy of the excited states obtained using the SUSY-QM charge operator is significantly higher, for a given basis set, than what is obtained variationally (or with QMC) from the original Hamiltonian [15, 32]. This procedure can be continued as follows. We define a sector-2 vector superpotential with components

$$W_{2\mu} = \frac{\partial}{\partial x_\mu} \ln \psi_{0\mu}^{(2)}. \tag{4.63}$$

Then it follows that

$$\vec{A}_2 \cdot \vec{\psi}_0^{(2)} = (\nabla + \vec{W}_2) \cdot \vec{\psi}_0^{(2)} = 0 \tag{4.64}$$

so we can write

$$\overleftrightarrow{H}_2 = \vec{A}_2^+ \vec{A}_2 + E_0^{(2)}\mathbf{I} \tag{4.65}$$

and Equation 4.59 is still satisfied. We form the scalar product of $\vec{A}_2$ with the first excited state Schrödinger equation to obtain

$$(\vec{A}_2 \cdot \vec{A}_2^+)\vec{A}_2 \cdot \vec{\psi}_1^{(2)} = E_1^{(2)}\vec{A}_2 \cdot \vec{\psi}_1^{(2)}. \tag{4.66}$$

Then we define the sector-3 *scalar* Hamiltonian by

$$H_3 = \vec{A}_2 \cdot \vec{A}_2^+ + E_0^{(2)} \tag{4.67}$$

with the ground-state wave equation

$$H_3\psi_0^{(3)} = E_0^{(3)}\psi_0^{(3)}. \tag{4.68}$$

It is easily seen that $E_0^{(3)} = E_1^{(2)} - E_0^2$. This procedure continues until all bound states of the original Hamiltonian are exhausted. It should also be clear that the sector-2 excited state wave function is obtained from the nodeless sector-3 ground state by applying $\vec{A}_2^+$ to it. Then the second excited state for sector-1 results from taking the scalar product of $\vec{A}_1^{\dagger}$ with $\vec{\psi}_1^{(?)}$. The approach thus leads to an alternating sequence of scalar and tensor Hamiltonians, *but in all cases we need only determine nodeless ground states.*

There are two additional aspects of the tensor sector problem that require discussion. First we consider the validity of the Rayleigh–Ritz variational principle. It is easily seen from Equation 4.56 that $\overleftrightarrow{H}_2$ is a Hermitian operator. Therefore, its eigenspectrum is real and its eigenvectors are complete. With these facts in hand, the proof of the variational principle follows the standard one in every detail. This is also true for the Hylleraas–Undheim theorem.

Second, the QMC method is also directly applicable to the tensor sector problem. For the example discussed above, we note that the energy is given by

$$E_{\text{trial}} = \frac{\int d\tau \vec{\psi}_{\text{trial}} \cdot \overleftrightarrow{H}_2 \cdot \vec{\psi}_{\text{trial}}}{\int d\tau \vec{\psi}_{\text{trial}} \cdot \vec{\psi}_{\text{trial}}}. \tag{4.69}$$

We next note that the integral can be expanded in terms of its components as

$$E_{\text{trial}} = \frac{\sum_{\mu\nu}\int d\tau(\psi_{\mu,\text{trial}}H_{2,\mu\nu}\psi_{\nu,\text{trial}})}{\sum_\mu \int d\tau(\psi_{\mu,\text{trial}})^2}. \tag{4.70}$$

It is then clear that each separate integral can be evaluated by QMC. For example, the $\mu \neq \nu$ cross-term is divided and multiplied by

$$\psi_{\mu,\text{trial}}\int d\tau\psi_{\nu,\text{trial}}\psi_{\mu,\text{trial}}. \tag{4.71}$$

Then the sampling is done relative to the mixed probability distribution,

$$P_{\mu\nu} = \frac{\psi_{\mu,\text{trial}}\psi_{\nu,\text{trial}}}{\int d\tau \psi_{\mu,\text{trial}}\psi_{\nu,\text{trial}}}. \tag{4.72}$$

A similar expression applies to each term in the energy expression and the evaluation would need to be performed self-consistently.

Thus far, we have developed a formalism that appears to be suitable for extending the SUSY-QM technique to higher-dimensional systems. We believe the approach we have outlined above will provide the mathematical basis for a number of potentially interesting theoretical results. Moreover, we anticipate that when combined with either variational or Monte Carlo methods, our multi-dimensional extension of SUSY-QM will facilitate the calculation of accurate excitation energies and excited state wave functions.

4.6 OUTLOOK

I presented a number of avenues we are actively pursuing with the goal of using SUSY-QM or SUSY-inspired-QM to solve problems that are difficult to solve using more conventional approaches. In addition to what I have discussed here we exploring the use of the Riccati equation to solve quantum scattering problems. It is as if one of the coauthors of this paper (DJK) has come full-circle since one of his first papers concerned solving the Hamilton–Jacobi equation for the action integral in quantum scattering [50],[*]

$$iS/\hbar = -\int_{r_o}^{r} W(r')dr'.$$

The integrand in this last equation is the SUSY superpotential. Furthermore, there is a connection between our work and the complex-valued quantum trajectories studied by Wyatt and Tannor and their respective coworkers.

ACKNOWLEDGMENTS

This work was supported in part by the National Science Foundation (ERB: CHE-0712981) and the Robert A. Welch foundation (ERB: E-1337, DJK: E-0608).

BIBLIOGRAPHY

1. E. Witten, *Nucl. Phys. B* (Proc. Supp.) **188**, 513 (1981).
2. E. Witten, *J. Different. Geom.* **17**, 661 (1982).
3. F. Cooper, A. Khare, and U. Sukhatme, *Phys. Rep.* **251**, 267 (1995).

---

[*] Coincidentally, Ref. [50] appeared in the *J. Chem. Phys.* issue immediately before the birthday of the other author of this paper. There appears to be some interesting Karma at work here.

4. E. A. Hylleraas and B. Undheim, *Z. Phys.* **65**, 759 (1930).
5. B. L. Hammond, W. A. Lester, and P. J. Reynolds, *Monte Carlo Methods in Ab Initio Quantum Chemistry*, vol. 1 of *World Scientific Lecture and Cousre Notes in Chemistry* (World Scientific Publishing, River Edge, NJ, 1994).
6. A. R. Porter, O. K. Al-Mushadani, M. D. Towler, and R. J. Needs, *J. Chem. Phys.* **114**, 7795 (2001), http://link.aip.org/link/?JCP/114/7795/1.
7. J. D. Doll, R. D. Coalson, and D. L. Freeman, *J. Chem. Phys.* **87**, 1641 (1987).
8. R. J. Needs, P. R. C. Kent, A. R. Porter, M. D. Trowler, and G. Rajagopal, *Int. J. Quantum Chem.* **86**, 218 (2001).
9. S. Sorella, *Phys. Rev. B* **71**, 241103 (2005).
10. D. Blume, M. Lewerenz, P. Niyaz, and K. B. Whaley, *Phys. Rev. E* **55**, 3664 (1997).
11. T. Bouabça, N. B. Amor, D. Maynau, and M. C. ffarel, *J. Chem. Phys.* **130**, 114107 (2009), http://link.aip.org/link/?JCP/130/114107/1.
12. X. Oriols, J. J. García, F. Martín, J. Suñé, T. Gonzàlez, J. Mateos, and D. Pardo, *App. Phys. Lett.* **72**, 806 (1998).
13. J. C. Light, I. P. Hamilton, and J. V. Lill, *J. Chem. Phys.* **82**, 1400 (1985), http://link.aip.org/link/?JCP/82/1400/1.
14. J. C. Light, in *Time-Dependent Quantum Molecular Dynamics*, edited by J. Broeckhove and L. Lathouwers (Plenum, New York, 1992), pp. 185–199.
15. D. J. Kouri, T. Markovich, N. Maxwell, and E. R. Bittner, *J. Phys. Chem. A* (2009), http://dx.doi.org/10.1021/jp905798m.
16. J. B. Maddox and E. R. Bittner, *J. Chem. Phys.* **119**, 6465 (2003), http://link.aip.org/link/?JCP/119/6465/1.
17. D. Bohm, *Phys. Rev.* **85**, 180 (1952).
18. P. R. Holland, *The Quantum Theory of Motion* (Cambridge University Press, Cambridge, 1993).
19. C. L. Lopreore and R. E. Wyatt, *Phys. Rev. Lett.* **82**, 5190 (1999).
20. E. R. Bittner and R. E. Wyatt, *J. Chem. Phys.* **113**, 8888 (2000).
21. R. E. Wyatt, c. L. Lopreore, and G. Parlant, *J. Phys. Chem.* **114**, 5113 (2001).
22. S. W. Derrickson and E. R. Bittner, *J. Phys. Chem. A* **110**, 5333 (2006), http://pubs.acs.org/cgi-bin/article.cgi/jpcafh/2006/110/i16/pdf/jp055889q.pdf.
23. S. W. Derrickson and E. R. Bittner, *J. Phys. Chem. A* **111**, 10345 (2007), http://pubs.acs.org/doi/pdf/10.1021/jp0722657, http://pubs.acs.org/doi/abs/10.1021/jp0722657.
24. L. D. Landau and E. M. Lifshitz, *Quantum Mechanics (Non-Relativistic Theory)*, vol. 3 of *Course of Theoretical Physics* (Pergammon, Oxford, 1974), 3rd edn.
25. E. R. Bittner, J. B. Maddox, and D. J. Kouri, *J. Phys. Chem. A* (2009), http://dx.doi.org/10.1021/jp9058017.
26. J. C. Light, I. P. Hamilton, and J. V. Lill, *J. Chem. Phys.* **82**, 1400 (1985).
27. H. Baer, A. Belyaev, T. Krupovnickas, and X. Tata, *Phys. Rev. D* **65**, 075024 (pages 8) (2002), http://link.aps.org/abstract/PRD/v65/e075024.
28. A. A. Andrianov, N. V. Borisov, M. V. Ioffe, and M. I. Eides, *Theoret. Math. Phys.* **61**, 965 (1984).
29. A. A. Andrianov, N. V. Borisov, M. I. Eides, and M. V. Ioffe, *Phys. Lett. A* **109**, 143 (1985).
30. A. A. Andrianov, N. V. Borisov, and M. V. Ioffe, *Phys. Lett. A* **105**, 19 (1984).
31. A. Gangopadhyaya, P. K. Panigrahi, and U. P. Sukhatme, *Phys. Rev. A* **47**, 2720 (1993).
32. E. R. Bittner, D. J. Kouri, K. Maji, and T. J. Markovich, *J. Phys. Chem.*, submitted (2010).
33. D. J. Kouri, T. Markovich, N. Maxwell, and B. G. Bodman, *J. Phys. Chem. A* **113**, 7698 (2009).

34. P. T. Leung, A. M. van den Brink, W. M. Suen, C. W. Wong, and K. Young, *J. Math. Phys.* **42**, 4802 (2001), http://link.aip.org/link/?JMP/42/4802/1.
35. R. de Lima Rodrigues, P. B. da Silva Filho, and A. N. Vaidya, *Phys. Rev. D* **58**, 125023 (pages 6) (1998), http://link.aps.org/abstract/PRD/v58/e125023.
36. A. Contreras-Astorga and D. J. F. C., AIP Conference Proceedings **960**, 55 (2007), http://link.aip.org/link/?APC/960/55/1.
37. A. A. Andrianov, N. V. Borisov, and M. V. Ioffe, *Theoret. Math. Phys.* **72**, 748 (1987).
38. A. A. Andrianov and M. V. Ioffe, *Phys. Lett. B* **205**, 507 (1988).
39. A. A. Andrianov, N. V. Borisov, and M. V. Ioffe, *Theoret. Math. Phys.* **61**, 1078 (1984).
40. A. A. Andrianov, N. V. Borisov, and M. V. Ioffe, *JETP Lett.* **39**, 93 (1984).
41. A. A. Andrianov, N. V. Borisov, and M. V. Ioffe, *Phys. Lett. B* **181**, 141 (1986).
42. F. Cannata, M. V. Ioffe, and D. N. Nishnianidze, *J. Phys. A* **35**, 1389 (2002), http://stacks.iop.org/0305-4470/35/1389.
43. A. Andrianov, M. Ioffe, and D. Nishnianidze, *Phys. Lett. A* **201**, 103 (2002).
44. A. Das and S. A. Pernice, arXiv:hep-th/9612125v1 (1996).
45. M. A. Gonzalez-Leon, J. M. Gullarte, and M. de la Torre Mayado, SIGMA **3**, 124 (2007).
46. A. A. Andrianov, M. V. Ioffe, and V. P. Spiridonov, *Phys. Lett. A* **174**, 273 (1993).
47. A. A. Andrianov, M. V. Ioffe, and D. N. Nishnianidze, *Theoret. Math. Phys.* **104**, 1129 (1995).
48. V. V. Kravchenko, *J. Phys. A* **38**, 851 (2005).
49. R. I. Dzhioev and V. L. Korenev, *Phys. Rev. Lett.* **99**, 037401 (2007), ISSN 0031-9007 (Print).
50. D. J. Kouri and C. F. Curtiss, *J. Chem. Phys.* **43**, 1919 (1965), http://link.aip.org/link/?JCP/43/1919/1.

5 Quantum Field Dynamics from Trajectories

Peter Holland

CONTENTS

5.1 INTRODUCTION

Transcending its origin in the debate on the interpretation of quantum theory, the continuous deterministic trajectory has proved a potent tool in computational quantum mechanics (an excellent introduction to numerical trajectory techniques is provided by Wyatt's text [1]). It is straightforward to develop a method to derive the time-dependent Schrödinger equation from a single-valued continuum of spacetime trajectories, and supply a corresponding exact formula for the wavefunction [2]. A natural language for this theory of evolution is offered by the hydrodynamic analogy, in which wave mechanics corresponds to the Eulerian picture (as shown by Madelung [3]) and the trajectory theory to the Lagrangian picture. The Lagrangian model for the quantum fluid may be developed from a variational principle involving a specific interaction potential (the quantum internal potential energy), and the Euler–Lagrange equations imply a nonlinear partial differential equation to calculate the trajectories of the fluid particles as functions of their initial coordinates and the initial wavefunction. The latter supplies two sets of data to the particle model: the initial density (via the amplitude) and the initial velocity (via the phase). The time-dependent wavefunction is then computed via the standard map between the Lagrangian coordinates and the Eulerian fields.

Much of the work in this area has concentrated on developing techniques in the arena of non-relativistic spin 0 quantum mechanics, using trajectory flows in spacetime. Our purpose here is to show how the scheme may be extended to include the derivation of the dynamics of quantum fields from a trajectory model. It turns out that a unified description embracing both bosons and fermions may be developed by generalizing the configuration space of the known scheme for first quantized systems to a Riemannian manifold. We shall illustrate this for the simplest non-trivial systems that exhibit the key algebraic features, namely, non-relativistic free spinless boson and fermion fields. In the fluid description the two types of statistics are distinguished by the choice of the configuration space and associated Riemannian metric, together with the potentials to which the flow is subject. Specifically, the trajectories in the case of bosons will be described by translational time-dependent oscillator coordinates, and an analogous system of rotational time-dependent coordinates is employed in the fermion case. We thus call attention to the value of considering flows in spaces other than physical (Euclidean) space.

In extending the continuum-mechanical construction of the wave equation to fermionic fields we shall exploit the fact that the anticommutative fermion algebra is essentially an algebra of $SU(2)$ angular momentum operators. It is necessary then to include spin $\frac{1}{2}$ in the hydrodynamical description. This problem has been addressed previously and indeed the inclusion of spin was the initial motivation for developing the Riemannian generalization that we use [4, 5]. In presenting this method here we generalize the previous treatment slightly to include external potentials, which are necessary for the application to both boson and fermion fields. The key point in the fermion case is that in the Lagrangian picture the fluid particles are treated as spin $\frac{1}{2}$ rotators described by sets of Euler angles. The necessity of this model arises because the usual textbook approach to spin employing spinors is not suitable for our purposes. We shall discuss this problem first.

5.2   SPIN AND TRAJECTORIES

An Eulerian hydrodynamic picture for spin $\frac{1}{2}$ follows in a fairly straightforward way for systems governed by the Pauli equation (e.g., Ref. [6] and references therein). However, the hydrodynamic equations in this approach to spin are rather complicated and difficult to interpret, at least in comparison with the simplicity of the spin 0 Madelung model, and there are ambiguities in identifying suitably defined phases of the spinor components. An additional problem in connection with our constructive program is that it is not clear how to develop a suitable Lagrangian-coordinate, or trajectory, version of the theory. We may see this with a simple example. For a spinor field $\psi^a(x), a = 1, 2$, the associated probability density is

$$\rho = \psi^\dagger \psi \tag{5.1}$$

and the flow lines are naturally defined as the integral curves of the velocity field

$$v_i(x) = \frac{\hbar}{2mi\rho} \left(\psi^\dagger \frac{\partial \psi}{\partial x_i} - \frac{\partial \psi^\dagger}{\partial x_i} \psi \right), \quad i = 1, 2, 3. \tag{5.2}$$

Undoubtedly, these paths provide considerable insight into spin phenomena [6]. The problem in the present context is that the ensemble of paths generated by varying the initial position does not contain sufficient information to construct the four real components of the spinor field according to the method set out above and described in [2]. To see the nature of the information that is missing, consider the following two wavefunctions (ψ' obeys the free Pauli equation if ψ does):

$$\psi(x,t) = \begin{pmatrix} \psi^1 \\ \psi^2 \end{pmatrix}, \quad \psi'(x,t) = \sigma_1 \psi(x,t) = \begin{pmatrix} \psi^2 \\ \psi^1 \end{pmatrix} \tag{5.3}$$

where we have used one of the Pauli matrices $\sigma_i, i = 1, 2, 3$ (any one would do in this example). These wavefunctions may be chosen to have non-trivially distinct time dependences. But the implied probability density and velocity are the same for each, given by the Equations 5.1 and 5.2. To capture the further information in the wavefunction not exhibited by the trajectories one needs to introduce additional hydrodynamic-like variables built from the spinor. The spin vector field $s_i(x,t) = (\hbar/2\rho)\psi^\dagger \sigma_i \psi$ plays such a role; denoting by s_i' the spin vector associated with ψ', the two states (Equation 5.3) are distinguished as follows:

$$s_1' = s_1, \quad s_2' = -s_2, \quad s_3' = -s_3. \tag{5.4}$$

Unfortunately, while such quantities suitably extend the Eulerian hydrodynamic description, this does not assist us since the additional functions do not have Lagrangian counterparts. That is, although one can potentially associate additional structure with the trajectories defined by Equation 5.2 by evaluating the functions along them, the trajectories exhibit no features that could be set in correspondence with the functions. One can partially ameliorate the situation by seeking to redefine the velocity so that it depends on the additional variables whilst corresponding to the same density. Just such an additional term is, in fact, required by relativistic considerations [7] and this leads to the following modified non-relativistic velocity:

$$v_i = \frac{\hbar}{2mi\rho} \left(\psi^\dagger \frac{\partial \psi}{\partial x_i} - \frac{\partial \psi^\dagger}{\partial x_i} \psi \right) + \frac{1}{m\rho} \sum_{j,l} \varepsilon_{ijl} \frac{\partial}{\partial x_j} (\rho s_l). \tag{5.5}$$

It remains the case, however, that we cannot uniquely infer from the modified trajectories the full time-dependent wavefunction.

The problem is compounded for many-body systems. As the number of particles increases, reproducing all the information in the wavefunction through additional hydrodynamic variables requires ever more complex quantities that lack physical interpretation and corresponding Lagrangian representatives.

It has been argued that the origin of the problems with the spin $\frac{1}{2}$ hydrodynamic theory in this context—its complexity and lack of an adequate Lagrangian formulation—is that the standard approach to spin on which it is based works with a (angular momentum) representation of the quantum theory in which the rotational freedoms appear as discrete indices in the wavefunction [6]. The local fluid quantities (such as density and velocity) are defined by summing, or "averaging," over these indices, and as a result essential information contained in the wavefunction is lost. The alternative procedure advocated in Reference [6] is to start from the angular coordinate representation in which the spin freedoms are represented as continuous parameters $(\alpha^1, \alpha^2, \alpha^3)$ (Euler angles) in the wavefunction, on the same footing as the spatial variables x: $\psi(x, \alpha, t)$. Specifically, we write

$$\psi(x, \alpha, t) = \sum_{a=1,2} \psi^a(x, t) u_a(\alpha), \tag{5.6}$$

where

$$\begin{aligned}
u_1(\alpha) &= (2\sqrt{2}\pi)^{-1} \cos(\alpha^1/2) e^{-i(\alpha^2 + \alpha^3)/2}, \\
u_2(\alpha) &= -i(2\sqrt{2}\pi)^{-1} \sin(\alpha^1/2) e^{i(\alpha^2 - \alpha^3)/2},
\end{aligned} \tag{5.7}$$

are spin $\frac{1}{2}$ basis functions with

$$\int u_a^*(\alpha) u_b(\alpha) d\Omega = \delta_{ab},$$

$$d\Omega = \sin \alpha^1 d\alpha^1 d\alpha^2 d\alpha^3, \quad \alpha^1 \in [0, \pi], \alpha^2 \in [0, 2\pi], \alpha^3 \in [0, 4\pi]. \tag{5.8}$$

This implies a physically clearer and simpler hydrodynamic model. The phase S of the wavefunction is immediately identifiable and the equations for the fluid paths are defined in terms of the gradient of S with respect to all the coordinates, an obvious generalization of the spin 0 theory. The paths thus comprise both spatial and angular components. The principal advantage of the angular coordinate approach for our considerations here is that it provides a sufficiently detailed Lagrangian picture that allows us to extend the constructive method to include the wavefunction (Equation 5.6). The enhanced detail is indicated, for example, by the fact that the velocity field (Equation 5.2) is the mean of the new velocity over the angles:

$$v_i(x, t) = (1/\rho) \int |\psi(x, \alpha, t)|^2 \frac{1}{m} \frac{\partial S(x, \alpha, t)}{\partial x_i} d\Omega. \tag{5.9}$$

Having obtained the wavefunction, the components of the time-dependent spinor field may be extracted by using Equation 5.8 to invert Equation 5.6:

$$\psi_a(x, t) = \int \psi(x, \alpha, t) u_a^*(\alpha) d\Omega. \tag{5.10}$$

In our application to fermions here the space variables are not relevant and we need consider only the angular freedoms.

5.3 TRAJECTORY CONSTRUCTION OF THE WAVE EQUATION IN A RIEMANNIAN MANIFOLD

5.3.1 LAGRANGIAN PICTURE

We describe here the Lagrangian-coordinate hydrodynamic derivation of the wave equation in an N-dimensional Riemannian manifold M with generalized coordinates

x^μ and (static) metric $g_{\mu\nu}(x)$, $\mu, \nu, \ldots = 1, \ldots, N$. In this space, the history of the fluid is encoded in the positions $\xi(\xi_0, t)$ of the distinct fluid elements at time t, each particle being distinguished by its position ξ_0 at $t = 0$. We assume that the mapping between these two sets of coordinates are single-valued and differentiable with respect to ξ_0 and t to whatever order is necessary, and that the inverse mapping $\xi_0(\xi, t)$ exists and has the same properties.

Let $P_0(\xi_0)$ be the initial density of some continuously distributed quantity in M (number of particles in our application) and $g = \det g_{\mu\nu}$. Then the quantity in an elementary volume $d^N \xi_0$ attached to the point ξ_0 is given by $P_0(\xi_0)\sqrt{|g(\xi_0)|}d^N\xi_0$. The conservation of this quantity in the course of the motion of the fluid element is expressed through the relation

$$P(\xi(\xi_0, t))\sqrt{|g(\xi(\xi_0, t))|}d^N\xi(\xi_0, t) = P_0(\xi_0)\sqrt{|g(\xi_0)|}d^N\xi_0 \qquad (5.11)$$

or

$$P(\xi_0, t) = D^{-1}(\xi_0, t)P_0(\xi_0) \qquad (5.12)$$

where

$$D(\xi_0, t) = \sqrt{|g(\xi)|/|g(\xi_0)|}J(\xi_0, t), \quad 0 < D < \infty, \qquad (5.13)$$

and J is the Jacobian of the transformation between the two sets of coordinates:

$$J = \frac{1}{N!} \sum_{\substack{\mu_1,\ldots,\mu_N \\ \nu_1,\ldots,\nu_N}} \varepsilon_{\mu_1\cdots\mu_N}\varepsilon^{\nu_1\cdots\nu_N}\frac{\partial\xi^{\mu_1}}{\partial\xi_0^{\nu_1}}\cdots\frac{\partial\xi^{\mu_N}}{\partial\xi_0^{\nu_N}}. \qquad (5.14)$$

In our later application where we take the limit $N \to \infty$ it is assumed that the determinant converges.

We assume that the Lagrangian for the set of fluid particles comprises a kinetic term, an internal quantum potential energy that represents a certain kind of particle interaction, and terms due to external scalar and vector potentials:

$$L = \int P_0(\xi_0)\left[\sum_{\mu,\nu}\left(\frac{1}{2}mg_{\mu\nu}(\xi)\frac{\partial\xi^\mu}{\partial t}\frac{\partial\xi^\nu}{\partial t} - g^{\mu\nu}(\xi)\frac{\hbar^2}{8m}\frac{1}{P^2}\frac{\partial P}{\partial\xi^\mu}\frac{\partial P}{\partial\xi^\nu}\right)\right.$$
$$\left. + \sum_\mu A_\mu(\xi)\frac{\partial\xi^\mu}{\partial t} - V(\xi)\right]\sqrt{|g(\xi_0)|}d^N\xi_0. \qquad (5.15)$$

Here P_0, A_μ, V, and $g_{\mu\nu}$ are prescribed functions, $\xi = \xi(\xi_0, t)$ and we substitute for P from Equation 5.12 and write

$$\frac{\partial}{\partial\xi^\mu} = J^{-1}\sum_\nu J_\mu^\nu\frac{\partial}{\partial\xi_0^\nu} \qquad (5.16)$$

where

$$J_\mu^\nu = \frac{\partial J}{\partial(\partial\xi^\mu/\partial\xi_0^\nu)} \qquad (5.17)$$

is the cofactor of $\partial \xi^\mu / \partial \xi_0^\nu$. The latter satisfies

$$\sum_\mu \frac{\partial \xi^\mu}{\partial \xi_0^\nu} J_\mu^\sigma = J \delta_\nu^\sigma. \tag{5.18}$$

It is assumed that P_0 and its derivatives take values such that the surface terms in the variational principle vanish. The parameter m will later be identified with the mass of the quantum system and should not be confused with the mass of a fluid particle, which is $mP\sqrt{|g|}d^N\xi$.

Varying the coordinates, the Euler–Lagrange equations of motion for the ξ_0th fluid particle take the form of Newton's second law in general coordinates:

$$\frac{\partial^2 \xi^\mu}{\partial t^2} + \sum_{\nu,\sigma} \begin{Bmatrix} \mu \\ \nu\sigma \end{Bmatrix} \frac{\partial \xi^\nu}{\partial t} \frac{\partial \xi^\sigma}{\partial t} = -\frac{1}{m} \sum_\nu g^{\mu\nu} \frac{\partial}{\partial \xi^\nu}(Q+V) + \frac{1}{m} \sum_\nu F^{\mu\nu} \frac{\partial \xi^\nu}{\partial t} \tag{5.19}$$

where $\begin{Bmatrix} \mu \\ \nu\sigma \end{Bmatrix} = \sum_\rho \frac{1}{2} g^{\mu\rho}(\partial g_{\sigma\rho}/\partial \xi^\nu + \partial g_{\nu\rho}/\partial \xi^\sigma - \partial g_{\nu\sigma}/\partial \xi^\rho)$, $F_{\mu\nu} = \partial A_\nu / \partial \xi^\mu - \partial A_\mu / \partial \xi^\nu$, and

$$Q = \sum_{\mu,\nu} \frac{-\hbar^2}{2m\sqrt{|g|}P} \frac{\partial}{\partial \xi^\mu} \left(\sqrt{|g|} g^{\mu\nu} \frac{\partial \sqrt{P}}{\partial \xi^\nu} \right) \tag{5.20}$$

is the quantum potential. Here we have written the force terms on the right-hand side of Equation 5.19 in condensed form and substituting for P from Equation 5.12 and for the derivatives with respect to ξ from Equation 5.16 we obtain a highly complex fourth-order (in ξ_0) local nonlinear partial differential equation. We shall see that from the solutions $\xi = \xi(\xi_0, t)$, subject to specification of $\partial \xi_0^\mu / \partial t$ whose determination is discussed next, we may derive solutions to Schrödinger's equation.

5.3.2 INITIAL CONDITIONS

To obtain a flow that is representative of Schrödinger evolution we choose the initial conditions of Equation 5.19 so that P_0 is the square of the amplitude of the initial wavefunction and the initial covariant components of the velocity field are given by

$$\sum_\nu g_{\mu\nu}(\xi_0) \frac{\partial \xi_0^\nu}{\partial t} = \frac{1}{m} \left(\frac{\partial S_0(\xi_0)}{\partial \xi_0^\mu} - A_\mu(\xi_0) \right) \tag{5.21}$$

where $S_0(\xi_0)$ is the phase of the initial wavefunction. To show that these assumptions imply motion characteristic of quantum evolution we first demonstrate that the form Equation 5.21 is preserved by the dynamical Equation 5.19. To this end, we use a generalization of the method employed in classical hydrodynamics based on Weber's transformation [8].

We first multiply Equation 5.19 by $\sum_\sigma g_{\sigma\mu} \partial \xi^\sigma / \partial \xi_0^\rho$ and integrate between the time limits $(0, t)$. The term involving the Christoffel symbols $\begin{Bmatrix} \mu \\ \nu\sigma \end{Bmatrix}$ drops out and we obtain

$$\sum_{\sigma,\mu} g_{\sigma\mu}(\xi(\xi_0,t)) \frac{\partial \xi^\sigma}{\partial \xi_0^\rho} \frac{\partial \xi^\mu}{\partial t} = \sum_\mu g_{\rho\mu}(\xi_0) \frac{\partial \xi_0^\mu}{\partial t} + \frac{1}{m} A_\rho(\xi_0) - \sum_\mu \frac{1}{m} A_\mu \frac{\partial \xi^\mu}{\partial \xi_0^\rho}$$

$$+ \frac{\partial}{\partial \xi_0^\rho} \int_0^t \left(\sum_{\mu,\nu} \frac{1}{2} g_{\mu\nu}(\xi(\xi_0,t)) \frac{\partial \xi^\mu}{\partial t} \frac{\partial \xi^\nu}{\partial t} \right.$$

$$\left. + \sum_\mu \frac{1}{m} A_\mu \frac{\partial \xi^\mu}{\partial t} - \frac{1}{m}(Q+V) \right) dt. \qquad (5.22)$$

Then, substituting Equation 5.21, we get

$$\sum_{\sigma,\mu} g_{\sigma\mu} \frac{\partial \xi^\sigma}{\partial \xi_0^\rho} \frac{\partial \xi^\mu}{\partial t} = \frac{1}{m} \left(\frac{\partial S}{\partial \xi_0^\rho} - \sum_\mu A_\mu \frac{\partial \xi^\mu}{\partial \xi_0^\rho} \right) \qquad (5.23)$$

where

$$S = S_0 + \int_0^t \left(\sum_{\mu,\nu} \frac{1}{2} mg_{\mu\nu} \frac{\partial \xi^\mu}{\partial t} \frac{\partial \xi^\nu}{\partial t} + \sum_\mu A_\mu \frac{\partial \xi^\mu}{\partial t} - Q - V \right) dt. \qquad (5.24)$$

The left-hand side of Equation 5.23 gives the covariant velocity components at time t with respect to the ξ_0-coordinates. To obtain the ξ components we multiply by $J^{-1} J_\nu^\rho$ and use Equations 5.16 and 5.18 to get, at time t,

$$\sum_\nu g_{\mu\nu} \frac{\partial \xi^\nu}{\partial t} = \frac{1}{m} \left(\frac{\partial S}{\partial \xi^\mu} - A_\mu(\xi) \right) \qquad (5.25)$$

where $S = S(\xi_0(\xi,t),t)$. We conclude that the covariant velocity of each particle retains forever the form Equation 5.21 if it possesses it at any moment.

5.3.3 EULERIAN PICTURE

We can transform the nonlinear dynamical equation into a linear one by changing the independent variable from the particle label ξ_0 to a fixed space point x. The link between the particle (Lagrangian) and wave-mechanical (Eulerian) pictures is defined by the following expression for the Eulerian density:

$$P(x,t)\sqrt{|g(x)|} = \int \delta(x - \xi(\xi_0,t)) P_0(\xi_0) \sqrt{|g(\xi_0)|} d^N \xi_0. \qquad (5.26)$$

The corresponding formula for the Eulerian velocity is contained in the expression for the current:

$$P(x,t)\sqrt{|g(x)|} v^\mu(x,t) = \int \frac{\partial \xi^\mu(\xi_0,t)}{\partial t} \delta(x - \xi(\xi_0,t)) P_0(\xi_0) \sqrt{|g(\xi_0)|} d^N \xi_0. \qquad (5.27)$$

These relations express both the change of picture and the temporal propagation of the system. Evaluating the integrals, Equations 5.26 and 5.27 are equivalent to the following local expressions

$$P(x,t)\sqrt{|g(x)|} = J^{-1}|_{\xi_0(x,t)} P_0(\xi_0(x,t))\sqrt{|g(\xi_0(x,t))|} \qquad (5.28)$$

$$v^\mu(x,t) = \left.\frac{\partial \xi^\mu(\xi_0,t)}{\partial t}\right|_{\xi_0(x,t)}. \qquad (5.29)$$

These formulas enable us to translate the Lagrangian flow equations into Eulerian language and solve the latter using the trajectories. Differentiating Equation 5.26 with respect to t and using Equation 5.27 we deduce the continuity equation

$$\frac{\partial P}{\partial t} + \frac{1}{\sqrt{|g|}} \sum_\mu \frac{\partial}{\partial x^\mu}(P\sqrt{|g|}v^\mu) = 0. \qquad (5.30)$$

Next, differentiating Equation 5.27 and using Equations 5.19 and 5.30 we get the analog of Euler's classical equation:

$$\frac{\partial v^\mu}{\partial t} + \sum_v v^v \frac{\partial v^\mu}{\partial x^v} + \sum_{v,\sigma} \begin{Bmatrix} \mu \\ v\sigma \end{Bmatrix} v^v v^\sigma = -\frac{1}{m} \sum_v g^{\mu v} \frac{\partial}{\partial x^v}(Q+V) + \frac{1}{m} \sum_v F^{\mu v} v^v$$

$$(5.31)$$

where Q is given by Equation 5.20 with ξ replaced by x. Finally, the Formula 5.25 for the velocity becomes

$$v^\mu = \frac{1}{m}\left(\sum_v g^{\mu v} \frac{\partial S(x,t)}{\partial x^v} - A^\mu(x)\right). \qquad (5.32)$$

Formulas 5.28 and 5.29 give the general solution of the coupled continuity and Euler Equations 5.30 and 5.31 in terms of the paths and initial wavefunction.

To establish the connection between the Eulerian equations and Schrödinger's equation we note that, using Equation 5.32, 5.31 implies

$$\frac{\partial S}{\partial t} + \frac{1}{2m} \sum_{\mu,v} g^{\mu v} \left(\frac{\partial S}{\partial x^\mu} - A_\mu\right)\left(\frac{\partial S}{\partial x^v} - A_v\right) + Q + V = 0 \qquad (5.33)$$

where we have absorbed a function of time in S. Combining Equation 5.33 with Equation 5.30 (where we substitute Equation 5.32) we find that the function $\Psi(x,t) = \sqrt{P}\exp(iS/\hbar)$ obeys the Schrödinger equation in external scalar and vector potentials in general coordinates:

$$i\hbar\frac{\partial \Psi}{\partial t} = \frac{-\hbar^2}{2m\sqrt{|g|}} \sum_{\mu,v}\left(\frac{\partial}{\partial x^\mu} - (i/\hbar)A_\mu\right)\left[\sqrt{|g|}g^{\mu v}\left(\frac{\partial \Psi}{\partial x^v} - (i/\hbar)A_v\Psi\right)\right] + V\psi.$$

$$(5.34)$$

We have thus deduced the linear wave equation from the nonlinear collective particle motion.

The explicit construction of the time-dependent wavefunction in terms of the trajectories follows straightforwardly from Equation 5.28 (which gives the amplitude) and Equations 5.29 and 5.32. The latter give $\partial S/\partial x^\mu$ and S is fixed up to an additive constant by Equation 5.33.

5.4   BOSONS

To make contact with the field theory for bosons we let $M = \mathbb{R} \otimes \cdots \otimes \mathbb{R}$ with ξ^μ a set of Cartesian coordinates and $g_{\mu\nu} = \delta_{\mu\nu}$. We assume moreover that the coordinate index is shorthand for three indices: $\mu \equiv k_i = (k_1, k_2, k_3)$ where $k_i = -2\pi n/L, \ldots, 2\pi n/L$, $i = 1, 2, 3$, with L the side of a box of volume L^3 and $N = (2n+1)^3$. We shall write k for k_i. Choosing $A_\mu = 0$ and $V = \sum_k \frac{1}{2} m \omega_k^2 \xi_k^2$ where $\omega_k = \hbar k^2/2m$, the equation of motion 5.19 of the ξ_{k0}th particle becomes

$$\frac{\partial^2 \xi_k}{\partial t^2} + \omega_k^2 \xi_k = -\frac{1}{m} \frac{\partial Q}{\partial \xi_k}, \quad Q = \sum_k \frac{-\hbar^2}{2m\sqrt{P}} \frac{\partial^2 \sqrt{P}}{\partial \xi_k^2}. \tag{5.35}$$

The fluid is therefore a continuum of linear oscillators coupled by quantum force terms. The corresponding classical system is obtained when the quantum force $-\partial Q/\partial \xi_k$ may be neglected.

For notational convenience we write q in place of x in passing to the Eulerian picture. The Schrödinger Equation 5.34 implied by Equation 5.35 is

$$i\hbar \frac{\partial \Psi}{\partial t} = \sum_k \left(\frac{-\hbar^2}{2m} \frac{\partial^2 \Psi}{\partial q_k^2} + \frac{1}{2} m \omega_k^2 q_k^2 \Psi \right). \tag{5.36}$$

Writing $p_k = -i\hbar \partial/\partial q_k$ and using the q_k, p_k commutation relations, the quantities

$$a_k = \left(\frac{1}{2} |k| q_k - p_k/i\hbar|k| \right), \quad a_k^\dagger = \left(\frac{1}{2} |k| q_k + p_k/i\hbar|k| \right) \tag{5.37}$$

define a set of bosonic creation and annihilation operators:

$$[a_k, a_{k'}^\dagger]_- = \delta_{kk'}, \quad [a_k, a_{k'}]_- = [a_k^\dagger, a_{k'}^\dagger]_- = 0. \tag{5.38}$$

Using these operators, the Hamiltonian in Equation 5.36 becomes

$$H = \sum_k E_k a_k^\dagger a_k \tag{5.39}$$

where $E_k = \hbar^2 k^2/2m$. To complete the derivation of the quantum field dynamics we take the limit $n \to \infty$ and define a complex field operator in space through the Fourier series

$$\psi(x) = L^{-3/2} \sum_k a_k e^{ik \cdot x}, \quad k_i = 2\pi n_i/L, n_i \in \mathbb{Z}, i = 1, 2, 3. \tag{5.40}$$

The relations Equation 5.38 imply that the field operators obey the commutation relations

$$[\psi(x), \psi^\dagger(x')]_- = \delta(x - x'), \quad [\psi(x), \psi(x')]_- = [\psi^\dagger(x), \psi^\dagger(x')]_- = 0 \tag{5.41}$$

and the Hamiltonian Equation 5.39 is that of a free boson field:

$$H = \frac{-\hbar^2}{2m} \int \psi^\dagger(x)\nabla^2\psi(x)d^3x. \tag{5.42}$$

We have thus derived the dynamics of a free quantum boson field from the trajectory theory of a continuum of harmonic oscillators coupled by forces derived from the quantum potential.

5.5 FERMIONS

In the case of fermions we make the following identification of the manifold supporting the flow: $M = SU(2) \otimes \cdots \otimes SU(2)$. Here the index $\mu = r, k_i$ where $r = 1, 2, 3$ and k_i is defined as in Section 5.4. The coordinates $\xi^{rk_i} = \sqrt{3/2}|k|^{-1}\theta^r_{k_i}$ where $\theta^r_{k_i} = (\theta^1_{k_i}, \theta^2_{k_i}, \theta^3_{k_i})$ are, for each k, a set of three Euler angles (for the definition see Ref. [6]), and the k-dependent coefficient is chosen in order to achieve a particular form for the moment of inertia. The metric on M is $g_{\mu\nu} = g_{rk_isk'_j}$ with $g_{rk_isk'_j} = G_{rs}(\theta_k)\delta_{k_ik'_j}$ and

$$G_{rs} = \begin{pmatrix} 1 & 0 & 0 \\ 0 & 1 & \cos\theta^1_k \\ 0 & \cos\theta^1_k & 1 \end{pmatrix}, \quad G^{rs} = \begin{pmatrix} 1 & 0 & 0 \\ 0 & \dfrac{1}{\sin^2\theta^1_k} & -\dfrac{\cos\theta^1_k}{\sin^2\theta^1_k} \\ 0 & -\dfrac{\cos\theta^1_k}{\sin^2\theta^1_k} & \dfrac{1}{\sin^2\theta^1_k} \end{pmatrix}. \tag{5.43}$$

Thus $g = \prod_k \det G_{rs}(\theta_k) = \prod_k \sin^2\theta^1_k$.

Instead of expressing the law of motion Equation 5.19 directly in terms of the Euler angles, we can more clearly appreciate the physical model if we use the angular velocity vector and other quantities carrying vector indices $i, j, \ldots$ with respect to which the metric is δ_{ij}. We may pass between the angle and vector descriptions using the matrices

$$X^i_r = \begin{pmatrix} -\cos\theta^2_k & 0 & -\sin\theta^1_k\sin\theta^2_k \\ \sin\theta^2_k & 0 & -\sin\theta^1_k\cos\theta^2_k \\ 0 & -1 & -\cos\theta^1_k \end{pmatrix},$$

$$X^r_i = \begin{pmatrix} -\cos\theta^2_k & \cot\theta^1_k\sin\theta^2_k & -\mathrm{cosec}\theta^1_k\sin\theta^2_k \\ \sin\theta^2_k & \cot\theta^1_k\cos\theta^2_k & -\mathrm{cosec}\theta^1_k\cos\theta^2_k \\ 0 & -1 & 0 \end{pmatrix}. \tag{5.44}$$

Then, for each k, the angular velocity is $\omega^i_k = \sum_r X^i_r \partial\theta^r_k/\partial t$ and the metrics are given by $G_{rs} = \sum_i X^i_r X^i_s$ and $\delta_{ij} = \sum_r X^r_i X^r_j$. Writing $A_{rk}(\theta_k) = -\sum_i \sqrt{mI_k}X^i_r B_{ik}(\theta_k)$ where $I_k = 3m/2k^2$ we have

$$\sum_{\mu} A_{\mu} \frac{\partial \xi^{\mu}}{\partial t} = -\sum_{i,k} I_k B_{ik} \omega_k^i. \tag{5.45}$$

Next, we introduce a set of angular momentum operators for each k,

$$\hat{M}_{ik} = -i\hbar \sum_r X_i^r \partial/\partial \theta_k^r, \quad i = 1,2,3, \tag{5.46}$$

so that

$$-\hbar^2 \sum_{\mu,v} g^{\mu v}(\xi) \frac{\partial P}{\partial \xi^{\mu}} \frac{\partial P}{\partial \xi^v} = -\hbar^2 \sum_{r,s,k} \left(\frac{2k^2}{3}\right) G^{rs}(\theta_k) \frac{\partial P}{\partial \theta_k^r} \frac{\partial P}{\partial \theta_k^s}$$

$$= \sum_{i,k} \left(\frac{2k^2}{3}\right) \hat{M}_{ik} P \hat{M}_{ik} P. \tag{5.47}$$

Putting all this together in Equation 5.15, the Lagrangian may be written

$$L = \int P_0(\theta_{k0}) \sum_{i,k} \left(\frac{1}{2} I_k \omega_k^i \omega_k^i + \frac{1}{8 I_k P^2} \hat{M}_{ik} P \hat{M}_{ik} P - I_k B_{ik}(\theta_k) \omega_k^i - V(\theta_k) \right) \prod_k d\Omega_{k0} \tag{5.48}$$

where $d\Omega_{k0} = \sin\theta_{k0}^1 d\theta_{k0}^1 d\theta_{k0}^2 d\theta_{k0}^3$. This is the Lagrangian of a collection of spherical rotators of moment of inertia I_k subject to external magnetic-type fields B_{ik} and a scalar potential V. Their quantum-mechanical character is expressed through the quadratic interaction potential and the constraints implied by the initial wavefunction. In terms of these variables the (final) Lorentz force-type term in Equation 5.19 contributes several terms including a Larmor-type torque and the equation of motion of the θ_{k0}th particle is

$$\frac{\partial \omega_k^i}{\partial t} = \sum_{j,l} \varepsilon_{ijl} B_{jk} \omega_k^l + \frac{1}{i\hbar I_k} \hat{M}_{ik}(Q + V) + \frac{1}{i\hbar} \sum_j \omega_k^j (\hat{M}_{ik} B_{jk} - \hat{M}_{jk} B_{ik}). \tag{5.49}$$

The fluid is therefore a continuum of spin $\frac{1}{2}$ rotators subject to Larmor (the first term on the right-hand side of Equation 5.49), quantum and external torques. The corresponding classical system is obtained when the quantum torque $\hat{M}_{ik} Q/i\hbar$ may be neglected.

The justification for this construction is seen on passing to the Eulerian picture where we write $x^{rk_i} = \sqrt{3/2}|k|^{-1}\alpha_{k_i}^r$. First, we note that the Schrödinger Equation 5.34 implied by Equation 5.49 describes a set of spherical rotators, in external "magnetic" and scalar potentials, in the angular coordinate representation:

$$i\hbar \frac{\partial \Psi}{\partial t} = \sum_{i,k} \frac{1}{2I_k} (\hat{M}_{ik} + I_k B_{ik})(\hat{M}_{ik} + I_k B_{ik})\Psi + V\Psi \tag{5.50}$$

where $\hat{M}_{ik}$ is given by Equation 5.46 with θ replaced by α. In this representation the wavefunction may be expanded in terms of a complete set of orthonormal spin $\frac{1}{2}$ basis functions Equation 5.7 for each k:

$$\Psi = \sum_{a_k=1,2} c_{\Pi_k a_k}(t) \prod_k u_{a_k}(\alpha_k) \tag{5.51}$$

where the product is antisymmetrized. To obtain the fermion field dynamics we need to choose special forms for the potentials, analogous to the oscillator potential in the boson case. We choose $A_k^r = (0, \sqrt{mI_k} B_k, 0) = \text{const.}$, where $B_k = E_k/\hbar$ so that $A_{rk} = \sqrt{mI_k} B_k (0, 1, \cos \theta_k^1)$, and $V = -\sum_k \frac{1}{2} I_k B_k B_k$. This implies a constant "magnetic" field $B_{ik} = (0, 0, B_k)$. With these assumptions, the rotator equation of motion 5.49 becomes

$$\frac{\partial \omega_k^i}{\partial t} = \sum_{j,l} \varepsilon_{ijl} B_{jk} \omega_k^l + \frac{1}{i\hbar I_k} \hat{M}_{ik} Q \tag{5.52}$$

and the Schrödinger Equation 5.50 is

$$i\hbar \frac{\partial \Psi}{\partial t} = \sum_{i,k} \frac{1}{2I_k} \hat{M}_{ik} \hat{M}_{ik} \Psi + B_k \hat{M}_{3k} \Psi. \tag{5.53}$$

It is a simple matter to map this theory into more familiar language. We note that the angular momentum operators obey the following commutation and anticommutation relations:

$$[\hat{M}_{ik}, \hat{M}_{jk}]_- = \sum_i i\hbar \varepsilon_{ijl} \hat{M}_{lk}, \quad [\hat{M}_{ik}, \hat{M}_{jk'}]_- = 0, \quad [\hat{M}_{ik}, \hat{M}_{jk}]_+ = 2(\hbar/2)^2 \delta_{ij}, k \neq k'. \tag{5.54}$$

Making a Jordan–Wigner transformation [9, 10], these relations may be expressed equivalently as an algebra of fermionic creation and annihilation operators. Thus, defining

$$a_k = (1/\hbar)\lambda_k(\hat{M}_{1k} - i\hat{M}_{2k}), \quad a_k^\dagger = (1/\hbar)\lambda_k(\hat{M}_{1k} + i\hat{M}_{2k}),$$

$$\lambda_k = \prod_{k'<k}(-2\hat{M}_{3k'}/\hbar), \tag{5.55}$$

the new operators obey the anticommutation relations that define a fermion theory:

$$[a_k, a_{k'}^\dagger]_+ = \delta_{kk'}, \quad [a_k, a_{k'}]_+ = [a_k^\dagger, a_{k'}^\dagger]_+ = 0. \tag{5.56}$$

The Hamiltonian in Equation 5.53 is given by Equation 5.39 and, using the definition Equation 5.40 of the field operator, the Hamiltonian takes the form Equation 5.42 but where the commutation relations Equation 5.41 are replaced by anticommutation relations. This completes the derivation of the dynamics of a free quantum fermion field from the trajectory theory of a continuum of spin $\frac{1}{2}$ Larmor rotators subject to torques derived from the quantum potential.

5.6 CONCLUSION

We have established that quantized fields admit two complementary descriptions which find a natural interpretation in the language of hydrodynamics: an Eulerian picture which corresponds to the Schrödinger picture of field theory, and a Lagrangian picture comprising a (continuously) many-"particle" system. In particular, we have shown how the Eulerian notion of evolution (the history of fields at a space point) may be derived from the Lagrangian one (the temporal sequence of fluid elements passing through that point). One may transform between the pictures using the Formulas 5.28 and 5.29 and their inverses. The dynamics of boson and fermion fields may be given a common description based on a generalized Riemannian geometry.

This and similar investigations suggest that the deterministic trajectory may be regarded as a foundational component of the quantum description and not merely an optional element of interpretation. We may regard the approach as an alternative method of "quantization," characterized as follows: starting from a single particle, pass to a continuum of particles and introduce an interparticle interaction (the second term in brackets in Equation 5.15). Next, generalize to a Riemannian space with external scalar and vector potentials. Finally, pass to an Eulerian description. The method attributes a fundamental formal significance to the quantum internal potential energy (the interaction term), beyond its original purely interpretational aspect [6]. This formalism has a certain universal character in that its Lagrangian technique of construction applies to a variety of quantum theories. Indeed, it may be applied to other field theories admitting representations in terms of conservation equations, such as Maxwell's equations [4]. A common feature in all these applications is the appearance of the quadratic interaction term, the different cases being distinguished by the choice of the configuration space and the external potentials. This approach thus brings to light a meaningful sense in which physical theories generally may be said to exhibit "wave–particle duality."

The spin $\frac{1}{2}$ rotator model was proposed originally as a solution to the problem of extending the de Broglie–Bohm theory to fields quantized according to fermion statistics, a problem that had hitherto been regarded as unsolvable [11]. In that case, one regards one set of angle variables θ_{k0}^r as preferential labels describing the actual state of the system (we have explained elsewhere why the de Broglie–Bohm model, although similar mathematically, should not be conflated with the constructive application of trajectories [2]). Conceptually, the model follows the lead of Bohm's treatment of the electromagnetic field in terms of oscillating normal coordinates (as described in the second of his classic papers [12]), an elementary version of which has been presented in Section 5.4. It thus has the benefit of establishing a fermionic analog of the normal mode decomposition of bosonic fields.

BIBLIOGRAPHY

1. R.E. Wyatt, *Quantum Dynamics with Trajectories* (Springer, New York, 2005).
2. P. Holland, *Ann. Phys. (NY)* **315**, 505 (2005).
3. E. Madelung, *Z. Phys.* **40**, 322 (1926).

4. P. Holland, *Proc. R. Soc. A* **461**, 3659 (2005).
5. P. Holland, *Found. Phys.* **36**, 369 (2006).
6. P.R. Holland, *The Quantum Theory of Motion* (Cambridge University Press, Cambridge, 1993).
7. P. Holland and C. Philippidis, *Phys. Rev. A* **67**, 062105 (2003).
8. H. Lamb, *Hydrodynamics*, 6th edition (Cambridge University Press, Cambridge, 1932).
9. P. Jordan and E.P. Wigner, *Z. Phys.* **47**, 631 (1928).
10. J.D. Bjorken and S.D. Drell, *Relativistic Quantum Fields* (McGraw-Hill, New York, 1966).
11. P.R. Holland, *Phys. Lett. A* **128**, 9 (1988).
12. D. Bohm, *Phys. Rev.* **85**, 180 (1952).

6 The Utility of Quantum Forces

Gary E. Bowman

CONTENTS

6.1 INTRODUCTION

Utility: The condition, quality, or fact of being useful or beneficial (*The Shorter Oxford English Dictionary*)

Bohmian mechanics—the causal interpretation of quantum mechanics first introduced in completed form by David Bohm in 1952 [1,2]—is generally believed to be empirically equivalent to "standard" quantum mechanics. In principle, any result obtainable in one should be obtainable in the other. It need not be true, however, that equivalent results may be obtained with equivalent ease in the two interpretations, or that they always render equally valuable physical insights.

Even *within* Bohmian mechanics there exist two formulations, utilizing different tools and concepts: one in terms of Hamilton–Jacobi theory, the other in terms of Newtonian classical mechanics. I will call these, respectively, the HJ (Hamilton–Jacobi) and QF (quantum force) formulations of Bohmian mechanics. Both must lead to equivalent physics, but—much like the contrast between Bohmian and standard quantum mechanics—this need not imply that they offer equivalent calculational tools or physical insights.

Herein I do not argue which interpretation of quantum mechanics offers the correct picture of the world, nor do I advocate one formulation of Bohmian mechanics over the other. My goal is less fundamental, more utilitarian: to illustrate that the QF formulation of Bohmian mechanics offers distinct mathematical and conceptual tools that are unavailable in either the HJ formulation or in standard quantum mechanics. Thus, the goal is not to show that the QF formulation is inherently superior to the HJ formulation, but to show that in some cases the former offers greater utility.

Section 6.2 shows how Bohmian mechanics arises from the Schrödinger equation. Section 6.3 discusses general conceptual and mathematical features of the QF formulation. Section 6.4 illustrates how the quantum force formulation may be applied to trajectories and quantum states. Section 6.5 presents a brief summary and conclusions.

6.2 FROM SCHRÖDINGER TO BOHM

The starting point for single-particle, non-relativistic Bohmian mechanics is the time-dependent Schrödinger equation:

$$i\hbar \frac{\partial \Psi(\mathbf{x}, t)}{\partial t} = \frac{-\hbar^2}{2m} \nabla^2 \Psi(\mathbf{x}, t) + V \Psi(\mathbf{x}, t). \tag{6.1}$$

Since $\Psi(\mathbf{x}, t)$ is a complex function, it may be written in polar form, $\Psi(\mathbf{x}, t) = R(\mathbf{x}, t) \exp(i S(\mathbf{x}, t)/\hbar)$, where the modulus, R, and the phase, $S/\hbar$, are real functions. Substituting this form into the Schrödinger equation, taking derivatives, and separating the real and imaginary parts, we obtain the following two coupled equations:

$$-\frac{\partial S}{\partial t} = \frac{(\nabla S)^2}{2m} + V - \frac{\hbar^2}{2m} \left(\frac{\nabla^2 R}{R} \right), \tag{6.2}$$

$$\frac{\partial (R^2)}{\partial t} = -\nabla \cdot \left(\left(\frac{R^2}{m} \right) \nabla S \right). \tag{6.3}$$

Note that $R^2 = |\Psi|^2$, the position probability distribution. Equation 6.3 is a continuity equation for probability, while Equation 6.2 closely resembles the classical Hamilton–Jacobi equation,

$$-\frac{\partial S_{CL}}{\partial t} = \frac{(\nabla S_{CL})^2}{2m} + V, \tag{6.4}$$

where S_{CL} denotes Hamilton's principal function [3].

Only now is Bohm's alternative interpretation introduced, by *choosing* to regard Equation 6.2 as analogous to the classical Equation 6.4. In classical Hamilton–Jacobi theory we obtain $\mathbf{p}_{CL} = \nabla S_{CL}$, where $\mathbf{p}_{CL}$ is the momentum of a classical particle. In Bohmian mechanics, then, we have the analogous *guidance condition*:

$$\mathbf{p} = \nabla S. \tag{6.5}$$

Here $\mathbf{p}$ describes the momentum of a particle on a Bohmian trajectory, and the quantum dynamics describes an ensemble of possible trajectories, corresponding to an ensemble of possible initial conditions, of a *real* particle with continuously existing and well-defined dynamical properties. Equation 6.5 is the defining equation of the HJ formulation.

The last term in Equation 6.2, the analog of which does not appear in Equation 6.4, is denoted Q,

$$Q = -\frac{\hbar^2}{2m}\left(\frac{\nabla^2 R}{R}\right). \tag{6.6}$$

In Bohmian mechanics Q is often regarded as an additional potential energy, and referred to as the *quantum potential*. We may then associate with Q a quantum force, denoted $\mathbf{F}_Q$:

$$\mathbf{F}_Q = -\nabla Q. \tag{6.7}$$

With Q and $\mathbf{F}_Q$ in hand, we may discuss Bohmian mechanics in terms of a Newtonian formulation, with the total (classical plus quantum) force given by:

$$\mathbf{F} = \mathbf{F}_{CL} + \mathbf{F}_Q = -\nabla(V + Q). \tag{6.8}$$

Equations 6.6, 6.7, and 6.8, along with Newton's second law of motion (into which $\mathbf{F}$ is substituted) form the basis for the QF formulation of Bohmian mechanics, or simply *QF mechanics*.

Forces act on particles, and in the ensemble of Bohmian trajectories that comprise a pure state, only one is occupied by an actual particle—although which one is unknown. Nevertheless, a Bohmian ensemble must evolve *as though* every trajectory is occupied by a particle subject to classical and quantum forces, evaluated at that trajectory. In what follows, then, we can think of forces on, or the momenta of, trajectories.

Is one formulation "correct" and the other "wrong?" Leaving aside precisely what such a question could even mean, the fact is that QF mechanics is acceptable as a means of obtaining results. As Bohm himself said [1, p. 170]: "use of the Hamilton–Jacobi equation in solving for the motion of the particle is only a matter of convenience ... in principle, we can always solve directly by using Newton's laws of motion and the correct boundary conditions."

6.3 FEATURES OF QUANTUM FORCE MECHANICS

6.3.1 GENERAL REMARKS

In Bohmian mechanics, the quantum state of interest is typically calculated by conventional means. The Bohm trajectory behavior may then be investigated: the squared wavefunction ($|\psi|^2$) is the position probability distribution, and thus the Bohm trajectory position distribution; the momentum distribution is obtained through Equation 6.5, the guidance condition. In this approach, QF mechanics plays a peripheral role, if any.

In some cases, however, QF mechanics may be used to *obtain* the time-dependent behavior of the quantum state and the Bohm trajectories. And even when it cannot, it can provide a means to think about quantum behavior in an approximate or qualitative sense; QF mechanics offers a conceptual—not just abstract and mathematical—means to understand a system's behavior.

I believe, and hope to convince you, that acquiring some familiarity with key features of QF mechanics can be worthwhile, quite apart from whether one's loyalties might otherwise lie with standard quantum mechanics or Bohm, with one form of Bohmian mechanics or the other.

First a qualitative remark: QF mechanics is expressed in terms of Newton's second law of motion, that is, in terms of concepts, tools, and language that most of us have been using since our earliest exposure to physics. These can be rather easily imported for use in QF mechanics.

6.3.2   Quantum and Classical Forces

Even in standard quantum mechanics, time evolution may be thought of as occurring due to both quantum-mechanical effects and the effects of the classical potential (i.e., due to classical forces). But there may be no way to separate these effects. Indeed, it is unclear what it could even mean to separate quantum and classical effects in standard quantum mechanics, where concepts such as particle, trajectory, and force—the fundamental stuff of classical mechanics—have no clear meaning.

It is a particular virtue of QF mechanics that we may disentangle classical and quantum effects. A remarkable feature is apparent in Equation 6.8: evidently the total force—which governs time evolution of the Bohm trajectories, and thus of the state—is separable into quantum and classical contributions. Another notable feature is evident in Equations 6.6 and 6.7: $\mathbf{F}_Q$ depends on the modulus, R, but not the phase, S.

Yet both the separability of forces and $\mathbf{F}_Q$'s phase independence hold only in a limited sense. A complicated interplay exists between classical and quantum forces. Over time, $\mathbf{F}_{CL}$ will alter the momentum distribution, and thus R, so $\mathbf{F}_Q$ at a given trajectory will differ from what it would have been if $\mathbf{F}_{CL} = 0$. In this sense $\mathbf{F}_Q$ *does* depend on $\mathbf{F}_{CL}$, despite separability. Similarly, $\mathbf{F}_Q$ is formally independent of the phase, but again, the momentum distribution (i.e., the phase) will alter R, and thus $\mathbf{F}_Q$, over time. (In Section 6.4 I consider cases where force separability and phase independence prove useful despite these complications.)

6.3.3   Nature of the Quantum Force

Newton's second law is a second-order ordinary differential equation, but to implement it in QF mechanics we must obtain $\mathbf{F}_Q$. From Equations 6.6 and 6.7, we have:

$$\mathbf{F}_Q = \frac{\hbar^2}{2mR} \left(\nabla^3 R - R^{-1} \nabla^2 R \nabla R \right). \tag{6.9}$$

Equation 6.6 may suggest that inspection of R's curvature easily yields information about $\mathbf{F}_Q$. Equation 6.9 suggests otherwise: R may be a highly complicated function, and $\mathbf{F}_Q$ is proportional to $\nabla^3 R - R^{-1}\nabla^2 R \nabla R$. Thus, even qualitative statements about $\mathbf{F}_Q$'s strength, or indeed just its direction, will likely require careful analysis; a cursory inspection of the wavefunction invites error.

In principle, QF mechanics offers one approach to numerical time evolution. But even simple, low-order derivatives can lead to unreliable numerical results, and Equation 6.9 includes ∇R, $\nabla^2 R$, and $\nabla^3 R$. Thus, while QF mechanics remains one option for numerical calculations, it does not seem a promising one.

6.4 STATES AND TRAJECTORIES FROM QUANTUM FORCES

For two simple potentials, QF mechanics easily leads to exact results. In what follows, these two cases will be presented in broad outline form. Detailed treatments are available elsewhere; the primary goal here is to illustrate the features discussed in Section 6.3.

6.4.1 UNIFORM FORCES

Forces that are uniform in space, that is, that have no spatial dependence, are easily treated using QF mechanics (for a full discussion, see Ref. [4]). In this case, $\mathbf{F}_{CL}$ imparts an identical momentum boost to each Bohmian trajectory. As a result, $\mathbf{F}_{CL}$ will translate the modulus R, but it cannot alter the momenta of the Bohmian trajectories relative to each other, and thus it cannot alter R's shape. Because $\mathbf{F}_Q$ depends only on R's shape, and $\mathbf{F}_{CL}$ cannot alter that shape, $\mathbf{F}_Q$ must remain identical for corresponding trajectories in the free and forced states.

The preceding argument relies only on the spatial uniformity of $\mathbf{F}_{CL}$. Thus our result still holds even if we introduce an arbitrary time dependence into our spatially uniform $\mathbf{F}_{CL}$. Thus, we can solve for the problem of an arbitrary initial state subject to an arbitrarily time-dependent, spatially uniform force, by solving the corresponding free-particle problem, and then simply "adding on" the classical translation and boost that the uniform force imparts.

While the foregoing is qualitative, we can nevertheless construct the formal expression for the state from the fact that in Bohmian mechanics the phase has a straightforward interpretation in terms of the momentum distribution of the trajectories. In particular, if $\Psi_f(\mathbf{x}, t)$ is the free-particle state obtained above, then the state subject to our uniform force, denoted $\Psi(\mathbf{x}, t)$, is simply

$$\Psi(\mathbf{x}, t) = \Psi_f(\mathbf{x} - \Delta\mathbf{x}_{CL}(t), t)\exp(i\,\Delta\mathbf{p}_{CL}(t) \cdot \mathbf{x}/\hbar)\exp(i\,S_0/\hbar), \qquad (6.10)$$

where $\Delta\mathbf{x}_{CL}$ and $\Delta\mathbf{p}_{CL}$ are the classical displacement and momentum change, respectively, due to $\mathbf{F}_{CL}$. (The exponent $S_0/\hbar$, which may depend on t but not $\mathbf{x}$, may be readily and explicitly calculated, as shown in Ref. [4].)

In sum, the only effects of an arbitrarily time-dependent, spatially uniform classical force are to translate and boost the corresponding free-particle quantum state as it would a classical particle.

6.4.2 ON GAUSSIANS

In Section 6.4.3 I focus on Gaussians subject to linear classical forces. Before doing so, it's worth considering Gaussians rather generally.

Consider a one-dimensional state with a Gaussian modulus centered at $x = x_0$, i.e., $\Psi(x) \propto e^{-\beta(x-x_0)^2}$. Here β characterizes the Gaussian's width; for a time-dependent width, $\beta = \beta(t)$. The quantum force is

$$F_Q(x) = \frac{4\hbar^2\beta^2}{m}(x - x_0), \qquad (6.11)$$

From Equations 6.6 and 6.7, F_Q is determined only by the modulus, and depends on position relative to the state itself, but not on its overall location, or its motion, or the momentum distribution.

Gaussians often seem particularly well suited to quantum-mechanical calculations. As one simple example, many introductory quantum textbooks solve *exactly* for the time evolution of a free-particle Gaussian—but not for other initial states.

Why not? Probably because it's hard, or impossible, to do so. And why is that? Why are Gaussians so "nice" for the free-particle, and for other cases as well? I am unaware of an answer—*except* in the context of QF mechanics, wherein we see that Gaussians possess a very special property: their quantum force is *linear*. This special property will play a crucial role in the next section.

6.4.3 LINEAR FORCES

Now consider a time-independent, spatially linear classical restoring force (a simple harmonic oscillator), and an initial state with a Gaussian modulus, and a phase at most quadratic in x. In this case *both* F_{CL} and F_Q, and thus the total force, F, are linear in x. As a result, our Gaussian will remain a Gaussian (as shown in Ref. [5]).

The Gaussian's peak motion will be exactly classical. This can be shown through explicit calculation or from Ehrenfest's theorem [6]. It also arises quite simply in QF mechanics.

Because our Gaussian remains a Gaussian, Equation 6.11 holds for all t, which in turn implies that $F_Q = 0$ at the peak ($x = x_0$) for all t. Thus the peak's motion, and therefore the state's overall motion, is strictly classical—determined by F_{CL} alone.

What of the Gaussian's width? Figure 6.1 illustrates the situation in QF mechanics. F_{CL} (dashed line) is a restoring force; from Equation 6.11, F_Q (dotted line) must be linear and expansive, i.e., an *anti-restoring* force. If, as in Figure 6.1, the slope of F (dotted-dashed line) is negative, then the Gaussian is subject to a net restoring force; if positive, it is subject to a net anti-restoring force.

The dynamics of the Gaussian's width depend on F, the total force, and thus (in contrast to its overall motion) on F_{CL} *and* F_Q. In particular, the *relative* behavior of two trajectories in the Gaussian is determined by the total force *difference* between the trajectories, and thus by the slope of F.

In addition to F, the width dynamics will depend on the state's initial (width) conditions, so let us assume it is narrowing at $t = 0$. This narrowing has no effect

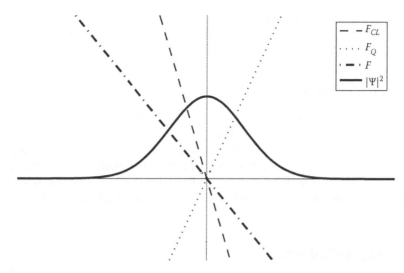

FIGURE 6.1 Classical, quantum, and total forces for a Gaussian centered at $x = 0$. Here the classical force is linear and restoring.

on F_{CL}. But β (in Equation 6.11) increases, so F_Q's slope, also, increases. As the state narrows this process continues, and F eventually becomes an anti-restoring force (its slope becomes positive). This eventually results in the state's expansion, with *decreasing* F_Q, until F once again becomes a *restoring* force. This, in turn, eventually leads to the packet again narrowing—and so the process repeats itself. It turns out that for the harmonic oscillator the width simply oscillates with half the classical period [5]. Note that though the dynamics of the width involve both F_Q and F_{CL}, the problem is essentially classical.

Now let our linear F_{CL} be either restoring or anti-restoring, with arbitrary time dependence (this case is discussed in Ref. [7]). Consider first the overall motion. F remains linear for all t, which is sufficient to insure that a Gaussian remains a Gaussian [5, 7]. Thus, Equation 6.11 still holds for all t, and $F_Q = 0$ at the peak for all t. So the state's overall motion is again classical—determined by F_{CL} alone (and again in accord with Ehrenfest's theorem).

As before, the width behavior is determined by the slope of F. Naturally, for F_{CL} restoring or anti-restoring and with arbitrary time dependence, we cannot expect simple behavior, as for the harmonic oscillator. Still, the basic method of solution, along with the accompanying conceptual picture, remain valid.

Evidently QF mechanics has reduced our Gaussian's time evolution to

- determination of the motion of the Gaussian's peak,
- time evolution of the Gaussian's width.

That is, we now have two *independent*, essentially *classical* problems.

In Ref. [7] we will show, through simple QF arguments, that the only effect of adding a uniform classical force to a linear classical force is to change the state's overall motion. This motion is simply the classical motion due to the sum of the uniform and linear forces; the addition of the uniform force has no other effect on the state's time evolution.

6.5   CONCLUSION

The quantum force formulation of Bohmian mechanics offers concepts, tools, and language that are unavailable in either standard quantum mechanics or in the Hamilton–Jacobi formulation of Bohmian mechanics. Though these features are unlikely to render easily soluble those problems that are intractable in standard quantum mechanics, or in the HJ formulation of Bohmian mechanics, the quantum force formulation remains inherently valuable for the unique methods and insights it makes available to us.

ACKNOWLEDGMENTS

Michelle McMillan carried out considerable work related to Section 6.4.3. I thank my friend and colleague, Ralph Baierlein, for critically reading the manuscript, and Pratim Chattaraj for the opportunity to contribute to this volume, as well as for his patience and understanding. I also thank Amy Caldwell, for giving me reason to go on.

This work is dedicated to the memory of my wife Katherine, through whom I learned the meaning of devotion.

BIBLIOGRAPHY

1. Bohm, D., 1952. A suggested interpretation of the quantum theory in terms of hidden variables, I. *Phys. Rev.* 85:166–79.
2. Bohm, D., 1952. A suggested interpretation of the quantum theory in terms of hidden variables, II. *Phys. Rev.* 85:180–93.
3. Goldstein, H., 1980. *Classical Mechanics*, 2nd edn. (Reading, MA: Addison-Wesley).
4. Bowman, G., 2006. Quantum-mechanical time evolution and uniform forces. *J. Phys. A* 39:157–62.
5. Bowman, G., 2002. Bohmian mechanics as a heuristic device: wave packets in the harmonic oscillator. *Am. J. Phys.* 70:313–18.
6. Messiah, A., 1964. *Quantum Mechanics* (Amsterdam: North-Holland).
7. McMillan, M. and Bowman, G., Gaussian wavepacket evolution in time-dependent quadratic potentials (in preparation).

7 Quantum Trajectories in Phase Space

Craig C. Martens, Arnaldo Donoso,
and Yujun Zheng

CONTENTS

7.1 INTRODUCTION

Since the time our prehistoric ancestors first portrayed motion by scratching a line in the dirt with the point of a spear, we have most naturally described dynamics with trajectories—paths though space parameterized by time. This eventually became formalized in the classical mechanics of particles by Newton, Lagrange, Hamilton, and others. Despite the formidable mathematical sophistication that classical dynamics can exhibit, the primordial intuition of matter being composed of things that can be found at a definite place at a given time remains at its foundation.

Quantum mechanics is the proper theoretical framework for describing the behavior of matter when it is composed of atoms and molecules [1, 2]. The unassailable successes of quantum mechanics in predicting the properties of matter at this scale are well-known. Equally well-known are the profound problems underlying the interpretation of the theory. The reconciliation of quantum effects such as the uncertainty principle, wave–particle duality, wave function collapse, entanglement, and others with the intrinsic classical perspective we view the world has a long and continuing history. A central part of this history is the desire to recover a "realistic" description of particle motion in quantum systems in terms of "hidden variables"—underlying classical-like paths that satisfy the desire for a description of nature that allow particles to always be somewhere definite. Developments along the way, in particular

Bell's theorem, have gone a long way in excluding hidden variables and enforcing the need to retain the nonclassical and nonintuitive elements of quantum mechanics. Nonetheless, efforts to understand quantum mechanics and the correspondence principle continue in corners of physical science and philosophy.

Most applications of theory—whether classical or quantum—to describing physical systems are made for practical rather than philosophical reasons. This is another area where classical mechanics often has an advantage over quantum mechanics. In this chapter we consider the problem of simulating quantum processes in molecular systems using classical trajectories and ensemble averaging. We mainly focus on the methodology rather than issues of interpretation.

For simple systems, a direct numerical solution of the time-dependent Schrödinger equation can be accomplished easily. For complex many-body problems the unfavorable scaling of computational cost of standard quantum methods with dimension and particle number make this approach intractable, and approximate methods must be employed. A broad range of such approaches have been developed, including mean-field methods, semiclassical and mixed classical–quantum methods, phenomenological reduced descriptions, and others.

One surprisingly effective approach in many cases is to simply *ignore* quantum effects altogether and use classical mechanics to describe the motion of atoms in molecular systems. The result is the method called classical molecular dynamics (MD) [3], a commonly used approach for studying many-particle systems where high temperatures, large masses, or other factors allow quantum effects in the atomic motion to be neglected. An MD simulation is performed by solving the appropriate Hamilton or Newton equations of motion given the forces of interaction and appropriate initial conditions. An individual classical trajectory for a multidimensional problem is much easier to integrate numerically than the time-dependent wave packet of the corresponding quantum system. Unless the anecdotal information revealed by a single trajectory is sufficient, however, significant numbers of trajectories—*ensembles*—must be employed. A distribution of trajectories evolving in phase space is the most direct classical analogue of an evolving quantum wave packet, and statistical averages of dynamical variables over the classical ensemble parallel the corresponding quantum expectation values of operators.

7.2 DYNAMICS IN PHASE SPACE

The state of a classical system is represented by a probability distribution $\rho(q, p, t)$ defined in the phase space (q, p) of the system. The evolution of $\rho(q, p, t)$ in phase space is governed by the classical Liouville equation [4]

$$\frac{\partial \rho}{\partial t} = \{H, \rho\}, \tag{7.1}$$

where we consider a system with one degree of freedom for simplicity; this can be easily generalized. The q and p are the canonical coordinate and momentum, respectively, $H(q, p) = p^2/2m + V(q)$ is the system Hamiltonian, where m is the mass and $V(q)$ is the potential energy function, and $\{H, \rho\}$ is the Poisson bracket of H and ρ, defined as

$$\{H, \rho\} \equiv \frac{\partial H}{\partial q}\frac{\partial \rho}{\partial p} - \frac{\partial \rho}{\partial q}\frac{\partial H}{\partial p}. \tag{7.2}$$

In classical MD, a solution of the Liouville equation is approximated by generating an ensemble of N distinct initial conditions $q_k(0)$ and $p_k(0)$ ($k = 1, 2, \ldots, N$) sampled from the given initial probability distribution $\rho(q, p, 0)$. Phase space trajectories are then determined by integrating the Hamilton equations,

$$\dot{q} = \frac{\partial H}{\partial p} \tag{7.3}$$

$$\dot{p} = -\frac{\partial H}{\partial q} \tag{7.4}$$

using the $q_k(0)$ and $p_k(0)$ as initial data. Then, aside from statistical error due to a finite N, $\rho(q, p, t)$ is given by the local phase space density of the evolving trajectories $(q_k(t), p_k(t))$ around the point (q, p). The relation between the evolution of the classical function $\rho(q, p, t)$ and the trajectory ensemble $(q_k(t), p_k(t))$ ($k = 1, 2, \ldots, N$) in phase space is illustrated schematically in the left panel of Figure 7.1.

Our goal is to describe *quantum mechanics* from a similar phase space trajectory ensemble perspective. To accomplish this, we adopt a phase space representation of quantum mechanics—the Wigner representation [5–8]. The quantum state of the system that was described using classical mechanics above is now represented by the wave function $\psi(q, t)$, which is a solution of the time-dependent Schrödinger equation [1]. An equivalent phase space description is given in terms of the Wigner

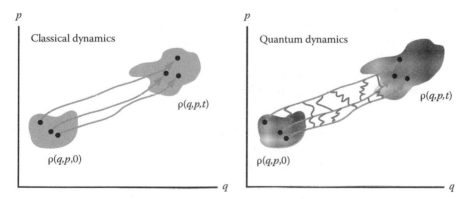

FIGURE 7.1 A pictorial representation of trajectory-based evolution of classical and quantum states in phase space. In the classical case (left), the individual trajectories evolve independently. In the quantum case (right), a trajectory-based representation unavoidably leads to a breakdown of the statistical independence of the ensemble members. A schematic representation of the resulting interactions is shown in the figure.

function $\rho_W(q, p, t)$ [5–8]. The Wigner function is related to the density operator $\hat{\rho}$ by

$$\rho_W(q, p, t) = \frac{1}{2\pi\hbar} \int_{-\infty}^{\infty} \left\langle q - \frac{y}{2} |\hat{\rho}(t)| q + \frac{y}{2} \right\rangle e^{ipy/\hbar} \, dy. \tag{7.5}$$

For a pure state with the time-dependent wave function $\psi(q, t)$ this becomes

$$\rho_W(q, p, t) = \frac{1}{2\pi\hbar} \int_{-\infty}^{\infty} \psi^* \left(q + \frac{y}{2}, t \right) \psi \left(q - \frac{y}{2}, t \right) e^{ipy/\hbar} \, dy. \tag{7.6}$$

The equation of motion for the Wigner function is

$$\frac{\partial \rho_W}{\partial t} = -\frac{p}{m} \frac{\partial \rho_W}{\partial q} + \int_{-\infty}^{\infty} J(q, p - \xi) \rho_W(q, \xi, t) d\xi \tag{7.7}$$

where

$$J(q, p) = \frac{i}{2\pi\hbar^2} \int_{-\infty}^{\infty} \left[V \left(\frac{q + y}{2} \right) - V \left(\frac{q - y}{2} \right) \right] e^{-ipy/\hbar} \, dy. \tag{7.8}$$

The integral in Equation 7.8 can be evaluated to give

$$J(q, \eta) = \frac{4}{\hbar^2} \text{Im}(\hat{V}(2\eta/\hbar) e^{-2i\eta q/\hbar}). \tag{7.9}$$

The Wigner representation is an exact and faithful representation of quantum mechanics, and so the Wigner function $\rho_W(q, p, t)$ contains the same information about observable quantities as does $\psi(q, t)$ for pure state systems. In order to treat mixed states, a wave function does not exist, and so the Wigner (or other density operator-based) representation must be used.

Equation 7.7 emphasizes the fundamental nonlocality of quantum mechanics: the time rate of change of ρ_W at point (q, p) depends on the value of ρ_W over a range of momentum values $\xi \neq p$.

For systems with potential $V(q)$ that have a power series expansion in q, the kernel of the integral $J(q, p)$ in Equation 7.8 becomes

$$J(q, p) = -V'(q) \delta'(p) + \frac{\hbar^2}{24} V'''(q) \delta'''(p) + \cdots, \tag{7.10}$$

giving the equation of motion as a power series in $\hbar$,

$$\frac{\partial \rho_W}{\partial t} = -\frac{p}{m} \frac{\partial \rho_W}{\partial q} + V'(q) \frac{\partial \rho_W}{\partial p} - \frac{\hbar^2}{24} V'''(q) \frac{\partial^3 \rho_W}{\partial p^3} + \cdots, \tag{7.11}$$

where the prime denotes the derivative with respect to q. The higher order terms not shown involve successively higher even powers of $\hbar$, odd derivatives of V with respect to q, and corresponding derivatives of ρ_W with respect to p. In the classical ($\hbar \to 0$)

limit, the $\hbar$-dependent terms vanish and the Wigner function becomes a solution of the classical Liouville equation for probability distributions in phase space.

Interpreting the Wigner function as a probability distribution is tempting, in analogy with the classical Liouville equation and phase space density. However, this is complicated by the fact that ρ_W, although always real, can assume negative values. Faithful representations of quantum mechanics that are built on positive probability distributions in phase space can be formulated. An example is the Husimi representation [6]. The Husimi distribution is constructed from the Wigner function by smoothing with a minimum uncertainty phase space Gaussian. We will explore this representation in more detail later in this chapter.

The nonlocality of quantum mechanics does not allow an arbitrarily fine subdivision of the quantum distribution into individual independent elements, as is possible in the connection between distribution functions and independent trajectories in classical mechanics. Rather, quantum mechanics insists that the entire state be propagated as a unified whole. If a trajectory ensemble representation of nonlocal quantum motion is to be achieved, the statistical independence of the trajectories must be abandoned and the individual members of the ensemble must *interact* with each other. This interdependence, or *entanglement*, of the trajectory ensemble is depicted schematically in the right panel Figure 7.1.

7.3 QUANTUM TRAJECTORIES

Recently, we have developed a method for solving the quantum Liouville equation in the Wigner representation in the context of a classical trajectory simulation [9–13]. In this approach, we represent the time-dependent state of the system $\rho(q, p, t)$ as an ensemble of trajectories. In classical mechanics, the ensemble members would evolve independently of each other under the Hamilton equations, as described above. For a quantum state, however, nonlocality prohibits an arbitrarily fine subdivision and independent treatment of its constituent parts—this would violate the uncertainty principle. We incorporate this non-classical aspect of quantum mechanics explicitly in our method *as a breakdown of the statistical independence of the members of the trajectory ensemble.* We derive non-classical forces acting *between* the ensemble members that model the quantum effects governing the evolution of the corresponding nonstationary wave packet.

A range of quantum trajectory methods have been pursued vigorously in recent years in the physics and chemistry literature. We direct the reader to the monograph by Wyatt for a review [14] and to the other chapters of this book for recent developments.

The continuous distribution function ρ is represented by a finite ensemble of N trajectories,

$$\rho(q, p, t) = \frac{1}{N} \sum_{j=1}^{N} \delta(q - q_j(t))\delta(p - p_j(t)). \tag{7.12}$$

We note here that this ansatz is an approximate one, as the exact Wigner function ρ_W can become negative. The assumed strictly positive form of the solution in

Equation 7.12 thus cannot capture the full quantum dynamics in the Wigner representation. A representation of quantum mechanics exists that is compatible with this ansatz, based on the Husimi distribution [6], a Gaussian smoothed Wigner function. Oscillations in ρ_W average out, resulting in a distribution function that has the desired non-negative property and can thus be interpreted probabilistically. In our method, we identify the continuous phase space function resulting from smoothing Equation 7.12 with an equivalent positive-definite smoothing of the underlying Wigner function ρ_W. There are a number of ways to implement the smoothing in practice.

Our trajectory representation of quantum mechanics includes quantum effects by *altering the motion of the trajectories themselves*. The instantaneous force acting on a particular member of the ensemble will thus depend on both the classical force $-V'(q)$ and on the phase space locations of all the other members of the ensemble. Their evolution will thus become mutually *entangled*.

Equations of motion for the trajectories can be derived based on principles of continuity and conservation of normalization. The phase space trace of the Wigner function is conserved: $\mathrm{Tr}\,\rho_W = \int \rho\,dqdp = 1$, a property shared by its approximation in Equation 7.12. In terms of the phase space flux $\vec{j} = \rho\vec{v}$, the ensemble must evolve collectively so that the continuity equation

$$\frac{\partial \rho}{\partial t} + \vec{\nabla} \cdot \vec{j} = 0 \tag{7.13}$$

is obeyed, where $\vec{\nabla}$ is the gradient in phase space. We exploit this continuity condition in our equations of motion by identifying the form of the current $\vec{j}$ in the Liouville equation, finding the corresponding vector field $\vec{v} = \vec{j}/\rho$, and then integrating the trajectories in phase space using $(\dot{q}, \dot{p}) = \vec{v}$.

We first consider the strict classical limit. Here, the $\hbar$-dependent terms in Equation 7.11 vanish, and the phase space density obeys the classical Liouville equation [4, 15]:

$$\frac{\partial \rho}{\partial t} = -\vec{\nabla} \cdot \vec{j} = \{H, \rho\}. \tag{7.14}$$

By noting that $\partial \dot{q}/\partial q + \partial \dot{p}/\partial p = 0$, we can identify the phase space current vector as

$$\vec{j} = \begin{pmatrix} \partial H/\partial p \\ -\partial H/\partial q \end{pmatrix} \rho. \tag{7.15}$$

Division by ρ then gives the familiar classical independent evolution of phase space trajectories under the conventional Hamiltonian equations $\dot{q} = v_q = \partial H/\partial p$, $\dot{p} = v_p = -\partial H/\partial q$. The density ρ cancels from the expression for the phase space vector field $\vec{v}$ when $\vec{j}$ in Equation 7.15 is divided by ρ.

We now turn to the quantum Liouville equation in the Wigner representation. The continuity condition involves the full equation of motion, Equation 7.11. Writing the divergence of the current as

$$\vec{\nabla} \cdot \vec{j} = \frac{\partial}{\partial q}\left(\frac{\partial H}{\partial p}\rho\right) + \frac{\partial}{\partial p}\left(-V'(q)\rho + \frac{\hbar^2}{24}V'''(q)\frac{\partial^2 \rho}{\partial p^2} + \cdots\right) \tag{7.16}$$

and dividing the corresponding current by ρ, we arrive at the equations of motion for the trajectory at point (q, p):

$$\dot{q} = v_q = \frac{p}{m}$$

$$\dot{p} = v_p = -V'(q) + \frac{\hbar^2}{24}V'''(q)\frac{1}{\rho}\frac{\partial^2\rho}{\partial p^2} + \cdots. \qquad (7.17)$$

Note that in this case ρ *does not cancel out of the equations*. In marked contrast with the classical Hamilton equations, the vector field now depends on the global state of the system as well as on the phase point (q, p).

A consequence of the additional ρ-dependent contribution to the force is that individual trajectory energies are not conserved.

$$\frac{dH}{dt} = \dot{q}\frac{\partial H}{\partial q} + \dot{p}\frac{\partial H}{\partial p} = \frac{p}{m}\left(\frac{\hbar^2}{24}V'''(q)\frac{1}{\rho}\frac{\partial^2\rho}{\partial p^2} + \cdots\right) \neq 0. \qquad (7.18)$$

This is acceptable—and in fact essential—if quantum effects are going to be represented by the method. Energy conservation is only required *on average*. It is straightforward to show from Equation 7.17 that the ensemble average $\langle\dot{p}\rangle = \text{Tr}(\dot{p}\rho) = -\langle V'\rangle$, and thus the method obeys Ehrenfest's theorem, while the average energy $\langle F\rangle = \text{Tr}(H\rho)$ is independent of time:

$$\left\langle\frac{dH}{dt}\right\rangle = \iint \rho\frac{dH}{dt}dq\,dp = \iint \frac{p}{m}\left(\frac{\hbar^2}{24}V'''(q)\frac{\partial^2\rho}{\partial p^2} + \cdots\right)dq\,dp = 0. \qquad (7.19)$$

The individual trajectories, however, can behave nonclassically—as they *must* if they are to capture the dynamics of quantum tunneling.

7.4 ENTANGLED TRAJECTORY MOLECULAR DYNAMICS

Our formalism can form the basis of a method for quantum dynamics in the context of a classical-like MD simulation. This is accomplished by generating an ensemble of initial conditions representing $\rho_W(q, p, 0)$ and then propagating the trajectory ensemble using Equation 7.17. In practice, the singular distribution ρ must be smoothed to allow a faithful representation of the analogously smoothed [6] quantum dynamics. The nonclassical ρ-dependent force is determined from a smooth local Gaussian representation of the instantaneous ensemble. In particular, the value of $\rho^{-1}\partial^2\rho/\partial p^2$ and terms involving higher derivatives at each phase space point (q_j, p_j) is calculated by assuming a local Gaussian approximation of ρ around $\vec{\Gamma}_j = (q_j, p_j)$:

$$\rho(q, p, t) \simeq \rho_o e^{-(\vec{\Gamma}-\vec{\Gamma}_j(t))\cdot\beta_j(t)\cdot(\vec{\Gamma}-\vec{\Gamma}_j(t))+\vec{\alpha}_j(t)\cdot(\vec{\Gamma}-\vec{\Gamma}_j(t))}. \qquad (7.20)$$

The state $\rho(t)$ at each trajectory location (q_j, p_j) in phase space is characterized by the time-dependent parameters in the matrix β_j and vector $\vec{\alpha}_j$. We determine these numerically in practice by calculating *local* moments of the ensemble around the reference point $\vec{\Gamma}_j$ [9,12]. These consist of sums of appropriate powers of the dynamical variables over the ensemble, weighted by a Gaussian cutoff $\phi(\vec{\Gamma}) = \exp(-\vec{\Gamma} \cdot \mathbf{h} \cdot \vec{\Gamma})$ centered at the point under consideration, where $\mathbf{h}$ is chosen to give a minimum uncertainty ϕ, consistent with the smoothing requirement for a positive quantum phase space distribution [6]. From this calculation, the parameters β_j and $\vec{\alpha}_j$ can be inferred at each point $\vec{\Gamma}_j = (q_j, p_j)$. The generator of modified moments is [12]

$$\tilde{I} = \int_{-\infty}^{\infty} \int_{-\infty}^{\infty} e^{-\beta_q \xi^2 - \beta_p \eta^2 - 2\beta_{qp}\xi\eta + \alpha_q\xi + \alpha_p\eta} \, \phi_{h_q, h_p}(\xi, \eta) \, d\xi d\eta, \tag{7.21}$$

where this includes a *local Gaussian window function* ϕ:

$$\phi_{h_q, h_p}(\xi, \eta) = \exp\left(-h_q \xi^2 - h_p \eta^2\right). \tag{7.22}$$

The *modified* mth, nth moment of ξ, η is then

$$\langle \xi^m \tilde{\eta}^n \rangle \equiv \frac{\langle \xi^m \eta^n \phi \rangle}{\langle \phi \rangle} = \frac{\iint \xi^m \eta^n \phi(\xi, \eta) \rho(\xi, \eta) d\xi d\eta}{\iint \phi(\xi, \eta) \rho(\xi, \eta) d\xi d\eta}. \tag{7.23}$$

For ρ a local Gaussian, these moments are generated by derivatives of $\tilde{I}$:

$$\langle \xi^m \tilde{\eta}^n \rangle = \frac{1}{\tilde{I}} \frac{\partial^{(m+n)}}{\partial \alpha_q^m \partial \alpha_p^n} \tilde{I}. \tag{7.24}$$

We define generalized variances and correlation:

$$\tilde{\sigma}_\xi^2 = \langle \tilde{\xi}^2 \rangle - \langle \tilde{\xi} \rangle^2 \tag{7.25}$$

$$\tilde{\sigma}_\eta^2 = \langle \tilde{\eta}^2 \rangle - \langle \tilde{\eta} \rangle^2 \tag{7.26}$$

$$\tilde{\sigma}_{\xi\eta}^2 = \langle \tilde{\xi}\tilde{\eta} \rangle - \langle \tilde{\xi} \rangle \langle \tilde{\eta} \rangle. \tag{7.27}$$

The *original* Gaussian parameters can then be reconstructed in terms of the generalized moments; for instance:

$$\alpha_p = \frac{\tilde{\sigma}_\xi^2 \langle \tilde{\eta} \rangle - \tilde{\sigma}_{\xi\eta}^2 \langle \tilde{\xi} \rangle}{\tilde{\sigma}_\xi^2 \tilde{\sigma}_\eta^2 - \tilde{\sigma}_{\xi\eta}^4} \tag{7.28}$$

$$\beta_p = \frac{\tilde{\sigma}_\xi^2}{2(\tilde{\sigma}_\xi^2 \tilde{\sigma}_\eta^2 - \tilde{\sigma}_{\xi\eta}^4)} - h_p. \tag{7.29}$$

The required modified moments can be calculated easily from the evolving ensemble:

$$\langle \xi^m \tilde{\eta}^n \rangle_k = \frac{\sum_{j=1}^{N} (q_j - q_k)^m (p_j - p_k)^n \phi(q_j - q_k, p_j - p_k)}{\sum_{j=1}^{N} \phi(q_j - q_k, p_j - p_k)}. \tag{7.30}$$

The local nature of the fit allows non-trivial densities with multiple maxima to be represented by the discrete ensemble in an accurate, efficient, and numerically stable manner.

We illustrate this general approach by considering a one-dimensional model of quantum mechanical tunneling [9, 12]. Using atomic units throughout, we treat a particle of mass $m = 2000$ moving on the potential

$$V(q) = \frac{1}{2} m \, \omega_o^2 q^2 - \frac{1}{3} b q^3, \tag{7.31}$$

where $\omega_o = 0.01$ and $b = 0.2981$. This system has a metastable potential minimum with $V = 0$ at $q = 0$ and a barrier to escape of height $V^{\ddagger} = 0.015$ at $q^{\ddagger} = 0.6709$. The parameters are chosen so that the system roughly mimics a proton bound with approximately two metastable bound states. The dynamics are thus expected to be highly quantum mechanical.

A series of minimum uncertainty quantum wave packets and corresponding trajectory ensembles are chosen as initial states. The mean momentum $\langle p \rangle = 0$ in all cases and the mean energy of the state is varied by selecting a range of initial average displacements. The trajectories are then propagated using the ensemble-dependent force given by Equation 7.17. For the potential in Equation 7.31, $V''' = -2b$ is constant, and the higher order terms in Equation 7.17 rigorously vanish. The force then becomes

$$\dot{p}_j = -V'(q_j) - \frac{\hbar^2 b}{12} \frac{\partial^2 \rho / \partial p^2 (q_j, p_j)}{\rho(q_j, p_j)} \tag{7.32}$$

for $j = 1, 2, \ldots, N$. The ρ-dependent factor depends on the parameters β_j and $\vec{\alpha}_j$, and thus involves summations over the entire trajectory ensemble. In terms of the local Gaussian parameters, the force becomes

$$\dot{p}_j = -V'(q_j) - \frac{\hbar^2 b}{12} (\alpha_{p,j}^2 - 2\beta_{p,j}), \tag{7.33}$$

where

$$\alpha_p = \frac{\tilde{\sigma}_\xi^{\,2} \langle \tilde{\eta} \rangle - \tilde{\sigma}_{\xi\eta}^{\,2} \langle \tilde{\xi} \rangle}{\tilde{\sigma}_\xi^{\,2} \tilde{\sigma}_\eta^{\,2} - \tilde{\sigma}_{\xi\eta}^{\,4}} \tag{7.34}$$

$$\beta_p = \frac{\tilde{\sigma}_\xi^{\,2}}{2(\tilde{\sigma}_\xi^{\,2} \tilde{\sigma}_\eta^{\,2} - \tilde{\sigma}_{\xi\eta}^{\,4})} - h_p. \tag{7.35}$$

In Figure 7.2, we show the time-dependent tunneling probabilities $P(t)$ for three initial conditions, numbered 1–3, each corresponding to an initial minimum uncertainty wave packet or ensemble. The entangled trajectory MD simulations are compared with purely classical results generated with the same number of trajectories but in the absence of the quantum force and the results of numerically exact quantum wave packet calculations performed using the method of Kosloff [16].

The trajectory results shown here correspond to ensembles containing $N = 900$ trajectories. The quantum reaction probability is defined at each time as the integral of $|\psi(q, t)|^2$ from $q^{\ddagger}$ to ∞, while the classical and entangled trajectory quantities are defined as the fraction of trajectories with $q > q^{\ddagger}$ at time t. The curves

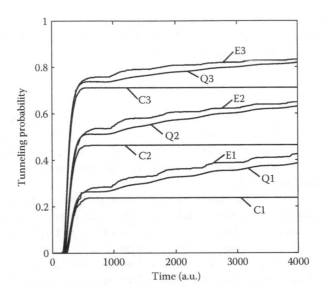

FIGURE 7.2 Time-dependent tunneling probabilities. Three initial wave packets or ensembles are considered, and the results of the entangled trajectory molecular dynamics simulations (E) are compared with purely classical (C) and exact quantum (Q) calculations. See the text for details.

labeled C, Q, and E indicate classical, quantum, and entangled trajectory ensemble results, respectively. Case 1 corresponds to a mean energy $E_o = \langle \psi | \hat{H} | \psi \rangle \simeq 0.75 \, V^{\ddagger}$. For case 2, $E_o \simeq 1.25 \, V^{\ddagger}$, while for case 3, $E_o \simeq 2.0 \, V^{\ddagger}$. Increasing the mean energy increases the short time transfer across the barrier, both classically and quantum mechanically. The classical reaction, however, ceases immediately after the first sharp rise, as the trajectories in the ensemble with energy below the barrier initially are trapped there for all time. The quantum wave packet, however, continues to escape from the metastable well by tunneling, and the reaction probability continues to grow slowly with time following the initial classical-like rise. This growth is modulated by the oscillations of the wave packet in the potential well. The entangled trajectory calculation tracks the exact quantum results quite well. Although these results slightly overestimate the exact instantaneous probability, the qualitative dynamics are described quite satisfactorily. In particular, the *purely nonclassical* longer time growth of the reaction probability is correctly described.

In Figure 7.3, we examine the agreement between quantum and entangled trajectory predictions of the non-classical tunneling dynamics in more detail. The decay of the survival probability $1 - P(t)$ at times longer than the initial rapid classical decay is fit to an exponential $\exp(-kt)$ and the tunneling rate constant k thus defined is plotted in the figure as a function of mean wave packet energy. The overall correspondence is very good, especially considering that a nonzero k is a quantity that results solely from the non-classical tunneling of the particle through the barrier.

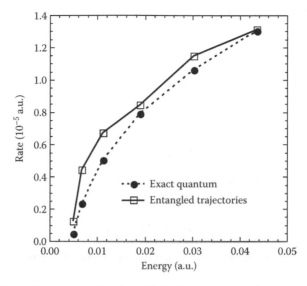

FIGURE 7.3 Tunneling rate as a function of initial mean wave packet energy, entangled trajectory molecular dynamics and exact quantum results.

7.5 HUSIMI REPRESENTATION

The above method employs local smoothing implicitly to formulate an entangled trajectory method in the Wigner representation. We now formalize this idea by a generalization of the approach that is based on a rigorous positive phase space representation of quantum mechanics: the Husimi representation [13].

The Husimi distribution is a locally smoothed Wigner function:

$$\rho_H(q, p) = \frac{1}{\pi\hbar} \int_{-\infty}^{\infty} \rho_W(q', p') e^{-\frac{(q-q')^2}{2\sigma_q^2}} e^{-\frac{(p-p')^2}{2\sigma_p^2}} \, dq' dp' \qquad (7.36)$$

where the smoothing is over a minimum uncertainty phase space Gaussian, satisfying

$$\sigma_q \sigma_p = \frac{\hbar}{2}. \qquad (7.37)$$

The smoothing can be represented using *smoothing operators* $\hat{Q}$ and $\hat{P}$:

$$\hat{Q} = e^{\frac{1}{2}\sigma_q^2 \frac{\partial^2}{\partial q^2}} \qquad (7.38)$$

$$\hat{P} = e^{\frac{1}{2}\sigma_p^2 \frac{\partial^2}{\partial p^2}}. \qquad (7.39)$$

The Husimi can then be written as a smoothed Wigner function as:

$$\rho_H(q, p) = \hat{Q}\hat{P}\rho_W(q, p). \qquad (7.40)$$

This is related to the interesting identity:

$$e^{-a(x-x')^2} = e^{\frac{1}{4a}\frac{\partial^2}{\partial x^2}}\, \delta(x - x').\tag{7.41}$$

We can consider the inverse *unsmoothing* operators $\hat{Q}^{-1}$ and $\hat{P}^{-1}$:

$$\hat{Q}^{-1} = e^{-\frac{1}{2}\sigma_q^2 \frac{\partial^2}{\partial q^2}}\tag{7.42}$$

$$\hat{P}^{-1} = e^{-\frac{1}{2}\sigma_p^2 \frac{\partial^2}{\partial p^2}}\tag{7.43}$$

so that the Wigner function can be written (at least formally) as an "unsmoothed" Husimi:

$$\rho_W(q, p) = \hat{Q}^{-1}\hat{P}^{-1}\rho_H(q, p).\tag{7.44}$$

We can then derive an equation of motion for the Husimi distribution

$$\frac{\partial \rho_H}{\partial t} = -\frac{1}{m}\hat{P}p\hat{P}^{-1}\frac{\partial \rho_H}{\partial q} + \int_{-\infty}^{\infty} \hat{Q}J(q,\eta)\hat{Q}^{-1}\rho_H(q, p + \eta, t)\, d\xi.\tag{7.45}$$

Note that there are no approximations; the Husimi representation provides an *exact* alternative description of quantum dynamics.

In the Husimi representation, powers of the coordinates and momenta become differential operators:

$$\hat{Q}q\hat{Q}^{-1} = q + \sigma_q^2 \frac{\partial}{\partial q}\tag{7.46}$$

$$\hat{P}p\hat{P}^{-1} = p + \sigma_p^2 \frac{\partial}{\partial p}\tag{7.47}$$

$$\hat{Q}q^2\hat{Q}^{-1} = q^2 + \sigma_q^2 + 2\sigma_q^2 q\frac{\partial}{\partial q} + \sigma_q^4 \frac{\partial^2}{\partial q^2}.\tag{7.48}$$

The Husimi equation of motion for the cubic system can then be written:

$$\frac{\partial \rho_H}{\partial t} = -\frac{1}{m}\hat{P}p\hat{P}^{-1}\frac{\partial \rho_H}{\partial q} + (m\omega_o^2 \hat{Q}q\hat{Q}^{-1} - b\hat{Q}q^2\hat{Q}^{-1})\frac{\partial \rho_H}{\partial p} + \frac{\hbar^2 b}{12}\frac{\partial^3 \rho_H}{\partial p^3}\tag{7.49}$$

where

$$\hat{Q}q\hat{Q}^{-1} = q + \sigma_q^2 \frac{\partial}{\partial q}\tag{7.50}$$

$$\hat{P}p\hat{P}^{-1} = p + \sigma_p^2 \frac{\partial}{\partial p}\tag{7.51}$$

$$\hat{Q}q^2\hat{Q}^{-1} = q^2 + \sigma_q^2 + 2\sigma_q^2 q\frac{\partial}{\partial q} + \sigma_q^4 \frac{\partial^2}{\partial q^2}.\tag{7.52}$$

Continuity conditions can be applied in the Husimi representation, which are now rigorously justified for a positive probability distribution:

$$\frac{\partial \rho_H}{\partial t} + \vec{\nabla} \cdot \vec{j}_H = 0. \tag{7.53}$$

This yields an expression for the divergence of the phase space current,

$$\vec{\nabla} \cdot \vec{j}_H = \frac{\partial}{\partial q}\left(\frac{p}{m}\rho_H\right) + \frac{\partial}{\partial p}\left(-V'(q)\rho_H + \frac{\hbar b}{2m\omega_o}\rho_H + \frac{\hbar b q}{m\omega_o}\frac{\partial \rho_H}{\partial q}\right.$$
$$\left. + \frac{\hbar^2 b}{4m^2\omega_o^2}\frac{\partial^2 \rho_H}{\partial q^2} - \frac{\hbar^2 b}{12}\frac{\partial^2 \rho_H}{\partial p^2}\right). \tag{7.54}$$

The phase space vector field can then be written as

$$\dot{q} = \frac{p}{m} \tag{7.55}$$

$$\dot{p} = -V'(q) + \frac{\hbar b}{2m\omega_o} + \frac{\hbar b q}{m\omega_o}\frac{1}{\rho_H}\frac{\partial \rho_H}{\partial q} + \frac{\hbar^2 b}{4m^2\omega_o^2}\frac{1}{\rho_H}\frac{\partial^2 \rho_H}{\partial q^2} - \frac{\hbar^2 b}{12}\frac{1}{\rho_H}\frac{\partial^2 \rho_H}{\partial p^2}. \tag{7.56}$$

The quantum force now contains additional terms not present in the Wigner representation quantum force. It is interesting to note that, because of the smoothing, the motion of a free particle is nonclassical in the Husimi representation,

$$\frac{\partial \rho_H}{\partial t} = -\frac{1}{m}\hat{P}p\hat{P}^{-1}\frac{\partial \rho_H}{\partial q} \tag{7.57}$$

or

$$\frac{\partial \rho_H}{\partial t} = -\frac{1}{m}p\frac{\partial \rho_H}{\partial q} - \frac{\sigma_p^2}{m}\frac{\partial^2 \rho_H}{\partial q \partial p}. \tag{7.58}$$

The extra terms are due to noncommutativity of classical time evolution and smoothing. (For a harmonic oscillator, the cross-terms resulting from the kinetic energy and potential energy cancel, leading to classical evolution.)

The MD methodology described above can be easily generalized to incorporate the additional terms in the equations of motion in the Husimi representation. When implemented, excellent agreement with exact results for model systems is again obtained [13]. In Figure 7.4, results for the cubic system are shown, indicating the level of agreement between exact quantum results, the Wigner-based method described above, and the implementation based on the positive Husimi representation.

7.6   INTEGRODIFFERENTIAL EQUATION FORM

The methods above have been based on an expansion of the quantum phase space density and potential in power series. When using a discrete representation of the density in terms of a sum of delta functions, Equation 7.12, the estimation of higher order

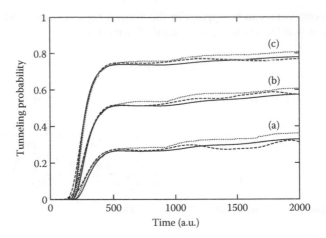

FIGURE 7.4 Tunneling probabilities versus time. Solid lines represent exact results. Fine dotted lines show results for the entangled trajectory method based on an $\hbar$ expansion of the Wigner function, as described in the text. Dashed lines present results performed in the Husimi representation. Cases a, b, and c refer to the initial placement of the center of the wave packet $q_0 = -0.2$ au, $q_0 = -0.3$ au, and $q_0 = -0.4$ au, respectively.

derivatives of ρ_W from this form is challenging. It is more natural to work directly with the integrodifferential equation form of the equations of motion, Equation 7.7, where the nonlocality of quantum mechanics is expressed explicitly as a convolution without any power series expansion and the discrete representation of ρ_W can be employed directly. With this in mind, we return to the Wigner equation of motion in the integrodifferential form, and try to solve it directly:

$$\frac{\partial \rho_W}{\partial t} = -\frac{p}{m}\frac{\partial \rho_W}{\partial q} + \int_{-\infty}^{\infty} J(q, p - \xi)\rho_W(q, \xi, t)d\xi. \qquad (7.59)$$

We write the divergence of the flux as:

$$\vec{\nabla} \cdot \vec{j}_W = \frac{\partial}{\partial q}\left(\frac{p}{m}\rho_W\right) - \int_{-\infty}^{\infty} J(q, \xi - p)\,\rho_W(q, \xi, t)\,d\xi. \qquad (7.60)$$

The momentum component of the flux divergence is then:

$$\frac{\partial}{\partial p}j_{W,p} = -\int_{-\infty}^{\infty} J(q, \xi - p)\,\rho_W(q, \xi, t)\,d\xi. \qquad (7.61)$$

Integrating, we obtain

$$j_{W,p} = -\int_{-\infty}^{\infty} \Theta(q, \xi - p)\,\rho_W(q, \xi, t)\,d\xi, \qquad (7.62)$$

where

$$\Theta(q, \xi - p) = \int_{-\infty}^{p} J(q, \xi - z)\, dz. \tag{7.63}$$

This can be written explicitly in terms of the potential $V(q)$:

$$\Theta(q, \xi - p) = \frac{1}{2\pi\hbar} \int_{-\infty}^{\infty} \left[V\left(q + \frac{y}{2}\right) - V\left(q - \frac{y}{2}\right) \right] \frac{e^{-i(\xi - p)y/\hbar}}{y}\, dy. \tag{7.64}$$

The quantum trajectory equations of motion then become

$$\dot{q} = \frac{p}{m} \tag{7.65}$$

$$\dot{p} = -\frac{1}{\rho_W(q, p)} \int \Theta(q, p - \xi) \rho_W(q, \xi)\, d\xi. \tag{7.66}$$

To proceed numerically, we write the Wigner function as a superposition of Gaussians:

$$\rho_W(q, p, t) = \frac{1}{N} \sum_{j=1}^{N} \phi(q - q_j(t), p - p_j(t)), \tag{7.67}$$

where

$$\phi(q, p) = \frac{1}{2\psi_q \sigma_p} \exp\left(-\frac{q^2}{2\sigma_q^2} - \frac{p^2}{2\sigma_p^2} \right). \tag{7.68}$$

After some algebra, we find

$$\dot{p}(q, p) = -\frac{\sum_{j=1}^{N} \phi_q(q - q_j) \Lambda(q - q_j, p - p_j)}{\sum_{j=1}^{N} \phi_q(q - q_j) \phi_p(q - q_j)} \tag{7.69}$$

where

$$\Lambda(q - q_j, p - p_j) = \int \frac{V(q + z/2) - V(q - z/2)}{z} \exp\left[i\frac{(p - p_j)z}{\hbar} - \frac{\sigma_p^2 z^2}{2\hbar^2} \right] dz. \tag{7.70}$$

This can be evaluated numerically for a given potential $V(q)$.

The results of this method are shown in Figure 7.5, and compared with both the exact quantum results and the Wigner method based on an expansion in $\hbar$, described above. The integrodifferential equation method yields better agreement with exact results, particularly at longer times. It should be noted that this implementation incorporates the purely classical part of the equations of motion in the kernel $\Lambda(q, p)$, giving a classical mechanics where the potential $V(q)$, rather than the force $-V'(q)$, appears in the equations of motion.

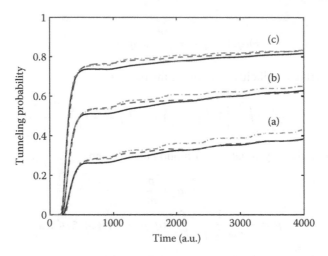

FIGURE 7.5 Time-dependent reaction probabilities for the cubic potential system with three different initial energies $E_0 = 0.75\ V_0$ (a), $E_0 = 1.25\ V_0$ (b), and $E_0 = 2.0\ V_0$ (c). Solid, dashed, and dash-dotted lines represent the results of an exact quantum calculation, entangled trajectory simulations using the integrodifferential equation formulation, Equation 7.69, and the results using the expansion in powers of $\hbar$, respectively.

7.7 GAUGE FREEDOM IN PHASE SPACE

We now briefly consider a gauge-like freedom in the definition of the vector field that phase space trajectories follow in our approach. We can express our continuity equation

$$\frac{\partial \rho_W}{\partial t} + \nabla \cdot \mathbf{j}_W = 0 \tag{7.71}$$

in terms of the components as

$$\frac{\partial}{\partial q}(\dot{q}\rho_W) = \frac{\partial}{\partial q}\left(\frac{p}{m}\rho_W + \theta_q \rho_W\right)$$
$$\frac{\partial}{\partial p}(\dot{p}\rho_W) = \frac{\partial}{\partial p}\left(-V'(q)\rho_W + \theta_p \rho_W\right) \tag{7.72}$$

which defines the "quantum vector field" $\theta = (\theta_q, \theta_p)$. We then consider the condition for the non-classical terms in the equations of motion:

$$\nabla \cdot (\theta \rho_W) = \frac{\hbar^2}{24} V'''(q) \frac{\partial^3 \rho_W}{\partial p^3}. \tag{7.73}$$

There is freedom in the solution of this differential equation for the vector θ. With the choice $\theta_q = 0$, Equation 7.73 can be integrated to yield

$$\theta_p = \frac{\hbar^2}{24} V'''(q) \frac{1}{\rho_W} \frac{\partial^2 \rho_W}{\partial p^2}. \tag{7.74}$$

This recovers the basis of the method described in this chapter [9–13]. Other choices are possible, however. For instance, we can choose $\theta_p = 0$, which then yields

$$\theta_q = \frac{\hbar^2}{24} \frac{1}{\rho_W} \int^q V'''(q') \frac{\partial^3 \rho_W(q', p)}{\partial p^3} \, dq'. \qquad (7.75)$$

This defines an alternative (and untested) quantum trajectory method. Many other divisions of the quantum vector field between its q and p components are possible, which all lead to the same quantum Liouville equation. In general, we can take a quantum trajectory method with non-classical term θ and add to the vector field any additional term ψ such that $\theta \to \theta + \psi$ and obtain an alternative quantum trajectory method, as long as the condition $\nabla \cdot \psi = 0$ is satisfied. This gauge-like freedom in the definition of quantum trajectories highlights their role simply as mathematical elements of the overall numerical method employed to represent the unified state of the system ρ_W, rather than as realistic descriptions of the actual paths of quantum particles—in other words, as hidden variables [14, 17–19].

7.8 DISCUSSION

The entangled trajectory formalism described in this chapter gives a unique but intuitively appealing picture of the quantum tunneling process. Rather than portraying tunneling as a "burrowing" through the barrier, trajectories that successfully surmount the obstacle do so by "borrowing" energy from their fellow ensemble members, and then going over the top in a classical-like manner. This energy loan is then paid back through the nonlocal inter-trajectory interactions, always keeping the mean energy of the ensemble a constant.

We have described an approach to the simulation of quantum processes using a generalization of classical MD and ensemble averaging. The general method was illustrated for the nonclassical phenomenon of quantum tunneling through a potential barrier. The basis of the method is the Liouville representation of quantum mechanics and its realization in phase space using the Wigner representation, or its generalization to strictly positive phase space densities, the Husimi representation. The evolution of the phase space functions is approximated by representing the distribution by a collection of trajectories, and then propagating equations of motion for the trajectory ensemble. In the classical limit, the members of the ensemble evolve independently under Hamilton's equations of motion. When quantum effects are included, however, the resulting quantum trajectories are no longer separable from each other. Rather, their statistical independence is destroyed by nonclassical interactions that reflect the nonlocality of quantum mechanics. Their time histories become interdependent and the evolution of the quantum ensemble must be accomplished by taking this entanglement into account.

BIBLIOGRAPHY

1. C. Cohen-Tannoudji, B. Diu, and F. Laloe, *Quantum Mechanics* (John Wiley, New York, 1977).

2. G. C. Schatz and M. A. Ratner, *Quantum Mechanics in Chemistry* (Prentice Hall, Englewood Cliffs, 1993).
3. M. P. Allen and D. J. Tildesley, *Computer Simulation of Liquids* (Clarendon Press, Oxford, 1987).
4. H. Goldstein, *Classical Mechanics* (Addison-Wesley, Reading, MA, 1980), 2nd ed.
5. E. P. Wigner, *Phys. Rev.* **40**, 749 (1932).
6. K. Takahashi, *Prog. Theor. Phys. Suppl.* **98**, 109 (1989).
7. H. W. Lee, *Phys. Rep.* **259**, 147 (1995).
8. S. Mukamel, *Principles of Nonlinear Optical Spectroscopy* (Oxford University Press, Oxford, 1995).
9. A. Donoso and C. C. Martens, *Phys. Rev. Lett.* **87**, 223202 (2001).
10. A. Donoso and C. C. Martens, *Int. J. Quantum Chem.* **87**, 1348 (2002).
11. A. Donoso and C. C. Martens, *J. Chem. Phys.* **116**, 10598 (2002).
12. A. Donoso, Y. Zheng, and C. C. Martens, *J. Chem. Phys.* **119**, 5010 (2003).
13. H. López, C. C. Martens, and A. Donoso, *J. Chem. Phys.* **1125**, 154111 (2006).
14. R. Wyatt, *Quantum Dynamics with Trajectories: Introduction to Quantum Hydrodynamics* (Springer, New York, 2005).
15. D. A. McQuarrie, *Statistical Mechanics* (HarperCollins, New York, 1976).
16. R. Kosloff, *Ann. Rev. Phys. Chem.* **45**, 145 (1994).
17. D. Bohm, *Phys. Rev.* **85**, 166 (1952).
18. D. Bohm, *Phys. Rev.* **85**, 180 (1952).
19. P. R. Holland, *The Quantum Theory of Motion* (Cambridge University Press, Cambridge, 1995).

8 On the Possibility of Empirically Probing the Bohmian Model in Terms of the Testability of Quantum Arrival/Transit Time Distribution

Dipankar Home and Alok Kumar Pan

CONTENTS

8.1   INTRODUCTION AND MOTIVATION

Born's interpretation of the squared modulus of a wave function ($|\psi|^2$) as the probability density of *finding* a particle within a specified region of space is a key ingredient of the standard framework of quantum mechanics, thereby implying that the standard interpretation of quantum mechanics is inherently *epistemological*. On the other hand, the possibility of an *alternative interpretation* of quantum mechanics by interpreting $|\psi|^2$ as the probability density of a particle *being* actually present within a specified region was first suggested by de Broglie [1]. Later, Bohm [2–4] developed the details of such an *ontological* model of quantum mechanics by using the notion of an *observer-independent spacetime trajectory* of an *individual particle* that is determined by its wave function through an equation of motion which is formulated in a way *consistent* with the Schrödinger time evolution. Bohm's model, thus, explicitly refuted the counterarguments (such as those put forward by Pauli [5] and von Neumann [6]) that claimed to have ruled out the formulation of such a model. Subsequently, much work has been done on various aspects of the Bohmian model [7–10]. That such a model is *not* unique has been elaborately discussed [11,12] while different versions of the ontological model of quantum mechanics have been proposed [13–21].

Although any such ontological model hinges on the notion of a definite *spacetime track* used to provide a description of the objective motion of a single particle, such trajectories are *not* directly measurable. Hence these trajectories have been essentially viewed as conceptual aids for understanding the various features of quantum mechanics. While recently a study has been reported which shows an application of such trajectories as computational aids for solving the time-dependent Schrödinger equation [22], here in this chapter we explore the question as to whether an ontological model such as the Bohmian one is empirically falsifiable.

The essence of the question as regards empirical equivalence between the standard and the Bohmian model can be seen as follows. Let us focus on the instant *when* an ensemble of particles *begins* to interact with a given potential. If at that *initial instant*, the position probability density of the ensemble is taken to be the *same* in both the standard and any ontological approach, then the time evolved position probability density of the final ensemble as calculated by the equation of motion specified by any ontological model is ensured to be the *same* as that obtained from the standard quantum mechanical formalism. The same equivalence holds good for any other dynamical observable represented by a Hermitian operator.

The above is the main reason why the ontological models have been dismissed by many as merely of metaphysical interest, having a "superfluous ideological superstructure" (see, for example, the criticism of the Bohmian model by Heisenberg [23]). In this chapter, we take a fresh look at the question of empirical falsifiability of an ontological model like the Bohmian one. The motivation underlying the present study stems from the type of question as was posed by Cushing [24]: Could it be that the additional interpretive ingredients (such as the notion of particles having objectively defined trajectories) in the Bohmian model might enable testable predictions in a suitable example where the standard version of quantum mechanics has an intrinsic *nonuniqueness* allowing for a number of calculational approaches, but, *prima facie*, none of these can be preferred over the others using a rigorous justification based on first principles? On this point one may recall that John Bell had once remarked that the Bohmian model of quantum mechanics is experimentally equivalent to the standard version "insofar as the latter is unambiguous" [25].

It is from this perspective that the kind of *nonuniqueness* inherent in the standard quantum mechanical calculation of time distributions seems to be a pertinent tool for exploring the possibility of subjecting an ontological model such as the Bohmian one to empirical scrutiny.

Time plays a peculiar role within the formalism of quantum mechanics—it differs fundamentally from all other dynamical quantities like position or momentum since it appears in the Schrödinger equation as a parameter, not as an operator. If the wave function $\psi(x, t)$ is the solution of Schrödinger's equation, then $\psi(x, t_1), \psi(x, t_2)$, $\psi(x, t_3)$, ... determine position probability distributions at the respective different instants t_1, t_2, t_3, ... for a fixed region of space, say, between x and $x + dx$. Now, if we fix the positions at X_1, X_2, X_3, ..., the question arises of whether the quantities $\psi(X_1, t), \psi(X_2, t), \psi(X_3, t)$, ... can specify the time probability distributions at respective various positions X_1, X_2, X_3, ...? However, one can easily see that, although $\int_{-\infty}^{+\infty} |\psi(x, t_i)|^2 dx = 1$, one would have in general, $\int_{-\infty}^{+\infty} |\psi(X_i, t)|^2 dt$ not normalisable. Hence, in order to quantum mechanically calculate the time probability distribution, unlike the position probability distribution, we do not readily have an unambiguously defined procedure.

The fundamental difficulty in constructing a self-adjoint time operator within the formalism of quantum mechanics was first pointed out by Pauli [26]. Another proof of the nonexistence of a time operator, specifically for the time-of-arrival operator, was given by Allcock [27–29]. Nevertheless, there were subsequent attempts to construct a suitable time operator. For instance, Grot et al. [30] and Delgado et al. [31] suggested a time-of-arrival operator for a free particle, and showed how the time probability distribution can be calculated using it; interestingly, such an operator has an orthogonal basis of eigenstates, although the operator is, in general, *not* self-adjoint.

In recent years there has been an upsurge of interest in analyzing the concept of an arrival time distribution in quantum mechanics; for useful reviews on this subject, see References [32, 33]. Here we note that a number of schemes have been analysed [34–61] for calculating what has been called the *arrival time distribution* in quantum mechanics, for example, the probability current density approach, using the path integral approach, the consistent history scheme, and by using the Bohmian trajectory model in quantum mechanics. However, since there is an inherent ambiguity within the standard formalism of quantum mechanics as regards calculating such a probability distribution, it remains an open question as to what extent these different approaches can be *empirically tested.*

There have been several specific toy models that have been suggested to investigate the feasibility of how the measurement of a transit time distribution can actually be performed in a way consistent with the basic principles of quantum mechanics. The earliest proposal for a model quantum clock in order to measure the time of flight of quantum particles was suggested by Salecker and Wigner [62], and later elaborated by Peres [63]. In effect, this model of quantum clock measures the change in the phase of a wave function over the duration to be measured. Such a model of quantum clock [63] can be used to calculate the expectation value for the transit time distribution of quantum particles passing through a given region of space. On the other hand, Azebel [64] has analyzed a process in which the thermal activation rate can serve as a clock. Applications of quantum clock models have also been studied for the motion of quantum particles in a uniform gravitational field by Davies [65] and others [53].

Against the backdrop of such studies, in the present chapter we proceed as follows. Let us first consider the following simple experimental arrangement. A particle moves in one dimension along the x-axis and a detector is placed at the position $x = X$. Let T be the time at which the particle is detected, which we denote as the *time of arrival* of the particle at X. Can we predict T from the knowledge of the state of the particle at the prescribed initial instant?

In classical mechanics, the time of arrival T at $x = X$ of an individual particle with the initial position x_0 and momentum p_0 is $T = t(t, x_0, p_0)$ which is fixed by the solution of the equation of motion of the particle concerned. But in quantum mechanics, this problem becomes nontrivial in terms of the probability distribution of times, say, $\Pi(T)$ over which the particles are registered at the detector location, say X, within the time interval, say, t and $t + dt$ so that $\int_{T_1}^{T_2} \Pi(T)dT$ is the probability that the particles are detected at $x = X$ between the instants T_1 and T_2. In other words, the relevant key question in quantum mechanics is how to determine $\Pi(T)dT$ at a specified position, given the wave function at the initial instant from which the propagation is considered.

An effort along this direction was made [54] by considering the measurement of arrival time using the emission of a first photon from a two-level system moving into a laser-illuminated region. The probability for this emission of the first photon was calculated for this purpose by specifically using the quantum jump approach. Subsequently, further work [55] was done on this proposal using Kijowski's distribution [61].

In this work we address this question of empirical verifiability from a new perspective so that starting from any axiomatically defined *transit/arrival time distribution* one can *directly* relate it to the actually testable results. For this, we first need to evaluate the transit time distribution $\Pi(t)$ by using one of the suggested approaches. Using such a calculated $\Pi(t)$, one can then derive a distribution of spin orientations $\Pi(\phi)$ along different directions for the spin-$\frac{1}{2}$ neutral particles emerging from a spin-rotator (SR) (which contains a constant magnetic field), ϕ being the angle by which the spin orientation of a spin-$\frac{1}{2}$ neutral particle (say, a neutron) is rotated from its initial spin-polarized direction. Note that this angle ϕ is determined by the transit time (t) within the SR. Thus, in our scheme, the SR serves as a *"quantum clock"* where the basic quantity is $\Pi(\phi)$ which determines the actual observable results corresponding to the probability distribution of spin orientations along different directions for the spin-$\frac{1}{2}$ neutral particles emerging from the SR. Such a calculated spin distribution function can be empirically tested by suitably using a Stern–Gerlach (SG) device, as explained later, thereby providing a test of $\Pi(t)$ based on which $\Pi(\phi)$ is calculated. In this chapter we particularly focus on analyzing the application of the Bohmian model in the context of such an example.

In order to recapitulate the essential aspects of this setup [49], we consider spin-$\frac{1}{2}$ neutral objects, specifically, spin-polarized neutrons. The initial quantum state is given by $\Psi(x, t = 0) = \psi(x, t = 0) \otimes \chi(t = 0)$ where $\chi(t = 0)$ is the initial spin state taken to be the *same* for *all* neutrons whose spins are oriented along the $\widehat{x}$-axis. The spatial part is taken to be a propagating Gaussian wave-packet (for simplicity, taken to be one dimensional) that moves along the $+\widehat{x}$-axis, and passes through a region containing constant magnetic field (**B**), confined between $x = 0$ and $x = d$ (Figure 8.1). The enclosed magnetic field is along the $+\widehat{z}$-axis, and is switched on *at the instant* (taken to be $t = 0$) *when* the peak of the wave-packet is at the entry point $x = 0$.

In passing through this bounded region, henceforth designated as SR, the spin variable of a neutron interacts with the constant magnetic field, this interaction being

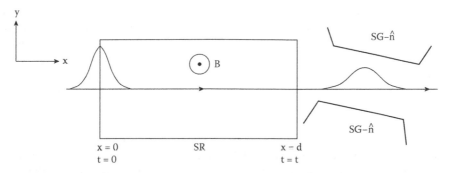

FIGURE 8.1 Spin-$\frac{1}{2}$ particles, say, neutrons with initial spin orientations polarized along the $+\widehat{\mathbf{x}}$-axis and associated with a localized Gaussian wave-packet (peaked at $x = 0$, $t = 0$) pass through a spin-rotator (SR) containing a constant magnetic field **B** directed along the $+\widehat{\mathbf{z}}$-axis. The particles emerging from the SR have a distribution of their spins oriented along different directions. Calculation of this distribution function is experimentally tested by measuring the spin observable along a direction $\widehat{n}(\theta)$ in the xy-plane making an angle θ with the initial spin polarized along the $+\widehat{\mathbf{x}}$-axis. This is done by suitably orienting the direction $\widehat{n}(\theta)$ of the inhomogeneous magnetic field in the Stern–Gerlach ($SG - \widehat{n}$) device.

mediated by the neutron's magnetic moment μ as the coupling constant. The angle ϕ by which the direction of the initial spin-polarization is rotated is determined by the time (τ) over which this interaction takes place. Now, if the spatial wave-packet is a superposition of energy eigenstates (plane waves), there will be a distribution ($\Pi(\tau)$) of times τ over which the interaction between neutron spins and the enclosed magnetic field takes place, corresponding to the relevant arrival/transit time distribution. Consequently, using the probability distribution function $\Pi(\tau)$, and an appropriate relation connecting the quantities ϕ and τ, one can calculate a distribution of spin orientations ($\Pi(\phi)$) for the neutrons emerging from the SR. It is, therefore, evident that for calculating $\Pi(\phi)$, the quantum mechanical evaluation of $\Pi(\tau)$ for the neutrons passing through the SR is a key issue.

We will first elaborate on the relevant setup by discussing how the determination of the quantity $\Pi(\phi)$ can actually be tested by using a SG device. Subsequently, we will discuss the key ingredients of the Bohmian procedure that can be adopted in the context of this setup for calculating $\Pi(\phi)$ in terms of $\Pi(t)$. The subtleties involved in such a procedure will be critically analyzed and directions for further studies will be indicated that are required to explore fully the potentiality of this example for subjecting the Bohmian model to empirical scrutiny.

8.2 THE SETUP

We consider an ensemble of spin-$\frac{1}{2}$ neutral particles, say, neutrons having magnetic moment μ. The spatial part of the total wave function is represented by a localized narrow Gaussian wave-packet $\psi(x, t = 0)$ (for simplicity, it is considered to be one dimensional) which is taken to be peaked at $x = 0$ at the initial instant $t = 0$ and moves with the group velocity u. Thus the initial total wave function is given by $\Psi = \psi(x, t = 0) \otimes \chi(t = 0)$ where $\chi(t = 0)$ is the initial spin state which is taken to be the *same* for all members of the ensemble oriented along the $\widehat{x}$-axis.

The SR used in our setup (Figure 8.1) has within it a constant magnetic field $\mathbf{B} = B\widehat{\mathbf{z}}$ directed along the $+\widehat{\mathbf{z}}$-axis, confined between $x = 0$ and $x = d$. We consider a specific situation where the magnetic field is turned on *while* the peak of the initial wave-packet *reaches* the position $x = 0$ at $t = 0$.

Application of the Larmor precession of spin in a magnetic field has earlier been discussed, for example, in the context of the scattering of a plane wave from a potential barrier [66, 67]. On the other hand, the scheme proposed here explores an application of Larmor precession such that one can empirically test *any* given quantum mechanical formulation for calculating the *arrival/transit time distribution* using a Gaussian wave-packet.

Now, for testing the scheme for calculating the probability density function $\Pi(\phi)$ of spin orientations of particles emerging from the SR, we consider the measurement of a spin variable, say $\widehat{\sigma}_\theta$, by a SG device (Figure 8.1) in which the inhomogeneous magnetic field is oriented along a direction $\widehat{n}(\theta)$ in the xy-plane making an angle θ with the initial spin-polarized direction ($+\widehat{x}$-axis) of the particles. The initial x-polarized spin state can be written in terms of the z-bases $|\uparrow\rangle_z$ and $|\downarrow\rangle_z$ as $\chi(0) = 1/\sqrt{2}\,(|\uparrow\rangle_z + |\downarrow\rangle_z)$. While passing through the SR, the spin-polarized state rotates only in the xy-plane by an angle ϕ with respect to the initial spin orientation along the $\widehat{x}$-axis. Such a rotated spin state in the xy-plane can be typically written as $\chi(\phi) = 1/\sqrt{2}\,(|\uparrow\rangle_z + e^{i\phi}|\downarrow\rangle_z)$. If one applies the SG-magnetic field along the direction $\widehat{n}(\theta)$, the relevant basis states corresponding to the spin operator $\widehat{\sigma}_\theta$ are respectively $|\uparrow\rangle_\theta = 1/\sqrt{2}\,(|\uparrow\rangle_z + e^{i\theta}|\downarrow\rangle_z)$ and $|\downarrow\rangle_\theta = 1/\sqrt{2}\,(|\uparrow\rangle_z - e^{i\theta}|\downarrow\rangle_z)$. Then for such spin measurements, the probabilities of getting $|\uparrow\rangle_\theta$ and $|\downarrow\rangle_\theta$ are $p_+(\theta) = |_\theta\langle\uparrow|\chi(\phi)\rangle|^2 = \cos^2(\theta - \phi)/2$ and $p_-(\theta) = |_\theta\langle\downarrow|\chi(\phi)\rangle|^2 = \sin^2(\theta - \phi)/2$ respectively.

8.3 THE TESTABLE PROBABILITIES

Since we are considering an ensemble of spin-$\frac{1}{2}$ neutrons passing through the SR where initially *all* members of this ensemble have their spins polarized along, say, the $+\widehat{x}$-axis, the spins of individual members of the ensemble evolve over *different times* (characterized by $\Pi(t)$, the *distribution of transit times* within the SR) of duration of the interaction with the constant magnetic field within the SR.

Thus, for the spins of the particles emerging from the SR polarized along different directions (with the respective probabilities $\Pi(\phi)$ of making angles ϕ with the $+\widehat{x}$-axis), the probabilities of finding the spin component along $+\theta$ direction and that along its opposite direction are respectively given by

$$P_+(\theta) = \int_0^{2n'\pi} \Pi(\phi)\cos^2\frac{(\theta - \phi)}{2}d\phi \qquad (8.1)$$

$$P_- (\theta) = \int_0^{2n'\pi} \Pi(\phi) \sin^2 \frac{(\theta - \phi)}{2} d\phi \qquad (8.2)$$

where $P_+ (\theta) + P_- (\theta) = 1$, and n' is any positive number.

It is these probabilities $P_+ (\theta)$ and $P_- (\theta)$ which constitute the actual *observable quantities* in our scheme which are determined by the distribution of spins $\Pi(\phi)$ of the particles emerging from the SR. The theoretical estimations of these probabilities crucially depend on *how* one calculates the quantity $\Pi(\phi)$ whose evaluation, in turn, is contingent on the approach adopted for calculating the relevant time distribution $\Pi(t)$ mentioned earlier. Now, the important point to stress here is that the specification of such a time distribution is *not* unique in quantum mechanics. For the setup indicated in Figure 8.1, $\Pi(t)$ represents the *arrival time distribution* at the exit point $(x = d)$ of the SR, which is *also* the *distribution of transit times(t)* as well as the *distribution of times of interaction with the magnetic field* within the SR. Thus, $\Pi(t)$ determining $\Pi(\phi)$ is the key quantity which needs to be evaluated first. In the following section we focus on discussing the Bohmian procedure for calculating $\Pi(t)$ leading to $\Pi(\phi)$.

8.4 THE EVALUATION OF $\Pi(t)$ AND $\Pi(\phi)$ USING THE BOHMIAN MODEL

In the Bohmian model [2–4, 7–10], each individual particle is assumed to have a definite position, irrespective of any measurement. The pre-measured value of position is revealed when an actual measurement is done. Over an ensemble of particles having the *same* wave function ψ, these ontological positions are distributed according to the probability density $\rho = |\psi|^2$ where the wave function ψ evolves with time according to the Schrödinger equation. For the purpose of the present chapter, the relevant key postulates of the Bohmian model can be expressed as follows:

(i) An individual "particle" embodying the innate attributes like mass, charge, and the magnetic moment has a *definite location* in space at *any* instant, irrespective of whether its position is measured or not. In particular, the pre-measurement value of position is taken to be the *same* as the value revealed by a measurement of position, whatever the measurement procedure may be. However, this feature does *not* hold for any other dynamical variable, such as the momentum or spin.

(ii) $|\psi|^2 dx$ is interpreted as the probability of a particle to be *actually present* within a region of space between x and $x + dx$, instead of Born's interpretation as the probability of *finding* the particle within the specified region.

(iii) Consistent with the Schrödinger equation and the equation of continuity being interpreted as corresponding to *actual flow of particles* with velocity $v(x, t)$, the equation of motion of any individual particle is determined by the following equation

$$v(x, t) = \frac{J(x, t)}{\rho(x, t)} = \frac{1}{m} \frac{\partial S(x, t)}{\partial t}. \qquad (8.3)$$

Here $J(x,t) = \rho(x,t)v(x,t)$ is interpreted as the probability current density of *actual particle flow* and the wave function is considered in the polar form $\psi = Re^{iS/\hbar}$ with $\rho = R^2$ where R and S are real functions of position and time. In the context of our SR example, coming to the question of whether there can be predictions obtained from the Bohmian model which may allow us to empirically discriminate it from the standard scheme, we analyze this issue as follows.

Here it should be noted that the Schrödinger expression for the probability current density is inherently *nonunique*. This is seen from the feature that if one adds any divergence-free or constant term to the expression for $J(x,t)$, the new expression also satisfies the same equation of continuity derived from the Schrödinger equation. However, it has been discussed by Holland [11, 12], followed by others [47–49], that the probability current density derived from the Dirac equation for any spin-$\frac{1}{2}$ particle is *unique*, and that this uniqueness is preserved in the *nonrelativistic limit* while containing a spin-dependent term in addition to the Schrödinger expression for $J(x,t)$. Accordingly, a modified Bohmian equation of motion for the spin-$\frac{1}{2}$ particle has been proposed [11, 12] which is of the form given by

$$\mathbf{v}(\mathbf{x},t) = \frac{1}{m}\left(\nabla S(x,t) + \nabla \log \rho(x,t) \times \mathbf{s(t)}\right) \qquad (8.4)$$

where $\mathbf{s}(t) = \frac{\hbar}{2}\chi(t)\widehat{\sigma}\chi^{\dagger}(t)$, $\chi(t)$ is the time evolving spin state and the components of $\widehat{\sigma}$ are the Pauli matrices. The modified Bohmian equation of motion given by Equation 8.4 reduces to the original Bohmian form given by Equation 8.3 if $|\nabla S| \gg \frac{\hbar}{2}|\nabla \log \rho|$, i.e., when the modulus of the spin-dependent term is negligibly small compared to the spin-independent term. In our setup, the pertinent parameters (viz. the magnitude of the magnetic field, the initial width and the peak velocity of the wave-packet) are taken to be such that this condition is satisfied. Hence, throughout this chapter, we will neglect any effect of the spin-dependent term in the modified Bohmian equation of motion.

Next, we recall the general question of equivalence between the standard version and the Bohmian model of quantum mechanics in predicting the observable results pertaining to any Hermitian operator which has been much discussed [2–4], along with some controversy [68–71]. Nonetheless, the central point we may stress is that, given any example, if and only if the procedure for calculating an observable quantity is *unambiguously defined* in *both* the standard and the Bohmian versions of quantum mechanics, the very formulation of the Bohmian model guarantees the equivalence (at least, in the non-relativistic domain) when the (common) formalism is applied.

Now, coming to our example, for the initial state given by $\Psi(x,t=0) = \psi(x,t=0) \otimes \chi(t=0)$, the time evolution within SR is subject to the interaction Hamiltonian $\mu\vec{\sigma}.B$. Note that as the spins of neutrons interact with the time-independent constant magnetic field, in order to conserve the energy, there will be changes in the momenta of the spatial parts associated with both the up and down spin components—these changes occur according to the potential energy of the spin–magnetic field interaction that develops corresponding to the up and down spin components respectively.

In the setup considered here, the relevant parameters (viz. the initial momentum, mass, magnetic moment, and the magnitude of the applied constant magnetic field) are chosen to be such that the magnitude of the potential energy ($\vec{\mu\sigma}.\mathbf{B}$) that arises because of the spin–magnetic field interaction is exceedingly *small* ($\approx 10^{-9}$ times) compared to the initial kinetic energy of the neutrons. This amounts to ensuring the magnitude of the momenta changes within the SR to be negligibly small compared to the initial momentum. Consequently, the time evolutions of the spatial and spin parts can be treated *independent* of each other, with the spatial wave function evolving *freely* within the SR (for the relevant justification based on a rigorous treatment, see Appendix A).

With respect to the setup shown in Figure 8.1, the initial wave function is given by

$$\psi(x, t = 0) = \frac{1}{\left(2\pi\sigma_0^2\right)^{1/4}} \exp\left[-\frac{x^2}{4\sigma_0^2} + ikx\right] \tag{8.5}$$

where the wave number $k = mu/\hbar$, u being the group velocity and σ_0 is the initial width of the wave-packet. Now, since the spatial wave function is considered to evolve *freely* within the SR, the freely propagating wave function at any instant t is given by

$$\psi(x, t) = \frac{1}{(2\pi A_t^2)^{1/4}} \exp\left[-\frac{(x - ut)^2}{4\sigma_0 A_t} + ik\left(x - ut/2\right)\right] \tag{8.6}$$

where

$$A_t = \sigma_0 \left(1 + \frac{i\hbar t}{2m\sigma_0^2}\right).$$

Writing Equation 8.6 in the polar form $\psi(x, t) = R(x, t)e^{iS(x,t)/\hbar}$, and using Equation 8.3, the Bohmian trajectory equation in this setup for the i-th individual particle having an initial position $x_i(0)$ is given by

$$x_i(t) = ut + x_i(0)\sqrt{1 + \beta t^2} \tag{8.7}$$

with

$$\beta = \frac{\hbar^2}{4m^2\sigma_0^4},$$

and the subscript "i" is used as the particle label.

The Bohmian velocity of the i-th individual particle can then be written as a function of the initial position by using Equation 8.7. The resulting expression is given by

$$v_i(x_{0i}, t) = \frac{dx_i(t)}{dt} = u + \frac{x_{0i}\beta t}{\sqrt{1 + \beta t^2}}. \tag{8.8}$$

From the above equation it follows that only those particles initially located around the peak position of the Gaussian wave-packet at $x_{0i} = 0$ follow the Newtonian equation of motion. On the other hand, the particles initially located within the front half ($x_{0i} = +$ve) are *all* accelerated, while the particles initially lying within the rear half ($x_{0i} = -$ve) are *all* decelerated [72]. It can then be seen that there will be turning points of the Bohmian trajectories provided the following relation is satisfied between the instant t' of the turning point and the Bohmian initial velocity u, given by

$$u = \frac{x_{0i}\beta t'}{\sqrt{1 + \beta(t')^2}} \tag{8.9}$$

where x_{0i} is essentially $-$ve. For the purpose of discussion in this chapter, we consider the relevant parameters (u and σ_0) to be such that the above condition is not satisfied, thereby ensuring that, in our example, the particles do *not* have any turning point [73,74]. Next, in order to proceed with the Bohmian calculation of $\Pi(\phi)$, we focus on the following two quantities pertaining to neutrons which are initially localized, say, within an arbitrarily chosen region between $x_i(0)$ and $x_i(0) + dx_i(0)$.

First, we note that the probability of neutrons to be actually localized between $x_i(0)$ and $x_i(0) + dx_i(0)$ is given by $|\psi(x_i(0), t = 0)|^2 dx_i(0)$. Secondly, we consider the probability of *these* neutrons to cross the region of the magnetic field enclosed in the SR within the time-interval, say, T and $T + dT$; this is denoted by $\Pi_B(T)dT$ where $\Pi_B(T)$ is the *arrival time distribution* at the *exit point* ($x = d$) of the SR, T being the time required by the i-th neutron to reach the exit point ($x = d$) of the SR.

Given the notion of trajectories in the Bohmian model, and taking that *all* the neutrons initially localized within $x_i(0)$ and $x_i(0) + dx_i(0)$ ultimately pass through the entire region of the SR (i.e., all of them cross the exit point ($x = d$) of the SR, and *none* of them re-enters), the above two quantities can be equated so that

$$\Pi_B(T)dT = |\psi(x_i(0), t = 0)|^2 dx_i(0) \tag{8.10}$$

whence

$$\Pi_B(T) = |\psi(x_i(0), t = 0)|^2 \frac{dx_i(0)}{dT}. \tag{8.11}$$

Note that Equation 8.11 is a very general relation that determines the arrival time distribution in any causal trajectory model, valid for any form of the initial wave-packet. In the particular case of a freely evolving Gaussian wave-packet we are considering here in the context of the Bohmian model, we proceed as follows:

By using Equation 8.7, and replacing t by T, the quantity $dx_i(0)/dT$ on the right hand side of Equation 8.11 is calculated by writing $x_i(0)$ in terms of $x_i(T)$. Further, by invoking the earlier outlined interpretive ingredients (*ii*) and (*iii*) of the Bohmian model, we write the quantity $|\psi(x_i(0), t = 0)|^2$ in terms of $x_i(T)$. This is done by using Equation 8.3 for $\psi(x, t = 0)$, replacing the argument "x" of this function by "$x_i(0)$" for varying "i" corresponding to different initial positions of the neutrons. Then, putting $x_i(T) = d$, using both Equation 8.7 and the Schrödinger expression for the probability current density $J(x,t)$, it can be verified that Equation 8.11 reduces to

$$\Pi_B(T) = J(d, T) \tag{8.12}$$

where

$$J(d, T) = \frac{1}{\left(2\pi\sigma_T^2\right)^{1/2}} \exp\left\{-\frac{(d - uT)^2}{2\sigma_T^2}\right\} \times \left\{u + \frac{(d - uT)\,\hbar^2 T}{4m^2\sigma_0^4 + \hbar^2 T^2}\right\} \tag{8.13}$$

is the current density of neutrons at the exit point($x = d$) of the SR. Here σ_T is the width of the wave-packet at the instant T. Note that $J(d,t)$ given by Equation 8.13 is essentially a +ve quantity. Thus, Equation 8.12 shows that the Bohmian *arrival time distribution* at $x = d$ is the same as that given by the probability current density (henceforth designated as PCD) approach [36]; i.e.,

$$\Pi_B(T) = \Pi_{PCD}(T) = J(d, T). \tag{8.14}$$

Then comes an important element in our example, viz. the consideration of the time evolution of the spin state according to an equation that is obtained by decoupling the position and the spin parts of the Pauli equation (adopting the "approximation" stated earlier whose justification is indicated in Appendix A), given by

$$\mu\vec{\sigma}.\mathbf{B}\,\chi = i\hbar\frac{\partial\chi}{\partial t}. \tag{8.15}$$

Now, we recall that in this particular example, the initial spin of any neutron is taken to be polarized along the $+\hat{\mathbf{x}}$-axis; i.e.,

$$\chi(0) = |\rightarrow\rangle_x = \frac{1}{\sqrt{2}}\left[|\uparrow\rangle_z + |\downarrow\rangle_z\right]. \tag{8.16}$$

Then, since the constant magnetic field $\mathbf{B} = B\hat{\mathbf{z}}$ within the SR, confined between $x = 0$ and $x = d$, is switched on at the instant ($t = 0$) the peak of the wave-packet reaches the entry point ($x = 0$) of the SR, the time evolved neutron spin state $\chi(\tau)$ under the interaction Hamiltonian $H = \mu\vec{\sigma}.\mathbf{B}$ is given by

$$\chi(\tau) = \exp\left(\frac{-iH\tau}{\hbar}\right)\chi(0) = \frac{1}{\sqrt{2}}e^{-i\omega\tau}\left[|\uparrow\rangle_z + e^{i2\omega\tau}|\downarrow\rangle_z\right] \tag{8.17}$$

where $\omega = \mu B/\hbar$ and τ is the spin–magnetic field interaction time; i.e., the time during which the neutron spin state evolves according to the spin–magnetic field interaction term. Equation 8.17 implies that after a time interval τ, the spin orientation of a neutron is rotated by an angle ϕ with respect to its initial spin-polarized direction, where

$$\phi = 2\omega\tau. \tag{8.18}$$

Subsequently, one may put $\tau = T$ in Equation 8.18, and use Equation 8.14 to obtain the probability distribution of spin orientations ($\Pi(\phi)$) of the neutrons emerging from the SR. Then, as applied to our example, the equivalence between the Bohmian scheme and the probability current density approach would seem to follow; i.e.,

$$\Pi_B(\phi) = \Pi_{PCD}(\phi) = J(d, \phi). \tag{8.19}$$

8.5   SUBTLETIES IN THE BOHMIAN CALCULATION OF $\Pi(t)$ AND $\Pi(\phi)$

A. Let us focus on a critical feature that is involved in obtaining the expression for $\Pi_B(T)$ given by Equation 8.12 within the Bohmian framework. This requires using the equality given by Equation 8.10, taking it to be true for *all* values of the initial positions $x_i(0)$ of the neutrons, including even for *those* neutrons initially located *outside* the SR. Consequently, $\Pi_B(T)$ given by Equation 8.12 is the *arrival time distribution* at $x = d$ (the exit point of the SR) pertaining to the *entire ensemble* of neutrons, including the neutrons initially located outside the SR.

Then, since writing the probability distribution $\Pi(\phi) = J(d, \phi)$ from $\Pi_B(T)$ of Equation 8.12 using Equation 8.18 is contingent upon taking $T = \tau$, this means regarding the *arrival time distribution* at $x = d$ to be essentially the *distribution of interaction times* τ over which the spin states of the neutrons belonging to the *entire ensemble* evolve under the interaction Hamiltonian $\mu\vec{\sigma}.\mathbf{B}$.

Now, if the above *proviso* is to be satisfied, this would imply that, within the Bohmian model, even if a neutron is located *outside* the region of the SR at the instant ($t = 0$) when the magnetic field within the SR is switched on, the spin state associated with *that* neutron would have to be regarded as evolving, under the spin–magnetic field interaction, from the instant $t = 0$ itself. However, a fundamental feature of the Bohmian model is that an individual particle such as a neutron is regarded as a *localized entity* in an objective sense, that has a *definite location* in space at *any* instant, and which embodies the innate properties such as the mass and the magnetic moment. Further, a relevant crucial point is that the spin–magnetic field interaction term $\mu\vec{\sigma}.\mathbf{B}$ has the *localized* particle attribute, magnetic moment (μ), as the coupling constant. It is, therefore, arguable that in applying the Bohmian scheme to our example, the time evolution of the spin state associated with an individual neutron is to be considered subject to the interaction $\mu\vec{\sigma}.\mathbf{B}$ *only* if the neutron is *inside* the SR within which the magnetic field is confined. Thus, from this point of view, the earlier mentioned *proviso* would be inadmissible for the neutrons which are *initially* located *outside* the SR. In other words, within the Bohmian model, this would mean taking the ontological position variable to be more fundamental than the global wave function in determining the instant from which the individual particle is subjected to the relevant interaction which is mediated through a localized property of the particle (in this case, the magnetic moment).

Consequently, it would not be legitimate to take the Bohmian arrival time distribution ($\Pi_B(T)$) at the exit point ($x = d$) of the SR to be the distribution of interaction times τ for *all* the neutrons belonging to the entire ensemble. This would then mean that the Bohmian probability distribution of spin orientations of neutrons emerging from the SR *cannot* be written as $J(d, \phi)$, thereby implying in the context of our setup, an alternative route for calculating the quantities $\Pi_B(t)$ and $\Pi_B(\phi)$.

Such an alternative Bohmian procedure for calculating $\Pi_B(t)$ and $\Pi_B(\phi)$ would entail the viewpoint that the spin state of an individual neutron which is initially located *outside* the SR would evolve under the spin–magnetic field interaction that acts (using the neutron's localized magnetic moment as the coupling constant) only *after* that neutron reaches the entry point ($x = 0$) of the SR. Therefore, for such a neutron, its arrival time at the exit point ($x = d$) of the SR would *not* be the time over which its spin–magnetic field interaction occurs in passing through the SR. On the other hand, for a neutron initially located *inside* the SR, its spin state would, of course, start to evolve from the instant $t = 0$ itself when the magnetic field is switched on.

Consequently, in this approach for calculating the Bohmian probability distribution $\Pi_B(\phi)$, the time over which the spin of a particle evolves under the magnetic field enclosed within the SR is essentially the particle's *transit time* within the SR (*not* the arrival time at $x = d$), which is the time taken by it to cross the region between $x = 0$ and $x = d$. Thus, in such a procedure, it is the computation of the *transit time distribution* within the SR (in contrast to the arrival time distribution at $x = d$ computed in the earlier discussed Bohmian procedure) that would provide the prediction of the Bohmian distribution $\Pi_B(\phi)$. *Prima facie*, it is thus an open question whether the quantitiy $\Pi'_B(\phi)$ calculated in this alternative way would agree with the result obtained (Equation 8.23) from the Bohmian calculational scheme discussed in the preceding section—a procedure which, in a sense, ascribes more fundamental ontological status to the global wave function associated with an individual particle than to its ontological position variable.

B. An important point to be stressed here is that the calculational scheme discussed in Section 8.4 is specifically couched in terms of a Gaussian wave-packet. In particular, the crucial expression given by Equation 8.12 is obtained from the Equations 8.10 and 8.11 in a way that involves an explicit use of the Gaussian wave-packet. Therefore, it is again another open question whether the equivalence between the Bohmian and the probability current density based approach in calculating the quantity $\Pi_B(t)$ would hold good if one considers non-Gaussian wave-packets. For example, one may consider a non-Gaussian wave function in the momentum space given by

$$\phi(k) = \frac{N}{\left(2\pi\sigma_k^2\right)^{1/4}} \exp\left[-\frac{(k - k_0)^2}{4\sigma_k^2}\right]\left(1 + \alpha \sin\left[\frac{k - k_0}{\beta\sigma_k}\right]\right), \tag{8.20}$$

where k is the wave number and $N = \left(1 + \frac{\alpha^2}{2}(1 - e^{-\frac{\pi^2}{8}})\right)^{-1}$ is the normalization constant. The above function contains four real parameters σ_k, k_0, α, and β among which σ_k is positive. The salient feature of the corresponding wave-packet, i.e., $|\phi(k)|^2$, is its infinite tail with the probability of finding particles negligibly small outside a bounded region determined by the parameter σ_k, while the wave-packet is asymmetric and it reduces to the Gaussian form upon continuous decrement of $|\alpha|$ to zero. Note that the infinite tail is in contradistinction to other non-Gaussian forms found in the literature which are generated by truncating the Gaussian distribution. The asymmetry of the wave-packet due to the sine function entails a difference between the mean and peak values of the wave-packet, and deprives k_0 of appropriating either of these values. Similarly, the parameter σ_k does not denote the standard deviation of $|\phi(k)|^2$.

Using Equation 8.20, the initial wave function in position space is just the Fourier transform of $\phi(k)$ and is given by

$$\psi(x, t = 0) \equiv \frac{1}{\sqrt{2\pi}} \int_{-\infty}^{\infty} \phi(k)e^{ikx}\, dk \tag{8.21}$$

$$= \frac{N}{(2\pi\sigma^2)^{1/4}} \exp\left[-\frac{x^2}{4\sigma^2} + ik_0 x\right]\left(1 + i\alpha e^{-\frac{\pi^2}{16}} \sinh\left[\frac{\pi x}{4\sigma}\right]\right)$$

where $\sigma = (2\sigma_k)^{-1}$.

It would thus be an interesting exercise to study the arrival/transit time distribution pertaining to the freely evolving wave-packet corresponding to Equation 8.21 and use the results to probe the equivalence between the Bohmian and the probability current density based scheme in calculating the empirically testable probabilities relevant to our setup.

8.6   SUMMARY AND OUTLOOK

The setup analyzed in this chapter seems to provide a particularly interesting example in the context of the foundations of quantum mechanics since it is capable of empirically discriminating between not only the various suggested standard quantum mechanical schemes for calculating the arrival/transit time distribution, but it may also enable us to empirically *verify* or *falsify* the specific *unique predictions* obtained from any *realist trajectory model* of quantum mechanics like that of the Bohmian type. This is irrespective of the question as regards which standard quantum mechanical scheme is the empirically correct one for calculating the observable probabilities in our example. While comprehensive computations are being pursued along the lines **A** and **B** indicated in Section 8.5, here it may be useful to summarize the implications of the different possible outcomes of the relevant experimental study of such a setup:

(a) If the experimental results turn out to corroborate the prediction obtained from $\Pi_B(\phi)(= \Pi_{PCD}(\phi))$ given by Equation 8.19, this would not only validate a particular quantum approach for calculating the arrival/transit time distribution with its justification provided by the Bohmian model, but it would also mean that if one analyzes our example within the Bohmian framework, one would have to infer that although the magnetic field is confined *within* the SR, it would have to affect, from $t = 0$, the time evolution of the spin state of even those individual neutrons that are initially *located outside* the SR. Such an effect within the Bohmian model is rather curious essentially because of the assumed objective (i.e., independent of measurement) localization of an individual particle, and because the neutron's spin–magnetic field interaction is mediated through the *localized* particle attribute, viz. the magnetic moment, as the coupling constant. Note that this type of question does not arise within the standard version of quantum mechanics because of the absence of the notion of objective localization of an individual particle.

(b) On the other hand, if the alternative Bohmian calculational procedure indicated in **A** of Section 8.5 yields the probability distribution $\Pi'_B(\phi)$ that is *different* from $\Pi_B(\phi)$, and if the experimental results are in conformity with $\Pi'_B(\phi)$, then, within the standard framework of quantum mechanics, this would pose a challenge to find out which particular quantum scheme, in this case, would yield results in agreement with the Bohmian prediction $\Pi'_B(\phi)$.

(c) Another possibility, apart from that in which the experimental results may *not* agree with either $\Pi_B(\phi)$ or $\Pi'_B(\phi)$ for the Gaussian wave-packet, stems from the line of study indicated in **B** of Section 8.5—i.e., if by using a non-Gaussian wave-packet, it is found that the Bohmian prediction in our example disagrees with that obtained from the probability current density based quantum approach, this would again require studying which particular quantum mechanical approach for calculating the arrival/transit time distribution, as applied to our example, would lead to results in agreement with the relevant Bohmian prediction using a non-Gaussian wave-packet.

Thus, whatever the experimental outcome, non-trivial questions would arise. Further, in view of a number of insightful variants [13–21] of the Bohmian model that have been proposed using the notion of objectively defined trajectories of individual particles, it should be instructive to analyze the example formulated in this chapter in terms of such models in order to find out the extent to which an empirical discrimination is possible. Such studies would reinforce the importance of pursuing a comprehensive experimental program in the direction that has been indicated here.

APPENDIX A

The Gaussian wave-packet considered in this chapter can be regarded as made up of plane wave components. The treatment given here pertains to an individual plane wave component.

Let the initial total wave function of a particle be represented by $\Psi_i = \psi_0 \otimes \chi$, where $\psi_0 = Ae^{ikx}$ is the spatial part which is taken to be a plane wave with wave number k, and $\chi = \frac{1}{\sqrt{2}}(\chi_{+z} + \chi_{-z})$ is the spin state polarized in the $+x$ direction where χ_{+z} and χ_{-z} are the eigenstates of σ_z. Now, we consider that the particle passes through a bounded region (called SR) that contains *constant* magnetic field directed along the $+\hat{z}$-axis.

The interaction Hamiltonian is $H_{int} = \mu\vec{\sigma}.\mathbf{B}$ where μ is the magnetic moment of the neutron, $\mathbf{B}$ is the enclosed magnetic field and $\vec{\sigma}$ is the Pauli spin vector. Then the time evolved total wave function at $t = \tau$ after the interaction of spins with the uniform magnetic field is given by

$$\Psi(\mathbf{x}, \tau) = \exp\left(-\frac{iH\tau}{\hbar}\right)\Psi(\mathbf{x}, 0)$$

$$= \frac{1}{\sqrt{2}}[\psi_+(\mathbf{x}, \tau) \otimes \chi_{+z} + \psi_-(\mathbf{x}, \tau) \otimes \chi_{-z}] \qquad (A.1)$$

where $\psi_+(\mathbf{x}, \tau)$ and $\psi_-(\mathbf{x}, \tau)$ are the two components of the spinor $\psi = \left(\begin{smallmatrix}\psi_+ \\ \psi_-\end{smallmatrix}\right)$ which satisfies the Pauli equation. We take the enclosed constant magnetic field as $\mathbf{B} = B\hat{z}$. The time evolutions of the position dependent amplitudes (ψ_+ and ψ_-) of the spin components of the wave function are thus given by

$$i\hbar\frac{\partial\psi_+}{\partial t} = -\frac{\hbar^2}{2m}\nabla^2\psi_+ + \mu B\psi_+ \qquad (A.2)$$

$$i\hbar\frac{\partial\psi_-}{\partial t} = -\frac{\hbar^2}{2m}\nabla^2\psi_- - \mu B\psi_-. \qquad (A.3)$$

Equations A.2 and A.3 imply that when a neutron having spin up interacts with the constant magnetic field within the SR, the associated spatial wave function (ψ_+) evolves under a *potential barrier* that has been generated due to the spin–magnetic field interaction, while for a neutron having spin down, the associated spatial wave function (ψ_-) evolves under a *potential well*.

Here we will specifically consider the situation where $E > |\mu B|$. This is because, even if one uses low-energy or ultra-cold neutrons having kinetic energy of the order of 5×10^{-7} eV, if the potential energy term ($|\mu B|$) has to exceed the kinetic energy term, one will require the magnitude of the magnetic field to be exceedingly high, to be of the order of 10 T. Magnetic fields of such high intensity are difficult to produce in the usual laboratory conditions, and therefore for all practical purposes, it suffices to treat the case where $E > |\mu B|$.

Now, since ψ_- evolves under a potential well confined between $x = 0$ and $x = d$ the reflected and transmitted parts are respectively given by

$$\psi_R^- = Ae^{-ikx}\frac{(k^2 - k_1^2)(1 - e^{2ik_1d})}{(k + k_1)^2 - (k - k_1)^2e^{2ik_1d}} \qquad (A.4)$$

$$\psi_T^- = Ae^{ikx}\frac{4kk_1e^{-ikd}e^{ik_1d}}{(k + k_1)^2 - (k - k_1)^2e^{2ik_1d}} \qquad (A.5)$$

where $k = \frac{\sqrt{2mE}}{\hbar}$, $k_1 = \frac{\sqrt{2m(E-\mu B)}}{\hbar}$, and d is the width of the SR arrangement which contains the uniform magnetic field.

On the other hand, for ψ_+ which evolves under a potential barrier, the expressions for the transmitted and the reflected parts are written by replacing all the k_1's in Equations A.4 and A.5 by k_2 where $k_2 = \frac{\sqrt{2m(E+\mu B)}}{\hbar}$

$$\psi_R^+ = Ae^{-ikx}\frac{(k^2 - k_2^2)(1 - e^{2ik_2d})}{(k + k_2)^2 - (k - k_2)^2 e^{2ik_2d}} \tag{A.6}$$

$$\psi_T^+ = Ae^{ikx}\frac{4kk_2 e^{-ikd}e^{ik_2d}}{(k + k_2)^2 - (k - k_2)^2 e^{2ik_2d}}. \tag{A.7}$$

Now, we note that the solutions in these cases, given our initial state, consist of a reflected part traveling in the $-\hat{x}$-direction and a transmitted part of the wave function, that travels in the $+\hat{x}$-direction. However, the reflected part of the wave function exists *only* to the left of the SR, and the transmitted part exists *only* to the right of the SR. Our objective here is to calculate the observable distribution of spins of the particles emerging from the SR. For this, we need to look at *only* the transmitted part of the wave function. Therefore, the final time evolved transmitted state that is relevant to our purpose is of the following form that represents an entangled state involving the spin and the spatial degrees of freedom

$$\Psi_f = \frac{N}{\sqrt{2}}(\psi_T^+ \chi_{+z} + \psi_T^- \chi_{-z}) \tag{A.8}$$

where N is the normalized constant that can be written as $N = \int_v (|\psi_T^+|^2 + |\psi_T^-|^2) dv$.

It is then seen that, in the regime $E > |\mu B|$, using Equations A.5 and A.7, one can rewrite Equation A.8 in the following form

$$\Psi_f = \frac{Ae^{ikx}}{\sqrt{2}}(Ce^{i\phi_2}\chi_{+z} + De^{i\phi_1}\chi_{-z}) \equiv \psi_0\chi(\phi) \tag{A.9}$$

whereby Equation A.9 is the form of the Larmor precession relation which is valid under the condition $E > |\mu B|$ that has been calculated in terms of the explicit time evolved solutions of Equations A.2 and A.3.

Here

$$C = \sqrt{\text{Re}(\psi_T^-)^2 + \text{Im}(\psi_T^-)^2} \tag{A.10}$$

$$D = \sqrt{\text{Re}(\psi_T^+)^2 + \text{Im}(\psi_T^+)^2} \tag{A.11}$$

$$\phi_1 = \tan^{-1}\frac{\text{Im}(\psi_T^-)}{\text{Re}(\psi_T^-)} \tag{A.12}$$

and

$$\phi_2 = \tan^{-1}\frac{\text{Im}(\psi_T^+)}{\text{Re}(\psi_T^+)}. \tag{A.13}$$

Note that, by using Equation A.5, one can find that

$$\text{Re}(\psi_T^-) = \frac{8kk_1(k^2 + k_1^2)\sin(kd)\sin(k_1d) + 16k^2k_1^2\cos(kd)\cos(k_1d)}{(k + k_1)^4 + (k - k_1)^4 - 2(k + k_1)^2(k - k_1)^2\cos(2k_1d)} \tag{A.14}$$

$$\text{Im}(\psi_T^-) = \frac{8kk_1(k^2 + k_1^2)\cos(kd)\sin(k_1d) - 16k^2k_1^2\sin(kd)\cos(k_1d)}{(k + k_1)^4 + (k - k_1)^4 - 2(k + k_1)^2(k - k_1)^2\cos(2k_1d)}. \tag{A.15}$$

Similarly, by using Equation A.7, one can obtain the expressions for $\text{Re}(\psi_T^+)$ and $\text{Im}(\psi_T^+)$ given by

$$\text{Re}(\psi_T^+) = \frac{8kk_2(k^2 + k_2^2)\sin(ka)\sin(k_2d) + 16k^2k_2^2\cos(kd)\cos(k_2d)}{(k + k_2)^4 + (k - k_2)^4 - 2(k + k_2)^2(k - k_2)^2\cos(2k_2d)} \quad (A.16)$$

$$\text{Im}(\psi_T^+) = \frac{8kk_2(k^2 + k_2^2)\cos(ka)\sin(k_2d) - 16k^2k_2^2\sin(kd)\cos(k_2d)}{(k + k_2)^4 + (k - k_2)^4 - 2(k + k_2)^2(k - k_2)^2\cos(2k_2d)}. \quad (A.17)$$

Now, let us examine in what limit one can obtain the standard expression for Larmor precession.

Considering the more stringent limiting condition $E \gg |\mu B|$, it can be seen that when the kinetic energy term of the Hamiltonian is much larger than the potential energy term, one can assume that the time evolution of the wave function occurs effectively due to a *very shallow well* and a *very low barrier*. This situation would correspond to the entire wave being transmitted while picking up just a phase.

From the expressions for k, k_1, k_2 given earlier, in the limit $E \gg |\mu B|$, we find that $k \approx k_1 \approx k_2$. Then, in order to get the standard expression for Larmor precession, one can first set $k = k_1 = k_2$ in Equations A.14, A.15, A.16, and in A.17, *except* when they appear inside the sine or cosine functions, since the latter terms are much more sensitive to the differences in the values of k, k_1, k_2. Equations A.14 and A.15 would then simplify to

$$\text{Re}(\psi_T^-) = \sin(kd)\sin(k_1d) + \cos(kd)\cos(k_1d) \quad (A.18)$$

$$\text{Im}(\psi_T^-) = \sin(k_1d)\cos(kd) - \cos(k_1d)\sin(kd). \quad (A.19)$$

Using the above expressions in Equations A.10 and A.12, we find that $C = 1$ and $\phi_1 = (k_1 - k)d$. Similarly, invoking Equations A.16 and A.17, and using Equations A.11 and A.13, we get $D = 1$ and $\phi_2 = (k_2 - k)d$. Therefore, Equation A.9 reduces to the form

$$\Psi_f = \frac{Ae^{ikx}}{\sqrt{2}}(e^{i(k_2 - k)d}\chi_{+z} + e^{i(k_1 - k)d}\chi_{-z}). \quad (A.20)$$

Now, remembering that

$$k = \frac{\sqrt{2mE}}{\hbar}, \quad k_1 = \frac{\sqrt{2m(E - \mu B)}}{\hbar} \quad \text{and} \quad k_2 = \frac{\sqrt{2m(E + \mu B)}}{\hbar},$$

we can binomially expand k_1 and k_2 around k, and keep terms up to the order of $\mu B/E$, since we have already stipulated the condition $k \approx k_1 \approx k_2$. Then, $(k_1 - k)d = -k\frac{\mu B}{2E}d$, and $v = \hbar k/m$, we can write $(k_1 - k)d = -\frac{\mu B}{\hbar}\frac{d}{v} = \omega t$.

Similarly, $(k_2 - k)d = \frac{\mu B}{\hbar}\frac{d}{v} = -\omega t$. Therefore, we can write Equation A.20 as

$$\Psi_f = \frac{Ae^{ikx}}{\sqrt{2}}\left(e^{-i\omega t}\chi_{+z} + e^{i\omega t}\chi_{-z}\right)$$

$$= \psi_0\frac{e^{-i\phi/2}}{\sqrt{2}}\left(\chi_{+z} + e^{i\phi}\chi_{-z}\right) \equiv e^{-i\phi/2}\psi_0\chi(\phi) \quad (A.21)$$

where $\phi = 2\omega t$. The spin part of this equation is Equation 8.17 given in the text, written by considering the time evolutions of the spatial and the spin parts to be *independent* of each other, with the spatial wave function evolving *freely* within the SR.

It is important to stress here that the rigorous treatment given above of the time evolution of the quantum state of a neutral spin-$\frac{1}{2}$ particle in a constant magnetic field reveals that the standard expression for Larmor precession given by Equation 8.17 in the text is essentially justified in the *limit* where the kinetic energy term is *much higher* than the potential energy term. Thus, subject to this condition being satisfied, the assumption of mutual independence of the evolutions of the spatial and the spin parts of the total wave function that has been used in the text is justified for the setup considered in this chapter.

ACKNOWLEDGMENTS

We thank Arka Banerjee for helpful interactions. AKP acknowledges useful discussions during his visits to the Perimeter Institute, Canada, Centre for Quantum Technologies, National University of Singapore, and Atominstitut, Vienna. DH thanks DST, Government of India, for the relevant project support. DH also thanks the Centre for Science and Consciousness, Kolkata. AKP acknowledges the Research Associateship of Bose Institute, Kolkata.

BIBLIOGRAPHY

1. L. de Broglie, *Electrons et Photons* (Gauthier-Villars, Paris, 1928), p. 105.
2. D. Bohm, *Phys. Rev.*, **85**, 166 (1952).
3. D. Bohm, *Phys. Rev.*, **85**, 180 (1952).
4. D. Bohm, *Phys. Rev.*, **89**, 458 (1953).
5. W.Pauli, in *Electrons et Photons* (Gauthier-Villars, Paris, 1928), pp. 280–282.
6. J. von Neumann, *Mathematische Grundlagen der Quantenmechanik* (Springer, Berlin, 1932); English Translation: *Mathematical Foundations of Quantum Mechanics* (Princeton University Press, Princeton, NJ, 1955), pp. 305–325.
7. D. Bohm and B.J. Hiley, *The Undivided Universe* (Routledge, London, 1993).
8. P.R. Holland, *The Quantum Theory of Motion* (Cambridge University Press, Cambridge, 1993).
9. K. Berndl, M. Daumer, D. Dürr, S. Goldstein and N. Zanghì, *Nuovo Cim.*, **B110**, 737 (1995).
10. J. T. Cushing, A. Fine, and S. Goldstein, *Bohmian Mechanics and Quantum Theory: An Appraisal* (Kluwer Academic Publishers, Dordrecht, 1996).
11. P. Holland, *Phys. Rev. A*, **60**, 4326 (1999).
12. P. Holland, *Ann. Phys. (Leipzig)*, **12**, 446(2003).
13. S.T. Epstein, *Phys. Rev.*, **89**, 319(1952).
14. S.T. Epstein, *Phys. Rev.*, **91**, 985(1953).
15. S.M. Roy and V. Singh, *Mod. Phys. Lett. A*, **10**, 709 (1995).
16. S.M. Roy and V. Singh, *Phys. Lett. A*, **255**, 201 (1999).
17. S.M. Roy and V. Singh, *Pramana—J. Phys.*, **59**, 337 (2002).

18. P. Holland, *Found. Phys.*, **28**, 881 (1998).
19. P. Holland, *Ann. Phys. (N.Y.)*, **315**, 503 (2005).
20. P. Holland, *Proc. Roy. Soc. A*, **461**, 3659 (2005).
21. E. Deotto and G.C. Ghirardi, *Found. Phys.*, **28**, 1 (1998).
22. P. Holland, *Ann. Phys.*, **315**, 505 (2005).
23. W. Heisenberg, in *Niels Bohr and the Development of Physics*, edited by W. Pauli (Pergamon Press, Oxford, 1955), pp. 17–19.
24. J.T. Cushing, *Quantum Mechanics* (The University of Chicago Press, Chicago, 1994), Ch. 4.
25. J.S. Bell, *Speakable and Unspeakable in Quantum Mechanics* (Cambridge University Press, Cambridge, 1987), pp. 111–116.
26. W. Pauli, in *Encyclopedia of Physics*, edited by S. Flugge (Springer, Berlin, 1958), vol. V/1, p. 60.
27. G.R. Allcock, *Ann. Phys. (N.Y.)*, **53**, 253 (1969).
28. G.R. Allcock, *Ann. Phys. (N.Y.)*, **53**, 286 (1969).
29. G.R. Allcock, *Ann. Phys. (N.Y.)*, **53**, 311 (1969).
30. N. Grot, C. Rovello, and R.S. Tate, *Phys. Rev. A*, **54**, 4676 (1996).
31. V. Delgado and J.G. Muga, *Phys. Rev. A*, **56**, 3425 (1997).
32. J.G. Muga and C.R. Leavens, *Phys. Rep.*, **338**, 353 (2000).
33. J.G. Muga, R. Sala Mayato, and I.L. Egusquiza (eds.), *Time in Quantum Mechanics* (Springer-Verlag, Berlin, 2008).
34. S. Brouard, R. Sala, and J.G. Muga, *Phys. Rev. A*, **49**, 4312 (1994).
35. N. Yamada and S. Takagi, *Prog. Theor. Phys.*, **85**, 599 (1991).
36. C.R. Leavens, *Phys. Lett. A*, **178**, 27 (1993).
37. J.G. Muga, S. Brouard, and D. Macias, *Ann. Phys.*, **240**, 351 (1995).
38. W.R. McKinnon and C.R. Leavens, *Phys. Rev. A*, **51**, 2748 (1995).
39. V. Delgado, *Phys. Rev. A*, **59**, 1010 (1999).
40. A.D. Baute, R.S. Mayato, J.P. Palao, J.G. Muga, and I.L. Egusquiza, *Phys. Rev. A*, **61**, 022118 (2000).
41. A.D. Baute, I.L. Egusquiza, J.G. Muga, and R. Sala-Mayato, *Phys. Rev. A*, **61**, 052111 (2000).
42. Y. Aharonov, J. Oppenheim, S. Popescu, B. Reznik, and W.G. Unruh, *Phys. Rev. A*, **57**, 4130 (1998).
43. J.G. Muga, S. Brouard, and D. Macias, *Ann. Phys.*, **240**, 351 (1995).
44. J.G. Muga, R. Sala Mayato, and J.P. Palao, *Superlattices and Microstructures*, **23**, 833 (1998).
45. V. Delgado, *Phys. Rev. A*, **59**, 1010 (1999).
46. J. Finkelstein, *Phys. Rev. A*, **59**, 3218 (1999).
47. Md.M. Ali, A.S. Majumdar, D. Home, and S. Sengupta, *Phys. Rev. A*, **68**, 042105 (2003).
48. S.V. Mousavi and M. Golshani, *J. Phys. A*, **41**, 375304 (2008).
49. A.K. Pan, Md.M. Ali, and D. Home, *Phys. Lett. A*, **352**, 296 (2006).
50. J.J. Halliwell and E. Zafiris, *Phys. Rev. D*, **57**, 3351 (1998).
51. J.J. Halliwell and E. Zafiris, *Phys. Rev. A*, **79**, 062101 (2009).
52. C.R. Leavens, *Phys. Rev. A*, **58**, 840 (1998).
53. Md.M. Ali, A.S. Majumdar, D. Home, and A.K. Pan, *Class. Quant. Grav.*, **23**, 6493 (2006).
54. J.M. Damborenea, I.L. Egusquiza, G.C. Hegerfeldt, and J.G. Muga, *Phys. Rev. A*, **66**, 052104 (2002).
55. G.C. Hagerfeldt, D. Seidel, and J.G. Muga, *Phys. Rev. A*, **68**, 022111 (2003).
56. S. Brouard, R. Sala Mayato, and J.G. Muga, *Phys. Rev. A*, **49**, 4312 (1994).

57. N. Yamada and S. Takagi, *Prog. Theor. Phys.*, **86**, 599 (1991).
58. D. Sokolovski and L. M. Baskin, *Phys. Rev. A*, **36**, 4604 (1987).
59. R. Giannitrapani, *Int. J. Theor. Phys.*, **36**, 1575 (1997).
60. Y. Aharonov and D. Bohm, *Phys. Rev.*, **122**, 1649 (1961).
61. J. Kijowski, *Rep. Math. Phys.*, **6**, 361 (1974).
62. H. Salecker and E.P. Wigner, *Phys. Rev.*, **109**, 571 (1958).
63. A. Peres, *Am. J. Phys.*, **48**, 552 (1980).
64. A. Azebel, *Phys. Rev. Lett.*, **68**, 98 (1992).
65. P.C.W. Davies, *Class. Quant. Grav.*, **21**, 2761 (2004).
66. A.I. Baz, *Sov. J. Nucl. Phys.*, **5**, 161 (1967).
67. M. Buttiker, *Phys. Rev. B*, **27**, 6178 (1983).
68. M. Golshani and O. Akhavan, *J. Phys. A: Math. Gen.*, **34**, 5259 (2001).
69. P. Ghose, *Pramana—J. Phys.*, **59**, 2 (2002).
70. G. Brida, E. Cagliero, G. Falzetta, M. Genovese, M. Gramegna, and C. Novero, *J. Phys. B. At. Mol. Opt. Phys.*, **35**, 4751 (2002).
71. W. Struyve, W. De Baere, J. De Neve, and S. De Weirdt, *J. Phys. A: Math. Gen.*, **36**, 1525 (2003).
72. P.R. Holland, *The Quantum Theory of Motion* (Cambridge University Press, Cambridge, 1993), pp. 161–162.
73. C. Leavens, *Phys. Rev. A*, **58**, 840 (1998).
74. S. Kreidl, G. Gruebl, and H.G. Embacher, *J. Phys. A: Math. Gen.*, **36**, 8851 (2003).

9 Semiclassical Implementation of Bohmian Dynamics

Vitaly Rassolov and Sophya Garashchuk

CONTENTS

9.1 INTRODUCTION

In this chapter we justify and explain the use of a semiclassical approach in quantum trajectory dynamics, then show how approximations can be balanced to make the dynamics stable in time. For simplicity, derivations of Sections 9.1 and 9.2 are given for one-spatial dimension x for a particle of mass m. Arguments of functions are omitted where unambiguous. Differentiation with respect to x is denoted as ∇. Differentiation with respect to a variable other than x is indicated with a subscript. The multidimensional generalization to an arbitrary coordinate system is given in Ref. [1].

In the Bohmian formulation [2] the usual time-dependent Schrödinger equation (TDSE),

$$\left(-\frac{\hbar^2}{2m}\nabla^2 + V\right)\psi(x,t) = \imath\hbar\frac{\partial\psi(x,t)}{\partial t}, \tag{9.1}$$

is transformed by representing the complex time-dependent wavefunction in polar form,

$$\psi(x,t) = A(x,t)\exp\left(\frac{\imath}{\hbar}S(x,t)\right), \tag{9.2}$$

where $A(x,t)$ and $S(x,t)$ are real functions. Identifying the gradient of the phase with the trajectory momentum,

$$p(x,t) = \nabla S(x,t), \tag{9.3}$$

and switching to the Lagrangian frame of reference,

$$\frac{d}{dt} = \frac{\partial}{\partial t} + \frac{p}{m}\nabla,$$ (9.4)

the TDSE (Equation 9.1) leads to the classical-like equations for a trajectory described by position x_t, momentum p_t and action function S_t,

$$\frac{dx_t}{dt} = \frac{p_t}{m}$$ (9.5)

$$\frac{dp_t}{dt} = -\nabla(V + Q)\big|_{x=x_t}$$ (9.6)

$$\frac{dS_t}{dt} = \frac{p_t^2}{2m} - (V + Q)\big|_{x=x_t}.$$ (9.7)

The term Q denotes the *quantum potential*,

$$Q = -\frac{\hbar^2}{2m}\frac{\nabla^2 A(x,t)}{A(x,t)},$$ (9.8)

entering Equations 9.6 and 9.7 on a par with the external "classical" potential V. Quantum effects are incorporated into the trajectory dynamics through the nonlocal quantum force, which depends on the wavefunction amplitude and its derivatives up to third order. With the exception of eigenstates, Q is singular at the wavefunction nodes, $A(x,t) = 0$, resulting in instability of the quantum trajectories describing quantum-mechanical (QM) interference. Thus, we pursue the idea of using the Approximate Quantum Potential (AQP), which captures the leading QM effects yet is cheap to compute. Calculation of the exact Q for general systems is expensive and inaccurate due to the inherent instability of exact quantum trajectories.

Evolution of the wavefunction density, $A^2(x,t)$, is defined by the continuity equation, a counterpart of Equation 9.7, and conserves the probability to find a particle in the volume element dx_t associated with each quantum trajectory. In other words, the trajectory "weight" $w(x_t)$ is constant in time [3],

$$w(x_t) = A^2(x_t)dx_t, \quad \frac{dw(x_t)}{dt} = 0.$$ (9.9)

In the AQP approach we use Equation 9.9 and do not explicitly compute $A(x,t)$. Once a wavefunction is represented in terms of trajectories, the expectation values of x-dependent operators can be readily computed,

$$\langle \hat{o}(x) \rangle_t = \int o(x)A^2(x,t)dx = \sum_i o(x_t^{(i)})w_i,$$ (9.10)

where the index i enumerates the trajectories. The wavefunction normalization is explicitly conserved. For time-independent V the total wavefunction energy is constant, but the energy of individual quantum trajectories changes in time due to the influence of Q.

In addition to x_t and p_t fully defining the quantum trajectory, we will also consider the *nonclassical* momentum $r(x,t)$,

$$r(x,t) = \frac{\nabla A(x,t)}{A(x,t)}.$$ (9.11)

Formally, it is complementary to the *classical* momentum $p(x,t)$, since both arise from the action of the QM momentum operator on the wavefunction in the polar form (Equation 9.2)

$$\hat{p}\psi = \left(-\imath\hbar\frac{\nabla A(x,t)}{A(x,t)} + \nabla S(x,t)\right)\psi = (-\imath\hbar r(x,t) + p(x,t))\,\psi.$$ (9.12)

Q expressed in terms of r becomes

$$Q = -\frac{\hbar^2}{2m}\left(r^2(x,t) + \nabla r(x,t)\right).$$ (9.13)

The equations of motion of the classical and nonclassical momenta computed along a trajectory have the same structure of the differential operator on the right-hand side

$$m\nabla V(x_t) + m\frac{dp_t}{dt} = \hbar^2\left(r_t + \frac{\nabla}{2}\right)\nabla r_t$$

$$-m\frac{dr_t}{dt} = \left(r_t + \frac{\nabla}{2}\right)\nabla p_t.$$ (9.14)

In order to solve TDSE using trajectories, or to visualize the wavefunction dynamics, an ensemble of trajectories is initialized at $t = 0$. For each initial position $x = x_0$, the trajectory weight, $w = A^2(x,0)dx_0$, and the classical momentum, $p_0 = \nabla S(x,0)$, are defined. The trajectories evolve according to Newton's Equations 9.5 and 9.6. Computation of the exact Q and its gradient requires derivatives of $A(x,t)$ through third order. This step is approximated in our semiclassical implementation.

In order to examine the semiclassical regime of the Bohmian trajectory method we have to define a classical limit of the quantum equations of motion. This is, in general, a complicated question [4]. In this chapter we take a practical approach and define the classical limit in terms of observables. For a particle of mass m, we define the classical limit of a quantum state $\psi(x)$ as $m \to \infty$ provided the energy, $\langle\psi|\hat{H}|\psi\rangle$, is constant. The averaging $\langle\ldots\rangle$ can be performed over any subset of coordinates (such as only for translational motion). This requires that the wavefunction density, $\psi^*\psi$, and the kinetic energy density, $-\hbar^2\nabla^2\psi(2m\psi)^{-1}$, remain unchanged as $m \to \infty$. The reactivity of chemical systems is typically governed by the distribution of energy in various degrees of freedom. It is natural to define the classical limit preserving this distribution.

The term "semiclassical" is widespread in chemistry and physics and generally means "quantum that is close to classical." The precise definition depends on the context. For a numerical procedure not based on the series expansion of the TDSE, it is natural to define the semiclassical method as an approximate method describing some quantum features, which becomes exact in the classical limit [5]. This is a stringent requirement, which in practice means that approximations must only be made for the quantities that are negligible in the $m \rightarrow \infty$ or $\hbar \rightarrow 0$ limits, and that the errors of the method can be controlled, at least in principle. In particular, removing singularities in the quantum potential by constraining the density, phase, or momentum would generally violate the semiclassical condition. In contrast, the AQP method constrains the functional form of the nonclassical momentum $r(x, t)$ appearing in Equation 9.12 with the $\hbar$ prefactor, while the wavefunction density itself is unconstrained. Thus, the AQP approach can be systematically improved if the linearization of the nonclassical momentum is accomplished over subspaces [6] or if $r(x, t)$ is represented in terms of a complete basis.

9.2 GLOBALLY DEFINED APPROXIMATE QUANTUM POTENTIAL (AQP)

The point of developing the AQP method is to have an inexpensive, robust semiclassical approach that can be applied to high-dimensional systems. "Inexpensive" means that the scaling with the system size is polynomial, ideally linear, such that the computation of the quantum force is a small addition to the classical trajectory propagation. The polynomial scaling allows the method to be applied to hundreds of degrees of freedom. "Robust" implies numerical stability. In particular, the numerical propagation should be insensitive to trajectory crossings—exact quantum trajectories do not cross (!)—and should switch to classical propagation if the quantum force becomes inaccurate. "Semiclassical" means that the method should describe the leading QM effects. However, it should not aim for exact QM dynamics—a necessary trade-off for a better than exponential scaling with system dimensionality. In addition, it is desirable for a method to have an accuracy criterion and a systematic procedure for achieving the exact QM description.

Analytical representation of the (approximate) quantum potential is essential for efficient and accurate computation of the quantum force. One physically appealing way of constructing the AQP is to fit the nonclassical component, $r(x, t)$, given by Equation 9.11. The representation of this single object within a basis $\vec{f}(x)$,

$$r(x, t) \approx \tilde{r}(x, t) = \vec{c}(t) \cdot \vec{f}(x), \tag{9.15}$$

can be used to obtain Q and its gradient by using $\tilde{r}$ instead of $r(x, t)$ in Equation 9.13. The vector $\vec{c}$ contains the time-dependent basis expansion coefficients found from the minimization of a functional

$$I = \int_{-\infty}^{\infty} (r - \tilde{r})^2 A^2(x, t) dx \tag{9.16}$$

with respect to the elements of $\vec{c}$. Integrating by parts and replacing integrals with sums over the trajectories, the solution to this linear minimization problem, $\nabla_c I = 0$, is

$$\vec{c} = -\frac{1}{2} \mathbf{S}^{-1} \vec{b}. \tag{9.17}$$

The matrix $\mathbf{S}$ contains the time-dependent basis function overlaps,

$$s_{jk} = \langle f_j | f_k \rangle = \sum_i w_i f_j f_k \Big|_{x=x_t^{(i)}} \tag{9.18}$$

and the vector $\vec{b}$ consists of averages of ∇f,

$$b_j = \langle \nabla f_j \rangle = \sum_i w_i \frac{df_j}{dx} \Big|_{x=x_t^{(i)}}. \tag{9.19}$$

As shown in Ref. [7], AQP evaluated for expansion coefficients given by Equation 9.17 conserves the total energy of a system.

The smallest physically meaningful basis $\vec{f}$ consists of just two functions $\vec{f} = (1, x)$, in which case, the approximation is particularly transparent,

$$\tilde{r} = -\frac{x - \langle x \rangle}{2(\langle x^2 \rangle - \langle x \rangle^2)}. \tag{9.20}$$

For normalized wavefunctions the nonclassical momentum is approximated with a linear function centered at the average position of the wavepacket with the slope inversely proportional to the variance, $\sigma = \langle x^2 \rangle - \langle x \rangle^2$. This functional form of $\tilde{r}$ determines the quadratic AQP and, consequently, the linear quantum force—thus the name Linearized Quantum Force (LQF). The force is inversely proportional to σ^2 and, therefore, quickly vanishes for spreading wavepackets. From the physical point of view, the linear nonclassical momentum exactly corresponds to a Gaussian wavepacket evolving in time in a locally quadratic potential. For general potentials, it can describe the dominant quantum effects in semiclassical systems, such as wavepacket bifurcation, wavepacket spread in energy, and moderate tunneling. Zero-point energy (ZPE) effects are reproduced for short times—a few vibrational periods and depending on the anharmonicity of the system, which is adequate for direct gas phase reactions.

For instance consider photodissociation of ICN described within the two-dimensional Beswick–Jortner model [8] (see Ref. [9] for details). A wavepacket representing the ICN molecule is promoted by a laser from the ground to the excited electronic state, where the molecule dissociates into I and CN. The initial wavefunction

$$\psi(x, y, 0) = \sqrt{\frac{2}{\pi}} (\alpha_{11}\alpha_{22} - \alpha_{12}^2)^{1/4} e^{-\alpha_{11}(y-\bar{y})^2 - \alpha_{22}(x-\bar{x})^2 + 2\alpha_{12}(y-\bar{y})(x-\bar{x})} \tag{9.21}$$

is defined as the lowest eigenstate of the ground electronic surface composed of the harmonic potentials in CN and CI stretches. Thus, $\psi(x, y, 0)$ is a correlated Gaussian wavepacket located on the repulsive wall of the excited surface. The photodissociation spectrum is computed from the Fourier transform of the wavepacket autocorrelation function $C(t)$,

$$I(E) = \int_{-\infty}^{\infty} C(t) e^{iEt} dt, \quad C(t) = \langle \psi(0) | \psi(t) \rangle.$$

For real $\psi(x, y, 0)$, $C(t)$ can be written as

$$C(t) = \langle\psi(-t/2)|\psi(t/2)\rangle = \langle\psi^*(t/2)|\psi(t/2)\rangle = \sum_i w_i \exp(2\imath S_{t/2}^{(i)}). \qquad (9.22)$$

Note, that in Equation 9.22 we need only half the propagation time, which improves accuracy.

The physical value of the repulsion parameter of the excited potential surface yields a rather simple dissociation dynamics; $C(t)$ decays on the time-scale of about one and a half oscillations of the CN stretch, and the LQF spectrum is in excellent agreement with the quantum result [10]. The repulsion parameter, three times larger than its physical value, leads to predissociation processes providing a more challenging test (system II in Ref. [9]). As detailed in Ref. [7], in two dimensions the nonclassical momentum is a vector, $\vec{r} = (r^{(x)}, r^{(y)})$. In the LQF treatment, both components of $\vec{r}$ are approximated in a linear basis $\vec{f} = (1, x, y)$ with the component-specific expansion coefficient vectors $\tilde{r}^{(x,y)} = \vec{f} \cdot \vec{c}^{(x,y)}$. The minimization procedure is analogous to the one-dimensional case,

$$\nabla_x I = \nabla_y I = 0, \quad I = \langle(r^{(x)} - \tilde{r}^{(x)})^2\rangle + \langle(r^{(y)} - \tilde{r}^{(y)})^2\rangle. \qquad (9.23)$$

The solution is Equation 9.17 with the vectors $\vec{c}$ and $\vec{b}$ replaced with matrices $\mathbf{C} = (\vec{c}^{(x)}, \vec{c}^{(y)})$, and $\mathbf{B} = (\nabla_x \vec{f}, \nabla_y \vec{f})$, respectively.

The wavefunction was represented by an ensemble of 167 trajectories propagated for three vibrational periods of the CN stretch. The LQF parameters describing wavepacket localization are shown in Figure 9.1a: $c_2^{(x)}$ represents the changing width of the wavepacket in the CN mode; $c_3^{(y)}$ describes the wavepacket spreading in the dissociation mode; $c_3^{(x)} = c_2^{(y)}$ reflects the correlation between the degrees of freedom. Figure 9.1b compares the semiclassical spectrum computed using Equation 9.22 with the result of the exact QM wavepacket evolution. The accuracy of the LQF correlation function deteriorates at long times, which is manifested in small errors of its

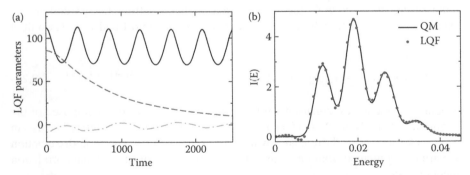

FIGURE 9.1 ICN dissociation. (a) The LQF wavepacket width parameters: $c_2^{(x)}$ (solid line), $c_3^{(y)}$ (dash), and $c_3^{(x)} = c_2^{(y)}$ (dot-dash). (b) The Fourier transform of the correlation function. The LQF result is computed using Equation 9.22.

Fourier transform for low energies. Nevertheless, the overall agreement of the spectra is quite good, demonstrating that LQF is sufficiently accurate for systems with fast dynamics.

There are several ways to improve the description of non-Gaussian wavefunctions evolving in general potentials. The LQF parameters can be optimized on subspaces representing, for example, the reaction channels [6]. More functions can be added to the basis $\vec{f}$. For example, the Chebyshev basis of up to six polynomials has been used in Ref. [11]. Small bases can be tailored to the system. For example, a two-function basis consisting of a constant and an exponent gives an exact description of r for the eigenstate of the Morse oscillator as used in Ref. [12] to describe the ZPE of H_2 in the reaction channel of the $O+H_2$ reaction. A mixed coordinate-space/polar wavefunction representation can be employed to describe hard QM effects, such as wavefunctions with nodes and nonadiabatic dynamics [13–15]. For condensed phase dynamics we developed a stabilized version of the LQF as it is cheap in many dimensions [16].

9.3 AQP DYNAMICS WITH BALANCED APPROXIMATION ERRORS

Most often a reaction occurring in a condensed phase is represented by a reactive coordinate coupled to a molecular environment or a "bath." The main QM effects that can be significant in such a system are (i) the motion along the reactive coordinate possibly influenced by QM tunneling and (ii) the ZPE—or, more generally, localization energy—in the reactive and bath degrees of freedom and energy flow among them. The energy changes in the reactive mode will obviously influence the probability of the reaction.

Formally, the ZPE is the energy of the lowest eigenstate. It is a sum of the kinetic and potential energy contributions due to localization, or "finite size," of the eigenstate. In the Bohmian formulation, the kinetic energy contribution to the ZPE is given by the expectation value of the quantum potential,

$$\langle Q \rangle = -\frac{\hbar^2}{2m}\langle A|\nabla^2 A\rangle = \frac{\hbar^2}{2m}\langle \nabla A|\nabla A\rangle = \frac{\hbar^2}{2m}\langle r^2 \rangle. \tag{9.24}$$

It can be called the "quantum" energy in contrast to the "classical" energy, $\langle p^2/2m + V\rangle$. The concept of "quantum" energy can be applied to any localized function, not just to the ground state. An efficient—scalable to high dimensionality—and stable description of the quantum energy, $\langle Q \rangle$, is the goal of this section.

The LQF approximation gives the exact quantum energy for all eigenstates and coherent states of the harmonic oscillator. In anharmonic systems, the LQF describes $\langle Q \rangle$ only on a short time-scale (depending on the anharmonicity). For an eigenfunction in the Bohmian formalism, the quantum force exactly cancels the classical one resulting in stationary trajectories. In the LQF, exact cancellation generally does not happen; a net force acting on the trajectories, representing the wavefunction tails, might be large causing these "fringe" trajectories to start moving. Sooner or later, the movement will affect the moments of the trajectory distribution resulting in a decoherence of the LQF trajectories and in a loss of the ZPE (or quantum energy) description. Since the total energy is conserved, the quantum energy will be transferred into the classical energy of the trajectory motion. Below, the LQF ideas are extended to prevent this "loss" of quantum energy for small anharmonicities, or more precisely for small nonlinearity of the classical and nonclassical momenta. Thus, a cheap and stable ZPE description over many oscillation periods is provided.

The derivation is given in atomic units ($\hbar = 1$) for a system described in N_{dim} Cartesian coordinates $\vec{x} = (x, y, \ldots)$ in vector notation. The classical and nonclassical momenta are vectors

$$\vec{p} = \nabla S, \quad \vec{r} = A^{-1}\nabla A. \tag{9.25}$$

In LQF, a linear approximation to each component of $A^{-1}\nabla A$ is made. The trajectory momenta are known but never used in the fitting procedure. The nonlinearity of $\vec{p}$, however, quickly translates into nonlinearity of $\vec{r}$. In this section, $\vec{r}$ and $\vec{p}$ are treated on an equal-footing. The deviations of both quantities from linearity are used to limit the accumulation of propagation errors with time.

In the multidimensional case, the time-evolution of $\vec{r}$ and $\vec{p}$ along a trajectory, given by Equation 9.14 in one dimension, is:

$$m\nabla V + m\frac{d\vec{p}}{dt} = (\vec{r} \cdot \nabla)\vec{r} + \frac{(\nabla \cdot \nabla)\vec{r}}{2}$$
$$-m\frac{d\vec{r}}{dt} = (\vec{r} \cdot \nabla)\vec{p} + \frac{(\nabla \cdot \nabla)\vec{p}}{2}. \tag{9.26}$$

For practical reasons the spatial derivatives of $\vec{r}$ and $\vec{p}$ in Equations 9.26 need to be determined from the global approximations to these quantities. The trajectory-specific quantities $\vec{r}$ and $\vec{p}$ are found by solving Equations 9.26. In general, the relations in Equation 9.25 are *not* fulfilled once approximations are made to the right-hand sides of Equations 9.26. As in LQF, $\vec{p}$ and $\vec{r}$ are expanded in a basis set of functions for the purpose of derivative evaluations. The expansion coefficients are determined from the minimization of the error functional using the total energy conservation as a constraint.

The total energy can be defined without spatial derivatives as

$$E = \frac{\langle \vec{p} \cdot \vec{p} \rangle}{2m} + \langle V \rangle + \frac{\langle \vec{r} \cdot \vec{r} \rangle}{2m}. \tag{9.27}$$

The energy conservation constraint will couple the fitting procedures of $A^{-1}\nabla A$ and $\vec{p}$. The linear basis $\vec{f} = (x, y, \ldots, 1)$ is considered here. Formulation for a general basis is given in Ref. [16]. The functions

$$\{\tilde{r}_x = \vec{c}_x^r \cdot \vec{f}, \ \tilde{r}_y = \vec{c}_y^r \cdot \vec{f}, \ldots\} \tag{9.28}$$

approximate components of the vector $A^{-1}\nabla A$. The functions

$$\{\tilde{p}_x = \vec{c}_x^p \cdot \vec{f}, \ \tilde{p}_y = \vec{c}_y^p \cdot \vec{f}, \ldots\} \tag{9.29}$$

approximate components of the vector $\vec{p}$. In order to express the energy conservation condition, $dE/dt = 0$, the fitting coefficients are arranged into matrices $\mathbf{C}^r$ and $\mathbf{C}^p$,

$$\mathbf{C}^r = [\vec{c}_x^{\,r}, \vec{c}_y^{\,r}, \ldots], \tag{9.30}$$

$$\mathbf{C}^p = [\vec{c}_x^{\,p}, \vec{c}_y^{\,p}, \ldots]. \tag{9.31}$$

Differentiating Equation 9.27 with respect to time and using Equations 9.26 with the derivatives obtained from Equations 9.28 and 9.29, the energy conservation condition becomes

$$\frac{dE}{dt} = \frac{\langle \vec{r}^{\,0} \cdot (\mathbf{C}^r \vec{p} - \mathbf{C}^p \vec{r}) \rangle}{m} = 0. \tag{9.32}$$

The quantity $\vec{r}^{\,0}$ denotes a vector of nonclassical momentum extended to the size of the basis $N_{\text{dim}} + 1$

$$\vec{r}^{\,0} = (r_x, r_y, \ldots, 0). \tag{9.33}$$

The matrices $\mathbf{C}^r$ and $\mathbf{C}^p$ are symmetric.

The least squares fit of $A^{-1}\nabla A$ and $\vec{p}$ in terms of a linear basis with the constraint Equation 9.32 is minimization of the functional

$$I = \langle \|A^{-1}\nabla A - \mathbf{C}^r \vec{f}\|^2 \rangle + \langle \|\vec{p} - \mathbf{C}^p \vec{f}\|^2 \rangle + 2\lambda \langle \vec{r}^{\,0} \cdot (\mathbf{C}^r \vec{p} - \mathbf{C}^p \vec{r}) \rangle \tag{9.34}$$

with respect to the fitting coefficients and with respect to the Lagrange multiplier λ. The optimal coefficients solve a system of linear equations,

$$\begin{pmatrix} \mathbf{M} & \mathbf{O} & \vec{\mathbf{D}}^p \\ \mathbf{O} & \mathbf{M} & \vec{\mathbf{D}}^r \\ \vec{\mathbf{D}}^p & \vec{\mathbf{D}}^r & 0 \end{pmatrix} \cdot \begin{pmatrix} \vec{\mathbf{C}}^r \\ \vec{\mathbf{C}}^p \\ \lambda \end{pmatrix} = \begin{pmatrix} \vec{\mathbf{B}}^r \\ \vec{\mathbf{B}}^p \\ 0 \end{pmatrix}. \tag{9.35}$$

In Equation 9.35 the following matrices and vectors are introduced: (i) $\mathbf{M}$ is the block-diagonal matrix of the dimensionality $N_{\text{dim}} N_b \times N_{\text{dim}} N_b$ with the basis function overlap matrix $\mathbf{S} = \langle \vec{f} \otimes \vec{f} \rangle$ as N_{dim} blocks on the diagonal and zeros otherwise; (ii) $\mathbf{O}$ is a zero matrix of the same size as $\mathbf{M}$; and (iii) the elements of the vectors $\vec{\mathbf{C}}^r$, $\vec{\mathbf{C}}^p$, $\vec{\mathbf{B}}^r$, $\vec{\mathbf{B}}^p$, $\vec{\mathbf{D}}^r$, and $\vec{\mathbf{D}}^p$ are the elements of the matrices $\mathbf{C}^r$, $\mathbf{C}^p$, $\mathbf{B}^r$, $\mathbf{B}^p$, $\mathbf{D}^r$, and $\mathbf{D}^p$, respectively, listed in column-by-column order. $\mathbf{C}^r$ and $\mathbf{C}^p$ are given by Equations 9.30 and 9.31. The remaining four matrices are defined as

$$\mathbf{B}^r = -\frac{1}{2}\langle (\nabla \otimes \vec{f})^T \rangle, \quad \mathbf{B}^p = \langle \vec{f} \otimes \vec{p} \rangle \tag{9.36}$$

$$\mathbf{D}^r = -\langle \vec{r}^{\,0} \otimes \vec{r} \rangle, \quad \mathbf{D}^p = \langle \vec{p}^{\,0} \otimes \vec{r} \rangle, \tag{9.37}$$

where $\vec{p}^{\,0}$ denotes a vector of classical momentum extended to the size of the basis

$$\vec{p}^{\,0} = (p_x, p_y, \ldots, 0). \tag{9.38}$$

Fitting of $A^{-1}\nabla A$ is the same as in the LQF procedure, except that now it is coupled to the least squares fit of $\vec{p}$. Formally, the total size of the matrix in Equation 9.35 is $2N_{\text{dim}}N_b+1$. Its structure allows one to invert the matrix on the left-hand side by performing a single matrix inversion of the block $\mathbf{S}$ of the size N_b [17]. Thus, the cost of the quantum force computation scales as $N_{\text{traj}}N_{\text{dim}}^2$. This is essential for efficient high-dimensional implementation.

From the conceptual point of view, the appealing features of the outlined approximation scheme are the energy conservation Equation 9.27 and an equal-footing treatment of $\vec{r}$ and $\vec{p}$. These features lead to a fuller utilization of trajectory information in the approximation. However, in the finite basis $\vec{f}$, Equations 9.26 are effectively truncated; for the linear basis the Laplacian terms are zeros. Another deficiency is that $\vec{r}$ computed along the trajectories is, in general, different from $A^{-1}\nabla A$ defined by the trajectory positions and Equation 9.9.

Approximation based on the linear basis $\vec{f}$ is cheap and gives exact dynamics in the important limit of Gaussian wavefunctions evolving in locally harmonic potentials. However, it results in a "cold" truncation of the time-evolution Equations 9.26 because the second derivatives of the basis functions are zeros. Such truncation of differential equations leads to dynamics which is unstable with respect to small deviations of $\vec{r}$ and $\vec{p}$ from nonlinearity. This can be compensated by introducing additional terms into Equations 9.26. These additional terms depend on the difference of the exact and approximated values of $\vec{p}$ and $\vec{r}$ and balance errors due to the linear basis in the first order of the nonlinearity parameters. The explicit form is determined from the analytical models and has no adjustable parameters.

Consider the lowest order nonlinearities of the classical and nonclassical momenta (in one dimension)

$$p = p_0 + p_1 x + \epsilon x^2, \quad r = \frac{\nabla A}{A}, \quad A = e^{-\alpha x^2}|1 + \delta \cdot (x - x_0)|. \tag{9.39}$$

Analysis of the short-time dynamics with spatial derivatives obtained from linearization of quantities given by Equations 9.39 has shown that the following approximate equations of motion

$$m\left(\frac{dp}{dt} + \nabla V\right) = r\nabla r + \frac{\nabla^2 r}{2} = r\nabla\tilde{r} + 2\nabla\tilde{r}\cdot(r - \tilde{r}) + O(\delta^4)$$

$$-m\frac{dr}{dt} = r\nabla p + \frac{\nabla^2 p}{2} = r\nabla\tilde{p} + 2\nabla\tilde{r}\cdot(p - \tilde{p}) + O(\epsilon\delta^3) \tag{9.40}$$

cancel the leading errors in the nonlinearity parameters δ and ϵ.

In the multidimensional case, the derivatives $\nabla\tilde{r}$ and $\nabla\tilde{p}$ of the approximate functions generalize into matrices $\mathbf{C}^r$ and $\mathbf{C}^p$, given by Equations 9.28 and 9.29. The approximate time-evolution equations become

$$-m\frac{d\vec{r}}{dt} = \mathbf{C}^p\vec{r} + 2\mathbf{C}^r(\vec{p} - \vec{p}^{\text{ fit}})$$

$$m\left(\frac{\vec{p}}{dt} + \nabla V\right) = \mathbf{C}^r\vec{r} + 2\mathbf{C}^r(\vec{r} - \vec{r}^{\text{ fit}}). \tag{9.41}$$

In Equation 9.30 the functions $(\tilde{r}_x, \tilde{r}_y, \ldots)$ approximate components of $A^{-1}\nabla A$ and their determination is coupled to the approximation of $\vec{p}$ in terms of $(\tilde{p}_x, \tilde{p}_y, \ldots)$ by the energy conservation condition. On the other hand, $\vec{r}^{\text{ fit}}$ approximates $\vec{r}$. In general, there is a difference between the fittings $(r_x^{\text{fit}}, r_y^{\text{fit}}, \ldots)$ and $(\tilde{r}_x, \tilde{r}_y, \ldots)$. The approximations $\vec{r}^{\text{ fit}}$ and $\vec{p}^{\text{ fit}}$ should be such that the stabilization terms do not contribute to the total energy of the trajectory ensemble and do not change the normalization of the wavefunction. The latter condition can be written as $\langle d\vec{r}/dt \rangle = 0$. Separate least squares fits of $\vec{r}$ and $\vec{p}$ in terms of the linear basis by minimizing $\langle \| \vec{r} - \vec{r}^{\text{ fit}} \|^2 \rangle$ and $\langle \| \vec{p} - \vec{p}^{\text{ fit}} \|^2 \rangle$ respectively, satisfy both of these requirements (the general basis is discussed in Ref. [16]). Physically, the stabilization terms provide "friction" opposing growth of the discrepancies between the trajectory-dependent $\vec{r}$ and $\vec{p}$ and their linear fits with time.

The first illustration of the stabilized dynamics is ZPE of a nonrotating hydrogen molecule, as in Ref. [6]. The classical potential V is the Morse oscillator with 17 bound states. The system is one-dimensional and is described in atomic units scaled by the reduced mass of H_2 to have $m = 1$. The initial wavepacket is a Gaussian wavefunction mimicking the ground state of H_2,

$$\psi(x, 0) = \left(\frac{2\alpha}{\pi}\right)^{1/4} \exp\left(-\alpha(x - x_m)^2\right), \tag{9.42}$$

where $\alpha = 9.33\ a_0^{-2}$ and $x_m = 1.4$ is the minimum of the Morse potential. Figure 9.2a shows positions of the trajectories obtained with the LQF method and with the stabilized AQP. The plot illustrates the effect of runaway trajectories leading to the trajectory "decoherence" and transfer of quantum potential energy into classical energy, which is described at the beginning of Section 9.3. Note the "fringe" LQF trajectory immediately leaving the trajectory ensemble toward the dissociation region. This behavior quickly drives the variance of the wavepacket and, thus, the quantum

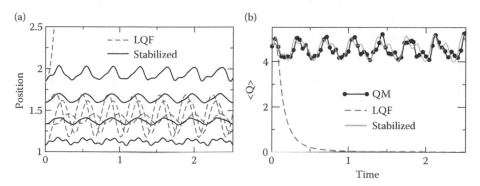

FIGURE 9.2 Quantum energy for the H_2 molecule. (a) Trajectory positions as functions of time obtained using the LQF and the stabilized dynamics. (b) Average quantum potential as a function of time obtained using the LQF (dash), the stabilized dynamics (solid line), and the exact QM propagation (circles).

potential and the quantum force to zero, as shown on the lower panel. Trajectories obtained with the stabilization procedure maintain their coherence and describe the quantum energy $\langle Q \rangle$ accurately as checked for 200 oscillation periods. The effects of stabilization terms can be seen in Figure 9.2a. The oscillatory behavior of the two central trajectories is a consequence of propagating the Gaussian function of Equation 9.42 rather than the eigenstate. These oscillations correlate with the oscillations in $\langle Q \rangle$. The behavior of the outlying trajectories of the stabilized AQP dynamics shows higher frequency oscillations superimposed on the oscillations of the central trajectories. These additional oscillations are due to the corrections of the "friction force" introduced into the equations of motion.

Application of approximate methods to high-dimensional systems must be validated by tests that can be compared to exact QM results, which generally implies separable Hamiltonians or harmonic potentials. For multidimensional testing of the stabilized AQP method, the average quantum energy has been computed for a model potential. The separable model potential consists of the Eckart barrier, centered at the zero of the reaction coordinate, and the Morse oscillators in the vibrational degrees of freedom. The vibrational degrees of freedom are the same as in the one-dimensional application above. The parameters of the barrier mimic the H+H$_2$ system and are given in Ref. [6]. The initial multidimensional wavepacket is defined as a direct product of a Gaussian in the reaction coordinate

$$\psi(x,0) = (2\alpha\pi^{-1})^{1/4} \exp\left(-\gamma(x - x_0)^2 + \imath p_0(x - x_0)\right) \tag{9.43}$$

with parameter values $\{\gamma = 6, \ x_0 = 4, \ p_0 = 6\}$, and Gaussian functions in the vibrational degrees of freedom, given by Equation 9.42. After the wavepacket in the reaction coordinate bifurcates, $\langle Q \rangle$ of the system becomes a sum of quantum energy of the vibrational modes.

Numerical performance of the stabilized dynamics has been tested for up to 40 dimensions with random Gaussian sampling of initial positions [17] using 2×10^4 trajectories. Calculation of the quantum potential and force is dominated by the computation of the moments of the trajectory distribution which scales as $N_{\text{traj}} N_{\text{dim}}^2$. Calculation of the global linearization parameters is performed at each time step for the ensemble of trajectories with the cost of N_{dim}^4. The average quantum potential divided by the number of the vibrational degrees of freedom, $\langle Q \rangle / (N_{\text{dim}} - 1)$, is shown in Figure 9.3. The difference in the quantum energy at short times is due to the quantum energy in the reactive coordinate which is not included in the QM calculation.

9.4   CONCLUSIONS

This chapter was focused on the semiclassical approximations to quantum dynamics of chemical systems based on the Madelung–de Broglie–Bohm formulation of time-dependent quantum mechanics. The appeal of the Bohmian formulation stems from the classical-like picture of quantum mechanical evolutions, which are interpreted using the trajectory language. The wavefunction is replaced by an ensemble of point particles that follow deterministic trajectories and obey a Newtonian equation of

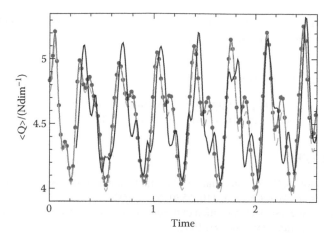

FIGURE 9.3 Average quantum potential per vibrational degree of freedom for a Gaussian wavepacket scattering on the Eckart barrier in the presence of $N_{dim} - 1$ Morse oscillators. Semiclassical results are shown for $N_{dim} = \{20, 40\}$ with circles and dash, respectively. The QM result for long times is shown with a solid line.

motion. The quantum effects are represented by a nonlocal quantum potential that enters the Newton equation and couples the evolution of different trajectories in the Bohmian ensemble.

The trajectory representation offers a number of advantages over the traditional grids and basis sets. Bohmian trajectories follow the quantum mechanical distributions, avoiding the unnecessary computational effort that is often taken in order to treat regions of very low quantum density. This feature of the Bohmian formulation allows one to eliminate the exponential scaling of the computational effort with system dimensionality, which is a common feature of the conventional approaches. Trajectories are easy to propagate using the tools developed for classical molecular dynamics. In contrast to the semiclassical techniques that also use the trajectory language, Bohmian mechanics is, in principle, exact and can be converged arbitrarily close to the accurate quantum mechanical answer.

At the same time, the exact Bohmian trajectory dynamics are expensive and unstable for general systems. This is due to difficulties in evaluating the quantum force on an unstructured trajectory grid, and because the behavior of the quantum trajectories is unstable in the presence of interference. The Bohmian formulation of quantum dynamics is most useful as a computational tool when it is implemented approximately to describe leading QM effects in semiclassical systems.

The AQP technique is designed to describe the dominant quantum effects with essentially linear scaling. Here, the quantum potential is defined from the moments of an ensemble of trajectories for the entire space or for a small number of subspaces. An AQP defined this way allows one to compute the quantum force analytically. Simple global approximations to quantum dynamics describe ZPE in anharmonic systems very well. Combined with the stabilization terms in the approximate equations of motion, the quantum effects in anharmonic systems can be described with semiclassical accuracy for essentially infinite propagation times.

ACKNOWLEDGMENTS

This research is based in part upon the work supported by the Chemistry Division of the National Science Foundation under Grant No. 0516889.

BIBLIOGRAPHY

1. V.A. Rassolov, S. Garashchuk, and G.C. Schatz. Quantum trajectory dynamics in arbitrary coordinates. *J. Phys. Chem. A*, 110: 5530–5536, 2006.
2. D. Bohm. A suggested interpretation of the quantum theory in terms of "hidden" variables, I and II. *Phys. Rev.*, 85: 166–193, 1952.
3. S. Garashchuk and V.A. Rassolov. Semiclassical dynamics based on quantum trajectories. *Chem. Phys. Lett.*, 364: 562–567, 2002.
4. P.R. Holland. *The Quantum Theory of Motion*. Cambridge University Press, Cambridge, 1993.
5. V.A. Rassolov and S. Garashchuk. Semiclassical nonadiabatic dynamics with quantum trajectories. *Phys. Rev. A*, 71: 032511, 2005.
6. V.A. Rassolov and S. Garashchuk. Bohmian dynamics on subspaces using linearized quantum force. *J. Chem. Phys.*, 120: 6815–6825, 2004.
7. S. Garashchuk and V.A. Rassolov. Energy conserving approximations to the quantum potential: Dynamics with linearized quantum force. *J. Chem. Phys.*, 120: 1181–1190, 2004.
8. J.A. Beswick and J. Jortner. Absorption lineshapes for the photodissociation of polyatomic molecules. *Chem. Phys.*, 24: 1–11, 1977.
9. R.D. Coalson and M. Karplus. Multidimensional variational Gaussian wave packet dynamics with application to photodissociation spectroscopy. *J. Chem. Phys.*, 93: 3919–3930, 1990.
10. S. Garashchuk and V.A. Rassolov. Quantum dynamics with Bohmian trajectories: Energy conserving approximation to the quantum potential. *Chem. Phys. Lett.*, 376: 358–363, 2003.
11. S. Garashchuk and V.A. Rassolov. Semiclassical nonadiabatic dynamics of NaFH with quantum trajectories. *Chem. Phys. Lett.*, 446: 395–400, 2007.
12. S. Garashchuk, V.A. Rassolov, and G.C. Schatz. Semiclassical nonadiabatic dynamics based on quantum trajectories for the $O(^3P,^1D)+H_2$ system. *J. Chem. Phys.*, 124: 244307, 2006.
13. S. Garashchuk and V.A. Rassolov. Modified quantum trajectory dynamics using a mixed wavefunction representation. *J. Chem. Phys.*, 121(18): 8711–8715, 2004.
14. S. Garashchuk, V.A. Rassolov, and G.C. Schatz. Semiclassical nonadiabatic dynamics using a mixed wave-function representation. *J. Chem. Phys.*, 123: 174108, 2005.
15. S. Garashchuk and T. Vazhappilly. Wavepacket approach to the cumulative reaction probability within the flux operator formalism. *J. Chem. Phys.*, 131(16): 164108, 2009.
16. S. Garashchuk and V.A. Rassolov. Stable long-time semiclassical description of zero-point energy in high-dimensional molecular systems. *J. Chem. Phys.*, 129: 024109, 2008.
17. W.H. Press, B.P. Flannery, S.A. Teukolsky, and W.T. Vetterling. *Numerical Recipes: The Art of Scientific Computing*, 2nd edition. Cambridge University Press, Cambridge, 1992.

10 Mixed Quantum/ Classical Dynamics: Bohmian and DVR Stochastic Trajectories

Christoph Meier, J. Alberto Beswick, and Tarik Yefsah

CONTENTS

10.1 INTRODUCTION

The dynamics of systems containing a large number of degrees of freedom is one of the most important challenges in contemporary theoretical chemistry. Full quantum mechanical wave-packet propagations in several degrees of freedom is a numerically demanding task, for which specialized methods like the multiconfiguration time-dependent Hartree (MCTDH) method [1,2] has proven to be a unique and particularly efficient tool. However, for processes like proton and electron transfer in isolated polyatomic molecules, liquids, interfaces and biological systems, intramolecular energy redistribution and unimolecular fragmentation, as well as interactions of atoms and molecules with surfaces [3], a full quantum treatment of all degrees of freedom is not feasible.

In many systems comprising a large number of particles, even though a detailed quantum treatment of all degrees of freedom is not necessary, there may exist subsets that have to be treated quantum mechanically under the influence of the rest of the system. If the typical time-scales between system and bath dynamics are very different,

Markovian models of quantum dissipation can successfully mimic the influence of the bath on the system dynamics [4]. However, in the femtosecond regime studied with ultrashort laser pulses, the so-called Markov approximation is not generally valid [5]. Furthermore, very often the bath operators are assumed to be of a special form (harmonic, for instance) which are sometimes not realistic enough.

Another class of approximate methods are hybrid quantum/classical schemes in which only the essential degrees of freedom are treated quantum mechanically while all others are described classically. The most popular of these mixed quantum/classical methods are the mean-field approximation [6], the surface hopping trajectories [7] or methods based on quantum/classical Liouville space representations [8–14].

In the mean-field treatment the force for the classical motion is calculated by averaging over the quantum wavefunction. The method is invariant to the choice of quantum representation (adiabatic or diabatic), it properly conserves total (quantum plus classical) energy, and it can provide accurate quantum transition probabilities when phase interference effects are unimportant and the energy difference of the relevant quantum states is negligible compared to the kinetic energy in the classical degrees of freedom. However, in cases where the system can evolve on several quite different quantum states, a situation that frequently occurs in photodissociation for instance, this method does not describe properly the correlation between classical and quantum motions.

In the surface hopping scheme the classical trajectories move according to a force derived from a *single quantum state* with the possibility of transitions to other states. In this method, the classical–quantum correlations are included by allowing a given trajectory to bifurcate properly into different branches, each governed by a particular quantum state and weighted by the amplitude of the state. The strengths and weaknesses of these two methods have been discussed in detail [15–19], including an exhaustive comparison between them by Tully [20].

In the surface hopping methods the quantum wavefunction is expanded in an appropriate basis, usually a finite (spectral) basis representation (FBR) for bound degrees of freedom. The coefficients in this basis are examined at specified intervals of time in order to determine if a state-to-state transition should occur. Bastida *et al.* [21] have proposed an alternative scheme (MQCS for *mixed quantum classical steps*) in which the quantum wavefunction is expanded in a discrete variable representation (DVR) [22] rather than in the usual FBR. This allows them to treat continuum states on a grid of points, as is usual in wave-packet dynamics, using well-known absorption techniques to avoid spurious reflections at the limit of the grid. The interpretation of the dynamics of the quantum degree of freedom is provided in terms of stochastic *hops* from one grid point to another, governed by the time evolution of the wave-packet.

Another way to mix quantum mechanics with classical mechanics, proposed in Refs. [23–29], is based on Bohmian quantum trajectories for the quantum/classical connection. Briefly, the quantum subsystem is described by a time-dependent Schrödinger equation that depends parametrically on classical variables. This is similar to the other approaches discussed above. The difference comes from the way the classical trajectories are calculated. In this approach, which was called mixed quantum/classical Bohmian (MQCB) trajectories, the wave-packet is used to define de Broglie–Bohm quantum *trajectories* [30–35] which in turn are used to calculate the force acting on the classical variables.

The main difference between MQCB and MQCS lies in the way the force on the classical variables is calculated: In the MQCS method, the force is calculated using different discrete points of the quantum degree of freedom, while in the MQCB method, it is calculated at the continuous points which follow the quantum trajectory.

Hence, the underlying question is to analyze the connection between the sequence of grid points obtained by stochastic hopping methods and the deterministic, continuous quantum trajectory. In his paper *Quantum mechanics in terms of discrete beables* [36], Jeroen C. Vink applied Bell's assumption [37] that all observables should be considered as discrete, i.e., only certain discrete values are possible *beables*, to the physical space described in a grid. From a discretized version of the Schrödinger equation, a rate equation can be written for the probability to find the system in a particular point of the grid. Trajectories are then defined by a stochastic treatment of this equation with jumps governed by Bell's choice [37]. Under certain conditions, Vink was able to show that in the limit of infinitesimal grid spacing, the dynamical equation gives the quantum trajectories. The proof was based in the three-point centered formula for the second derivative of the wavefunction.

This chapter is organized as follows. In Sections 10.2 and 10.3, we give a brief review of the MQCB and MQCS methods; in Section 10.4 we address the problem of representing a quantum wave-packet by stochastic hops between discrete points in space, and under which conditions the sequence of hopping trajectories converge toward a deterministic quantum trajectory. Finally Section 10.5 gives a general discussion of the relationship between all these treatments.

10.2 BOHMIAN TRAJECTORIES AND THE MQCB METHOD

Since the MQCB method of mixing quantum and classical mechanics can be considered as an approximate method derived from the de Broglie–Bohm formulation of quantum mechanics, this completely equivalent perspective of quantum mechanics will be briefly reviewed.

To this end, we consider a two-dimensional Hilbert space. Note that considering two dimensions is no restriction to what will be shown below; actually, x, X can be viewed as *collective variables*, one of which will comprise all quantum degrees of freedom while the other all classical ones. Writing the wavefunction as $\psi(x, X, t) = R(x, X, t)\exp(iS(x, X, t)/\hbar)$, with R, S being real, the Schrödinger equation can be recast in terms of a continuity equation,

$$\frac{\partial R^2}{\partial t} + \frac{1}{m}\frac{\partial S}{\partial x}\frac{\partial R^2}{\partial x} + \frac{1}{M}\frac{\partial S}{\partial X}\frac{\partial R^2}{\partial X} = -R^2\left(\frac{1}{m}\frac{\partial^2 S}{\partial x^2} + \frac{1}{M}\frac{\partial^2 S}{\partial X^2}\right) \tag{10.1}$$

and a quantum Hamilton–Jacobi equation:

$$\frac{\partial S}{\partial t} + \frac{1}{2m}\left(\frac{\partial S}{\partial x}\right)^2 + \frac{1}{2M}\left(\frac{\partial S}{\partial X}\right)^2 + V(x, X) + Q(x, X, t) = 0 \qquad (10.2)$$

where $Q(x, X, t)$ is the so-called *quantum potential* [30]

$$Q(x, X, t) = -\frac{\hbar^2}{2m}\frac{1}{R}\frac{\partial^2 R}{\partial x^2} - \frac{\hbar^2}{2M}\frac{1}{R}\frac{\partial^2 R}{\partial X^2}. \qquad (10.3)$$

Hence one sees that the phase of $\psi(x, X, t)$ can be viewed as an action function, a solution to the Hamilton–Jacobi equation with an additional potential term, the quantum potential. This observation led to the definition of *trajectories* $(\mathbf{x}(t), \mathbf{X}(t))$ [30], the conjugate momenta of which are given by the derivative of $S(x, X, t)$:

$$\mathbf{p} = m\dot{\mathbf{x}} = \left.\frac{\partial S}{\partial x}\right|_{x=\mathbf{x}(t), X=\mathbf{X}(t)} \quad ; \quad \dot{\mathbf{p}} = -\frac{\partial}{\partial x}(V + Q) \qquad (10.4)$$

$$\mathbf{P} = M\dot{\mathbf{X}} = \left.\frac{\partial S}{\partial X}\right|_{x=\mathbf{x}(t), X=\mathbf{X}(t)} \quad ; \quad \dot{\mathbf{P}} = -\frac{\partial}{\partial X}(V + Q). \qquad (10.5)$$

Thus, within the Bohmian formulation of quantum mechanics, *quantum trajectories* move according to the usual Hamilton's equations, subject to the additional quantum potential defined in Equation 10.3. An ensemble of *quantum particles* at positions $(\mathbf{x}(t), \mathbf{X}(t))$ distributed initially according to

$$P([x, x + dx]; [X, X + dX]) = |\psi_0(x, X)|^2 \, dx \, dX \qquad (10.6)$$

and propagated alongside using Equations 10.4 and 10.5, will represent the probability distribution of the quantum mechanical wavefunction at any time [30].

From Equations 10.4 and 10.5 one sees that whenever the additional force due to the quantum potential is negligible, one has a purely classical motion. Thus, the limit from quantum theory to classical mechanics appears naturally within this theory.

In order to establish the mixed quantum–classical method based on Bohmian trajectories (MQCB) [23], we take the same approach as in the de Broglie–Bohm formulation of quantum mechanics, as detailed above. Hence we start from the same, full dimensional initial wavefunction $\psi_0(x, X)$ alongside an ensemble of trajectories at initial positions $(\mathbf{x}(t = 0), \mathbf{X}(t = 0))$ distributed according to $R^2(x, X, t = 0) = |\psi_0(x, X)|^2$. After taking derivatives with respect to x and X of Equation 10.2 we neglect the term involving the second derivative of S with respect to X. In addition, we neglect the second derivative of S with respect to X in Equation 10.1 and the second derivative of R with respect to X in Equation 10.3. Considering the simplest case of a free two-dimensional Gaussian wave-packet, one sees that these terms describe the dispersion in the X-direction. In the limit of large M the wave-packet behaves classically and does *not* show much dispersion in the X-direction. Hence neglecting these terms should be a good approximation to the real quantum dynamics. In this sense, X will from now on be called the *classical* degree of freedom. Note that this approximation cannot be made in the original Schrödinger Equation 10.1 but *only* in the equations for the amplitude and phase! We then have from Equation 10.1:

$$\frac{\partial \tilde{R}^2}{\partial t} + \frac{\partial}{\partial x}\left(\tilde{R}^2 \frac{1}{m}\frac{\partial \tilde{S}}{\partial x}\right) + \frac{1}{M}\frac{\partial \tilde{S}}{\partial X}\frac{\partial \tilde{R}^2}{\partial X} = 0 \tag{10.7}$$

and from Equation 10.2:

$$\frac{\partial}{\partial t}\left(\frac{\partial \tilde{S}}{\partial x}\right) + \left(\frac{1}{m}\frac{\partial \tilde{S}}{\partial x}\right)\left(\frac{\partial^2 \tilde{S}}{\partial x^2}\right) + \left(\frac{1}{M}\frac{\partial \tilde{S}}{\partial X}\right)\left(\frac{\partial^2 \tilde{S}}{\partial x \partial X}\right) = -\frac{\partial}{\partial x}(V + \tilde{Q}) \tag{10.8}$$

$$\frac{\partial}{\partial t}\left(\frac{\partial \tilde{S}}{\partial X}\right) + \left(\frac{1}{m}\frac{\partial \tilde{S}}{\partial x}\right)\left(\frac{\partial^2 \tilde{S}}{\partial X \partial x}\right) = -\frac{\partial}{\partial X}(V + \tilde{Q}) \tag{10.9}$$

where $\tilde{Q} = -(\hbar^2/2m\tilde{R})\partial^2\tilde{R}/\partial x^2$ and the tilde quantities stand for the approximate solutions. As in the usual hydrodynamic formulation detailed above, the Bohmian trajectories associated with these approximate equations are:

$$\mathbf{p} = m\dot{\mathbf{x}} = \left.\frac{\partial \tilde{S}}{\partial x}\right|_{x=\mathbf{x}(t), X=\mathbf{X}(t)} \tag{10.10}$$

$$\mathbf{P} = M\dot{\mathbf{X}} = \left.\frac{\partial \tilde{S}}{\partial X}\right|_{x=\mathbf{x}(t), X=\mathbf{X}(t)} \tag{10.11}$$

together with the *same initial conditions* as in the hydrodynamic formulation of quantum mechanics. This does not pose any problem, since within the MQCB method, the full-dimensional initial wavefunction is supposed to be known. Hence the initial values $(\mathbf{x}(t=0), \mathbf{X}(t=0))$ are chosen according to the distribution as above:

$$P([x, x+dx]; [X, X+dX]) = |\psi_0(x, X)|^2\, dx\, dX. \tag{10.12}$$

The next step consists in evaluating Equations 10.7 and 10.8 at $X = \mathbf{X}(t)$. Using Equation 10.11 one gets:

$$\frac{d\tilde{R}^2}{dt} + \frac{\partial}{\partial x}\left(\tilde{R}^2\frac{1}{m}\frac{\partial \tilde{S}}{\partial x}\right) = 0 \tag{10.13}$$

$$\frac{d}{dt}\left(\frac{\partial \tilde{S}}{\partial x}\right) + \left(\frac{1}{m}\frac{\partial \tilde{S}}{\partial x}\right)\left(\frac{\partial^2 \tilde{S}}{\partial x^2}\right) = -\frac{\partial}{\partial x}(V + \tilde{Q}) \tag{10.14}$$

where d/dt stands for $d\tilde{f}/dt = \partial\tilde{f}/\partial t + \dot{\mathbf{X}}\cdot\left(\partial\tilde{f}/\partial X\right)_{X=\mathbf{X}(t)}$.

The important observation at this point is that Equations 10.13 and 10.14 are rigorously equivalent to a quantum problem in the x subspace with $\mathbf{X}$ being a time-dependent parameter. Thus the approximate wavefunction $\widetilde{\psi}(x, \mathbf{X}(t), t) = \widetilde{R}(x, \mathbf{X}(t), t) \exp(i\widetilde{S}(x, \mathbf{X}(t), t)/\hbar)$ obeys the Schrödinger equation

$$i\hbar \frac{d\widetilde{\psi}(x, \mathbf{X}(t), t)}{dt} = \left(-\frac{\hbar^2}{2m} \frac{\partial^2}{\partial x^2} + V(x, \mathbf{X}(t)) \right) \widetilde{\psi}(x, \mathbf{X}(t), t). \tag{10.15}$$

Note the appearance of the total derivative in the left-hand side of this equation. As will be shown below, this is important when an adiabatic basis set is used for solving the quantum problem. Since we have only approximated the *equations of motion* and supposed that the initial wavefunction $\psi_0(x, X)$ is known, we have

$$\widetilde{\psi}(x, \mathbf{X}(t = 0), t = 0) = \psi_0(x, X)|_{X = \mathbf{X}(t = 0)} \tag{10.16}$$

as the initial wavefunction in the quantum subspace.

A consistent equation of motion for the classical degrees of freedom is obtained by taking the total derivative with respect to time of Equation 10.11, noting that this leads to a term involving the second derivative of S with respect to X, which we have assumed to be small. Hence, we can use Equation 10.9 to write

$$\dot{\mathbf{P}} = -\frac{1}{M} \frac{\partial \left(V(x, X) + \widetilde{Q}(x, X) \right)}{\partial X} \Bigg|_{x = \mathbf{x}(t), X = \mathbf{X}(t)}. \tag{10.17}$$

The fact that at this level of approximation, the quantum potential corresponding to the quantum subsystem remains in the classical equation of motion, is somewhat reminiscent of the Pechukas force in the surface hopping method [7]. In both cases, the classical degrees of freedom are directly affected by changes in the quantum subspace. In practice, however, since we do not solve Equations 10.8 and 10.9 directly but only follow a specific trajectory $\mathbf{X}(t)$, we additionally neglect the quantum potential in the classical degree of freedom. At this level of approximation, the MQCB method is identical to the one proposed independently by Prezhdo *et al.* [27–29].

10.3 STOCHASTIC DVR AND MQCS

Considering the same two-dimensional Hamiltonian as above, we choose a one-dimensional (diabatic) DVR basis $|x_n\rangle$ with eigenvalue x_n of the position operator, for the quantum degree of freedom, and expand the total wavefunction according to:

$$\psi(x, \mathbf{X}(t), t) = \sum_n c_n(t) |x_n\rangle, \tag{10.18}$$

where the wave-packet $\Psi(x, t)$ propagates with time along the x coordinate, and is a DVR basis function with eigenvalue x_n of the position operator, such that $c_n(t) = \langle x_n|\psi(t)\rangle = \psi(x_n, \mathbf{X}(t), t)$. The values of the time-dependent coefficients can be obtained by solving the time-dependent Schrödinger equation using any conventional method (split operator, etc.).

Initially the wave-packet is a coherent superposition of DVR states. The time evolution of the coefficients of the wave-packet in these DVR states is given by

$$\frac{dc_n}{dt} = -\frac{i}{\hbar} \sum_m c_m(t)\langle x_n|H|x_m\rangle = -\frac{i}{\hbar}\left[\sum_m c_m(t)\langle x_n|T|x_m\rangle + V(x_n, \mathbf{X}(t))\right]$$

(10.19)

where we have used the fact that the nondiagonal matrix elements of the Hamiltonian reduce to the matrix elements of the kinetic energy operator T in the DVR basis and the potential energy matrix is diagonal.

In the MQCS method [21], the classical degrees of freedom are calculated using the force evaluated at one single DVR point,

$$\dot{\mathbf{P}} = -\frac{1}{M}\frac{\partial\,(V(x, X))}{\partial X}\bigg|_{x=\mathbf{x}_n(t), X=\mathbf{X}(t)}.$$

(10.20)

The dynamics in the quantum subspace is accounted for by the possibility of quantum jumps between these points. Hence, $\mathbf{x}_n$ is a sequence of DVR points, forming a *stochastic quantum trajectory*, defined as follows: Equation 10.19 determines the transition probabilities between DVR states. Let us introduce the diagonal density matrix elements $P_n = |c_n|^2 = |\psi(x_n, \mathbf{X}(t), t)|^2$ representing the probability for the system to be in state $|x_n\rangle$ at time t. The time evolution of $P_n(t)$ is deduced from Equation 10.19:

$$\frac{dP_n}{dt} = \sum_m b_{nm}(t)$$

(10.21)

where b_{nm} is given by

$$b_{nm} = \frac{2}{\hbar}\Im\left(c_n^* c_m\langle x_n|T|x_m\rangle\right).$$

(10.22)

In Equation 10.21, the b_{nm} coefficients can be interpreted as the population flux from state $|x_n\rangle$ to $|x_m\rangle$, with the property that $b_{nm}(t) = -b_{mn}(t)$. Hence if the system is in state $|x_n\rangle$ at time t, the probability to jump to state $|x_m\rangle$ at time $t + \delta t$ is given by

$$\mathcal{P}_{n\to m} = \frac{b_{nm}}{P_n}\delta t.$$

(10.23)

In the MQCS method [21], this equation is used as follows. An initial state $|x_n\rangle$ is selected at random, following a probability distribution given by $P_n(t = 0) \equiv |\langle x_n|\Psi(x, 0)\rangle|^2$. For each initial state a sequence of events is selected by testing at each time step the probability for transition to any other state $|x_m\rangle$. Suppose that the sequence of events has led to state $|x_n\rangle$ at time t. A random number g is generated between 0 and 1. If it is smaller than b_{n1}, then at time $t + \delta t$ the state will be 1; if $b_{n1} < g < b_{j1} + b_{n2}$ then the state will be state 2, etc.; and if $\sum_{n\neq m} b_{nm} < g$, then the state at time $t + \delta t$ will still be $|x_n\rangle$. At the end of the propagation ($t = t_f$), the recombination of the final states of all these chains of events gives the wave-packet at time t_f.

This approach is similar to the way the quantum potential energy surface for the classical particle dynamics is selected in Tully's fewest switches method [17] for mixed quantum–classical dynamics. The MQCS method can thus be thought of as the implementation of the surface hopping method in the DVR, rather than in the usual FBR.

Equation 10.21 can be rewritten to give:

$$\frac{dP_n}{dt} = \sum_m (T_{nm} P_m - T_{mn} P_n) \tag{10.24}$$

with

$$T_{nm} = \frac{\hbar}{M} D^{(2)}_{nm} \Im \left\{ \frac{c_m^* c_n}{|c_m|^2} \right\} \quad \text{if } T_{nm} > 0 \tag{10.25}$$

where we have denoted by $D^{(2)}_{nm} = \langle x_n | \partial^2 / \partial x^2 | x_m \rangle$ the matrix elements of the second derivative.

10.4   DISCRET BEABLES

In this section, we discuss the connection between a continuous quantum trajectory and stochastic hopping between discrete grid points first proposed by Vink in 1993 [36]. To this end, it is enough to consider only one quantum degree of freedom.

10.4.1   Vink's Formulation

We consider a particle of mass M in a one-dimensional space extending over $-\infty < x < +\infty$. This space is discretized as $x_n = \epsilon n$; $n = -\infty, +\infty$ where ϵ is the grid spacing. If we denote by $P_n = |\psi(x_n, t)|^2 = |\psi_n|^2$ the probability to find the particle at x_n, the evolution in time of P_n is given by

$$\frac{\partial P_n}{\partial t} = -\frac{\hbar}{M} \Im \left\{ \psi_n^* \frac{\partial^2 \psi}{\partial x^2} \Big|_{x_n} \right\}. \tag{10.26}$$

Using the three-point centered expansion of the second derivative,

$$\frac{\partial^2 \psi_n}{\partial x^2} \Big|_{x_n} = \frac{\psi_{n-1} - 2\psi_n + \psi_{n+1}}{\epsilon^2} \tag{10.27}$$

it is found that

$$\frac{\partial P_n}{\partial t} = -\frac{\hbar}{M\epsilon^2} \Im \left\{ \psi_n^* \psi_{n-1} + \psi_n^* \psi_{n+1} \right\}. \tag{10.28}$$

This equation can be written as

$$\frac{\partial P_n}{\partial t} = \sum_m J_{nm} \tag{10.29}$$

with

$$J_{nm} = -\frac{\hbar^2}{M\epsilon^2}\Im\{\psi_n^*\psi_{n+1}\delta_{m,n+1} + \psi_n^*\psi_{n-1}\delta_{m,n-1}\}. \tag{10.30}$$

Writing

$$\psi_{n\pm1} = \psi_n \pm \epsilon\psi_n' \tag{10.31}$$

where the prime denotes a first derivative with respect to x, and using the form $\psi_n = A_n \exp(i S_n/\hbar)$ with A_n and S_n real, we obtain

$$J_{nm} = \frac{\hbar}{M\epsilon}\left(S_m' P_m\delta_{m,n-1} - S_m' P_m\delta_{m,n+1}\right) \tag{10.32}$$

and we note that $J_{nm} = -J_{mn}$. Comparing Equation 10.29 with the rate equation

$$\frac{\partial P_n}{\partial t} = \sum_m (T_{nm} P_m - T_{mn} P_n) \tag{10.33}$$

and using Bell's minimal choice [37], we get for $n \neq m$,

$$T_{nm} = \frac{J_{nm}}{\hbar P_m}; \quad \text{if } J_{nm} \geq 0 \tag{10.34}$$

and 0 otherwise. Thus using Equation 10.32, we obtain, for $n \neq m$,

$$T_{nm} = \frac{S_m'}{M\epsilon}\delta_{m,n-1}; \quad \text{if } S_m' \geq 0$$

$$T_{nm} = -\frac{S_m'}{M\epsilon}\delta_{m,n+1}; \quad \text{if } S_m' \leq 0. \tag{10.35}$$

Finally T_{nn} is determined by the normalization

$$T_{nn} dt = 1 - \sum_{n \neq m} T_{nm} dt \tag{10.36}$$

with

$$dt \leq \frac{M\epsilon}{|S_{n-1}' - S_{n+1}'|} \tag{10.37}$$

in order to have $T_{nn} dt \geq 0$. Vink has shown [36] that if one uses Equations 10.35 and 10.36 it is possible to recover the causal Bohmian trajectories through a stochastic algorithm with minimal choice for the transition density. The proof is straightforward. For a trajectory starting in point n at $t = 0$ the probability to jump to the position $n + 1$ is, if we assume that $S_n' \neq 0$,

$$\mathcal{P}_{n+1} = \frac{S_n'}{M\epsilon} dt. \tag{10.38}$$

The mean average displacement after dt will then be given by $dx = \epsilon\mathcal{P}_{n+1}$ which using Equation 10.38 gives in the limit $\epsilon \to 0$

$$\frac{\partial x}{\partial t} = \frac{1}{M}\frac{\partial S}{\partial x} \tag{10.39}$$

which is precisely the Bohmian equation for the quantum trajectories.

What remains to be done is to show that the second moment of the distribution, i.e., the dispersion in the average displacement, vanishes in the limit $\epsilon \to 0$. The mean-squared dispersion for one jump is given by

$$\sigma^2 = \epsilon^2 \mathcal{P}_{n+1} - (\epsilon \mathcal{P}_{n+1})^2 = \epsilon \frac{S_n'}{M} dt \left(1 - \frac{S_n'}{M} dt\right) \tag{10.40}$$

which indeed vanishes for $\epsilon \to 0$.

10.4.2 N-POINT FORMULA

If instead of a three-point finite difference formula for the second derivative as in Equation 10.27, we use a five-point centered expression [38], we shall have

$$\left.\frac{\partial^2 \psi_n}{\partial x^2}\right|_{x_n} = \frac{-\psi_{n-2} + 16\,\psi_{n-1} - 30\,\psi_n + 16\,\psi_{n+1} - \psi_{n+2}}{12\epsilon^2} \tag{10.41}$$

and we get

$$J_{nm} = -\frac{\hbar^2}{12M\epsilon^2}\Im\{-\psi_n^*\psi_{n-2}\delta_{m,n-2} + 16\psi_n^*\psi_{n-1}\delta_{m,n-1} \tag{10.42}$$

$$+ 16\psi_n^*\psi_{n+1}\delta_{m,n+1} - \psi_n^*\psi_{n+2}\delta_{m,n+2}\}. \tag{10.43}$$

Writing

$$\psi_{n\pm v} = \psi_n \pm v\epsilon\psi_n' \tag{10.44}$$

and following the line of reasoning of the three-point case, we finally get, for $n \neq m$,

$$T_{nm} = \frac{S_m'}{6M\epsilon}(\delta_{m,n+2} + 8\delta_{m,n-1}); \quad \text{if } S_m' \geq 0$$

$$T_{nm} = -\frac{S_m'}{6M\epsilon}(\delta_{m,n-2} + 8\delta_{m,n+1}); \quad \text{if } S_m' \leq 0. \tag{10.45}$$

The difference with respect to the three-point case is that now, even for a given sign of S', we can have jumps to $m \pm 1$ and also to $m \mp 2$ with probabilities

$$\mathcal{P}_{n\pm1} = \frac{4S_n'}{3M} dt; \quad \mathcal{P}_{n\pm2} = \frac{S_n'}{6M} dt. \tag{10.46}$$

However if we calculate the mean average distance for a dt we get

$$dx = \epsilon\mathcal{P}_{n+1} - 2\epsilon\mathcal{P}_{n-2} = \frac{3}{4}\epsilon\mathcal{P}_{n+1} = \frac{S_n'}{M} dt \tag{10.47}$$

i.e., the same Bohmian trajectory equation. As for the mean-squared dispersion

$$\sigma^2 = \epsilon^2 \mathcal{P}_{n+1} + (-2\epsilon)^2 \mathcal{P}_{n-2} - \left(\frac{3}{4}\epsilon\mathcal{P}_{n+1}\right)^2 = 2\epsilon^2 \frac{S_n'}{M} dt \left(1 - \frac{S_n'}{2M} dt\right) \tag{10.48}$$

which again vanishes for $\epsilon \to 0$.

These results can be generalized to an arbitrary number of points. According to Fornberg [38], the expression for the second derivative in the $(p + 1)$-point centered finite difference approximation is

$$\left.\frac{\partial^2 \psi_n}{\partial x^2}\right|_{x_n} = \frac{1}{\epsilon^2} \sum_{j=-p/2}^{p/2} C_j(p)\psi_{n+j} \tag{10.49}$$

with

$$C_{j\neq 0}(p) = \frac{(-)^{j+1}2(p/2!)^2}{j^2(p/2+j)!\,(p/2-j)!}; \quad C_0(p) = -2\sum_{k=1}^{p/2}\frac{1}{k^2}. \tag{10.50}$$

We note that in the limit $p \to \infty$

$$C_{j\neq 0}(\infty) = \frac{(-)^{j+1}2}{j^2}; \quad C_0(\infty) = -\frac{\pi^2}{3}. \tag{10.51}$$

Using these expressions and Equation 10.44 we obtain for $S'_m > 0$ and $n \neq m$ the jumping density

$$T_{nm} = \frac{S'_m}{M\epsilon} \sum_{j=-p/2}^{p/2} jC_j(p)\delta_{m,n-j} = \frac{S'_m}{M\epsilon} \sum_{j=-p/2}^{p/2} \frac{(-)^{j+1}2(p/2!)^2}{j(p/2+j)!\,(p/2-j)!}\delta_{m,n-j}. \tag{10.52}$$

Using this equation we get

$$\mathcal{P}_{n+j} = \frac{S'_n}{M\epsilon} jC_j(p)\,dt = \frac{S'_n}{M}\frac{(-)^{j+1}2(p/2!)^2}{j(p/2+j)!\,(p/2-j)!}\,dt \tag{10.53}$$

from which

$$dx = \epsilon \sum_j j\mathcal{P}_{n+j} = \frac{S'_n}{M\epsilon}\,dt\,2\sum_j \frac{(-)^{j+1}(p/2!)^2}{(p/2+j)!\,(p/2-j)!} \tag{10.54}$$

and since

$$\sum_{k=1}^{n}\frac{(-)^{k+1}(n!)^2}{(n+k)!\,(n-k)!} = 1/2 \tag{10.55}$$

the Bohmian equation is found.

Similarly, the mean-squared dispersion is given by

$$\sigma^2 = \epsilon^2\left(\sum_j j^2\mathcal{P}_{n+j} - \sum_{j,l\neq j} jl\mathcal{P}_{n+j}\mathcal{P}_{n+l}\right) \tag{10.56}$$

which presumably also vanishes for $\epsilon \to 0$.

10.4.3   DVR POINTS

If in Equation 10.25, the linearization approximation Equation 10.44 is used, we can write

$$T_{nm} = \frac{(x_n - x_m)D^{(2)}_{nm}}{M} \Im \left\{ \frac{c^*_m c'_m}{|c_m|^2} \right\} \quad \text{if } T_{nm} > 0 \tag{10.57}$$

and we note that imposing $c_m(t) = R_m(t) \exp i S_m(t)/\hbar$, we have

$$\Im \{ c^*_m c'_m / |c_m|^2 \} = S'_m \tag{10.58}$$

and therefore Equation 10.57 becomes

$$T_{nm} = \frac{(x_n - x_m)D^{(2)}_{nm}}{M} D^{(2)}_{nm} S'_m \quad \text{if } T_{nm} > 0. \tag{10.59}$$

Comparing this expression with those found in the N-point formulation we note that indeed there is a clear correspondence between them. They are all proportional to S'_m/M and if in the limit of infinite number of DVR points

$$\sum_n (x_n - x_m)^2 D^{(2)}_{nm} = 1 \tag{10.60}$$

they all give the Bohmian trajectories.

10.5   DISCUSSION

Comparing MQCB and MQCS, we see that both methods calculate the back-reaction of the quantum system onto the classical system not by averaging the force over the quantum wavefunction (as does the mean-field method), but by choosing to evaluate this force only at one single point. The time evolution of this point, however, is different in both methods: the MQCB method uses the deterministic Bohm trajectory associated with the quantum wavefunction; the MQCS method generated these points by stochastic jumps between discrete DVR points of an underlying DVR basis. On the other hand, generalizing Vink's formulation, we have shown that Bohmian trajectories can be obtained by stochastic jumps between discrete points in the limit of a large number of points. Clearly, the MQCS method is based on the same idea except that the discrete points are chosen according to the DVR representation. In the limit of a large number of DVR points the MQCS trajectories should converge toward the Bohmian MQCB trajectories.

BIBLIOGRAPHY

1. H.-D. Meyer, U. Manthe, and L.S. Cederbaum, *Chem. Phys. Lett.*, **165**, 73 (1990).
2. H.-D. Meyer, F. Gatti, and G.A. Worth, Eds., *Multidimensional Quantum Dynamics: MCTDH Theory and Applications*, Wiley-VCH, Weinheim (2009).

3. B. Berne, G. Cicotti, and D. Coker, Eds., *Classical and Quantum Dynamics in Condensed Phase Simulations*, World Scientific, Singapore (1998).
4. U. Weiss, Ed., *Quantum Dissipative Systems*, World Scientific, Singapore (1993).
5. C. Meier and D. Tannor, *J. Chem. Phys.*, **111**, 3365 (1999).
6. G. Billing, *Int. Rev. Phys. Chem.*, **13**, 309 (1994).
7. J.C. Tully, *Int. J. Quantum Chem.*, **25**, 299 (1991).
8. D.A. Micha and B. Thorndyke, *Adv. Quantum Chem.*, **47**, 292 (2004).
9. B. Thorndyke and D.A. Micha, *Chem. Phys. Lett.*, **403**, 280 (2005).
10. A. Donoso and C.C. Martens, *J. Chem. Phys.*, **102**, 4291 (1998).
11. R. Kapral and G. Cicotti, *J. Chem. Phys.*, **110**, 8919 (1999).
12. S. Nielsen, R. Kapral, and G. Cicotti, *J. Chem. Phys.*, **115**, 5805 (2001).
13. I. Horenko, C. Salzmann, B. Schmidt, and C. Schütte, *J. Chem. Phys.*, **117**, 11075 (2002).
14. I. Horenko, M. Weiser, B. Schmidt, and C. Schütte, *J. Chem. Phys.*, **120**, 8913 (2004).
15. J.C. Tully, Nonadiabatic dynamics. In *Modern Methods for Multidimensional Dynamics Computations in Chemistry*, edited by D.L. Thompson, World Scientific, Singapore (1998), p. 34.
16. U. Müller and G. Stock, *J. Chem. Phys.*, **107**, 6230 (1997).
17. J.C. Tully, *J. Chem. Phys.*, **93**, 1061 (1990).
18. C.C. Martens and J.-Y. Fang, *J. Chem. Phys.*, **106**, 4918 (1997).
19. G.D. Billing, *J. Chem. Phys.*, **107**, 4286 (1997).
20. J.C. Tully, *Faraday Discuss.*, **110**, 1 (1998).
21. A. Bastida, J. Zuniga, A. Requena, N. Halberstadt, and J.A. Beswick, *Phys. Chem. Comm.*, **7**, 29 (2000).
22. Z. Bačić and J.C. Light, *J. Chem. Phys.*, **85**, 4594 (1986).
23. E. Gindensperger, C. Meier, and J.A. Beswick, *J. Chem. Phys.*, **113**, 9369 (2000).
24. E. Gindensperger, C. Meier, and J.A. Beswick, *Adv. Quantum Chem.*, **47**, 331 (2004).
25. E. Gindensperger, C. Meier, and J.A. Beswick, *J. Chem. Phys.* **116**, 8 (2002).
26. E. Gindensperger, C. Meier, J.A. Beswick, and M.-C. Heitz, *J. Chem. Phys.*, **116**, 10051 (2002).
27. O.V. Prezhdo and C. Brooksby, *Phys. Rev. Lett.*, **86**, 3215 (2001).
28. L.L. Salcedo, *Phys. Rev. Lett.*, **90**, 118901 (2003).
29. O.V. Prezhdo and C. Brooksby, *Phys. Rev. Lett.*, **90**, 118902 (2003).
30. P.R. Holland, *The Quantum Theory of Motion*, Cambridge University Press, Cambridge (1993).
31. L. de Broglie, *C. R. Acad. Sci. Paris*, **183**, 447 (1926).
32. L. de Broglie, *C. R. Acad. Sci. Paris*, **184**, 273 (1927).
33. D. Bohm, *Phys. Rev.*, **85**, 166 (1952).
34. D. Bohm, *Phys. Rev.*, **85**, 180 (1952).
35. R.E. Wyatt, *Quantum Dynamics with Trajectories*, Springer, New York (2005).
36. J.C. Vink, *Phys. Rev. A*, **48**, 1808 (1993).
37. J.S. Bell, *Speakable and Unspeakable in Quantum Mechanics*, Cambridge University Press, Cambridge (1987).
38. B. Fornberg, *A Practical Guide to Pseudospectral Methods*, Cambridge University Press, Cambridge (1998).

11 A Hybrid Hydrodynamic–Liouvillian Approach to Non-Markovian Dynamics

Keith H. Hughes and Irene Burghardt

CONTENTS

11.1 INTRODUCTION

Trajectory based approaches have proved invaluable in most areas of dynamics. In chemistry, trajectory based classical molecular dynamics has made it possible to study the dynamics of up to 10^3–10^4 atoms, and its success in this field is well documented. In fluid dynamics, Lagrangian trajectory based methods have also proved popular and are the method of choice for many fields in the fluid dynamics community. Trajectory based approaches have the advantages that they provide an intuitive insight into a dynamical process and, as witnessed in classical molecular dynamics, trajectory based approaches can be very computationally efficient.

Against this backdrop of success of trajectory based approaches it is no surprise that there has been a drive toward developing such approaches in quantum mechanics.

Probably the most well-known (although not exclusive) quantum trajectory approach in physics and chemistry is based on the quantum hydrodynamic, or Bohmian mechanical, approach [1–7]. The literature is rich with discussions of the interpretation of the hydrodynamic approach to quantum mechanics [5,6,8] and over the last decade there has been a surge of interest in developing numerical quantum trajectory techniques [9–18]. Most quantum trajectory based approaches have focused on pure states, i.e., wavefunctions. However, the central theme of this chapter is the quantum hydrodynamics of mixed states.

In the quantum dynamical treatment of mixed states the equations of motion are generally formulated for the density matrix ρ or, by a suitable transformation, for a phase-space density $\rho_W(q, p)$. Both these representations require $2N$ dimensions to describe an N-dimensional system. The hydrodynamic description of mixed states is an alternative approach that involves a reduction of the $2N$-dimensional space to N dimensions. The hydrodynamics approach may be derived from either the phase-space (q, p) representation or density matrix in the positional (x, x') representation. The hydrodynamic formulation of mixed quantum states dates back to the work of Takabayasi [19], Moyal [20] and Zwanzig [21] in the late 1940s and early 1950s and has subsequently been investigated in a number of studies [22–25]. Although the mixed state hydrodynamical theory has been developed since the 1940s, relatively little connection to the corresponding pure state Bohmian mechanics was made [19,26–28]. In the last two decades, work by Muga and collaborators [27] as well as our work [26,28–31] contributed to establishing this connection explicitly.

Recently, we also formulated a novel hybrid hydrodynamic-Liouvillian approach that is ideally suited for developing a mixed quantum-classical scheme [32–34]. The method is known as the *quantum-classical moment* (QCM) approach and involves a partition of the global system into a hydrodynamic part that remains quantum mechanical and a Liouville phase-space part that involves making a classical approximation to the Liouville sector. The classical approximation involves ignoring all $\hbar$ terms in the phase-space version of the quantum Liouville equation and the resulting mixed quantum-classical description is then equivalent to the quantum-classical Liouville equation [35–40]. However, if the potentials for the Liouville sector are no more than quadratic polynomials, i.e., harmonic, then the equations of motion for the Liouville sector contain no $\hbar$ terms and the QCM approach is then quantum mechanically exact in such cases. For the Hamiltonians described in this chapter the potentials for the Liouville sector are harmonic and so no classical approximations are made to the equations of motion.

The hybrid hydrodynamic-Liouville equations are defined in terms of a particular type of moments, obtained by integrating over the momentum p of the quantum part of the Wigner phase-space distribution $\rho_W(q, p; Q, P)$ of the composite system spanned by q, p, Q, P, i.e., $\langle \mathcal{P}^n \rho \rangle_{qQP} = \int dp\, p^n \rho_W(q, p; Q, P)$, see Section 11.2.6. We will refer to these moment quantities as *partial* moments. Exact equations of motion for the moments are then derived before transforming the equations of motion to a Lagrangian trajectory framework. The resulting trajectory equations for the dynamical q, Q, P variables involve a q, Q, P dependent generalized quantum force. The approach was demonstrated for a completely harmonic composite system by Burghardt and Parlant [32], and later for a double well potential system by

Hughes *et al.* [34]. In this study we extend the approach to simulate non-Markovian system-bath dynamics of open quantum systems.

In the following sections a non-Markovian approach to dissipative dynamics is described which is based on the construction of a hierarchy of effective environmental modes that is terminated by coupling the final member of the hierarchy to a Markovian bath. The effective modes in question can be understood as collective coordinates, or generalized Brownian oscillator modes [41–43] by which the dynamics is successively unraveled as a function of time. The definition of these modes emerges naturally from spin-boson type models and linear vibronic coupling models for nonadiabatic dynamics [44]. It has been shown [45–49] that the short-time dynamics of the system is entirely accounted for by the first member of the effective-mode hierarchy. Inclusion of higher order members of the chain captures the dynamics for longer times.

For the hybrid hydrodynamic-Liouville approach described in this chapter, the effective modes are harmonic and are described in Liouville phase-space. The system part is described hydrodynamically. Due to the harmonicity of the effective-mode environment the equations of motion in the Liouville phase-space contain no $\hbar$ terms and so the equations are equivalent to the corresponding classical Liouville phase-space description

11.2 THEORY

We begin the following theory section by describing the Hamiltonian associated with the non-Markovian system-bath dynamics (Section 11.2.1). It is then shown how the Hamiltonian may be transformed to a chain like representation where each member of the chain is bilinearly coupled to its nearest-neighbor (Sections 11.2.2–11.2.4). To ensure irreversible dissipative behavior the final member of the chain is coupled to a Markovian bath. However, the dynamics of the reduced system is strongly non-Markovian.

In the following sections the equations of motion associated with the composite chain Hamiltonian are initially formulated in Liouville phase-space, in terms of the Wigner function (Section 11.2.5). Subsequently, a transformation to the hybrid hydrodynamic-Liouville representation is made (Section 11.2.6). Mass-weighted coordinates are used throughout this chapter.

11.2.1 SYSTEM-BATH HAMILTONIAN

The starting point is the familiar partition of the Hamiltonian into a system H_S, bath H_B, and a system-bath coupling part H_{SB}

$$H = H_S + H_{SB} + H_B. \tag{11.1}$$

Here, the system is taken to be one-dimensional, with the following Hamiltonian in mass-weighted coordinates,

$$H_S = \frac{1}{2}p_0^2 + V(x_0) \tag{11.2}$$

where $p_0 = -i\hbar\partial/\partial x_0$ and $V(x_0)$ is the system potential energy. The bath consists of a number N of harmonic oscillators, with $N \to \infty$ to represent truly irreversible dynamics,

$$H_B = \frac{1}{2}\sum_{n=1}^{N}(p_n^2 + \omega_n^2 x_n^2) \tag{11.3}$$

where ω_n is the frequency of the nth-harmonic oscillator. The bath is assumed to be bilinearly coupled to the system coordinate x_0

$$H_{SB} = x_0\sum_{n=1}^{N}c_n x_n \tag{11.4}$$

where c_n are the system-bath coupling coefficients.

The combination of H_B of Equation 11.3 and H_{SB} of Equation 11.4 defines the spectral density function $J(\omega)$ which reflects the spectral distribution of the bath, weighted by the system-bath couplings,

$$J(\omega) = \frac{\pi}{2}\sum_{n=1}^{N\to\infty}\frac{c_n^2}{\omega_n}\delta(\omega - \omega_n). \tag{11.5}$$

11.2.2 EFFECTIVE-MODE REPRESENTATION

The strategy as described in Refs. [48, 49] is to generate a series of approximate spectral densities from reduced-dimensional analogs of the system-bath Hamiltonian equation 11.1. These reduced-dimensional models are obtained from a transformed version of the Hamiltonian equation 11.1,

$$H = H_S + H_{SB} + H_{\text{eff}} + H_{\text{eff},B'} + H_{B'} \tag{11.6}$$

where the transformation $H_{SB} + H_B \to H_{SB} + H_{\text{eff}} + H_{\text{eff},B'} + H_{B'}$ is based upon a transformed set of bath modes comprising an effective-mode X_1 and a set of residual modes $\{X_2, \ldots, X_N\}$.

In the transformed representation H_{SB} is represented in terms of a single collective mode X_1

$$H_{SB} = Dx_0 X_1 \tag{11.7}$$

with the effective-mode Hamiltonian H_{eff} given by

$$H_{\text{eff}} = \frac{1}{2}(P_1^2 + \Omega_1^2 X_1^2). \tag{11.8}$$

The effective mode is coupled to the residual bath modes by

$$H_{\text{eff},B'} = X_1\sum_{n=2}^{N}d_{1,n}X_n. \tag{11.9}$$

The residual bath part $H_{B'}$ is given by

$$H_{B'} = \frac{1}{2} \sum_{n=2}^{N} (P_n^2 + \Omega_n^2 X_n^2) + \sum_{n=2}^{N} \sum_{m>n}^{N} d_{nm} X_n X_m. \tag{11.10}$$

The total transformed Hamiltonian H has now been partitioned into the following form

$$H = H_S + H_{SB} + H_{\text{eff}} + H_{\text{eff},B'} + H_{B'}. \tag{11.11}$$

The transformation from Equation 11.1 to Equation 11.11 inevitably transforms the diagonal frequency matrix $\boldsymbol{\omega}$ to a non-diagonal symmetric form $\boldsymbol{\omega}^2 \to \boldsymbol{T}^\dagger \boldsymbol{\omega}^2 \boldsymbol{T} = \boldsymbol{W}$ that contains the following elements,

$$\Omega_n^2 = \sum_{m=1}^{N} \omega_m^2 t_{mn}^2, \quad d_{nm} = \sum_{k=1}^{N} \omega_k^2 t_{kn} t_{km} \tag{11.12}$$

where t_{mn} are the elements of the transformation matrix $\boldsymbol{T}$. In Equation 11.11 the effective-mode couples to the system and to all the residual bath modes via the d_{nm} terms. The couplings $\{d_{nm}\}$ within the residual bath part $H_{B'}$ can be chosen in various ways, since there is some freedom in defining the transformation $\boldsymbol{T}$. In particular, the couplings between the residual modes may be completely eliminated while all modes couple to the effective-mode X_1 via $H_{\text{eff},B'}$ of Equation 11.9. Alternatively, the coupling matrix $\boldsymbol{d}$ may be cast into a chain-type, band-diagonal form [48,50–53] with nearest-neighbor couplings within the residual bath space, along with a nearest-neighbor (X_1-X_2) coupling in $H_{\text{eff},B'}$,

$$H_{\text{eff},B'} = d_{12} X_1 X_2$$

$$H_{B'} = \frac{1}{2} \sum_{n=2}^{N} (P_n^2 + \Omega_n'^2 X_n^2) + \sum_{n=2}^{N} d_{n\,n+1} X_n X_{n+1}. \tag{11.13}$$

This representation has been referred to as the hierarchical electron–phonon (HEP) model [50,51].

11.2.3 EFFECTIVE-MODE CHAIN INCLUDING MARKOVIAN CLOSURE

In the following non-Markovian model, a band-diagonal (tri-diagonal) form of the residual bath couplings is adopted [50,52,53] in conjunction with a Markovian closure acting only on the Mth final member of the Mori-type effective-mode chain. In the following discussion the Caldeira–Leggett model [54] is used to describe the influence of the bath on X_M. To connect to the Caldeira–Leggett model, a potential renormalization counter-term for X_M

$$H_c = X_M^2 \sum_{n=M+1}^{N} \frac{d_{M,n}^2}{2\Omega_n^2} \tag{11.14}$$

is included in the Hamiltonian to give

$$H = H_S + Dx_0 X_1 + \sum_{n=1}^{N} \frac{1}{2}(P_n^2 + \Omega_n^2 X_n^2)$$

$$+ \sum_{n=2}^{M} d_{n-1,n} X_{n-1} X_n + X_M \sum_{n=M+1}^{N} d_{M,n} X_n + X_M^2 \sum_{n=M+1}^{N} \frac{d_{M,n}^2}{2\Omega_n^2}. \tag{11.15}$$

In Equation 11.15 the second summation represents the coupling between successive modes of the chain. The third summation represents the coupling between the remaining bath modes and X_M is the final member of the chain.

11.2.4 SPECTRAL DENSITIES

In the Caldeira–Leggett model the coupling between the bath modes and X_M is defined by an Ohmic spectral density $J(\Omega)$. In mass-weighted coordinates $J(\Omega)$ is given by

$$J(\Omega) = \frac{\pi}{2} \sum_n \frac{d_{M,n}^2}{\Omega_n} \delta(\Omega - \Omega_n) = 2\gamma\Omega \exp(-\Omega/\Lambda) \tag{11.16}$$

where γ is a frequency independent friction coefficient and Λ is a cut-off frequency for the bath modes. Although the dynamics of the simple Caldeira–Leggett model of a system coupled to a bath is Markovian (within the high-temperature approximation), for the hierarchical representation presented here, where the coupling between the system and the Mth mode is ohmic, the overall system-bath dynamics is strongly non-Markovian. For the case where the chain is truncated at the first order $M = 1$ effective-mode level it was demonstrated by Garg *et al.* [43] that the transformation from the Hamiltonian of Equations 11.1 through 11.15 has a spectral density given (in mass-weighted coordinates) by

$$J_{\text{eff}}^{(1)}(\omega) = \frac{2\gamma\omega D^2}{(\Omega_1^2 - \omega^2)^2 + 4\gamma^2\omega^2}. \tag{11.17}$$

Here, D is the transformed coupling coefficient of Equation 11.7, Ω_1 is the effective-mode frequency $\Omega_1^2 = \sum_{m=1}^{N} \omega_m^2 t_{m1}^2$ of Equation 11.8, and γ is the friction coefficient that defines the Ohmic spectral density of Equation 11.16 for the residual bath modes. It is shown in Refs. [48, 49] that truncation at a higher order for the chain enables the construction of more complicated spectral densities that could provide suitable models for reproducing various experimental spectral densities.

If the residual bath conforms to an Ohmic distribution, so that the Hamiltonians of Equations 11.7 through 11.10 and Equations 11.3 through 11.4 are related by a unitary transformation, Equation 11.17 is obtained under the condition that the cut-off Λ of Equation 11.16 is large, see Refs. [43,48,49]. (Another caveat again concerns the fact that this is strictly valid only in the high-temperature limit, where the residual bath is memory-free.)

The generalization of the construction of the spectral density to higher orders has been described in detail in Ref. [48]. Here, we summarize the key aspects of the derivation. For a bilinear $x_0 X_1$ coupling and a zeroth-order system Hamiltonian of the form $H_S = p_0^2/2 + V(x_0)$, the spectral function may be derived from the Fourier–Laplace transform of the equations of motion for the dynamical variables $\{x_0, X_1 \ldots X_M\}$. In the transformed form the equation of motion for $x_0(z)$ is given by [43,48]

$$\hat{K}^{(M)}(z)\hat{x}_0(z) = -V_z'(x_0) \tag{11.18}$$

where

$$\hat{x}_0(z) = \int_0^\infty x_0(t) \exp(-izt)dt, \quad \mathrm{Im}z < 0 \tag{11.19}$$

for $z = \omega - i\epsilon$. In Equation 11.18, $V_z'(x_0)$ is the Fourier–Laplace transform of $V'(x_0)$, the spatial derivative of the system potential.

It is shown in Ref. [48] that $\hat{K}^{(M)}(z)$ is given by the continued fraction

$$\hat{K}^{(M)}(z)\hat{x}_0(z)$$

$$= -z^2\hat{x}_0(z) - \cfrac{D^2}{\Omega_1^2 - z^2 - \cfrac{d_{1,2}^2}{\Omega_2^2 - z^2 - \cdots \cfrac{d_{M-2,M-1}^2}{\Omega_{M-1}^2 - z^2 - \cfrac{d_{M-1,M}^2}{\Omega_M^2 - z^2 + i2\gamma z}}}}\hat{x}_0(z) \tag{11.20}$$

where the $\{\Omega_n\}$ and $\{d_{nm}\}$ are the diagonal and off-diagonal components of the transformed frequency matrix, see Equation 11.13. The spectral density $J_{\mathrm{eff}}^{(M)}(\omega)$ associated with the Mth-order model Equations 11.3 through 11.4 is related to $\hat{K}^{(M)}(z)$ by Refs [43,55]

$$J_{\mathrm{eff}}^{(M)}(\omega) = \lim_{\epsilon \to 0^+} \mathrm{Im}\hat{K}^{(M)}(\omega - i\epsilon). \tag{11.21}$$

11.2.5 REDUCED-DIMENSIONAL EFFECTIVE-MODE CHAIN INCLUDING MARKOVIAN CLOSURE

In line with the above form of the Mth order spectral density, the explicit representation of the bath modes $(M + 1, \ldots, N)$ in Equation (11.15) can be replaced by a formulation of the dynamics in terms of the following master equation:

$$\frac{\partial \rho^{(M)}(t)}{\partial t} = -\frac{i}{\hbar}[H^{(M)}, \rho^{(M)}] + \mathcal{L}_{\mathrm{diss}}^{(M)}\rho^{(M)}. \tag{11.22}$$

Here, $\rho^{(M)}$ is a reduced-density matrix comprising all modes of the effective-mode hierarchy that are included explicitly in the dynamics, i.e., $\rho^{(M)} = \mathrm{Tr}_{n>M}\{\rho\}$ where ρ is the density matrix of the full-dimensional system comprising N bath modes.

The dissipative Liouvillian $\mathcal{L}_{\text{diss}}^{(M)}$ acting at the Mth order of the hierarchy chosen as a Caldeira–Leggett [54] form is given by

$$\mathcal{L}_{\text{diss}}^{(M)} \rho^{(M)} = -\frac{i\gamma}{\hbar}[X_M, [P_M, \rho^{(M)}]_+] - \frac{2\gamma E_{\text{th}}}{\hbar^2}[X_M, [X_M, \rho^{(M)}]] \qquad (11.23)$$

where γ is the friction coefficient and $E_{\text{th}} = \hbar\Omega_M/2 \coth[\hbar\Omega_M/(2k_B T)]$ is the mean equilibrium energy of the Mth harmonic mode; k_B is Boltzmann's constant and T is the temperature. Here, the Caldeira–Leggett equation has been formally extended from the high-temperature limit (where the equation is rigorously valid) to low temperatures and zero temperature. This extension is consistent with a quantum fluctuation-dissipation theorem [58,59], but does not correctly account for non-exponential effects that are typical of the low-temperature regime. Still, remarkably good agreement is often observed with explicit quantum-dynamical calculations at zero temperature, and with predictions of zero-temperature decoherence rates [60, 61]. This trend is also confirmed by the results of Refs. [48, 49].

In Section 11.2.6 the hybrid hydrodynamic-Liouville approach is described for the non-Markovian situation described in the previous sections. For the hybrid hydrodynamic-Liouville approach the equations are more easily derived from a Liouville phase-space representation for the system and effective modes. For the phase-space distribution function $\rho_W(q, p, Q, P)$, where q, p are the position and momentum coordinates for the subsystem parts and Q, P are the M-dimensional position and M-dimensional momenta for the effective modes of the chain, the equation of motion that corresponds to Equation 11.22 with the dissipative form of Equation 11.23 is given in Liouville phase-space by

$$\frac{\partial \rho_W}{\partial t} = -p\frac{\partial \rho_W}{\partial q} + \sum_{\substack{k=1 \\ \text{odd}}}^{\infty} \frac{1}{k!}\left(\frac{\hbar}{2i}\right)^{k-1}\frac{\partial^k V}{\partial q^k}\frac{\partial^k \rho_W}{\partial p^k} + DQ_1\frac{\partial \rho_W}{\partial p} + Dq\frac{\partial \rho_W}{\partial P_1}$$

$$+ \sum_{m=1}^{M}\left(\Omega_m^2 Q_m\frac{\partial \rho_W}{\partial P_m} - P_m\frac{\partial \rho_W}{\partial Q_m}\right) + \sum_{m=2}^{M} d_{m-1,m}\left(Q_m\frac{\partial \rho_W}{\partial P_{m-1}} + Q_{m-1}\frac{\partial \rho_W}{\partial P_m}\right)$$

$$+ \gamma\frac{\partial P_M \rho_W}{\partial P_M} + \gamma E_{\text{th}}\frac{\partial^2 \rho_W}{\partial P_M^2}. \qquad (11.24)$$

This equation of motion for the Wigner function is exact for the bath model that we are considering here. If the bath was anharmonic, the quantum-classical Liouville equation would be employed [35–40] which disregards the $\hbar$ correction terms in the classical sector.

11.2.6 PARTIAL MOMENTS

Central to the formulation of the hybrid hydrodynamic-Liouville approach is the construction of *partial* hydrodynamic moments which may be derived from the density operator in the coordinate representation $\rho(x_0, x_0', X, X')$ $(X = X_1 \ldots X_M)$ or by an integration over the momentum part of the quantum subsystem for the complete phase-space,

$$\langle \mathcal{P}^n \rho \rangle_{qQP} = \int dp \; p^n \rho_W(q, p; Q, P) \tag{11.25}$$

where $q = 1/2(x_0 + x_0')$ is associated with the hydrodynamic space. In the coordinate representation the moments may also be defined by differentiation of $\rho(x_0, x_0', X, X')$ with respect to the difference coordinate $r = x_0 - x_0'$

$$\langle \mathcal{P}^n \rho \rangle_{qQP} = \frac{1}{2\pi\hbar} \int dR \; \exp(-i P R/\hbar) \left(\frac{\hbar}{i}\right)^n \frac{\partial^n}{\partial r^n} \rho(q, r; Q, R) \Big|_{r=0}. \tag{11.26}$$

The coordinate-space density plays the role of a moment-generating function where the hydrodynamic moments correspond to the coefficients of the Taylor expansion of the coordinate-space density with respect to the coordinate r,

$$\rho(q, r; Q, P) = \sum_n \frac{1}{n!} \langle \mathcal{P}^n \rho \rangle_{qQP} \left(\frac{ir}{\hbar}\right)^n. \tag{11.27}$$

The equations of motion for the partial moments may be obtained from the coordinate representation of the density matrix or from the phase-space distribution function $\rho_W(q, p; Q, P)$. The phase-space representation presents the easiest option for deriving the equations of motion, and these are obtained by applying Equation 11.25 to Equation 11.24. The resulting equations of motion for the partial moments are given by

$$
\begin{aligned}
\frac{\partial \langle \mathcal{P}^n \rho \rangle_{qQP}}{\partial t} = & -\frac{\partial}{\partial q} \langle \mathcal{P}^{n+1} \rho \rangle_{qQP} - n\left(\frac{\partial V(q)}{\partial q} + DQ_1\right) \langle \mathcal{P}^{n-1} \rho \rangle_{qQP} \\
& - \sum_{\substack{l_1=3 \\ \text{odd}}}^{n} \binom{n}{l_1} \left(\frac{\hbar}{2i}\right)^{l_1-1} \frac{\partial^{l_1} V(q)}{\partial q^{l_1}} \langle \mathcal{P}^{n-l_1} \rho \rangle_{qQP} \\
& + \sum_{m=1}^{M} \left(\Omega_m^2 Q_m \frac{\partial}{\partial P_m} - P_m \frac{\partial}{\partial Q_m}\right) \langle \mathcal{P}^n \rho \rangle_{qQP} \\
& + Dq \frac{\partial}{\partial P_1} \langle \mathcal{P}^n \rho \rangle_{qQP} \\
& + \sum_{m=2}^{M} d_{m-1,m} \left(Q_m \frac{\partial}{\partial P_{m-1}} + Q_{m-1} \frac{\partial}{\partial P_m}\right) \langle \mathcal{P}^n \rho \rangle_{qQP} \\
& + \gamma \frac{\partial P_M \langle \mathcal{P}^n \rho \rangle_{qQP}}{\partial P_M} + \gamma E_{\text{th}} \frac{\partial^2 \langle \mathcal{P}^n \rho \rangle_{qQP}}{\partial P_M^2}.
\end{aligned}
\tag{11.28}
$$

The equations of motion display coupling to both higher and lower moments—the up-coupling result from the kinetic energy-term and the down-coupling results from the terms involving the potential energy. The implication of the coupling is that an infinite hierarchy of equations of motion is obtained. For the general case of mixed states the hierarchy is non-convergent and moment closure schemes [62–65] need to be implemented to terminate the hierarchy. However, for Gaussian densities the moment hierarchy can be terminated analytically by representing the third moment in terms of the lower moments [66–68]

$$\langle \mathcal{P}^3 \rho \rangle_{qQP} = \bar{p}^3(q) \langle \rho \rangle_{qQP} + 3\bar{p}_{qQP} \left[\langle \mathcal{P}^2 \rho \rangle_{qQP} - \bar{p}_{qQP}^2 \langle \rho \rangle_{qQP} \right] \qquad (11.29)$$

where $\bar{p}_{qQP}^2$ is defined in the next section. In this chapter we focus on Gaussian densities in harmonic systems where it is well-known that Gaussian densities maintain their Gaussian form throughout a propagation.

11.2.7 LAGRANGIAN TRAJECTORIES

The equation of motion for the partial moments defined in Equation 11.28 was formulated for an Eulerian framework. To analyze trajectories for the partial moments the equation of motion needs to be transformed to a Lagrangian framework. In a Lagrangian framework the coordinates q, Q, P become dynamical variables and are interpreted as fluid particles that evolve along hydrodynamic-Liouville space trajectories. For the non-dissipative case the trajectory equations of motion $\dot{q}, \dot{Q}, \dot{P}$ can be derived by interpreting the zeroth moment as a continuity equation in the hybrid hydrodynamic-Liouville space,

$$\frac{\partial \langle \rho \rangle_{qQP}}{\partial t} = -\nabla_{qQP} \cdot \mathbf{j}_{qQP} \qquad (11.30)$$

with $\nabla_{qQP} = (\partial/\partial q, \partial/\partial Q, \partial/\partial P)$ and the current

$$\frac{\mathbf{j}_{qQP}}{\langle \rho \rangle_{qQP}} = \begin{pmatrix} \dot{q} \\ \dot{Q} \\ \dot{P} \end{pmatrix} = \begin{pmatrix} \bar{p}_{qQP} \\ (\partial H/\partial P) \\ -(\partial H/\partial Q) \end{pmatrix} \qquad (11.31)$$

where the momentum field $\bar{p}_{qQP}$ is given by

$$\bar{p}_{qQP} = \frac{\langle \mathcal{P} \rho \rangle_{qQP}}{\langle \rho \rangle_{qQP}}. \qquad (11.32)$$

The quantity $\bar{p}_{qQP}$ represents the *average* momentum derived from the underlying Wigner distribution for a given combination of independent variables (q, Q, P). At the Mth truncation of the chain the trajectories are explicitly expressed as

$$\dot{q} = \bar{p}_{qQP}$$

$$\dot{\bar{p}}_{qQP} = -\frac{\partial}{\partial q} V(q) + DQ_1 + F_{\text{hyd}}$$

$$\dot{Q}_1 = P_1$$

$$\dot{P}_1 = -\Omega_1^2 Q_1 - Dq - d_{12} Q_2$$

$$\cdots$$

$$\dot{Q}_M = P_M$$

$$\dot{P}_M = -\Omega_M^2 Q_M - d_{M-1,M} Q_{M-1} \qquad (11.33)$$

where the hydrodynamic force $F_{hyd}(q, Q, P)$ is given by

$$
\begin{aligned}
F_{hyd}(q, Q, P) &= \frac{1}{\langle \rho \rangle_{qQP}} \frac{\partial}{\partial q} \sigma_{qQP} \\
&= \frac{1}{\langle \rho \rangle_{qQP}} \frac{\partial}{\partial q} \left(\langle \mathcal{P}^2 \rho \rangle_{qQP} - \bar{p}_{qQP}^2 \langle \rho \rangle_{qQP} \right).
\end{aligned}
\tag{11.34}
$$

For the dissipative case described in this chapter a Langevin description is implemented for the Mth order effective mode [69]

$$
\dot{P}_M = -\Omega_M^2 Q_M - d_{M-1,M} Q_{M-1} - \gamma P_M + R(t) \tag{11.35}
$$

where $R(t)$ is a random force distributed according to the Gaussian [70]

$$
\text{Prob}[R(t)] = \frac{\exp(0.5 R^2(t)/\langle R^2 \rangle)}{\sqrt{2\pi \langle R^2 \rangle}} \tag{11.36}
$$

and satisfying $\langle R(t) \rangle = 0$, $\langle R(t)R(t') \rangle = 2\gamma k_B T \delta(t - t')$.

11.3   RESULTS AND DISCUSSION

In the following sections trajectories generated from Equations 11.33 and 11.35 are illustrated for the reduced dimensional effective-mode chain model of Sections 11.2.5 and 11.2.6. For the examples demonstrated, the Markovian closure is implemented at the $M = 1$ level of the hierarchy. The overall picture for this model is that of a subsystem bilinearly coupled to a single effective mode that is itself Ohmically coupled to a Markovian bath. Examples using higher order effective-mode hierarchies are described in Refs. [48,49]. The system-bath spectral density $J^{(1)}(\omega)$ at the $M = 1$ level is given by Equation 11.17 and corresponds to the form described by Garg et al. [43]. In the following examples the influence of the residual bath on the trajectories of the effective mode and the system are investigated in terms of the friction strength γ and temperature.

The initial condition for the Gaussian density given by

$$
\langle \rho \rangle_{qQP} = \frac{\sqrt{\pi \omega_0}}{\pi^2} \exp(-\omega_0(q - q_0)^2 - \Omega_1(Q - Q_0)^2 - (P - P_0)^2/\Omega^2) \tag{11.37}
$$

(at time $t = 0$, $\langle \mathcal{P}\rho \rangle_{qQP} = 0$) are summarized in Table 11.1. Also shown in Table 11.1 are the frequency and coupling parameters used in the examples of this chapter. Unless otherwise stated atomic units (a.u.) are used throughout. The initial condition was a coherent state Gaussian displaced to $q_0 = 47.14$ along the subsystem coordinate and $Q_0 = P_0 = 0$ for the effective mode.

For the parameters defined in Table 11.1 the spectral density is depicted in Figure 11.1a for various strengths of γ. For the smallest value of $\gamma = 5 \times 10^{-5}$, $J_{eff}^{(1)}(\omega)$ is quite sharply peaked around Ω_1. Increasing γ leads to a broadening of the peak, and for a large friction term $\gamma = 5 \times 10^{-3}$ the peak is shifted to lower frequencies

TABLE 11.1

Parameters, Quoted in Atomic Units, Used for the Examples Described in Section 11.3

ω_0	Ω_1	$D(\times 10^7)$	q_0	Q_0	P_0
0.0018	0.002	4.0	47.14	0	0

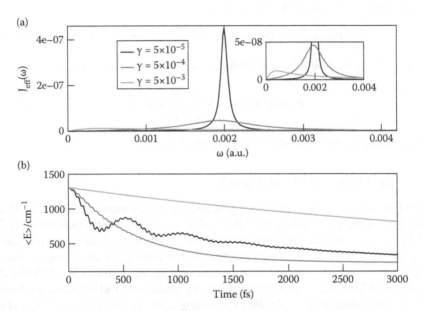

FIGURE 11.1 (a) Spectral density $J_{eff}^{(1)}(\omega)$ for three values of the friction $\gamma = 5 \times 10^{-5}$, 5×10^{-4} and 5×10^{-3}. Also depicted in the inset is a magnification of $J_{eff}^{(1)}(\omega)$. (b) Energy relaxation for the spectral densities depicted in (a) for a temperature $T = 0$ K and parameters defined in Table 11.1.

and the distribution becomes increasingly skewed (see the inset of Figure 11.1a). The corresponding relaxation of the system mean energy $\langle E \rangle$ for temperature $T = 0$ K is depicted in Figure 11.1b. For $\gamma = 5 \times 10^{-5}$, $\langle E \rangle$ displays strong recurrences as a result of energy transfer between $q \leftrightarrow Q$. The amplitude of the recurrences diminishes over time. The rate of energy transfer depends on the amplitude of the oscillation dynamics along the Q coordinate. A large amplitude oscillation along Q facilitates greater transfer of energy. This is clearly illustrated by the trajectories depicted in Figure 11.2b for the Q coordinate, where the trajectories oscillate with an increasingly larger amplitude up until around 250–300 fs. The amplitude of oscillation then diminishes until around 550 fs where there is only a small amplitude along Q. For longer timescales the oscillations along Q are damped by the coupling of the effective mode

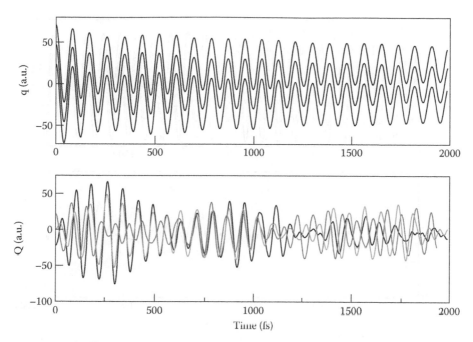

FIGURE 11.2   For temperature $T = 0$ K and friction $\gamma = 5 \times 10^{-3}$: (a) trajectories along the subsystem q coordinate for initial condition $Q = P = 0$; (b) trajectories along the effective-mode coordinate for initial condition $q = P = 0$.

to the residual bath. Throughout the propagation the hydrodynamic trajectories along q maintain a coherent oscillatory pattern with a gradual damping of the oscillation.

For a larger friction of $\gamma = 5 \times 10^{-4}$, $\langle E \rangle$ relaxes with no observable back transfer of energy from the effective mode to the system. Here, any oscillation along Q is rapidly damped by the strong coupling of the effective mode to the residual bath. For the large $\gamma = 5 \times 10^{-3}$, the Q mode becomes strongly overdamped and the relaxation rate of $\langle E \rangle$ reduces significantly. The Q trajectories are depicted in Figure 11.3b and due to the increasing influence of the random force in the fluctuation term in Equation 11.35, the trajectories display stochastic motion. In Figure 11.3a the q trajectories still maintain a coherent oscillatory pattern but due to the overdamping of the effective mode there is hardly any noticeable damping of the amplitude of oscillation along q.

The stochastic nature of the trajectories for the effective-mode sector is more prominent in the P trajectories. As depicted in Figure 11.4 fluctuations in the momentum trajectories are manifested in the smaller $\gamma = 5 \times 10^{-5}$ case. The trajectories oscillate with the frequency of the effective mode but the influence of the random force is also discernable. For the larger γ values the P oscillation is damped so rapidly that no large amplitude oscillation occurs, only stochastic motion around the mean $\langle P \rangle = 0$. Due to the rapid damping of P, there is no discernable Q vibration which is required to facilitate dissipation for the subsystem part.

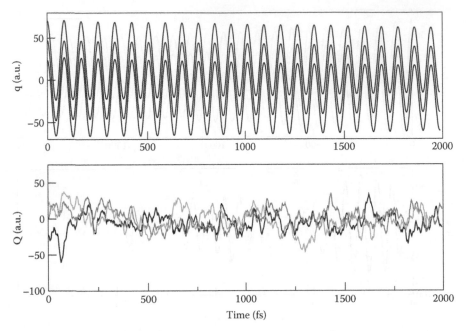

FIGURE 11.3 For temperature $T = 0$ K and friction $\gamma = 5 \times 10^{-3}$: (a) trajectories along the subsystem q coordinate for initial condition $Q = P = 0$; (b) trajectories along the effective-mode coordinate for initial condition $q = P = 0$.

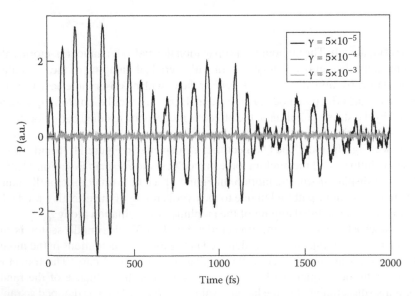

FIGURE 11.4 For temperature $T = 0$ K and the three γ terms shown the effective-mode momentum trajectories are illustrated. For all three examples the initial condition $q = Q = P = 0$ is used.

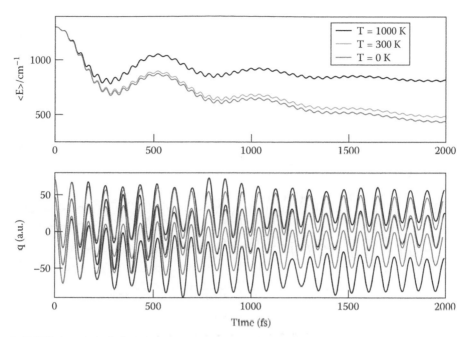

FIGURE 11.5 (a) Energy relaxation at different temperatures computed using a fixed value of $\gamma = 5 \times 10^{-5}$. (b) Trajectories along the subsystem q coordinate for initial condition $Q = P = 0$ at $T = 1000$ K (black). Also shown for comparison are the corresponding trajectories (grey) for the $T = 0$ K.

The influence of temperature on the dynamics is depicted in Figure 11.5 for fixed $\gamma = 5 \times 10^{-5}$ and temperatures of $T = 0$, 300 and 1000 K. In Figure 11.5a, the relaxation of $\langle E \rangle$ is depicted for the three temperatures. In all three cases the influence of the effective mode on $\langle E \rangle$ is manifested by the transfer of energy between the $q \leftrightarrow Q$ modes. Due to the coupling to the residual bath the subsystem eventually relaxes to $\langle E \rangle = E_{\text{th}}$. The width of the relaxed equilibrium density $\rho_{\text{eq}}(qQP)$ increases with temperature and this is manifested in the trajectories. As depicted in Figure 11.5b the trajectories for the q sector become more widely separated for the higher temperature residual bath; thus indicating a wider distribution for $\rho_{\text{eq}}(qQP)$.

As a final example we consider the case where the subsystem frequency is resonant with the effective-mode frequency. In Figure 11.6a the relaxation of $\langle E \rangle$ is depicted for the case where ω_0 is (i) resonant with Ω_1 and (ii) off-resonant with Ω_1. The presence of resonances increases the overall rate of energy dissipation to the bath. From the pre-transformed direct system-bath perspective of Section 11.2.1, the spectral density $J_{\text{eff}}^{(1)}(\omega)$, which is a measure of the strength of the system-bath coupling, is sharply peaked at Ω_1, and so it is expected that the rate of energy dissipation is greatest when ω_0 is resonant with the peak in $J_{\text{eff}}^{(1)}(\omega)$. From a post-transformed effective-mode interpretation (which has been the main focus of this chapter) the presence of resonances enhances the energy transfer between the $q \leftrightarrow Q$ modes. With more

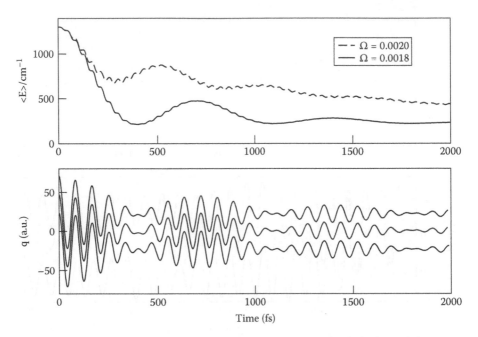

FIGURE 11.6   (a) Energy relaxation dynamics $\langle E \rangle$. In one example (dashed curve) the system harmonic frequency $\omega_0 = 0.0018$ is off-resonant with the effective-mode frequency Ω_1; in the other example (solid curve) ω_0 is resonant with Ω_1. (b) Trajectories along q for the resonant case.

energy being transferred to the effective mode the overall rate of energy dissipation increases as a result of the effective mode to the residual bath coupling.

The q trajectories for the resonant case are depicted in Figure 11.6b. The amplitude of oscillation along q is strongly correlated to the rate of energy transfer. In Figure 11.6 the amplitude of trajectory oscillation along q varies synchronously with the rate of energy transfer—the dips in $\langle E \rangle$ coincide with very small amplitude oscillations along q.

11.4   CONCLUSION

In this chapter a hybrid hydrodynamic phase-space approach to non-Markovian dynamics was described. The non-Markovian approach involves a hierarchical chain of coupled effective modes that is terminated by coupling the final Mth member of the chain to a Markovian bath. This Mori-type approach gives rise to a hierarchy of spectral densities that are defined in Equation 11.21 as the imaginary part of a continued fraction obtained from the classical equation of motion for the subsystem coordinate variable x_0 (see Ref. [48] for a detailed derivation).

In the reduced dimensional model obtained from the effective-mode construction, the subsystem part was described in a hydrodynamic setting and the effective modes were represented in a Liouville phase-space setting. Starting from a $2M$ phase-space, a reduction involving the partial moment construction of Equation 11.25 was carried out, resulting in a hierarchy of equations of motion that are obtained for these moment quantities. This hierarchy results from the up and down coupling between the moments and is unrelated to the effective-mode hierarchy. The effective-mode hierarchy determines the dimensionality of the reduced model while the hydrodynamic hierarchy affects the number of coupled equations of motion for Equation 11.28.

For Gaussian densities of the completely harmonic reduced dimensional model described here, the hydrodynamic hierarchy terminates at the second-order level. However, for anharmonic systems the hierarchy can only be terminated by implementing an approximate closure scheme, such as the Gauss–Hermite approach recently developed by Hughes et al. [62]. A challenge in developing a closure scheme for quantum hydrodynamics is finding a low-order closure to the hierarchy. A large number of moments are generally required to account for quantum effects. For one-dimensional problems a high-order closure to the hierarchy poses no computational challenges. However, for large dimensional problems the number of moments grows considerably and soon becomes computationally challenging.

The equations of motion could well have been formulated completely in a hydrodynamic setting; however, the advantage of a partial hydrodynamic representation becomes clear when considering anharmonic systems that require approximate moment closure schemes. For a one-dimensional subsystem the number of moments involved that includes the effective modes in a completely hydrodynamic setting is $\sum_{n=0}^{N}(M+1)^n$ where M is the number of effective modes in the reduced dimensional model and N is the order of the highest moment in the closure. For the hybrid hydrodynamic phase-space approach described here only N moments would be required for the closure.

For the examples described in this chapter the effective-mode hierarchy was truncated at the $M = 1$ level. Higher order models have been discussed in Refs. [48, 49] where it was shown that reconstruction of the spectral density from the effective-mode chain representation can account for quite complicated spectral densities.

The trajectories formulated in Equations 11.33 and 11.35 capture an intuitive picture of the non-Markovian behavior of the system-bath dynamics. From the transformed effective-mode representation the trajectories clearly displayed the energy transfer between the modes. The dissipative influence of the bath was buffered, or delayed, by the coupling to the effective mode and this was also manifested in the trajectories.

ACKNOWLEDGMENTS

We thank Klaus B. Møller for useful discussions.

BIBLIOGRAPHY

1. L. de Broglie, *Compt. Rend.*, **183**, 447(1926).
2. L. de Broglie, *Tentative d'interprétation causale et non-linéaire de la mécanique ondulatoire*, Gauthier-Villars, Paris (1956).

3. D. Bohm, *Phys. Rev.*, **85**, 166 (1952).
4. D. Bohm, *Phys. Rev.*, **85**, 180 (1952).
5. J.S. Bell, *Speakable and Unspeakable in Quantum Mechanics*, Cambridge University Press, Cambridge (1989).
6. P.R. Holland, *The Quantum Theory of Motion*, Cambridge University Press, New York (1993).
7. R.E. Wyatt, *Quantum Dynamics with Trajectories: Introduction to Quantum Hydrodynamics*, Springer, Heidelberg (2005).
8. V. Allori, V. Dürr, S. Goldstein, and N. Zanghì, *J. Opt. B*, **4**, 482 (2002).
9. C.L. Lopreore and R.E. Wyatt, *Phys. Rev. Lett.*, **82**, 5190 (1999).
10. R.E. Wyatt, *Chem. Phys. Lett.*, **313**, 189 (1999).
11. R.E. Wyatt and E.R. Bittner, *J. Chem. Phys.*, **113**, 8898 (2000).
12. B.K. Dey, A. Askar and H.A. Rabitz, *J. Chem. Phys.*, **109**, 8770 (1998).
13. C.J. Trahan, K.H. Hughes and R.E. Wyatt, *J. Chem. Phys.*, **118**, 9911 (2003).
14. B.K. Kendrick, *J. Chem. Phys.*, **119**, 5805 (2003).
15. J.B. Maddox and E.R. Bittner, *J. Phys. Chem. B*, **106**, 7981 (2002).
16. V.A. Rassolov and S. Garashchuk, *J. Chem. Phys.*, **121**, 8711 (2004).
17. Y. Zhao and N. Makri, *J. Chem. Phys.*, **119**, 60 (2003).
18. Y. Goldfarb, I. Degani and D.J. Tannor, *J. Chem. Phys.*, **125**, 231103 (2006).
19. T. Takabayasi, *Prog. Theoret. Phys.*, **11**, 341 (1954).
20. J.E. Moyal, *Proc. Cambridge Philos. Soc.*, **45**, 99 (1949).
21. J.H. Irving and R.W. Zwanzig, *J. Chem. Phys.*, **19**, 1173 (1951).
22. H. Fröhlich, *Physica*, **37**, 215 (1967).
23. J. Yvon, *J. Phys. Lettres*, **39**, 363 (1978).
24. M. Ploszajczak and M.J. Rhoades-Brown, *Phys. Rev. Lett.*, **55**, 147 (1985).
25. L.M. Johansen, *Phys. Rev. Lett.*, **80**, 5461 (1998).
26. I. Burghardt and L.S. Cederbaum, *J. Chem. Phys.*, **115**, 10303 (2001).
27. J.G. Muga, R. Sala and R.F. Snider, *Physica Scripta*, **47**, 732 (1993).
28. I. Burghardt and K.B. Møller, *J. Chem. Phys.*, **117**, 7409 (2002).
29. I. Burghardt and L.S. Cederbaum, *J. Chem. Phys.*, **115**, 10312 (2001).
30. I. Burghardt, K.B. Møller, G. Parlant, L.S. Cederbaum and E.R. Bittner, *Int. J. Quant. Chem.*, **100**, 1153 (2004).
31. I. Burghardt, K.B. Møller and K.H. Hughes, Quantum hydrodynamics and a moment approach to quantum-classical theory. In *Quantum Dynamics of Complex Molecular Systems*, edited by D.A. Micha and I. Burghardt, Springer, Berlin (2006).
32. I. Burghardt and G. Parlant, *J. Chem. Phys.*, **120**, 3055 (2004).
33. I. Burghardt, *J. Chem. Phys.*, **122**, 094103 (2005).
34. K.H. Hughes, S.M. Parry, G. Parlant and I. Burghardt, *J. Phys. Chem. A*, **111**, 10269 (2009).
35. C.C. Martens and J.-Y. Fang, *J. Chem. Phys.*, **106**, 4918 (1996).
36. A. Donoso and C.C. Martens, *J. Phys. Chem.*, **102**, 4291 (1998).
37. R. Kapral and G. Ciccotti, *J. Chem. Phys.*, **110**, 8919 (1999).
38. S. Nielsen, R. Kapral and G. Ciccotti, *J. Chem. Phys.*, **115**, 5805 (2001).
39. J. Caro and L.L. Salcedo, *Phys. Rev. A*, **60**, 842 (1999).
40. I. Horenko, M. Weiser, B. Schmidt and C. Schütte, *J. Chem. Phys.*, **120**, 8913 (2004).
41. S. Mukamel, *Principles of Nonlinear Optical Spectroscopy*, Oxford University Press, New York (1995).
42. Y. Tanimura and S. Mukamel, *J. Phys. Soc. Jpn.*, **163**, 66 (1994).
43. A. Garg, J.N. Onuchic and V. Ambegaokar, *J. Chem. Phys.*, **83**, 4491 (1985).

44. H. Köppel, W. Domcke and L.S. Cederbaum, *Adv. Chem. Phys.*, **57**, 59 (1984).
45. L.S. Cederbaum, E. Gindensperger and I. Burghardt, *Phys. Rev. Lett.*, **94**, 113003 (2005).
46. E. Gindensperger, I. Burghardt and L.S. Cederbaum, *J. Chem. Phys.*, **124**, 144103 (2006).
47. E. Gindensperger, I. Burghardt and L.S. Cederbaum, *J. Chem. Phys.*, **124**, 144104 (2006).
48. K.H. Hughes, I. Burghardt and C.D. Christ, *J. Chem. Phys.*, **131**, 024109 (2009).
49. K.H. Hughes, I. Burghardt and C.D. Christ, *J. Chem. Phys.*, **131**, 124108 (2009).
50. H. Tamura, E.R. Bittner and I. Burghardt, *J. Chem. Phys.*, **127**, 034706 (2007).
51. H. Tamura, J.G.S. Ramon, E.R. Bittner and I. Burghardt, *J. Phys. Chem. B*, **112**, 495 (2008).
52. I. Burghardt and H. Tamura, Non-Markovian dynamics at a conical intersection: effective-mode models for the short-time dynamics and beyond. In *Dynamics of Open Quantum Systems*, edited by K.H. Hughes, CCP6, Daresbury (2007).
53. E. Gindensperger, H. Köppel and L.S. Cederbaum, *J. Chem. Phys.*, **126**, 034106 (2007).
54. A.O. Caldeira and A.J. Legget, *Physica A*, **121**, 587 (1983).
55. A. Leggett, *Phys. Rev. B*, **30**, 1208 (1984).
56. H. Mori, *Prog. Theoret. Phys.*, **33**, 423 (1965).
57. H. Mori, *Prog. Theoret. Phys.*, **34**, 399 (1965).
58. U. Weiss, *Quantum Dissipative Systems*, World Scientific, Singapore (1999).
59. P. Hänggi and G. Ingold, *Chaos*, **15**, 026105 (2005).
60. J.P. Paz, S. Habib and W.H. Zurek, *Phys. Rev. D*, **47**, 488 (1993).
61. I. Burghardt, M. Nest and G.A. Worth, *J. Chem. Phys.*, **119**, 5364 (2003).
62. K.H. Hughes, S.M. Parry and I. Burghardt, *J. Chem. Phys.*, **130**, 054115 (2009).
63. P. Degond and C. Ringhofer, *J. Stat. Phys.*, **112**, 587 (2003).
64. C.D. Levermore, *J. Stat. Phys.*, **83**, 1021 (1996).
65. M. Junk, *Math. Models Methods Appl. Sci.*, **10**, 1001 (2000).
66. I. Burghardt and L.S. Cederbaum, *J. Chem. Phys.*, **115**, 10303 (2001).
67. J.B. Maddox, E.R. Bittner and I. Burghardt, *Int. J. Quant. Chem.*, **89**, 313 (2002).
68. I. Burghardt and K.B. Møller, *J. Chem. Phys.*, **117**, 7409 (2002).
69. K.B. Møller and I. Burghardt, Dissipative quantum dynamics with trajectories. In *Dynamics of Open Quantum Systems*, edited by K.H. Hughes, CCP6, Daresbury (2006).
70. J.M. Thijssen, *Computational Physics*, Cambridge University Press, Cambridge (2000).

12 Quantum Fluid Dynamics within the Framework of Density Functional Theory

Swapan K. Ghosh

CONTENTS

12.1 INTRODUCTION

The concepts of quantum mechanics (QM) are so radically different from those of classical mechanics that soon after Schrödinger proposed the concept of quantum mechanical wavefunction, there were attempts to provide a classical interpretation of QM. In fact, the success of Madelung [1] in transforming the Schrödinger equation into a pair of hydrodynamical equations consisting of the continuity equation and an Euler type equation of hydrodynamics, giving birth to so-called quantum fluid dynamics (QFD) [2–5], in 1926, itself is testimony to the strong urge to have a classical interpretation of the strange world of QM. The concept was resurrected with full vigor more than 25 years later through the pioneering works of Bohm [6] and Takabayashi [7]. In the meantime, Wigner [8] introduced, in 1932, the concept of the quantum mechanical phase-space distribution function, another quantity that possesses a classical flavor and has also provided an alternative route [9] to the formulation of QFD. Besides the phase-space function, the density matrix has also been used [10] to derive the equations of QFD, both these quantities having the additional advantage of encompassing the mixed states in contrast to the wavefunction-based approach of Madelung [1] which can take care of only the pure states of QM. All the early attempts towards the formulation of QFD, however, dealt with only a single quantum particle, and thus the corresponding QFD that resulted was automatically in three-dimensional (3-D) space.

For many-particle systems, Madelung's wavefunction-based approach leads to QFD in configuration space [11]. Therefore, there have been attempts to reduce or convert them to QFD equations in 3-D space. Quite naturally, the framework of single-particle theories, where the many-particle wavefunction is expressed in terms of single-particle orbitals, has been chosen to provide the obvious route. Thus, the Hartree, Hartree–Fock and even natural-orbital based approaches have been employed [11–13] for this purpose. The QFD approach that emerges from these self-consistent field theories mainly differs from the single-particle QFD in that the overall velocity field becomes irrotational, in contrast to the single-particle case where it is rotational, although individual orbital contributions to the velocity field are irrotational here as well. Another important aspect is that the many-particle fluid is a composite one consisting of fluid-component mixtures which move in a self-consistent field consisting of forces resulting from many-body effects in addition to the external potential. The quantum potential, the main ingredient of QFD, arises in this case too, although it consists of individual orbital contributions, consistent with the picture of a fluid mixture.

The most remarkable single-particle theory, which is also exact in principle, is the so-called density functional theory (DFT) [14–16], a well-known theoretical approach that uses the single-particle electron density [17] as the basic variable and provides a conceptually simple and computationally economic route to the description of many-particle systems within a single-particle framework. DFT thus qualifies to be an ideal vehicle to travel to the land of QFD in 3-D space.

Although originated [14] as a ground-state theory for time-independent systems, DFT has been extended to include excited states [18] as well. Its extension and generalization to time-dependent (TD) situations, initially for periodic time-dependence [19–21] and later for more arbitrary time-dependence [22] and inclusion of magnetic fields [23] in addition to the electric field, have been instrumental in the application of TD density functional theory (TDDFT) for the transcription to QFD in 3-D space. TDDFT employs, in addition to the electron density, another density quantity, viz. the current density and both TDDFT and the corresponding QFD framework can efficiently handle the TD many-electron correlations, an important issue in many-body physics.

As is well-known, the implementation of DFT lies in suitable approximation of the exchange-correlation (XC) and the kinetic energy (KE) functionals. The so-called local-density approximation (LDA) for the KE with neglect of XC altogether corresponds to the well-known Thomas–Fermi (TF) [24] theory, the predecessor of the modern DFT which was introduced soon after QM came into being. Also, as early as 1933, the TD generalization of the TF theory was introduced by Bloch [25] who applied the corresponding QFD framework for the study of photoabsorption. Many years later, Mukhopadhyay and Lundqvist [26] developed the QFD corresponding to the more generalized DFT which included, in principle, the XC contributions. This route to QFD, where the net-density quantities (and not the orbital densities) are involved, has also been of considerable interest although the approach is somewhat phenomenological in nature. A quantum fluid density functional theory (QFDFT) has also been developed [27] and implemented through a generalized nonlinear Schrödinger equation.

The TDDFT corresponding to the orbital based Kohn–Sham version has also been used extensively [20] for the development of QFD in 3-D space. The main difference again lies in the interpretation of a single pure fluid as against the mixture of fluid components. In the TDDFT for arbitrary electric and magnetic fields developed by Ghosh and Dhara [23], the TDDFT and its QFD version have been extensively integrated and unified.

The QFD equations correspond to a continuous fluid that can however be discretized to have a discrete-particle picture which, although fictitious, can lead to the concept of quantum trajectory [28] in analogy to the well-established concept of a classical trajectory. The concept of quantum trajectories within QFD, their numerical evaluation through solution of the QFD equations, and the corresponding application areas have recently been discussed in detail by Wyatt [5].

In this work, our objective is to provide a brief account of QFD within the level of a single-particle theory, viz. the TDDFT. In what follows, we first present QFD for a single-particle system in Section 12.2, followed by a very brief discussion on many-particle systems in Section 12.3. We then discuss the single-particle framework of DFT and TDDFT in Section 12.4 and its application to QFD in 3-D space in Section 12.5. Some miscellaneous aspects of DFT and QFD based theories are discussed in Section 12.6. Finally, in Section 12.7, we offer a few concluding remarks.

12.2　QUANTUM FLUID DYNAMICAL TRANSCRIPTION OF QUANTUM MECHANICS (QM) OF A SINGLE PARTICLE

The quantum mechanical description of a single-electron moving in a scalar potential $V(\mathbf{r}, t)$, is based on the TD Schrödinger equation given by

$$D^2\nabla^2\psi(\mathbf{r}, t) - (1/2m)V(\mathbf{r}, t)\psi(\mathbf{r}, t) = -iD(\partial/\partial t)\psi(\mathbf{r}, t), \qquad (12.1)$$

where $D = \hbar/2m$. Madelung [1] was the first to use the polar form for the complex wavefunction $\psi(\mathbf{r}, t) = R(\mathbf{r}, t)\exp[S(\mathbf{r}, t)]$ in terms of the real quantities, viz. the amplitude $R(\mathbf{r}, t)$ and the phase $S(\mathbf{r}, t)$ and obtained the two fluid dynamical equations given by the continuity equation

$$(\partial/\partial t)\rho(\mathbf{r}, t) + \nabla.[\rho(\mathbf{r}, t)\mathbf{v}(\mathbf{r}, t)] = 0 \qquad (12.2)$$

and the Euler equation

$$(d/dt)\mathbf{v}(\mathbf{r}, t) \equiv [(\partial/\partial t) + \mathbf{v}(\mathbf{r}, t).\nabla]\mathbf{v}(\mathbf{r}, t) = -(1/m)\nabla[V(\mathbf{r}, t) + Q(\mathbf{r}, t)], \quad (12.3)$$

where $\rho(\mathbf{r}, t) = R^2(\mathbf{r}, t)$ is the density variable, $\mathbf{v}(\mathbf{r}, t) = 2D\nabla S(\mathbf{r}, t)$ the velocity field and $\mathbf{j}(\mathbf{r}, t) = \rho(\mathbf{r}, t)\mathbf{v}(\mathbf{r}, t)$ is the current density. These equations of QFD denote the motion of the electron fluid in the force field of the classical potential $V(\mathbf{r}, t)$, augmented by the force arising from the quantum potential $Q(\mathbf{r}, t)$, given by

$$Q(\mathbf{r}, t) = -2mD^2\nabla^2 R(\mathbf{r}, t)/R(\mathbf{r}, t). \tag{12.4}$$

The transformation of the Schrödinger equation into the pair of QFD equations is thus clear. It is, however, possible to consider the reverse transformation as well and, as has been shown [29] recently, for arbitrary values of the constant D (other than $D = \hbar/2m$ assumed here) and the presence of a potential of the form of Equation 12.4 as arising here, the fluid dynamical equations lead to a macroscopic Schrödinger-like equation.

The velocity field $\mathbf{v}(\mathbf{r}, t)$, as defined here, is in fact only the real part of a more general complex velocity field [3, 28] expressed as

$$\mathbf{u}(\mathbf{r}, t) = -2iD\nabla \ln \psi(\mathbf{r}, t) \tag{12.5}$$

which is also irrotational in nature.

The QFD formalism corresponding to the Schrödinger equation in a vector potential $\mathbf{K}(\mathbf{r}, t)(= (e/mc)\mathbf{A}(\mathbf{r}, t)$ for an electromagnetic field), viz.

$$[(1/2)(-2iD\nabla - \mathbf{K}(\mathbf{r}, t))^2 + (1/m)V(\mathbf{r}, t)]\psi(\mathbf{r}, t) = 2iD(\partial/\partial t)\psi(\mathbf{r}, t), \tag{12.6}$$

has also been developed through the use of the polar form of the wavefunction, leading to the continuity Equation 12.2 with a modified expression for the velocity field $\mathbf{v}(\mathbf{r}, t) = 2D\nabla S(\mathbf{r}, t) - \mathbf{K}(\mathbf{r}, t)$ and the Euler equation given by

$$[(\partial/\partial t) + \mathbf{v}.\nabla]\mathbf{v}(\mathbf{r}, t) = -(\partial\mathbf{K}/\partial t) + \mathbf{v} \times \text{curl } \mathbf{K} - (1/m)\nabla[V(\mathbf{r}, t) + Q(\mathbf{r}, t)]. \tag{12.7}$$

For an electromagnetic field, the velocity field consists of paramagnetic and diamagnetic contributions and the terms $-(\partial\mathbf{K}/\partial t) = (-(e/mc)(\partial\mathbf{A}/\partial t))$ and $\mathbf{v} \times \text{curl } \mathbf{K}$ correspond respectively to the magnetic contribution to the TD electric field $\mathbf{E}(\mathbf{r}, t) = -(1/c)(\partial\mathbf{A}/\partial t) - \nabla V(\mathbf{r}, t)$ and the magnetic force $(e/c)\mathbf{v} \times \text{curl } \mathbf{A}$. The QFD equations in this situation also are again classical-like and are governed by the quantum force in addition to the classical force.

12.3 QUANTUM FLUID DYNAMICAL EQUATIONS FOR MANY-PARTICLE SYSTEMS IN CONFIGURATION SPACE

For an N-electron system, with its electrons moving in the field of their mutual repulsion as well as the nuclei and external fields due to scalar as well as vector potentials, the corresponding TD Schrödinger equation is given by

$$\left[(1/2)\sum_j(-2iD\nabla_j - \mathbf{K}(\mathbf{r}_j, t))^2 + (1/m)V(\mathbf{r}^N, t)\right]\psi(\mathbf{r}^N, t) = 2iD(\partial/\partial t)\psi(\mathbf{r}^N, t)$$

$$\tag{12.8}$$

with the potential $V(\mathbf{r}^N, t) = \sum_j [V_0(\mathbf{r}_j) - e\phi(\mathbf{r}_j, t)] + U(\mathbf{r}^N, t)$, consisting of contributions from the potential V_0 due to the nuclei, mutual inter-electronic Coulomb repulsion $U(\mathbf{r}^N, t)$, and the external TD scalar potential $\phi(\mathbf{r}, t)$. For convenience, we will be using $D = \hbar/2m$ and the vector potential $\mathbf{K}(\mathbf{r}, t) = (e/mc)\mathbf{A}(\mathbf{r}, t)$ for an electromagnetic field in all the discussion that will now follow. The TD Schrödinger equation can now be transformed in a manner similar to that for a single particle, i.e., by using the polar form $\psi(\mathbf{r}^N, t) = R(\mathbf{r}^N, t)\exp[S(\mathbf{r}^N, t)]$ to obtain [11] the similar looking QFD equations, viz. the continuity equation

$$(\partial/\partial t)\rho^N(\mathbf{r}^N, t) + \sum_k^N \nabla_k \cdot \mathbf{j}_k(\mathbf{r}^N, t) = 0 \tag{12.9}$$

and an Euler equation that can be written [11] as

$$(\partial/\partial t)\mathbf{v}_k(\mathbf{r}^N, t) + \sum_j (\mathbf{v}_j(\mathbf{r}^N, t).\nabla_k)\mathbf{v}_j(\mathbf{r}^N, t)$$

$$+ \sum_j (1 - \delta_{jk})\mathbf{v}_j(\mathbf{r}^N, t) \times (\nabla_k \times \mathbf{v}_j(\mathbf{r}^N, t))$$

$$= -(e\mathbf{E}(\mathbf{r}_k, t) + (e/c)\mathbf{v}_k(\mathbf{r}^N, t) \times \mathbf{B}(\mathbf{r}_k, t))$$

$$- (1/m)\nabla[V_0(\mathbf{r}^N, t) + U(\mathbf{r}^N, t) + Q(\mathbf{r}^N, t)]. \tag{12.10}$$

Here, the notation $\mathbf{r}^N$ denotes all the N-electron coordinates and ∇_k represents the gradient operator involving the coordinate $\mathbf{r}_k$ of the k-th electron, and the external electric and magnetic fields corresponding to the TD scalar and vector potentials are given by $\mathbf{E}(\mathbf{r}, t) = -\nabla\phi(\mathbf{r}, t) - (1/c)(\partial/\partial t)\mathbf{A}(\mathbf{r}, t)$ and $\mathbf{B}(\mathbf{r}, t) = \text{curl }\mathbf{A}(\mathbf{r}, t)$.

The resulting QFD equations are, however, in configuration space, thus involving the N-particle density $\rho^N(\mathbf{r}^N, t) = R^2(\mathbf{r}^N, t)$ and the configuration space current density $\mathbf{j}_k(\mathbf{r}^N, t) = \rho^N(\mathbf{r}^N, t)\mathbf{v}_k(\mathbf{r}^N, t)$, with the $3N$-D velocity field (corresponding to the k-th electron) given by

$$\mathbf{v}_k(\mathbf{r}^N, t) = (\hbar/m)\nabla_k S(\mathbf{r}^N, t) - (e/mc)\mathbf{A}(\mathbf{r}_k, t). \tag{12.11}$$

Although these density variables in $3N$-dimensional configuration space can easily be projected to 3-D space to obtain the single-particle electron density $\rho(\mathbf{r}, t)$ and the current density $\mathbf{j}(\mathbf{r}, t)$ quantities defined as

$$\rho(\mathbf{r}, t) = \int d\mathbf{r}^N \rho^N(\mathbf{r}^N, t) \sum_k \delta(\mathbf{r} - \mathbf{r}_k) \tag{12.12}$$

$$\mathbf{j}(\mathbf{r}, t) = \int d\mathbf{r}^N \sum_k \delta(\mathbf{r} - \mathbf{r}_k)\mathbf{j}_k(\mathbf{r}^N, t) \tag{12.13}$$

and the continuity Equation 12.9 can also easily be transformed into 3-D space to obtain

$$(\partial/\partial t)\rho(\mathbf{r}, t) + \nabla.\mathbf{j}(\mathbf{r}, t) = 0, \tag{12.14}$$

the same is not true for the Euler Equation 12.10 which is in configuration space. Here, the integrations correspond to $3N$-dimensional integrals and the delta function in Equations 12.12 and 12.13 helps one to essentially represent the $(N − 1)$-electron coordinate integral and multiplication by N.

The QFD equations are, however, appealing only if they are in 3-D space in terms of the basic variables $\rho(\mathbf{r}, t)$ and $\mathbf{j}(\mathbf{r}, t)$. For many-electron systems, to obtain an Euler equation of QFD in 3-D space, one can resort to the framework of single-particle theories which employ an orbital partitioning of the electron-density and the current-density variables.

As has already been mentioned in the Introduction, DFT provides a rigorous single-particle based framework for the description of many-particle systems in 3-D space. With its TD generalization, it now enables one to investigate the dynamical properties, thus extending the scope of DFT enormously. Since TDDFT already involves the current density variable, in addition to the electron density, it is ideally suited for providing a route to QFD. In the following section, we discuss the framework of DFT and TDDFT in particular and explore, in the next section, their versatility for obtaining the QFD in 3-D space.

12.4 DENSITY FUNCTIONAL THEORY OF MANY-ELECTRON SYSTEMS: A SINGLE-PARTICLE FRAMEWORK FOR QFD IN 3-D SPACE

As is well-known, DFT has established itself as a versatile tool for describing many-particle systems within a single-particle framework. Unlike other single-particle theories like Hartree or Hartree–Fock theories, DFT is formally exact and can incorporate the important effect of many-electron correlations effectively and economically (also in principle, exactly). Although it existed earlier in the form of TF theory and its variants, its rigorous foundation was laid by the pioneering Hohenberg–Kohn theorem [14], demonstrating a unique mapping between the electron density $\rho(\mathbf{r})$ of a many-electron system in its ground state and the external potential $V_0(\mathbf{r})$ (say, due to the nuclei) which characterizes the system. In the Kohn–Sham version, the ground-state DFT consists of a set of single-particle Schrödinger-like equations, the so-called Kohn–Sham equations, expressed as

$$\{-(\hbar^2/2m)\nabla^2 + V_0(\mathbf{r}) + v_{\text{SCF}}(\mathbf{r}; [\rho])\}\psi_k(\mathbf{r}) = \varepsilon_k \psi_k(\mathbf{r}), \tag{12.15}$$

which are to be solved self-consistently to obtain the one-electron orbitals $\{\psi_k(\mathbf{r})\}$, which in turn, determine the density as $\rho(\mathbf{r}) = \sum_k \psi_k^*(\mathbf{r})\psi_k(\mathbf{r})$ and an effective self-consistent potential given by

$$v_{\text{SCF}}(\mathbf{r}; [\rho]) = V_{\text{COUL}}(\mathbf{r}) + v_{\text{XC}}(\mathbf{r}); \quad V_{\text{COUL}} = (\delta U_{\text{int}}/\delta\rho); \quad V_{\text{XC}} = (\delta E_{\text{XC}}/\delta\rho). \tag{12.16}$$

Here, $U_{\text{int}}[\rho] = (1/2)e^2 \iint \rho(\mathbf{r})\rho(\mathbf{r}')/|\mathbf{r} − \mathbf{r}'|$ is the classical Coulomb energy and $E_{\text{XC}}[\rho]$ is the XC energy, components of the well-known Hohenberg–Kohn–Sham energy density functional

$$E[\rho] = T_s[\rho] + \int d\mathbf{r} v_0(\mathbf{r})\rho(\mathbf{r}) + U_{int}[\rho] + E_{XC}[\rho], \qquad (12.17)$$

where $T_s[\rho](= -(\hbar^2/2m)\sum_k\langle\psi_k^*(\mathbf{r})|\nabla^2|\psi_k(\mathbf{r})\rangle)$ represents the KE of a fictitious non-interacting system of a density the same as that of the actual interacting system. Clearly, the framework of the Kohn–Sham version of DFT employs an orbital partitioning of the density and the picture that emerges is that of a fictitious non-interacting system of particles moving in the field of the external potential $V_0(\mathbf{r})$ and an additional density-dependent effective potential $V_{SCF}(\mathbf{r}; [\rho])$. For practical implementation, the XC energy density functional $E_{XC}[\rho]$ is usually to be approximated since an exact expression for this functional is still unknown.

The ground-state DFT was later extended to excited states, and oscillating TD potentials and, mainly through the work of Peuckert [19], Bartolotti [21] and Deb and Ghosh [20], a formal TDDFT was born for oscillating time-dependence. A generalized TDDFT for an arbitrary TD scalar potential was subsequently developed by Runge and Gross [22] which was later extended to TD electric and magnetic fields with arbitrary time-dependence by Ghosh and Dhara [23]. For many-electron systems characterized by a potential $V_0(\mathbf{r})$ due to the nuclei and subjected to an additional oscillating external potential $v_{ext}(\mathbf{r}, t)$, a Hohenberg–Kohn like theorem was proved by using the Hamiltonian for the steady state, viz. $(H(t) - i\hbar(\partial/\partial t))$ and the minimal property of a quasi-energy quantity defined as a time-average of its expectation value over a period. The TD analog of the Kohn–Sham equation, as derived by them, is given by

$$\{-(\hbar^2/2m)\nabla^2 + V_0(\mathbf{r}) + V_{ext}(\mathbf{r}, t) + V_{SCF}(\mathbf{r}, t)\}\psi_k(\mathbf{r}, t) = i\hbar(\partial/\partial t)\psi_k(\mathbf{r}, t) \quad (12.18)$$

which is to be solved for the one-electron orbitals $\psi_k(\mathbf{r}, t)$ self-consistently, with the electron-density and the current-density obtained by using the relations

$$\rho(\mathbf{r}, t) = \sum_k \psi_k^*(\mathbf{r}, t)\psi_k(\mathbf{r}, t); \qquad (12.19)$$

$$\mathbf{j}(\mathbf{r}, t) = -(i\hbar/2m)\sum_k[\psi_k^*(\mathbf{r}, t)\nabla\psi_k(\mathbf{r}, t) - \psi_k(\mathbf{r}, t)\nabla\psi_k^*(\mathbf{r}, t)]. \qquad (12.20)$$

The effective potential $V_{SCF}(\mathbf{r}, t)$, like its time-independent counterpart (Equation 12.16), consists of the classical Coulomb as well as XC contributions which now depend on the TD density. The explicit form of the XC functional may, however, be different and may also be current-density dependent as well. An elegant analysis of the Floquet formulation of TDDFT for oscillating time-dependence has recently been discussed by Samal and Harbola [30].

The restriction to the case of oscillating TD scalar potentials was overcome first by Runge and Gross [22] who formulated a TDDFT for arbitrary TD scalar potentials [22] leading to TD Kohn–Sham equations very similar to the ones described above. Their approach was later used by Ghosh and Dhara who formulated [23] a TDDFT for TD scalar as well as TD vector potentials, with arbitrary time-dependence. This broadened the scope and applicability of TDDFT to include the study of the interaction with electromagnetic radiation, a magnetic field, and various other problems.

Considering the TD-Schrödinger equation $(i\hbar(\partial/\partial t) - \hat{H})\psi = 0$ for an N-electron system moving initially in the field of an external potential $V_0(\mathbf{r})$ due to the nuclei besides their mutual Coulomb interaction and then subjected to an additional TD-scalar potential $\phi(\mathbf{r}, t)$ and a TD-vector potential $\mathbf{A}(\mathbf{r}, t)$ and the Taylor series expansion of these two potentials with respect to time around the initial time, it was proved that both the potentials are uniquely (apart from only an additive TD function) determined by the single-particle current density $\mathbf{j}(\mathbf{r}, t)$ of the system. It is also obvious from the continuity equation that the current density $\mathbf{j}(\mathbf{r}, t)$ uniquely determines the electron density as well. This enables one to treat the energy and other quantities for the TD problems as functionals of the current density or more conveniently as functionals of both $\mathbf{j}(\mathbf{r}, t)$ and $\rho(\mathbf{r}, t)$, in contrast to the stationary ground state where one has only the electron-density functionals.

Based on the mapping to a fictitious system of non-interacting particles and hence an orbital partitioning along the lines of the Kohn–Sham [15] version of time-independent DFT, Ghosh and Dhara [23] showed that the TD one-electron orbitals $\psi_k(\mathbf{r}, t)$ obtained through self-consistent solution of the effective one-particle TD Schrödinger-like equations (TD Kohn–Sham type equations) given by

$$\left\{\frac{1}{2m}\left(-i\hbar\nabla + \frac{e}{c}\mathbf{A}_{\text{eff}}(\mathbf{r}, t)\right)^2 + v_{\text{eff}}(\mathbf{r}, t)\right\}\psi_k(\mathbf{r}, t) = \left(i\hbar\frac{\partial}{\partial t}\right)\psi_k(\mathbf{r}, t), \quad (12.21)$$

determine the exact densities $\rho(\mathbf{r}, t)$ and $\mathbf{j}(\mathbf{r}, t)$ through the relations

$$\rho(\mathbf{r}, t) = \sum_k n_k \psi_k^*(\mathbf{r}, t)\psi_k(\mathbf{r}, t) \tag{12.22}$$

and

$$\mathbf{j}(\mathbf{r}, t) = -\frac{i\hbar}{2m}\sum_k n_k[\psi_k^*(\mathbf{r}, t)\nabla\psi_k(\mathbf{r}, t) - \psi_k(\mathbf{r}, t)\nabla\psi_k^*(\mathbf{r}, t)] + \frac{e}{mc}\rho(\mathbf{r}, t)\mathbf{A}_{\text{eff}}(\mathbf{r}, t),$$

$$(12.23)$$

where $\{n_k\}$ represents the occupation numbers of the orbitals and is unity for occupied and zero for unoccupied orbitals at zero temperature. The finite temperature effects can, however, be incorporated by assuming the n_k's to be governed by the Fermi–Dirac distribution. The picture that emerges here is that the density and current density of the actual system of interacting electrons characterized by the given external potentials are obtainable by calculating the same for a system of non-interacting particles moving in the field of effective scalar and vector potentials $V_{\text{eff}}(\mathbf{r}, t)$ and $\mathbf{A}_{\text{eff}}(\mathbf{r}, t)$ respectively, which can be expressed as

$$V_{\text{eff}}(\mathbf{r}, t) = V_0(\mathbf{r}) - e\phi(\mathbf{r}, t) + \frac{c}{e}\left(\frac{\delta U_{\text{int}}}{\delta\rho}\right) + \frac{c}{e}\left(\frac{\delta E_{\text{XC}}}{\delta\rho}\right) + \left(\frac{e^2}{2mc^2}\right)(A_{\text{eff}}^2 - A^2)$$

$$(12.24)$$

$$\mathbf{A}_{\text{eff}}(\mathbf{r}, t) = \mathbf{A}(\mathbf{r}, t) + \frac{c}{e}\left(\frac{\delta U_{\text{int}}}{\delta\mathbf{j}}\right) + \frac{c}{e}\left(\frac{\delta E_{\text{XC}}}{\delta\mathbf{j}}\right), \tag{12.25}$$

having contributions from the external potentials augmented by internal contributions determined by the density variables. The energy quantities $U_{int}[\rho, \mathbf{j}]$ and $E_{XC}[\rho, \mathbf{j}]$ again represent the classical Coulomb energy, and the XC energy density functional respectively, and are in general both electron-density and current-density dependent.

While the basic equations of DFT and TDDFT are exact in principle, for practical implementation, the XC energy functional has to be approximated using a suitable form. The standard LDA or other gradient corrected forms or non-local approaches are often used widely since an exact form for this quantity is still not in sight. For TD potentials, the approximations to suitable XC potentials has however been less well-developed and further work in this area is essential.

12.5 QUANTUM FLUID DYNAMICAL EQUATIONS WITHIN A DENSITY FUNCTIONAL FRAMEWORK

The discussion on DFT and TDDFT makes it clear that one can solve the TD Kohn–Sham like Equations 12.21 self-consistently to obtain the one-electron orbitals and hence the orbital densities and current densities as well as the corresponding total densities. This approach, based on an orbital partitioning scheme, can be used to derive the QFD equations by transforming Equation 12.21 through the use of the polar form of each orbital wavefunction, viz. $\psi_k(\mathbf{r}, t) = R_k(\mathbf{r}, t) \exp[S_k(\mathbf{r}, t)]$. The real and imaginary parts of the resulting equation can then be separated and after algebraic manipulations, one obtains, for individual orbitals (k-th orbital), the continuity equation

$$(\partial/\partial t)\rho_k(\mathbf{r}, t) + \nabla \cdot \mathbf{j}_k(\mathbf{r}, t) = 0 \tag{12.26}$$

and the Euler equation

$$
\begin{aligned}
(d/dt)\mathbf{v}_k(\mathbf{r}, t) &\equiv [(\partial/\partial t) + \mathbf{v}_k(\mathbf{r}, t) . \nabla]\mathbf{v}_k(\mathbf{r}, t) \\
&= -(e/m)[\mathbf{E}_{eff}(\mathbf{r}, t) + (1/c)\mathbf{v}_k(\mathbf{r}, t) \times \mathbf{B}_{eff}(\mathbf{r}, t)] \\
&\quad - (1/m)\nabla[V_{eff}(\mathbf{r}, t) + Q_k(\mathbf{r}, t)],
\end{aligned}
\tag{12.27}
$$

where the effective electric and magnetic fields are given respectively by

$$\mathbf{E}_{eff}(\mathbf{r}, t) = -\nabla\phi(\mathbf{r}, t) - (1/c)(\partial/\partial t)\mathbf{A}_{eff}(\mathbf{r}, t) \tag{12.28}$$

and

$$\mathbf{B}_{eff}(\mathbf{r}, t) = \operatorname{curl} \mathbf{A}_{eff}(\mathbf{r}, t), \tag{12.29}$$

and the quantum potential is given by

$$
\begin{aligned}
Q_k(\mathbf{r}, t) &= (-\hbar^2/2m)\nabla^2 R_k(\mathbf{r}, t)/R_k(\mathbf{r}, t) \\
&= (\hbar^2/8m)\nabla\rho_k(\mathbf{r}, t).\nabla\rho_k(\mathbf{r}, t)/\rho_k^2(\mathbf{r}, t) - (\hbar^2/4m)\nabla^2\rho_k(\mathbf{r}, t)/\rho_k(\mathbf{r}, t),
\end{aligned}
\tag{12.30}
$$

which is clearly orbital-dependent.

Here, the orbital density and current density (for the k-th orbital) are respectively given by

$$\rho_k(\mathbf{r}, t) = R_k^2(\mathbf{r}, t) \tag{12.31}$$

and

$$\mathbf{j}_k(\mathbf{r}, t) = \rho_k(\mathbf{r}, t)\mathbf{v}_k(\mathbf{r}, t), \tag{12.32}$$

with the corresponding velocity field expressed as

$$\mathbf{v}_k(\mathbf{r}, t) = (\hbar/m)\nabla S_k(\mathbf{r}, t) + (e/mc)\mathbf{A}_{\text{eff}}(\mathbf{r}, t). \tag{12.33}$$

The net electron and current densities are respectively given by $\rho(\mathbf{r}, t) = \sum_k n_k \rho_k(\mathbf{r}, t)$ and $\mathbf{j}(\mathbf{r}, t) = \sum_k n_k \mathbf{j}_k(\mathbf{r}, t)$. It may be noted that the picture is that of a fluid mixture with components corresponding to each orbital, moving in the force field of effective electric and magnetic fields which are the same for each fluid component and an additional quantum force which is orbital dependent and is thus different for different fluid components, as is obvious from Equations 12.28 through 12.30.

It may be noted that while the continuity equations for the individual orbitals can easily be summed up to result in the overall continuity Equation 12.14 given by $(\partial/\partial t)\rho(\mathbf{r}, t) + \nabla.\mathbf{j}(\mathbf{r}, t) = 0$, one cannot express the Euler Equation 12.27 in terms of the overall density quantities. The Euler Equation 12.27 can, however, be recast into other standard forms of fluid dynamical equations, such as the Navier–Stokes equation, given by

$$(\partial/\partial t)\mathbf{j}_k(\mathbf{r}, t) = -(e/m)[\rho_k(\mathbf{r}, t)\mathbf{E}_{\text{eff}}(\mathbf{r}, t) + (1/c)\mathbf{j}_k(\mathbf{r}, t)\times\mathbf{B}_{\text{eff}}(\mathbf{r}, t)]$$
$$- (1/m)\rho_k(\mathbf{r}, t)\nabla V_{\text{eff}}(\mathbf{r}, t) + \nabla.\mathbf{T}_k(\mathbf{r}, t), \tag{12.34}$$

where the stress tensor $\mathbf{T}_k(\mathbf{r}, t)$ which consists of quantum contribution [3, 31] corresponding to the quantum potential $Q_k(\mathbf{r}, t)$ as well as contributions from the current density of the k-th orbital is expressed as

$$\mathbf{T}_k(\mathbf{r}, t) = (\hbar/2m)^2\nabla\nabla\rho_k(\mathbf{r}, t)$$
$$+ (1/\rho_k(\mathbf{r}, t))[\mathbf{j}_k(\mathbf{r}, t)\mathbf{j}_k(\mathbf{r}, t) - (\hbar/2m)^2(\nabla\rho_k(\mathbf{r}, t))(\nabla\rho_k(\mathbf{r}, t))]. \tag{12.35}$$

Thus, in this QFD formulated through orbital based TDDFT, the electron fluid is interpreted as a mixture of N components and each fluid component is described by Euler type equations characterized by common effective electric and magnetic fields, and an orbital-dependent quantum force or stress tensor.

While the hydrodynamical scheme mentioned above involves the orbital partitioning of the density quantities, it may be noted that besides the orbital based DFT, there is also the net-density based orbital-free DFT, the TD version of which can also be used for the description of QFD. In this scheme, one has the continuity Equation 12.14 and the Euler equation expressed as

$$(\partial/\partial t)\mathbf{j}(\mathbf{r}, t) = \mathbf{P}_{\{V\}}[\rho(\mathbf{r}, t), \mathbf{j}(\mathbf{r}, t)] \tag{12.36}$$

where the vector $\mathbf{P}_{\{V\}}[\rho(\mathbf{r}, t), \mathbf{j}(\mathbf{r}, t)]$ is a functional of the two densities $\rho(\mathbf{r}, t)$ and $\mathbf{j}(\mathbf{r}, t)$ for a specified external potential denoted by $\{V\}$, and can be written formally as

$$\mathbf{P}_{\{V\}}[\rho(\mathbf{r}, t), \mathbf{j}(\mathbf{r}, t)] = (i\hbar)^{-1} \langle \psi(t)|[\hat{\mathbf{j}}_0, \hat{H}_0]|\psi(t) \rangle$$
$$- (1/m)\rho(\mathbf{r}, t)[\nabla V_0(\mathbf{r}) + e\mathbf{E}(\mathbf{r}, t)] - (e/mc)\mathbf{j}(\mathbf{r}, t) \times \mathbf{B}(\mathbf{r}, t)$$
$$(12.37)$$

where $\mathbf{E}(\mathbf{r}, t)$ and $\mathbf{B}(\mathbf{r}, t)$ denote the TD electric and magnetic fields defined by the scalar and vector potentials as $\mathbf{E}(\mathbf{r}, t) = -\nabla\phi(\mathbf{r}, t) - (1/c)(\partial/\partial t)\mathbf{A}(\mathbf{r}, t)$ and $\mathbf{B}(\mathbf{r}, t) = \text{curl}\, \mathbf{A}(\mathbf{r}, t)$. The operators $\hat{\mathbf{j}}_0$ and $\hat{H}_0$ are the paramagnetic current operator and unperturbed Hamiltonian operator respectively. In Equation 12.37, the total force is contributed by the classical forces (last two terms on the right-hand side of the equation) due to various external potentials, as well as the forces of quantum origin (the first term on the right-hand side) which includes the XC contribution, which are to be expressed as density functionals of $\rho(\mathbf{r}, t)$ and $\mathbf{j}(\mathbf{r}, t)$ so that the hydrodynamical Equations 12.36 through 12.37 can be used along with the continuity Equation 12.14 for the direct calculation of these density and current-density quantities.

Explicit forms of the functional $P_{\{V\}}[\rho(\mathbf{r}, t), \mathbf{j}(\mathbf{r}, t)]$ have been employed in Equations 12.36 and 12.37 by Deb and coworkers [32] and Chattaraj and coworkers [33], leading to the so-called QFDFT [27] and a generalized nonlinear Schrödinger-type equation. Applications to collision processes and interaction of atomic and molecular systems with radiation fields have been considered by them in detail.

12.6 MISCELLANEOUS ASPECTS OF THE QUANTUM FLUID DYNAMICAL APPROACH

The QFD of many-electron systems in 3-D space presented here corresponds to the TDDFT framework. The exact XC potential corresponding to TDDFT is, however, unknown although various approximations have been introduced. Other routes [3] to QFD via the phase space distribution function or the reduced density matrix have become available. Also discussed are the QFD for non-local potentials [3] as well as for mixed states, dissipation, and corresponding to a new hybrid quantum-classical approach [34].

The picture of the electron fluid has been further strengthened through a local thermodynamic transcription [35] within a DFT based framework. Its TD generalization has also been proposed [36]. Other equations of the equilibrium properties for the electron fluid have been derived [37]. Various concepts of an interpretive nature such as local temperature, local hardness and softness, local virial and hypervirial relations, have been introduced within the DFT framework. Various other local force laws have also been obtained [38]. Different applications of QFD and quantum potential to quantum transport [39], tunneling, interaction with intense laser fields [40] and various strong field processes [41] have been reported. Various TD processes have also been studied [42] through QFDFT. It will be of interest to develop the QFD equations of change [43] within the framework of DFT using the transition density.

12.7 CONCLUDING REMARKS

For many-electron systems such as atoms, molecules and solids under the action of TD external fields, which may include interaction with radiation and thus correspond to both scalar and vector potentials, QFD is a versatile approach for the description of equilibrium as well as dynamical properties. In this short review, a brief overview of the underlying principles of the quantum fluid dynamical approach has been presented. The emphasis has been on the formal aspects of QFD in 3-D space in the presence of external TD electric and magnetic fields with arbitrary time-dependence. This formalism is suitable for the treatment of interactions of radiation with atomic and molecular systems. The Kohn–Sham like TD equations of TDDFT have been used to derive the QFD equations and the basic picture that emerges is that of a multi-component fluid mixture moving in common effective electric and magnetic fields and component-specific quantum potentials.

The QFD-based approach within the TDDFT framework as presented here is a versatile tool for the investigation of dynamical properties of many-electron systems. The limitation on practical applications of TDDFT and hence QFD based on this theory due to lack of exact and accurate XC functionals of density and current density is gradually being overcome through ongoing research in the area of energy density functional development. The formalism presented here has neglected the spin polarization, although it can be incorporated easily. The latter is particularly important in view of the many recent developments in the areas of magnetism and spintronics. Also, only selected aspects of QFD have been covered here to provide a glimpse of the basic formalism in 3-D space through a single-particle framework like TDDFT, although recent years have witnessed many new developments in this exciting area of research.

ACKNOWLEDGMENTS

I am indebted to Professor B.M. Deb for introducing me to the subject of quantum fluid dynamics and density functional theory. It is a pleasure to thank Professor P.K. Chattaraj for inviting and encouraging me to write this chapter and also for his patience.

BIBLIOGRAPHY

1. Madelung, E., *Z. Phys.*, **1926**, 40, 332.
2. Ghosh, S.K. and Deb, B.M., *Phys. Rep.*, **1982**, 92, 1.
3. Deb, B.M. and Ghosh, S.K., in *The Single-Particle Density in Physics and Chemistry*, Ed. N.H. March and B.M. Deb, Academic Press, New York, **1987**.
4. Holland, P.R., *The Quantum Theory of Motion*, Cambridge University Press, Cambridge, **1993**.
5. Wyatt, R.E., *Quantum Dynamics with Trajectories: Introduction to Quantum Hydrodynamics*, Springer, New York, **2005**.
6. Bohm, D., *Phys. Rev.*, **1952**, 85, 166, 180.
7. Takabayashi, T., *Prog. Theor. Phys.*, **1954**, 11, 341.
8. Wigner, E.P., *Phys. Rev.*, **1932**, 40, 749.

9. Irving, J.H. and Zwanzig, R.W., *J. Chem. Phys.*, **1951**, 19, 1173; Mazo, R.M. and Kirkwood, J.G., *J. Chem. Phys.*, **1958**, 28, 644.
10. Frolich, H., *Physica*, **1967**, 37, 215.
11. Ghosh, S.K. and Deb, B.M., *Int. J. Quant. Chem.*, **1982**, 22, 871.
12. Wong, C.Y., *J. Math. Phys.*, **1976**, 16, 1008.
13. Ghosh, S.K., *Curr. Sci.*, **1983**, 52, 769.
14. Hohenberg, P. and Kohn, W., *Phys. Rev.*, **1964**, 136, B864.
15. Kohn, W. and Sham, L.J., *Phys. Rev.*, **1965**, 140, A1133.
16. Parr, R.G. and Yang, W., *Density Functional Theory of Atoms and Molecules*, Oxford University Press, New York, **1989**.
17. March, N.H. and Deb, B.M., Editors, *The Single-Particle Density in Physics and Chemistry*, Academic Press, New York, **1987**.
18. Theophilou, A.K., *J. Phys.*, **1979**, C12, 5419.
19. Peuckert, V., *J. Phys.*, **1978**, C11, 4945.
20. Deb, B.M. and Ghosh, S.K., *J. Chem. Phys.*, **1982**, 77, 342.
21. Bartolotti, L.J., *Phys. Rev.*, **1981**, A24, 1661.
22. Runge, E. and Gross, E.K.U., *Phys. Rev. Lett.*, **1984**, 52, 997.
23. Ghosh, S.K. and Dhara, A.K., *Phys. Rev. A*, **1988**, 38, 1149.
24. Thomas, L.H., *Proc. Camb. Phil. Soc.*, **1926**, 23, 542; Fermi, E., *Z. Phys.*, **1928**, 48, 73. For a review, see March, N.H., *Adv. Phys.*, **1957**, 6, 1.
25. Bloch, F., *Z. Phys.*, **1933**, 81, 363.
26. Mukhopadhyay, G. and Lundqvist, S., *Nuovo Cimento*, **1975**, 27B, 1.
27. Giri, S., Duley, S., Khatua, M., Sarkar, U. and Chattaraj, P.K., *Quantum Fluid Density Functional Theory and Chemical Reactivity Dynamics*, This Volume, Chapter 14, **2010**.
28. Hirschfelder, J.O., Christoph, A.C. and Palke, W.E., *J. Chem. Phys.*, **1974**, 61, 5435.
29. Nottale, L., *J. Phys. A*, **2009**, 42, 275306.
30. Samal, P. and Harbola, M.K., *Chem. Phys. Lett.*, **2006**, 433, 204.
31. Bamzai, A.S. and Deb, B.M., *Rev. Mod. Phys.* **1981**, 53, 96.
32. Deb, B.M. and Chattaraj, P.K., *Phys. Rev. A*, **1989**, 39, 1696; Deb, B.M., Chattaraj, P.K. and Mishra, S., *Phys. Rev. A*, **1991**, 43, 1248; Dey, B.K. and Deb, B.M., *Int. J. Quantum Chem.*, **1995**, 56, 707.
33. Chattaraj, P.K., *Int. J. Quantum Chem.* **1992**, 41, 845; Chattaraj, P.K. and Sengupta, S., *Phys. Lett. A*, **1993**, 181, 225; Sengupta, S. and Chattaraj, P.K. *Phys. Lett. A*, **1996**, 215, 119; Chattaraj, P.K., Sengupta, S. and Poddar, A., *Int. J. Quantum Chem.*, **1998**, 69, 279; Chattaraj, P.K., Sengupta, S. and Maiti, B. *Int. J. Quantum Chem.*, **2004**, 100, 254.
34. Burghardt, I., Moller, K.B., Parlant, G., Cederbaum, L.S. and Bitner, E.R., *Int. J. Quant. Chem.*, **2004**, 100, 1153.
35. Ghosh, S.K., Berkowitz, M. and Parr, R.G., *Proc. Natl. Acad. Sci. USA*, **1984**, 81, 8028.
36. Nagy, A., *J. Mol. Struc. THEOCHEM*, **2009**, 943, 48.
37. Ghosh, S.K. and Berkowitz, M., *J. Chem. Phys.*, **1985**, 83, 2976.
38. Ghosh, S.K., *J. Chem. Phys.*, **1987**, 87, 3513.
39. Asenov, A., Watling, J.R., Brown, A.R. and Ferry, D.K., *J. Computat. Electron.*, **2002**, 1, 1572.
40. Wadehra, A. and Deb, B.M., *J. Chem. Sci.*, **2007**, 119, 335.
41. Roy, A.K. and Chu, Shih-I, *Phys. Rev. A*, **2002**, 65, 043402.
42. Chattaraj, P.K., Sengupta, S. and Poddar, A., *Int. J. Quant. Chem.*, **1998**, 69, 279.
43. Hirschfelder, J.O., *J. Chem. Phys.*, **1978**, 68, 5151.

13 An Account of Quantum Interference from a Hydrodynamical Perspective

Ángel S. Sanz and Salvador Miret-Artés

CONTENTS

13.1 INTRODUCTION

Quantum interference is ubiquitous to many (if not all) quantum processes and phenomena, for it is a direct manifestation of their inherent coherence. Experiments based on matter wave interferometry [1] constitute very nice empirical evidences of this property. Though the first diffraction experiments were carried out with electrons [2], nowadays the (matter) particles utilized range from large atoms and single molecules [3–6] to very complex, large molecules [7] and Bose–Einstein condensates (BECs) [8]. This kind of experiments, apart from confirming the wave nature of matter, has also given rise to important practical applications. For example, sensing, metrology or quantum information processing are potential applications which have motivated in the last few years the remarkable development of BEC interferometry [9–11]. This can be observed in the increasing amount of (theoretical and

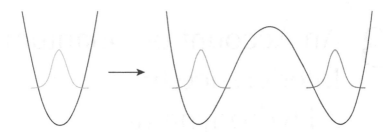

FIGURE 13.1   Basic scheme illustrating the essentials of BEC interferometry (see text for details).

experimental) work appearing in the literature [12–18] since the first experimental evidence of interference between two freely expanding BECs [19]. The basic scheme upon which this type of interferometry is based is depicted in Figure 13.1. In the first stage (left), the atomic cloud is cooled down in a magnetic trap until condensation takes place. Then (right), the BEC is split up coherently by means of radio-frequency [16] or microwave fields [17]. Finally, the double-well-like trapping potential [20] is switched off, releasing the two resulting BECs, which will interfere by free expansion (e.g., see Figure 6 of Ref. [18] as an illustration of a real experimental outcome).

Now, despite all the experimental and theoretical work grounded on the concept of quantum interference, very little attention beyond the implications of the superposition principle has been devoted to better understanding this phenomenon at a more fundamental level. In this regard, Bohmian mechanics [21,22] constitutes a very convenient tool to explore and understand the quantum system dynamics as well as the associated processes and phenomena, as pointed out by John S. Bell in his already famous book *Speakable and Unspeakable in Quantum Mechanics* [23]. Although Bohmian mechanics is at the same predictive level as standard quantum mechanics regarding statistical properties (*observables*) [24], the former provides a more "intuitive" insight into the problem under study since it allows us to monitor the probability "flow" with (quantum) trajectories—this is exactly the same as when tiny charcoal particles are released on the surface of a fluid in order to determine the flow of the fluid stream.

As an *analytical* approach, Bohmian mechanics has been applied, for example, to the study of systems, processes and phenomena with interest in surface physics [25–30] or quantum chemistry [31, 32]. As a *synthetic* approach, the *quantum trajectory method* [33] has been developed as a computational implementation to the hydrodynamic formulation of quantum mechanics [34] to generate the wave function by evolving ensembles of quantum trajectories [35]. Furthermore, based on the complex version of Bohmian mechanics [36], the problems of stationary states [37–39] or wave-packet scattering dynamics [40–43] have also been treated. Here, we present an analytical study of quantum interference from the viewpoint of Bohmian mechanics (or, in terms of standard quantum mechanics, of *quantum probability fluxes*), which will render some interesting properties that go beyond what standard quantum mechanics textbooks teach us about this phenomenon [44]. Accordingly, we have organized the chapter as follows. In Section 13.2 we introduce the basic theoretical ingredients of quantum interference, Bohmian mechanics and quantum fluxes, as well as several illustrative examples. In Section 13.3 we provide a formal discussion of interference in terms of fluxes, showing how the Bohmian *non-crossing property* already appears in standard quantum mechanics. As a consequence of this property, any interference process can be understood as the action of some effective potential on one of the interfering beams, as explained in Section 13.4. To be self-contained, in Section 13.5 we provide a discussion of quantum interference when it is considered within the complex plane, which requires a complexification of Bohmian mechanics (a further development can be found in Chapter 18). Finally, the main conclusions derived from this chapter are briefly summarized in Section 13.6.

13.2 A FIRST LOOK INTO QUANTUM INTERFERENCE

13.2.1 SOME BASIC THEORETICAL BACKGROUND

Due to the linearity of the Schrödinger equation, any solution will satisfy the superposition principle. That is, if ψ_1 and ψ_2 are solutions of this equation,

$$\Psi(\mathbf{r}, t) = c_1 \psi_1(\mathbf{r}, t) + c_2 \psi_2(\mathbf{r}, t) = c_1 \left[\psi_1(\mathbf{r}, t) + \sqrt{\alpha} \psi_2(\mathbf{r}, t) \right] \tag{13.1}$$

will also be a solution, with $\alpha = (c_2/c_1)^2$ (there should also be an additional phase factor $e^{i\delta}$ multiplying ψ_2, but we will consider $\delta = 0$, for simplicity). The particular interest in solutions like Equation 13.1 relies on the fact that they express pretty well the essence of interference, more apparent through the associated probability density,

$$\rho = |\Psi|^2 = c_1^2 \left[\rho_1 + \alpha \rho_2 + 2\sqrt{\alpha}\sqrt{\rho_1 \rho_2} \cos \varphi \right], \tag{13.2}$$

where both ψ_1 and ψ_2 are expressed in polar form ($\psi_i = \rho_i^{1/2} e^{iS_i/\hbar}$, $i = 1, 2$) and $\varphi = (S_2 - S_1)/\hbar$. The presence of the oscillatory term in Equation 13.2 constitutes *observable* evidence of interference, possible because ψ_1 and ψ_2 form a *coherent superposition*. As can be noticed, ρ does not satisfy the superposition principle, as also happens with the associated probability density current,

$$\mathbf{J} = \frac{1}{m} \text{Re} \left[\Psi^* \hat{\mathbf{p}} \Psi \right] = -\frac{i\hbar}{2m} \left[\Psi^* \nabla \Psi - \Psi \nabla \Psi^* \right], \tag{13.3}$$

where $\hat{\mathbf{p}} = -i\hbar\nabla$ is the momentum (vector) operator. This vector field provides information about the flow of ρ in configuration space, according to the *continuity equation*,

$$\frac{\partial \rho}{\partial t} + \nabla \mathbf{J} = 0. \tag{13.4}$$

Substituting Equation 13.1 into Equation 13.3, with ψ_1 and ψ_2 in polar form, yields

$$\mathbf{J} = \frac{c_1^2}{m} \Big[\rho_1 \nabla S_1 + \alpha \rho_2 \nabla S_2 + \sqrt{\alpha}\sqrt{\rho_1 \rho_2} \nabla(S_1 + S_2) \cos \varphi$$
$$+ \hbar\sqrt{\alpha}(\rho_1^{1/2}\nabla\rho_2^{1/2} - \rho_2^{1/2}\nabla\rho_1^{1/2}) \sin \varphi \Big], \tag{13.5}$$

which, effectively, does not satisfy the superposition principle.

Within the hydrodynamical picture of quantum mechanics provided by ρ and $\mathbf{J}$ one can further proceed and determine probability streamlines (lines along which the probability flows), as in classical hydrodynamics. Consider the total wave function in polar form, $\Psi = \rho^{1/2}e^{iS/\hbar}$. After its substitution into the Schrödinger equation, we find the continuity equation 13.4, with $\mathbf{J} = \rho\nabla S/m$, and the so-called *quantum Hamilton–Jacobi equation*,

$$\frac{\partial S}{\partial t} + \frac{(\nabla S)^2}{2m} + V + Q = 0, \tag{13.6}$$

where

$$Q = -\frac{\hbar^2}{2m}\frac{\nabla^2\rho^{1/2}}{\rho^{1/2}} \tag{13.7}$$

is the *quantum potential*. According to Equation 13.6, quantum mechanics can be understood in terms of the eventual, well-defined (in space and time) trajectories that a quantum system can follow (as in classical mechanics). These trajectories are obtained by integrating the equation of motion

$$\dot{\mathbf{r}} = \frac{\mathbf{J}}{\rho} = \frac{\nabla S}{m} = -\frac{i\hbar}{2m}\frac{\Psi^*\nabla\Psi - \Psi\nabla\Psi^*}{\Psi^*\Psi}$$

$$= \frac{1}{m}\frac{\rho_1\nabla S_1 + \alpha\rho_2\nabla S_2 + \sqrt{\alpha}\sqrt{\rho_1\rho_2}\nabla(S_1 + S_2)\cos\varphi}{\rho_1 + \alpha\rho_2 + 2\sqrt{\alpha}\sqrt{\rho_1\rho_2}\cos\varphi}$$

$$+ \sqrt{\alpha}\frac{\hbar}{m}\frac{(\rho_1^{1/2}\nabla\rho_2^{1/2} - \rho_2^{1/2}\nabla\rho_1^{1/2})\sin\varphi}{\rho_1 + \alpha\rho_2 + 2\sqrt{\alpha}\sqrt{\rho_1\rho_2}\cos\varphi}. \tag{13.8}$$

In order to reproduce the same results as in standard quantum mechanics it is only necessary to consider a large ensemble of initial positions distributed according to the initial probability density, $\rho_0(\mathbf{r})$.

From Equation 13.8, one can notice that Bohmian trajectories never cross the same (configuration) space point at the same time—unlike classical trajectories, for which this restriction takes place in phase space. This is the so-called *non-crossing property* of Bohmian mechanics. Suppose that two velocities are assigned to the same point $\mathbf{r}$, $v_1(\mathbf{r})$ and $v_2(\mathbf{r})$, with $v_1 \neq v_2$. According to Equation 13.8, these velocities will be related to some wave functions $\psi_1(\mathbf{r})$ and $\psi_2(\mathbf{r})$, respectively. If both wave functions represent the same problem, they can only differ in a constant or time-dependent phase at $\mathbf{r}$, i.e.,

$$S_2(\mathbf{r}, t) = S_1(\mathbf{r}, t) + \phi(t). \tag{13.9}$$

Now, since the velocities are the gradients of these functions at $\mathbf{r}$, v_2 and v_1 must be equal. This means that, given a superposition, there will be as many *domains* (regions which cannot be crossed) as interfering waves; trajectories started within one of the domains will remain confined within it at any time. Next, we illustrate this property as well as some interesting effects derived from it by means of some examples (for specific details, the reader can go to the corresponding references).

13.2.2 THE TWO-SLIT EXPERIMENT

Consider a simple two-slit diffraction experiment described in terms of a superposition, like Equation 13.1, which represents the interference of the outgoing diffracted beams. In principle, the position of the interference maxima is independent of the diffraction process (they only depend on the zeros of the cosine, according to Equation 13.2). However, the envelope of the interference pattern in the *far-field* or *Fraunhofer region* [25, 26] does depend on the transmission function associated with the slits and, therefore, on the kind of diffraction process that takes place. For example, let us assume two identical slits with total transmission along the whole distance covered by each slit, in one case, and Gaussian transmission in the other. In the Fraunhofer region, the envelope is given by the Fourier transform of the transmission function, which in our case will correspond to a sinc-function and a Gaussian, respectively, as shown in Figure 13.2a and b.

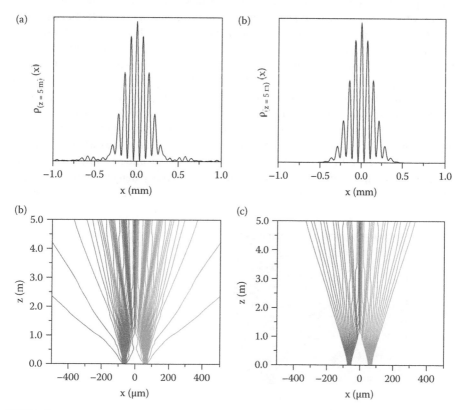

FIGURE 13.2 **Top:** Two-slit diffraction patterns observed along x (at a distance $z = 5$ m from the two slits), arising when the slit transmission function is: (a) a "hat"-function and (b) a Gaussian. **Bottom:** Quantum trajectories illustrating the probability flow from the slits to the detection screen of the cases displayed in the corresponding upper panels. The results shown here describe a typical cold neutron diffraction experiment [26].

The associated Bohmian trajectories will show the flow of the probability density from the slits to the detection screen, placed in the Fraunhofer region. The transmission function will also influence the topology of the trajectories, which spread out faster (see Figure 13.2c) or slower (see Figure 13.2d) within the *near-field* or *Fresnel region* [25, 26] in order to satisfy the requirement that, in the detection region, the swarm of trajectories has to cover the whole range covered by $\rho_0(x)$. Regarding interference, we note that the trajectories start to group in bundles as they evolve from the Fresnel to the Fraunhofer region, each bundle contributing to a particular interference intensity maximum. However, the most interesting feature is that, as mentioned above, trajectories arising from one of the slits, i.e., starting with the domain of one of the interfering waves, will remain within the same domain all the time, without crossing the borderline that separates them; because of symmetry, here both domains are equal and the trajectories present mirror symmetry.

The previous description of the two-slit experiment is a quite simplified one which makes possible the derivation of analytical results. However, a more realistic simulation of the experiment requires us to consider an appropriate potential model to describe the interaction between the incident particles and the atoms constituting the material where the slits are "carved" [45]. These models are usually obtained from *ab initio* calculations or from reasonable fitted functions, whose validity is tested by running some type of dynamical calculation (either molecular dynamics for many degree-of-freedom systems or wave packets for simpler systems), and then comparing with the corresponding experimental data. In Figure 13.3 we have considered a purely repulsive potential energy surface, which describes the scattering/diffraction of an incident electron beam (here, simulated by a quasi-plane wave) when it interacts with two slits [26, 29]. The backward-scattered and transmitted (diffracted) beams are represented in Figure 13.3a in terms of the probability density, while the angular intensity distribution is displayed in Figure 13.3b. In Figure 13.4 we show the

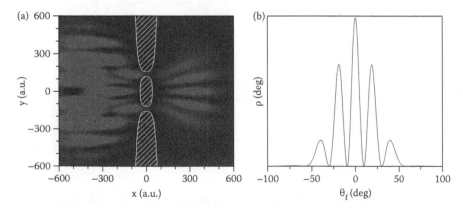

FIGURE 13.3   (a) Probability density after an electron beam interacts with two slits simulated by a soft, repulsive potential energy surface [26, 29]. (b) Angular intensity distribution after diffraction.

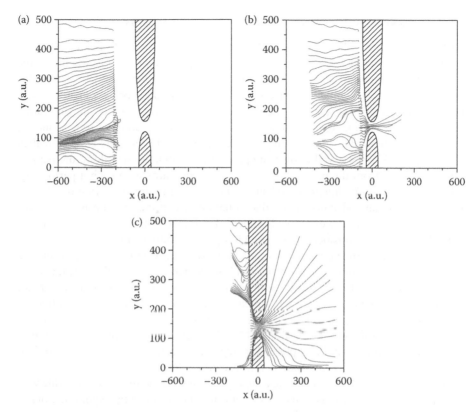

FIGURE 13.4 Quantum trajectories illustrating the probability flow associated with the scattering process represented in Figure 13.3. In each frame, the trajectories have been launched from the same x_0 initial position (varying y_0), which has been taken at the rearmost (a), central (b) and foremost (c) positions on the initial wave function. For the sake of clarity, we have only plotted the trajectories in the region $y > 0$ and, in cases (a) and (b), only their scattered parts.

trajectory picture of this process, where only the region $y > 0$ has been plotted due to symmetry with respect to the borderline at $y = 0$. The sets of trajectories displayed are all associated with the same initial wave function, but selected according to their x_0 initial condition; trajectories within the same frame have all the same x_0 value, which has been chosen at the rearmost (a), central (b) and foremost (c) part of the incident wave function. Apart from the non-crossing rule, we also observe that, depending on the particular initial conditions, trajectories display very different dynamics. The trajectories associated with the rearmost part of the wave function (see Figure 13.4a) "bounce" backwards quite far from the physical potential, on a sort of *effective potential energy surface* [26–28], while those in the foremost part (see Figure 13.4c) are the only ones (at the energy considered) which can pass through the two slits.

13.2.3 Many Slits: The Talbot Effect and Multimode Cavities

As mentioned above, the number of slits determines the number of domains. Now, as the number of slits, N, increases, a certain well-organized structure starts to emerge in the Fresnel region, whose extension also increases. In the limit $N \to \infty$, this structure covers the whole subspace behind the grating and forms a "carpet" with the particularity that it is periodic along both the direction parallel to the slit grating (the periodicity is equal to the spacing between the slits or grating period, d) and the direction perpendicular to the grating, as seen in Figure 13.5a. This is the so-called *Talbot effect* [30], very well known in optics. The periodicity (d) along the parallel direction arises from the presence of well-defined domains with such a periodicity, which gives rise to a *channeling* of the quantum flux (as shown in Figure 13.5b in terms of Bohmian trajectories), similar to having it confined within a waveguide. This is a physical picture of the well-known Bloch theorem and Born–von Karman periodic conditions considered when one has periodic lattices in solid state physics, for example. Along the perpendicular direction, however, we find a first recurrence at $z_T = d^2/\lambda$, known as the *Talbot distance* (with λ being the wavelength of the incoming particle beam), which is π-shifted with respect to $\rho_0(x)$. Then, a second recurrence is found at $2z_T$, which is an exact replica of $\rho_0(x)$ (i.e., the wave function that we have just behind the grating) at $2z_T$. At other fractional distances of z_T, there are composite, fractional (including fractal-like [46]) replicas of ρ_0 instead of identical ones. Talbot carpets can be explained taking into account the periodic conditions imposed by the grating period and, therefore, can be related to multimode interference processes [29, 30, 46]: by generating a superposition with the modes compatible with such periodic conditions (as is also done with superpositions of the eigenfunctions of a square box or a harmonic oscillator).

Talbot-like structures are not only observable with slits, but they also appear when we have periodic surfaces illuminated by a coherent incident beam. In this latter case,

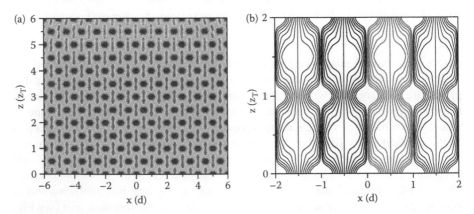

FIGURE 13.5 (a) Talbot carpet arising from the interference of outgoing He-atom beams with an initial Gaussian profile [30]. (b) Enlargement of the associated Bohmian trajectories, showing the periodicity of the quantum flux along both x and z.

due to the effect of the attractive and/or repulsive forces mediating the interaction between the incident beam and the surface, the Talbot pattern will present a certain distortion near the classical interaction region due to the *Beeby effect*. This is what we have called the *Talbot–Beeby effect* [30].

13.2.4 INTERFERENCE IN TWO-DIMENSIONS: VORTICAL DYNAMICS

Finally, consider a flat (or almost flat) surface upon which there is an obstacle, such an adsorbed particle. Due to the presence of the obstacle, an interference-like structure emerges. The frames shown in Figure 13.6 illustrate the appearance of such a structure when an incoming quasi-plane wave function (which represents an incident He-atom beam) approaches an almost flat Pt surface with a CO molecule adsorbed on it [26, 27]. As a result of the still incoming wave and the outgoing semicircular wave formed by the adsorbate, a web of vortices appears—far from the adsorbate, the

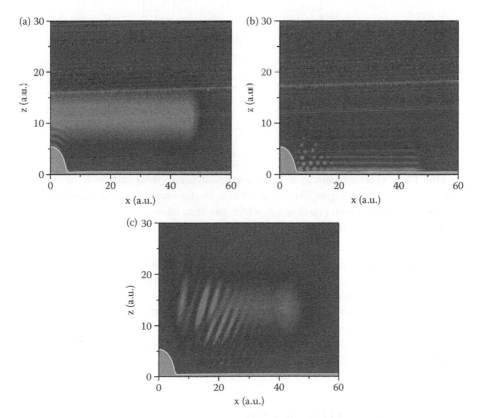

FIGURE 13.6 From (a) to (c), formation of a vortical web when an extended wave function approaches an imperfection on a surface (adsorbate) [26, 27]. The well-defined vortical web (b) arises as a consequence of the superposition of the outgoing circular wavefronts and the (still) incoming wavefront.

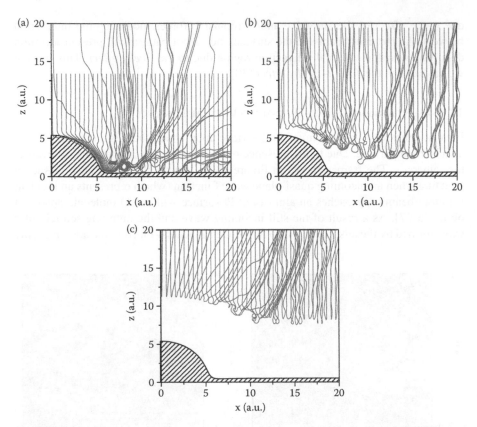

FIGURE 13.7 Quantum trajectories illustrating the dynamics associated with the diffracted processes represented in Figure 13.6. In each frame, the trajectories have been launched from the same z_0 initial position (varying x_0), which has been taken at the foremost (a), central (b) and rearmost (b) positions on the initial wave function.

outgoing wavefront is almost flat and, therefore, we only appreciate the formation of parallel ripples. By inspecting the associated trajectory dynamics (see Figure 13.7), we find that a vortical dynamics develops around the adsorbate, with quantum trajectories displaying temporary loops around different vortices. Moreover, as in the case of the soft potential modeling the two slits, here we also find that trajectories started at the regions on the initial wave function further from the adsorbate will not be able to reach it physically, but will feel a sort of effective potential energy surface [26–28].

13.3   QUANTUM FLUX ANALYSIS

Consider the coherent superposition described by Equation 13.1, with ψ_1 and ψ_2 given by normalized Gaussian wave packets, which propagate initially in opposite, converging directions and are far enough in order to minimize the (initial) overlap between them, i.e., $\rho_{0,1}(\mathbf{r})\rho_{0,2}(\mathbf{r}) \approx 0$. The dynamics observed will depend on the ratio between the propagation velocity v_p and the *spreading* velocity v_s associated with the interfering wave functions [44] as well as on their relative weighting factor, α. If $v_p \gg v_s$, the interference process is temporary and localized spatially; this *collision-like* behavior is typical, for example, of interferometry experiments. On the other hand, if $v_p \ll v_s$, an interference pattern appears asymptotically and remains stationary after some time (at the Fraunhofer region); this *diffraction-like* behavior is the one observed in slit diffraction experiments, for example.

As can be noticed in Equations 13.5 and 13.8, there are two well-defined contributions related to the effects caused by the exchange of ψ_1 and ψ_2 on the particle motion after interference has taken place. The first contribution is *even* after only exchanging the modulus or only the phase of the wave packets; the second one changes its sign with these operations. From the terms that appear in each contribution, it is apparent that the first contribution is associated with the evolution of each individual flux as well as with their combination. Thus, it provides information about both the asymptotic behavior of the quantum trajectories and also about the interference process (whenever the condition $\rho_1(\mathbf{r}, t)\rho_2(\mathbf{r}, t) \approx 0$ is not satisfied). On the other hand, the second contribution describes interference effects connected with the asymmetries or differences of the wave packets. Therefore, it will vanish at $x = 0$ when the wave packets are identical although their overlap might be non-zero.

Consider the symmetric collision-like case [44]. From now on, we will refer to the domains associated with ψ_1 and ψ_2 as I and II, respectively. Usually, after the wave packets have maximally interfered at a given time $t_{\max}^{\text{int}}$ (see panel for $t = 0.3$, on the left, in Figure 13.8), it is commonly assumed that ψ_1 moves to the domain II and ψ_2 to I. However, if the process is considered from the viewpoint of quantum fluxes or Bohmian trajectories (see right-hand-side panel in Figure 13.8), the non-crossing property forbids such a possibility. As infers from Equation 13.8, for identical, counter-propagating wave packets the velocity field is zero along $x = 0$ at any time, i.e., there cannot be any probability density (or particle) flow from domain I to II, or vice versa. Therefore, after interference (see panels for $t > 0.3$, on the left, in Figure 13.8), each outgoing wave packet can be somehow connected (through the quantum flux or, equivalently, the corresponding quantum trajectories) with the initial wave packet within its domain and the whole process can be understood as a sort of bouncing effect undergone by the wave packets after reaching maximal interference (maximal "approaching").

Although the trajectory bundles do not cross each other, they behave asymptotically as if they did [44]. From a hydrodynamic viewpoint, this means that the sign of the associated velocity field will change after the collision—before the collision its sign points inwards (toward $x = 0$); at $t_{\max}^{\text{int}}$ it does not point anywhere, but remains *steady*; and, after the collision, it points outwards (diverging from $x = 0$). This conciliates with the standard description of wave-packet crossing (based on a literal interpretation of the superposition principle), which can be understood as a "transfer" of the probability functions describing the fluxes from one domain to the other. That is, in analogy to classical particle–particle elastic collisions, where particles remain

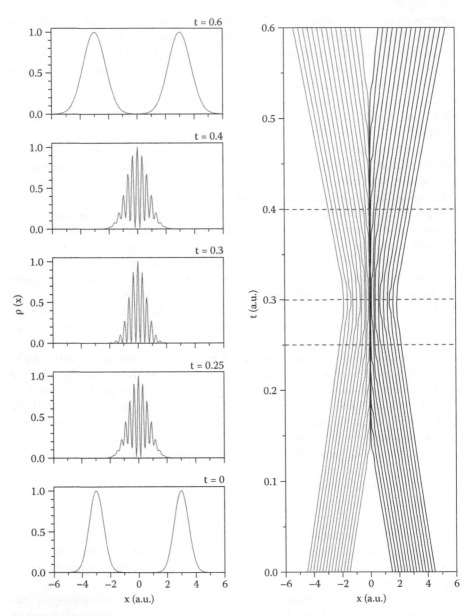

FIGURE 13.8 **Left:** From bottom to top, the frame sequence shows the appearance of interference in a collision-like process ($v_p \gg v_s$) through the probability density ρ [44]. Times are indicated on the top-right corner of each frame. **Right:** Quantum trajectories illustrating the probability flow as the two wave packets approach and then separate again after the "collision." The horizontal dashed lines denote the intermediate stages shown in the left column.

as such (i.e., the collision does not give rise to the formation of new particles) though their momentum is exchanged, here the particle swarms exchange their probability distributions "elastically." This process can be described as follows. Initially, depending on the trajectory domain, they can be described approximately by the equations of motion $\dot{r}_I \approx \nabla S_1/m$ or $\dot{r}_{II} \approx \nabla S_2/m$. Now, asymptotically (i.e., for $t \gg t_{max}^{int}$), Equation 13.8 can be expressed as

$$\dot{r} \approx \frac{1}{m} \frac{\rho_1 \nabla S_1 + \rho_2 \nabla S_2}{\rho_1 + \rho_2}, \qquad (13.10)$$

where we have assumed $\rho_1(x,t)\rho_2(x,t) \approx 0$. Note that this approximation also means that $\rho = \rho_1 + \rho_2$ is non-vanishing only on those space regions covered either by ρ_1 or by ρ_2. Specifically, in domain I, $\rho \approx \rho_2$, and in II, $\rho \approx \rho_1$. Inserting this result into Equation 13.10, we obtain that $\dot{r}_I \approx \nabla S_2/m$ and $\dot{r}_{II} \approx \nabla S_1/m$, which reproduce (and explain) the asymptotic dynamics. This trajectory picture thus provides us with a totally different interpretation of interference processes: although probability distributions transfer, particles always remain within the domain where they started from.

In diffraction-like cases, the spreading is faster than the propagation, this being the reason why we observe the well-known quantum trajectories of a typical two-slit experiment in the *Fraunhofer region* [25,26]. It is relatively simple to show [30] that, in this case, the asymptotic solutions of Equation 13.8 are

$$x(t) \approx 2\pi n \frac{\sigma_0}{x_0} (v_s t), \qquad (13.11)$$

with $n = 0, \pm 1, \pm 2, \ldots$. That is, there are bundles of quantum trajectories whose slopes are quantized quantities (through n) proportional to v_s. This means that, when Equation 13.8 is integrated exactly, one will observe quantized trajectory bundles which, on average, are distributed around the value given by Equation 13.11.

13.4 EFFECTIVE INTERFERENCE POTENTIALS

Consider now the problem of a wave packet scattered off an impenetrable potential wall, with $v_p > v_s$ [44]. At the time of maximal interference between its still incident (forward, f) part and the already outgoing (backward, b) part (see Figure 13.6b, for example), the corresponding wave function can be expressed as

$$\Psi = \psi_f + \psi_b \sim e^{imv_px/\hbar} + e^{-imv_px/\hbar}, \qquad (13.12)$$

where considerations of the shape of each wave packet are neglected for simplicity. From Equation 13.12, we find $\rho(x) \sim \cos^2(mv_px/\hbar)$, with the distance between two consecutive minima being $w_0 = \pi\hbar/mv_p$, which is the same distance between two consecutive minima in the collision-like wave-packet interference process described above. Hence, although each process (barrier scattering and wave-packet collision) has a different physical origin, the observable effect is similar when the interference patterns from both processes are superposed and compared. The only difference consists of a $\pi/2$-shift in the position of the corresponding maxima/minima, which arises from the constraint imposed by the impenetrable wall in the case of barrier scattering: the impenetrable wall forces the wave function to have a node at $x = 0$. This problem can be solved taking into account flux considerations. Since quantum trajectories do not cross, half of the trajectories contributing to the central intensity interference peak will arise from a different wave packet. In terms of potentials, this is equivalent to having a series of peaks with the same width and one, the closest one to the wall, with half-width, $w \sim \pi\hbar/2p = w_0/2$. Based on boundary conditions and the forward–backward interference discussed above, this peak cannot arise from interference, but from another mechanism: a *resonance process*. Thus, in order to establish a better analogy between the two problems, a potential well has also to be considered, whose width should be, at least, of the order of w in order to support a resonance or quasi-bound state. In standard quantum mechanics [47], the presence of bound states is usually connected with a relationship, such as

$$V_0 a^2 = n \frac{\hbar^2}{2m}, \tag{13.13}$$

between the half-width of the well ($a = w/2$) and its well depth (V_0), and where n is an integer number. From Equation 13.13, assuming $n = 1$, the estimate we obtain for the well depth is

$$V_0 = \frac{16}{\pi^2} \frac{p^2}{2m}. \tag{13.14}$$

With this, the effective interference potential is given by an impenetrable well with a short-range attractive well,

$$V(x) = \begin{cases} 0 & x < -w \\ -V_0 & -w \le x \le 0 \\ \infty & 0 < x, \end{cases} \tag{13.15}$$

which allows us to find an excellent match between the interference peak widths in both problems, as can be seen in Figure 13.9a. Of course, a better matching could be achieved by means of more sophisticated methods, similar to *ab initio* methods [31] and that could be based on the similarity of the associated quantum trajectories (see Figure 13.9b) to check the feasibility of the corresponding potentials.

The equivalence between the two-wave-packet collision problem and the scattering with a potential barrier can also be extended to the diffraction-like case [44], as shown in Figure 13.10. However, now the central diffraction maximum increases its width with time, which means that the width of the corresponding effective potential well has also to increase in time. To determine this "dynamical" or time-dependent potential function, we proceed as follows. First, we note that the two-wave-packet collision problem can be expressed as a superposition of two Gaussians [44] as

$$\Psi \sim e^{-(x+x_t)^2/4\tilde{\sigma}_t\sigma_0+ip(x+x_t)/\hbar+iEt/\hbar} + e^{-(x-x_t)^2/4\tilde{\sigma}_t\sigma_0-ip(x-x_t)/\hbar+iEt/\hbar}, \tag{13.16}$$

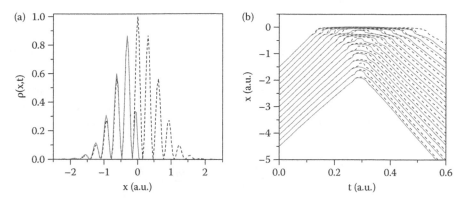

FIGURE 13.9 (a) Probability density at $t = 0.3$ for the collision of a Gaussian wave packet with the external potential described by Equation 13.15 (solid line) and another (identical) Gaussian wave packet under collision like conditions (dashed line) [44]. To compare, the maxima at $x \approx -0.32$ of both probability densities are normalized to the same height. (b) Bohmian trajectories illustrating the dynamics associated with the two cases displayed in part (a).

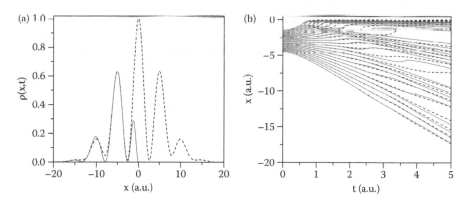

FIGURE 13.10 (a) Probability density at $t = 5$ for the collision of a Gaussian wave packet with the external, time-dependent potential described by Equation 13.24 (solid line) and another (identical) Gaussian wave packet under diffraction-like conditions (dashed line) [44]. To compare, the maxima at $x \approx -5$ of both probability densities are normalized to the same height. (b) Bohmian trajectories illustrating the dynamics associated with the two cases displayed in part (a).

where $x_t = x_0 - v_p t$ (for simplicity, the time-dependent prefactor is neglected, since it does not play any important role regarding either the probability density or the quantum trajectories). The probability density associated with Equation 13.16 is

$$\rho(x, t) \sim e^{-(x+x_t^2)/2\sigma_t^2} + e^{-(x-x_t^2)/2\sigma_t^2} + 2e^{-(x^2+x_t^2)/2\sigma_t^2} \cos[f(t)x], \qquad (13.17)$$

with

$$f(t) \equiv \frac{\hbar t}{2m\sigma_0^2} \frac{x_t}{\sigma_t^2} + \frac{2p}{\hbar}. \tag{13.18}$$

As can be noticed, Equation 13.17 is maximum when the cosine is $+1$ (constructive interference) and minimum when it is -1 (destructive interference). The first minimum (with respect to $x = 0$) is then reached when $f(t)x = \pi$, i.e.,

$$x_{\min}(t) = \frac{\pi}{\frac{2p}{\hbar} + \frac{\hbar t}{2m\sigma_0^2} \frac{x_t}{\sigma_t^2}} = \frac{\pi\sigma_t^2}{\frac{2p\sigma_0^2}{\hbar} + \frac{\hbar t}{2m\sigma_0^2}x_0}, \tag{13.19}$$

for which

$$\rho[x_{\min}(t)] \sim 4e^{-(x_{\min}^2 + x_t^2)/2\sigma_t^2} \sinh\left(\frac{x_{\min}x_t}{2\sigma_t^2}\right), \tag{13.20}$$

which is basically zero if the initial distance between the two wave packets is relatively large compared with their spreading.

In Figure 13.11a, we can see the function $x_{\min}(t)$ for different values of the propagation velocity v_p. As seen, $x_{\min}(t)$ decreases with time up to a certain value, and then increases again, reaching linear asymptotic behavior. From Equation 13.19 we find that the minimum value of $x_{\min}(t)$ is reached at

$$t_{\min} = \frac{4m\sigma_0^4}{\hbar^2 x_0}\left[-p + \sqrt{p^2 + p_s^2\left(\frac{x_0}{\sigma_0}\right)^2}\right]. \tag{13.21}$$

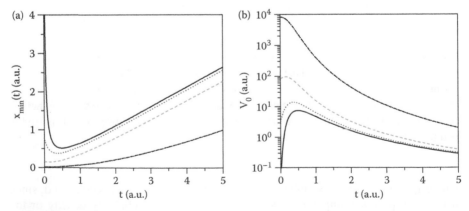

FIGURE 13.11 (a) $x_{\min}$ as a function of time for different values of the propagation velocity: $v_p = 0.1$ (solid line), $v_p = 2$ (dotted line), $v_p = 10$ (dashed line) and $v_p = 100$ (dash-dotted line). (b) V_0 as a function of time for the same four values of v_p considered in panel (a). In all cases, $v_s = 1$.

The linear time-dependence at long times is characteristic of the Fraunhofer regime, where the width of the interference peaks increases linearly with time. On the other hand, the fact that, at $t = 0$, $x_{min}(t)$ increases as v_p decreases (with respect to v_s) could be understood as a "measure" of the coherence between the two wave packets, i.e., how important the interference among them is when they are far apart (remember that, despite their initial distance, there is always an oscillating term in between due to their coherence [42]). In connection with the standard quantum–mechanical argumentation, the diffraction-like features can be associated with the wave nature of particles, while collision-like ones will be related to their corpuscular nature (they behave more classical-like). Thus, as the particle becomes more "quantum–mechanical," the initial reaching of the "effective" potential well should be larger, whereas, as the particle behaves in a more classical fashion, this reach should decrease and be only relevant near the scattering or interaction region, around $x = 0$. From Equation 13.19, two limits are thus worth discussing. In the limit $p \sim 0$,

$$x_{min}(t) \approx \frac{\pi \sigma_t^2}{x_0} \frac{\tau}{t} \tag{13.22}$$

and $t_{min} \approx \tau$. In the long-time limit, this expression becomes $x_{min}(t) \approx (\pi \hbar / 2m)$ (t/x_0), i.e., x_{min} increases linearly with time, as mentioned above. On the other hand, in the limit of large σ_0 (or, equivalently, $v_p \gg v_s$),

$$x_{min}(t) \approx \frac{\pi \hbar}{2p} \tag{13.23}$$

and $t_{min} \approx 0$. That is, the width of the "effective" potential barrier remains constant in time, thus justifying our former hypothesis above, in the scattering-like process, when we considered $w \sim \pi \hbar / 2p$.

After Equation 13.19, the time-dependent "effective" potential barrier is defined as Equation 13.15,

$$V(t) = \begin{cases} 0 & x < x_{min}(t) \\ -V_0[x_{min}(t)] & x_{min}(t) \leq x \leq 0 \\ \infty & 0 < x, \end{cases} \tag{13.24}$$

with the (time-dependent) well depth being

$$V_0[x_{min}(t)] = \frac{2\hbar^2}{m} \frac{1}{x_{min}^2(t)}. \tag{13.25}$$

The variation of the well depth with time is plotted in Figure 13.11a for the different values of v_p considered in Figure 13.11b. As seen, the well depth increases with v_p (in the same way that its width, x_{min}, decreases with it) and decreases with time. For low values of v_p, there is a maximum, which indicates the formation of the quasi-bound state that will give rise to the innermost interference peak (with half the width of the remaining peaks, as shown in Figure 13.10a). Note that, despite the time-dependence of the well depth, in the limit $v_p \gg v_s$, we recover Equation 13.14.

13.5 QUANTUM INTERFERENCE IN THE COMPLEX PLANE

A complex version of Bohmian mechanics arises when the wave function is expressed as $\Psi(x,t) = e^{iS(x,t)/\hbar}$, with $S(x,t)$ being a complex-valued phase. Substitution of this form into the time-dependent Schrödinger equation gives rise to the *complex-valued quantum Hamilton–Jacobi* (CQHJ) *equation*,

$$\frac{\partial S}{\partial t} + \frac{(\nabla S)^2}{2m} + V - \frac{i\hbar}{2m}\nabla^2 S = 0, \tag{13.26}$$

where the last term plays the role of a complex, non-local quantum potential. Due to the one-to-one correspondence between Ψ and S, both functions provide exactly the same information and Equation 13.26 can be regarded as the "logarithmic" form of the time-dependent Schrödinger equation. Equation 13.26 can be further generalized by analytic continuation assuming that both S and Ψ are complex-valued functions of a complex (space) variable $z = x + iy$, i.e., $\bar{S} \equiv S(z,t)$ and $\bar{\Psi} \equiv \Psi(z,t)$. Then, analogously to standard Bohmian mechanics, complex quantum trajectories can be obtained by integrating the equation of motion

$$\bar{p} \equiv m\dot{z} = \nabla\bar{S} = -i\hbar\nabla\ln\bar{\Psi}, \tag{13.27}$$

where $\bar{p}$ is the *quantum momentum function* (QMF). As shown elsewhere [43] (see also Chapter 18), two kind of singularities are especially relevant: (i) *nodes* of the wave function, which correspond to *poles* of the QMF, and (ii) *stagnation points* [39], which occur where the QMF is zero and correspond to points where the first derivative of the wave function is also zero. In addition, *caustics* are related to free wave-packet propagation [42].

The relationship between the complex quantum trajectories obtained from Equation 13.26 and the standard Bohmian trajectories has to be understood as follows. A complex quantum trajectory starting from any point on the complex space will cross the real axis at a certain time. All complex trajectories crossing the real axis at the same time are *isochronous* and, therefore, form a family; the initial positions of all these trajectories arise from an *isochrone curve* [39], which coincides with the real axis at some time. A standard Bohmian trajectory, therefore, consists of a series of crossings of different isochronous trajectories [42]. In Figure 13.12a, a set of quantum trajectories representative of the dynamics associated with a complex collision-like interference process [43] is displayed. The isosurfaces $|\bar{\Psi}(z,t)| = 0.053$ and $|\partial\bar{\Psi}(z,t)/\partial z| = 0.106$ are also displayed in a sort of three-dimensional Argand plot. As seen, tubular shapes develop around nodes (lighter "tubes") and stagnation points (darker "tubes"), which alternate with each other and whose centers correspond to vortical and stagnation curves, respectively. The sharp features and well defined vertical tubes observed in Figure 13.12a, reminiscent of *stalactites* and *stalagmites*, led us [43] to call this kind of plots *quantum caves*. The signature of quantum interference in the complex domain is, therefore, the formation of quantum caves, as can also be noticed in diffraction-like interference processes (see Figure 13.12b). The complex quantum trajectories in Figure 13.12 display counterclockwise helical wrapping around the stagnation

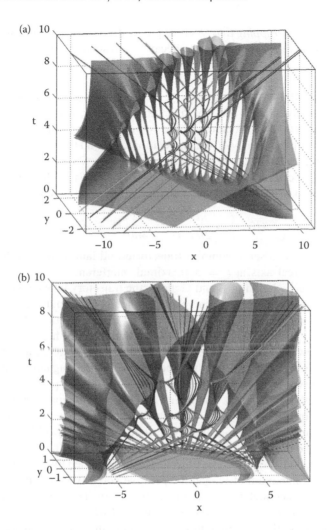

FIGURE 13.12 Complex quantum trajectories and quantum cave formation for two-wave-packet interference in the collision-like (a) and diffraction-like (b) cases. In the collision-like case, the caves are formed with the isosurfaces $|\bar{\Psi}(z,t)| = 0.053$ and $|\partial\bar{\Psi}(z,t)/\partial z| = 0.106$. In the diffraction-like case, the caves are formed with the isosurfaces $|\bar{\Psi}(z,t)| = 0.16$ (pink sheets) and $|\partial\bar{\Psi}(z,t)/\partial z| = 0.23$ (violet sheets). In both cases, the complex quantum trajectories have been launched from two branches of the isochrone that reaches the real axis at $t = 5$. (For color details, please, consult Ref. [43].)

tubes, while they are hyperbolically deflected or "repelled" when vortical tubes are approached. This intricate motion depicts the probability density flow around the vortical and stagnation tubes. Trajectories launched from different initial positions may "wrap" around the same stagnation curve and remain trapped for a certain time interval. This can be understood as a kind of *resonance*, the time lasting during the wrapping being the *wrapping time*. As time proceeds, these trajectories separate from the stagnation curves in analogy to the decay of a resonant state. Now, as one goes from collision-like processes to diffraction-like ones, the temporary trapping becomes permanent.

Due to the QMF two-dimensionality ($\bar{p} = \bar{p}_x + i\,\bar{p}_y$), one can compute its divergence and vorticity along a complex quantum trajectory, which describe the *local* expansion/contraction and rotation of the (complex) quantum fluid, respectively. Thus, by means of the Cauchy–Riemann relations, we find that the QMF first derivative can be expressed as $\partial\bar{p}/\partial z = (\Gamma + i\Omega)/2$, where $\Gamma \equiv \nabla\cdot\bar{p} = \partial\bar{p}_x/\partial x + \partial\bar{p}_y/\partial y$ is the QMF divergence and $\Omega \equiv |\nabla\times\bar{p}| = (\partial\bar{p}_y/\partial x - \partial\bar{p}_x/\partial y)$ is the vorticity. In terms of these magnitudes, the complex quantum potential from Equation 13.26 reads as

$$\bar{Q}(z,t) = \frac{\hbar}{2mi}\frac{\partial\bar{p}}{\partial z} = \frac{\hbar}{4mi}\,(\Gamma + i\Omega)\,. \tag{13.28}$$

Figure 13.13a shows a set of complex trajectories, all launched from the isochrone that reaches the real axis at $t = 5$ (maximal interference), while Figure 13.13b presents the time evolution of Γ and Ω along the trajectory labeled as 1. When this trajectory approaches the vortical curve at a, it undergoes a repulsive force driven by the QMF pole and, therefore, displays hyperbolic deflection; then, Γ and Ω show a sudden spike, as shown in Figure 13.13b. The same behavior happens when the trajectory moves away from position e, where it is again acted upon by a repulsive force. After a (and before e), the trajectory becomes trapped by the stagnation curve between two vortical curves. When the trajectory approaches turning points (b, c or d), the QMF undergoes rapid changes, this giving rise to sharp fluctuations in Γ and Ω; as inferred from Equation 13.28, $\bar{Q}$ becomes larger near these positions. The whole process indicates important dynamical activity, which is lacking within the real-valued version of this problem, where no divergence or vorticity can be defined [42].

The wrapping time for a specific trajectory can be defined by the interval between the first and last minimum of Ω, and the positive vorticity within this time interval describes the counterclockwise twist of the trajectory. The sign of Γ indicates the local expansion or contraction of the quantum fluid when it approaches or leaves a turning point, respectively. Within this time interval, the particle obviously feels the presence of stagnation points and nodes, and the trajectory displays the interference dynamics. From Figure 13.13b, the wrapping process lasts from $t \approx 3.7$ to $t \approx 6.9$. In addition, trajectories 1 and 2 wrap around the same stagnation curve with different wrapping times and numbers of loops. The wrapping time around a stagnation curve is determined by Γ and Ω, which are used to characterize the turbulent flow. Thus, the average wrapping time for those trajectories reaching the real axis at the time of maximal interference can be used to define the *life-time* for the interference process observed on the real axis.

Trajectories 2 and 3 start from the isochrone with the initial separation $\Delta z_0 \approx 0.3$, wrap around different stagnation curves, and then end with the separation at $t = 10$, $\Delta z \approx 0.8$. These two trajectories avoid the vortical curve and this greatly increases the separation between them. This behavior is consistent with what one observes when looking at the quantum flow in real space: the trajectory distribution is sparse near

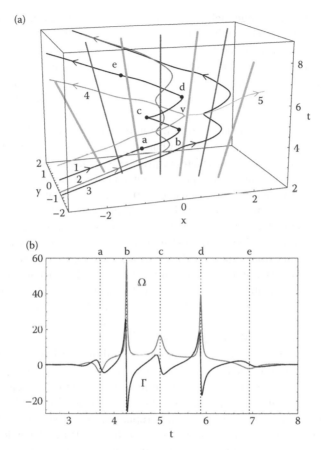

FIGURE 13.13   (a) Trajectories 1, 2, 3, 4 and 5, launched from the isochrone which reaches the real axis at $t = 5$, display hyperbolic deflection around the vortical curves (lighter straight lines) and helical wrapping around the stagnation curves (darker straight lines). Trajectories 4 and 5 diverge at v, near the vortical curve. (b) QMF divergence and vorticity along trajectory 1.

nodes of the wave function and dense between two consecutive nodes. In addition, trajectories 4 and 5 start with slightly different initial positions, $\Delta z_0 = 0.01$, and they suddenly separate at position v, near the vortical curve. This leads to the continuously increasing separation between them and a positive Lyapunov exponent, analogous to the case reported in real space [29]. The alternating structure for the vortical and stagnation tubes thus leads to divergent trajectories and may generate chaos [48].

13.6   CONCLUDING REMARKS

Here, we have seen that there are at least two important properties which are usually masked when quantum interference is studied from a standard viewpoint: (a) it is always possible to determine uniquely the departure region of the particles giving rise to an interference fringe, and (b) interference processes can be described as the effect of a particle interacting with an effective potential. Though Bohmian mechanics allows us to notice these properties, they can also be inferred within standard quantum mechanics through considerations of the quantum probability density current, as shown here. This leads us to a very different conception of quantum phenomena and processes, where none of the fundamental principles of quantum mechanics is violated.

Properties (a) and (b) lead to a sort of distinguishability, which should not be confused with that associated with identical particles (i.e., fermion or boson statistics). For example, in collision-like interference processes, a classical-like description can be considered. If the probability is equi-distributed between the two wave packets, the effect is similar to classical elastic particle–particle collisions, where only the (average) momentum is transferred through the probability density current, as indicated by the constant number of trajectories which remain associated with each wave packet. If the probabilities are asymmetrically distributed, there is a probability transfer which translates into a trajectory transfer, where the well-defined non-crossing line disappears; this behavior is similar to a classical inelastic particle–particle collision. These ideas can help importantly in the design, development, improvement and/or implementation of trajectory-based algorithms [35], where the way to handle (spatially) separate wave packets and the nodes emerging from their interference constitutes one of the main drawbacks of their application.

In connection with property (b), we find a straightforward equivalence with the scenario of classical collisions, where a two-body problem can be replaced by a (center of mass) one-body problem acted on by an (effective) interaction central potential. As shown here, any quantum interference process (collision-like or diffraction-like) can also be rearranged in a similar fashion, replacing the two-wave-packet problem by the scattering of a single wave packet with an effective (time-dependent) potential barrier. In this regard, we would like to note that this property can be considered as a precursor of more refined and well-known *ab initio* many-body methods [31], where the many degrees of freedom involved are replaced by effective interaction potentials. Furthermore, we have shown that, in order to describe properly the interference fringe associated with the non-crossing border, such effective potentials have to support temporary bound states or resonances.

Finally, the complex version of Bohmian mechanics provides an insightful alternative to the traditional analysis of quantum interference phenomena in real space. As seen, complex quantum trajectories displaying helical wrapping and hyperbolic deflection undergo turbulent flow in the complex plane. This counterclockwise circulation of trajectories launched from different positions around the same stagnation tubes can be viewed as a resonance process in the sense that during interference some trajectories keep circulating around the tubes for finite times and then escape as time progresses. On the other hand, in conventional quantum mechanics, the interference pattern transiently observed on the real axis is attributed to constructive and destructive interference between components of the total wave function. In contrast, within the complex quantum trajectory formalism, two counter-propagating wave packets are *always* interfering with each other in the complex plane. This leads to a persistent pattern of nodes and stagnation points which is a signature of the "quantum coherence" demonstrating the connection between both wave packets before or after interference fringes are observed on the real axis. The interference features observed on the real axis are connected to the rotational dynamics of the nodal line in the complex plane. Therefore, the average wrapping time for trajectories and the rotation rate of the nodal line in the complex plane provide two methods to define the *interference life-time* observed on the real axis.

ACKNOWLEDGMENTS

The Authors acknowledge interesting and fruitful discussions with Chia-Chun Chou and Robert E. Wyatt (authors of Chapter 18), which led to the works indicated in Reference [43], summarized in the Section 13.5 of this chapter. This work has been supported by the Ministerio de Ciencia e Innovación (Spain) under Projects FIS2007-62006 and SB2006-0011. A.S. Sanz thanks the Consejo Superior de Investigaciones Científicas for a JAE-Doc Contract.

BIBLIOGRAPHY

1. P. R. Berman (Ed.), *Atom Interferometry* (Academic Press, San Diego, 1997)
2. C. J. Davisson and L. H. Germer, *Phys. Rev.* **30**, 705, 1927.
3. W. Schöllkopf and J. P. Toennies, *Science* **266**, 1345, 1994.
4. M. S. Chapman, C. R. Ekstrom, T. D. Hammond, J. Schmiedmayer, B. E. Tannian, S. Wehinger, and D. E. Pritchard, *Phys. Rev. A* **51**, R14, 1995.
5. D. Leibfried, E. Knill, E. Seidelin, J. Britton, R. B. Blakestad, J. Chiaverini, D. B. Hume, W. M. Itano, J. D. Jost, C. Langer, R. Ozeri, R. Reichle, and D. J. Wineland, *Nature* **438**, 639, 2005.
6. J. O. Cáceres, M. Morato, and A. González-Ureña, *J. Phys. Chem. A* **110**, 13643, 2006; A. González-Ureña, A. Requena, A. Bastida, and J. Zúñiga, *Eur. Phys. J. D* **49**, 297, 2008.
7. M. Arndt, O. Nairz, J. Vos-Andreae, C. Keller, G. van der Zouw, and A. Zeilinger, *Nature* **401**, 680, 1999; O. Nairz, B. Brezger, M. Arndt, and A. Zeilinger, *Phys. Rev. Lett.* **87**, 160401, 2001; B. Brezger, L. Hackermüller, S. Uttenthaler, J. Petschinka, M. Arndt, and A. Zeilinger *Phys. Rev. Lett.* **88**, 100404, 2002; L. Hackermüller, S. Uttenthaler, K. Hornberger, E. Reiger, B. Brezger, A. Zeilinger, and M. Arndt, *Phys. Rev. Lett.* **91**, 090408, 2003.
8. L. Deng, E. W. Hagley, J. Denschlag, J. E. Simsarian, M. Edwards, C. W. Clark, K. Helmerson, S. L. Rolston, and W. D. Phillips, *Phys. Rev. Lett.* **83**, 5407, 1999.
9. J. Javanainen and S. M. Yoo, *Phys. Rev. Lett.* **76**, 161, 1996.
10. Y. Castin and J. Dalibard, *Phys. Rev. A* **55**, 4330, 1997.
11. C. J. Pethick and H. Smith, *Bose–Einstein Condensation in Dilute Gases* (Cambridge University Press, Cambridge, 2002).
12. Y. Shin, M. Saba, T. A. Pasquini, W. Ketterle, D. E. Pritchard, and A. E. Leanhardt, *Phys. Rev. Lett.* **92**, 050405, 2004.
13. M. Zhang, P. Zhang, M. S. Chapman, and L. You, *Phys. Rev. Lett.* **97**, 070403, 2006.
14. L. S. Cederbaum, A. I. Streltsov, Y. B. Band, and O. E. Alon, *Phys. Rev. Lett.* **98**, 110405, 2007.
15. T. J. Haigh, A. J. Ferris, and M. K. Olsen, arXiv:0907.1333v1 (2009).

16. T. Schumm, S. Hofferberth, L. M. Andersson, S. Wildermuth, S. Groth, I. Bar-Joseph, J. Schmiedmayer, and P. Krüger, *Nature Physics* **1**, 57, 2005.

17. P. Böhl, M. F. Reidel, J. Hoffrogge, J. Reichel, T. W. Hänsch, and P. Treutlein, *Nature Phys.* **5**, 592, 2009.

18. R. J. Sewell, J. Dingjan, F. Baumgärtner, I. Llorente-García, S. Eriksson, E. A. Hinds, G. Lewis, P. Srinivasan, Z. Moktadir, C. O. Gollash, and M. Kraft, arXiv:0910.4547v1 (2009).

19. M. R. Andrews, C. G. Townsend, H.-J. Miesner, D. S. Durfee, D. M. Kurn, and W. Ketterle, *Science* **275**, 637, 1997.

20. W. Hänsel, J. Reichel, P. Hommelhoff, and T. W. Hänsch, *Phys. Rev. Lett.* **86**, 608, 2001; *Phys. Rev. A* **64**, 063607, 2001; E. A. Hinds, C. J. Vale, and M. G. Boshier, *Phys. Rev. Lett.* **86**, 1462, 2001; E. Andersson, T. Calarco, R. Folman, M. Andersson, B. Hessmo, and J. Schmiedmayer, *Phys. Rev. Lett.* **88**, 100401, 2002; K. T. Kapale and J. P. Dowling, *Phys. Rev. Lett.* **95**, 173601, 2005.

21. D. Bohm, *Phys. Rev.* **85**, 166, 1952; **85**, 180, 1952.

22. P. R. Holland, *The Quantum Theory of Motion—An Account of the de Broglie–Bohm Causal Interpretation of Quantum Mechanics* (Cambridge University Press, Cambridge, 1993).

23. J. S. Bell, *Speakable and Unspeakable in Quantum Mechanics* (Cambridge University Press, Cambridge, 1987).

24. M. Born, *Z. Physik* **37**, 863, 1926; *Z. Physik* **38**, 803, 1926.

25. A. S. Sanz, F. Borondo, and S. Miret-Artés, *Phys. Rev. B* **61**, 7743, 2000.

26. A. S. Sanz, F. Borondo, and S. Miret-Artés, *J. Phys.: Condens. Matter* **14**, 6109, 2002; A. S. Sanz and S. Miret-Artés, Atom-Surface Diffraction: A Quantum Trajectory Description, in *Quantum Dynamics of Complex Molecular Systems*, Eds. D. A. Micha and I. Burghardt (Springer, Berlin, 2007).

27. A. S. Sanz, F. Borondo, and S. Miret-Artés, *J. Chem. Phys.* **120**, 8794, 2004; *Phys. Rev. B* **69**, 115413, 2004.

28. A. S. Sanz and S. Miret-Artés, *J. Chem. Phys.* **122**, 014702, 2005; A. S. Sanz and S. Miret-Artés, *Phys. Rep.* **451**, 37, 2007.

29. R. Guantes, A. S. Sanz, J. Margalef-Roig, and S. Miret-Artés, *Surf. Sci. Rep.* **53**, 199, 2004.

30. A. S. Sanz and S. Miret-Artés, *J. Chem. Phys.* **126**, 234106, 2007.

31. A. S. Sanz, X. Giménez, J. M. Bofill, and S. Miret-Artés, *Time-Dependent Density Functional Theory from a Bohmian Perspective*, in *Theory of Chemical Reactivity: A Density Functional View*, Ed. P. K. Chattaraj (Taylor and Francis, New York, 2009).

32. A. S. Sanz, X. Giménez, J. M. Bofill and S. Miret-Artés, *Chem. Phys. Lett.* **478**, 89, 2009; *Chem. Phys. Lett.* **488**, 235, 2010 (Erratum).

33. C. L. Lopreore and R. E. Wyatt, *Phys. Rev. Lett.* **82**, 5190, 1999.

34. E. Madelung, *Z. Phys.* **40**, 332, 1926.

35. R. E. Wyatt, *Quantum Dynamics with Trajectories* (Springer, New York, 2005).

36. R. A. Leacock and M. J. Padgett, *Phys. Rev. Lett.* **50**, 3, 1983; *Phys. Rev. D* **28**, 2491, 1983.

37. M. V. John, *Found. Phys. Lett.* **15**, 329, 2002; *Ann. Phys.* **324**, 220, 2009.

38. C. D. Yang, *Ann. Phys. (N.Y.)* **319**, 444, 2005; *Ann. Phys. (N.Y.)* **321**, 2876, 2006; *Chaos Soliton Fract.* **30**, 342, 2006.

39. C. C. Chou and R. E. Wyatt, *Phys. Rev. A* **76**, 012115, 2007; *J. Chem. Phys.* **128**, 154106, 2008.

40. Y. Goldfarb, I. Degani, and D. J. Tannor, *J. Chem. Phys.* **125**, 231103, 2006; A. S. Sanz and S. Miret-Artés, *J. Chem. Phys.* **127**, 197101, 2007; Y. Goldfarb, I. Degani, and D. J. Tannor, *J. Chem. Phys.* **127**, 197102, 2007; Y. Goldfarb, J. Schiff, and D. J. Tannor, *J. Phys. Chem. A* **111**, 10416, 2007; Y. Goldfarb and D. J. Tannor, *J. Chem. Phys.* **127**, 161101, 2007.

41. B. A. Rowland and R. E. Wyatt, *J. Phys. Chem. A* **111**, 10234, 2007; *J. Chem. Phys.* **127**, 164104, 2007; *Chem. Phys. Lett.* **461**, 155, 2008; R. E. Wyatt and B. A. Rowland, *J. Chem. Phys.* **127**, 044103, 2007; *J. Chem. Theory Comput.* **5**, 443, 2009; *J. Chem. Theory Comput.* **5**, 452, 2009.
42. A. S. Sanz and S. Miret-Artés, *Chem. Phys. Lett.* **458**, 239, 2008.
43. C. C. Chou, A. S. Sanz, S. Miret-Artés, and R. E. Wyatt, *Phys. Rev. Lett.* **102**, 250401, 2009; *Ann. Phys. (N.Y.)* (2010), doi: 10.1016/j.aop.2010.05.009.
44. A. S. Sanz and S. Miret, *J. Phys. A* **41**, 435303, 2008.
45. R. E. Grisenti, W. Schöllkopf, J. P. Toennies, G. C. Hegerfeldt, and T. Köhler, *Phys. Rev. Lett.* **83**, 1755, 1999; R. E. Grisenti, W. Schöllkopf, J. P. Toennies, J. R. Manson, T. A. Savas, and H. I. Smith, *Phys. Rev. A* **61**, 033608, 2000.
46. A. S. Sanz, *J. Phys. A* **38**, 6037, 2005.
47. L. I. Schiff, *Quantum Mechanics* (McGraw-Hill, Singapore, 1968) 3rd Ed.
48. C. Efthymiopoulos, C. Kalapotharakos, and G. Contopoulos, *J. Phys. A* **40**, 12945, 2007.

14 Quantum Fluid Density Functional Theory and Chemical Reactivity Dynamics

Santanab Giri, Soma Duley, Munmun Khatua,
Utpal Sarkar, and Pratim K. Chattaraj

CONTENTS

14.1 QUANTUM FLUID DENSITY FUNCTIONAL THEORY

Ever since the advent of quantum mechanics attempts have been made to express it in terms of 3D classical-like quantities. According to density-functional theory (DFT) the single-particle density $\rho(\mathbf{r})$ contains all information of an N-electron system in its ground state [1]. It is defined as

$$\rho(\mathbf{r}) = N \int \cdots \int \Psi^*(\mathbf{r}, \mathbf{r}_2, \ldots, \mathbf{r}_N) \Psi(\mathbf{r}, \mathbf{r}_2, \ldots, \mathbf{r}_N) d\mathbf{r}_2 \cdots d\mathbf{r}_N \qquad (14.1)$$

where $\Psi(\mathbf{r}_1, \mathbf{r}_2, \ldots, \mathbf{r}_N)$ is the N-electron wavefunction of the system with external potential $\upsilon(\mathbf{r})$ and total electronic energy E. The map between $\upsilon(\mathbf{r})$ and $\rho(\mathbf{r})$ is invertible up to a trivial additive constant and all ground state properties are unique functionals of $\rho(\mathbf{r})$ which is obtainable through the solution of the following Euler–Lagrange equation,

$$\frac{\delta E[\rho]}{\delta \rho} = \mu. \qquad (14.2)$$

In Equation 14.2, $E[\rho]$ is the energy functional and μ is the chemical potential which is the Lagrange multiplier associated with the normalization condition,

$$\int \rho(\mathbf{r}) d\mathbf{r} = N. \qquad (14.3)$$

A time dependent (TD) variant of DFT is also available [2] which allows one to invert the mapping between the TD external potential and the TD density up to an additive TD function. All TD properties are unique functionals of density $\rho(\mathbf{r}, t)$ and current density $\mathbf{j}(\mathbf{r}, t)$. The TDDFT may be used through the solution of a set of TD Kohn–Sham type equations [2,3]. Alternatively, one may resort to a many-particle version of Madelung's quantum fluid dynamics (QFD) [4] to obtain the basic QFD variables $\rho(\mathbf{r}, t)$ and $\mathbf{j}(\mathbf{r}, t)$ and in turn all other TD properties. The backbone of QFD comprises two equations, viz., an equation of continuity,

$$\frac{\partial \rho(\mathbf{r}, t)}{\partial t} + \nabla \cdot \mathbf{j}(\mathbf{r}, t) = 0 \tag{14.4a}$$

and an Euler type equation of motion,

$$\frac{\partial \mathbf{j}(\mathbf{r}, t)}{\partial t} = P_{\upsilon(\mathbf{r},t)}[\rho(\mathbf{r}, t), \mathbf{j}(\mathbf{r}, t)]. \tag{14.4b}$$

Unfortunately TDDFT does not provide the exact form of the functional $P_{\upsilon(\mathbf{r},t)}[\rho(\mathbf{r}, t), \mathbf{j}(\mathbf{r}, t)]$. A quantum fluid density functional theory (QFDFT) [5] has been developed to tackle this problem by introducing an approximate version of this functional, writing the many-particle QFD equations as

$$\frac{\partial \rho}{\partial t} + \nabla \cdot (\rho \nabla \xi) = 0 \tag{14.5a}$$

and

$$\frac{\partial \xi}{\partial t} + \frac{1}{2}(\nabla \xi)^2 + \frac{\delta G[\rho]}{\delta \rho} + \int \frac{\rho(\mathbf{r}', t)}{|\mathbf{r} - \mathbf{r}'|} d\mathbf{r}' + \upsilon_{\text{ext}}(\mathbf{r}, t) = 0. \tag{14.5b}$$

In the above equations ξ is the velocity potential and the universal Hohenberg–Kohn functional $G[\rho]$ is composed of kinetic and exchange-correlation energy functionals, while $\upsilon_{\text{ext}}(\mathbf{r}, t)$ contains any TD external potential apart from $\upsilon(\mathbf{r})$. These two equations and a 3D complex-valued hydrodynamical function $\Phi(\mathbf{r}, t)$ provide the QFDFT whose basic equation is the following generalized nonlinear Schrödinger equation (GNLSE):

$$\left[-\frac{1}{2}\nabla^2 + \upsilon_{\text{eff}}(\mathbf{r}, t) \right] \Phi(\mathbf{r}, t) = i \frac{\partial \Phi(\mathbf{r}, t)}{\partial t}, \quad i = \sqrt{-1} \tag{14.6a}$$

where

$$\Phi(\mathbf{r}, t) = \rho(r, t)^{\frac{1}{2}} \exp[i\xi(\mathbf{r}, t)] \tag{14.6b}$$

$$\rho(\mathbf{r}, t) = \Phi^*(\mathbf{r}, t)\Phi(\mathbf{r}, t) \tag{14.6c}$$

$$j(\mathbf{r}, t) = [\Phi_{\text{re}}\nabla\Phi_{\text{im}} - \Phi_{\text{im}}\nabla\Phi_{\text{re}}] = \rho\nabla\xi \tag{14.6d}$$

and the effective potential $\upsilon_{\text{eff}}(\mathbf{r}, t)$ is given by

$$\upsilon_{\text{eff}}(\mathbf{r}, t) = \frac{\delta T_{\text{NW}}[\rho]}{\delta \rho} + \frac{\delta E_{\text{xc}}[\rho]}{\delta \rho} + \int \frac{\delta \rho(\mathbf{r}, t)}{|\mathbf{r} - \mathbf{r}'|} d\mathbf{r}' + \upsilon_{\text{ext}}(\mathbf{r}, t), \tag{14.6e}$$

$T_{\text{NW}}[\rho]$ and $E_{\text{xc}}[\rho]$ being respectively the non-Weizsäcker contribution to the kinetic energy and the exchange-correlation energy functionals.

The phase of the hydrodynamical function $\Phi(\mathbf{r}, t)[\xi(\mathbf{r}, t)]$, Equation 14.6b, may be used to define a velocity (Equation 14.6d) as

$$\dot{\mathbf{r}} = \nabla\xi(\mathbf{r}, t)|_{\mathbf{r}=\mathbf{r}(t)} \tag{14.7}$$

which will govern the motion of a point-particle guided by a pilot wave described by $\Phi(\mathbf{r}, t)$, like the wavefunction $\Psi(\mathbf{r}, t)$, within the quantum theory of motion (QTM) [6]. Now an ensemble of particle motions may be generated by considering different initial positions of the particles guided by the same wave and satisfying the constraint that $\rho(\mathbf{r}, t)d\mathbf{r}$ gives the probability that a particle belongs to an ensemble distributed between $\mathbf{r}$ and $\mathbf{r} + d\mathbf{r}$ at time t, where $\rho(\mathbf{r}, t)$ is given by $|\Phi(\mathbf{r}, t)|^2$ (Equation 14.6c). Thus the solution of Equation 14.7 or equivalently the associated Newtonian equation of motion with different initial positions would yield the so-called quantum or Bohmian trajectories [7].

The particle motion in QTM and the fluid motion in QFD take place under the influence of the external classical potential augmented by a quantum potential $V_{qu}(\mathbf{r}, t)$ written as

$$V_{qu}(\mathbf{r}, t) = -\frac{1}{2}\frac{\nabla^2\rho^{\frac{1}{2}}(\mathbf{r}, t)}{\rho^{\frac{1}{2}}(\mathbf{r}, t)}. \tag{14.8}$$

The quantum domain behavior of classically chaotic systems like anharmonic oscillators, field induced barrier penetration in a double well potential, chaotic ionization in Rydberg atoms, etc. has been analyzed [8] using QFD and QTM mainly in terms of the distance between two initially close quantum (Bohmian) trajectories and the related quantum Lyapunov exponent and Kolmogorov–Sinai entropy.

14.2 CHEMICAL REACTIVITY DYNAMICS

Chemical reactivity can be analyzed through various global and local reactivity descriptors defined within a conceptual DFT framework [9]. The chemical potential (Equation 14.2) may be expressed as [10],

$$\mu = -\chi = \left(\frac{\partial E}{\partial N}\right)_{v(\mathbf{r})} \tag{14.9}$$

where χ is the electronegativity. Correspondingly, the second derivative defines the hardness [11], viz.,

$$\eta = \left(\frac{\partial^2 E}{\partial N^2}\right)_{v(\mathbf{r})} = \left(\frac{\partial\mu}{\partial N}\right)_{v(\mathbf{r})} \tag{14.10}$$

or equivalently [12]

$$\eta = \frac{1}{N}\iint \eta(\mathbf{r}, \mathbf{r}')f(\mathbf{r}')\rho(\mathbf{r})d\mathbf{r}d\mathbf{r}' \tag{14.11}$$

where $\eta(\mathbf{r}, \mathbf{r}')$ is the hardness kernel [12] and $f(\mathbf{r})$ is the Fukui function [13]. Electrophilicity is defined as [14]

$$\omega = \frac{\mu^2}{2\eta} = \frac{\chi^2}{2\eta}. \tag{14.12}$$

Several electronic structure principles have been introduced to understand these descriptors better. During the molecule formation process electronegativity gets equalized. According to the electronegativity equalization principle [15] the molecular electronegativity is roughly the geometric mean of the corresponding isolated atom values. During an acid–base reaction, hard likes hard and soft likes soft as per the statement of the hard–soft acid–base principle [16]. Moreover, the maximum hardness principle (MHP) dictates that [17], "there seems to be a rule of nature that molecules arrange themselves so as to be as hard as possible." Owing to the inverse relationship [18] between hardness and polarizability a minimum polarizability principle has also been proposed [19] as, "the natural direction of evolution of any system is toward a state of minimum polarizability." A minimum electrophilicity principle is also known [20].

In order to understand the dynamical implications of these electronic structure principles the GNLSE has been solved for two TD processes and the time evolution of various reactivity parameters has been analyzed [21]. The chosen TD processes include: (a) ion–atom collision and (b) atom–field interaction, known as two of the most popular models to simulate a chemical reaction.

A TD chemical potential may be defined as [22]

$$\mu(t) = \frac{\delta E(t)}{\delta \rho} = \frac{1}{2}|\nabla \xi|^2 + \frac{\delta T}{\delta \rho} + \int \frac{\rho(\mathbf{r}',t)}{|\mathbf{r} - \mathbf{r}'|}d\mathbf{r}' + \frac{\delta E_{XC}}{\delta \rho} + v_{ext}(\mathbf{r}',t) \quad (14.13)$$

which is equal to the total electrostatic potential at a point r_c (a measure of the covalent radii of atoms in the ground state [23]) where the functional derivative of the total kinetic energy balances that of the exchange-correlation energy, viz.,

$$\left[\frac{1}{2}|\nabla \xi|^2 + \frac{\delta T}{\delta \rho} + \frac{\delta E_{XC}}{\delta \rho}\right]\Bigg|_{r=r_c} = 0. \quad (14.14)$$

Therefore,

$$\mu(t) = -\chi(t) = \int \frac{\rho(\mathbf{r},t)}{|\mathbf{r}_c - \mathbf{r}|}d\mathbf{r} + v_{ext}(\mathbf{r}_c,t). \quad (14.15)$$

In the ion–atom collision problem, the dynamic μ profile (Figure 14.1) divides the whole collision process into three distinct regimes, viz., approach, encounter, and departure [21a]. The qualitative features of these profiles for various electronic states (^1S, ^1P, ^1D, ^1F) of different helium isoelectronic systems (He, Li$^+$, Be^{2+}, B^{3+}, C^{4+}, only two states for C^{4+}) remain the same in the scale we plot. In the encounter regime, where the actual chemical reaction takes place, hardness gets maximized [21a] (Figure 14.2) and polarizability gets minimized [21a] (Figure 14.3). Moreover, for a hard–hard interaction (say between He and H$^+$) the maximum hardness value is the largest and the minimum polarizability value is the smallest when compared with the other similar interactions [21a]. These results justify the validity of the associated electronic principles in a dynamical context.

During the interaction between an atom and an external axial electric field there starts a competition between the atomic nucleus and the external field to govern the electron-density distribution. Figure 14.4 depicts the external fields of different colors

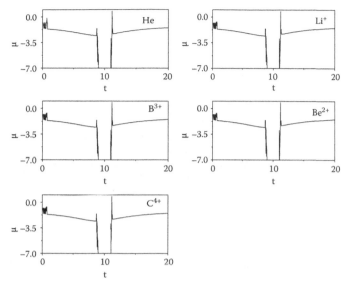

FIGURE 14.1 Time evolution of chemical potential (μ, au) during a collision process between an X-atom/ion (X = He, Li$^+$, Be^{2+}, B^{3+}, C^{4+}) in its ground state and a proton. (Reprinted with permission from Chattaraj, P. K. and Maiti, B., *J. Am. Chem. Soc.*, 125, 2705, 2003. Copyright 2003 American Chemical Society.)

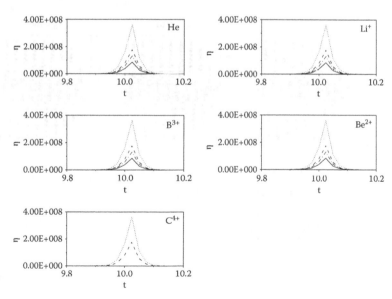

FIGURE 14.2 Time evolution of hardness (η, au) during a collision process between an X-atom/ion (X = He, Li$^+$, Be^{2+}, B^{3+}, C^{4+}) in various electronic states (^{1}S, ^{1}P, ^{1}D, ^{1}F) and a proton. ($\cdots\cdots$) ^{1}S; (− − −) ^{1}P; (−·−·−·−) ^{1}D; (——) ^{1}F. (Reprinted with permission from Chattaraj, P. K. and Maiti, B., *J. Am. Chem. Soc.*, 125, 2705, 2003. Copyright 2003 American Chemical Society.)

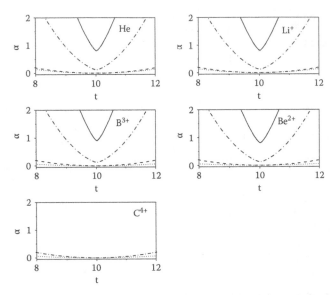

FIGURE 14.3 Time evolution of polarizability (α, au) during a collision process between an X-atom/ion and a proton. See the caption of Figure 14.2 for details. (Reprinted with permission from Chattaraj, P. K. and Maiti, B., *J. Am. Chem. Soc.*, 125, 2705, 2003. Copyright 2003 American Chemical Society.)

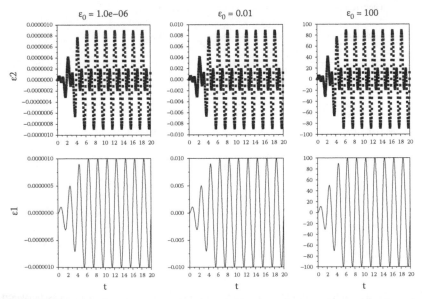

FIGURE 14.4 Time evolution of the external field: $\epsilon1$ (—) monochromatic pulse; $\epsilon2$ (— –) bichromatic pulse. Maximum amplitudes: $\epsilon_0 = 10^{-6}$, 0.01, 100; $\omega_0 = \pi$, $\omega_1 = 2\omega_0$. (Reprinted with permission from Chattaraj, P. K. and Maiti, B., *J. Phys. Chem. A*, 105, 169, 2001. Copyright 2001 American Chemical Society.)

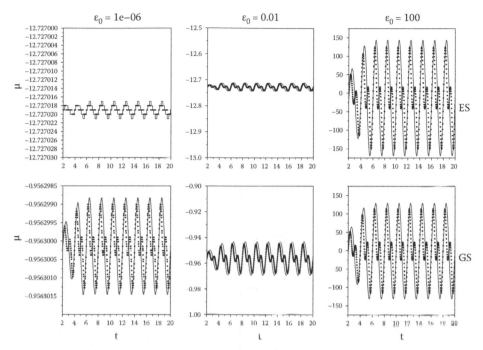

FIGURE 14.5 Time evolution of Chemical potential (μ, au) when a helium atom is subjected to external electric fields (GS, ground state; ES, excited state): (——) monochromatic pulse; ($\cdots$) bichromatic pulse. Maximum amplitudes: $\varepsilon_0 = 10^{-6}$, 0.01, 100; $\omega_0 = \pi$, $\omega_1 = 2\omega_0$. (Reprinted with permission from Chattaraj, P. K. and Maiti, B., *J. Phys. Chem. A*, 105, 169, 2001. Copyright 2001 American Chemical Society.)

and intensities. After the initial transients die out the TD chemical potential oscillates in phase (Figure 14.5) with the external field. However, hardness being a second order response requires a stronger field to exhibit in-phase oscillations (Figure 14.6). In general the ground state hardness is larger than the corresponding excited state value while electrophilicity shows the reverse trend [21c,d], as expected. These descriptors may be used in analyzing the chaotic ionization in Rydberg atoms [24]. The in-phase oscillations in chemical potential continues (Figure 14.7) even in the presence of a super-intense field (intensity $= 3.509 \times 10^{24}$ W/cm^2) which is, however, getting disturbed in the case of hardness (Figure 14.8).

For a weak external field the central Coulomb field due to the nucleus will provide a pulsating atom with roughly a spherical electron-density distribution. However, as the field intensity increases there will be a cylindrical density distribution and the atom will behave as an oscillating dipole which will emit radiation including higher order harmonics [21d].

In summary, conceptual DFT provides quantitative definitions for different reactivity descriptors whose time evolution during a TD process like an ion–atom collision or an atom–field interaction may be analyzed within a QFDFT framework. The

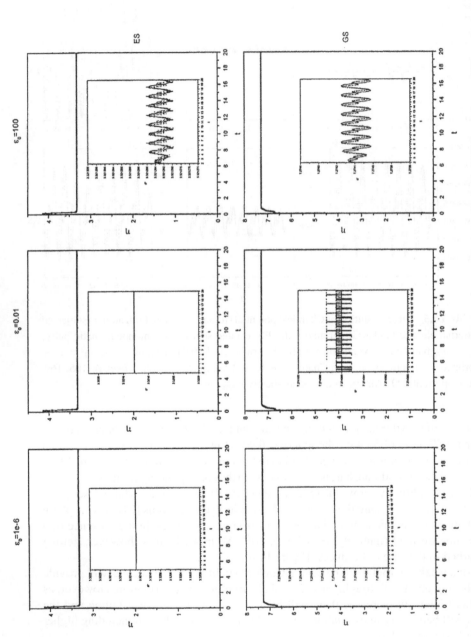

FIGURE 14.6 Time evolution of hardness (η) when a helium atom is subjected to external electric fields (GS, ground state; ES, excited state): (–) monochromatic pulse; (···) bichromatic pulse. Maximum amplitudes: $\varepsilon_0 = 10^{-6}$, 0.01, 100; $\omega_0 = \pi$, $\omega_1 = 2\omega_0$. (Reprinted with permission from Chattaraj, P. K. and Maiti, B., *J. Phys. Chem. A*, 105, 169, 2001. Copyright 2001 American Chemical Society.)

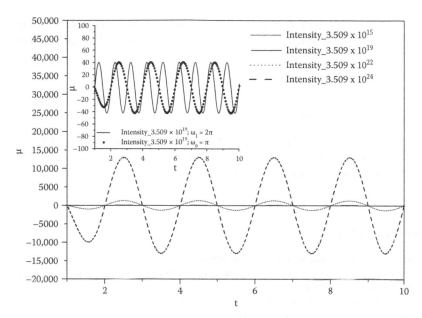

FIGURE 14.7 Time evolution of chemical potential (μ, au) when a helium atom in its ground state is subjected to external electric fields: Intensities: $I = 3.509 \times 10^{15}$ W/cm^2, 3.509×10^{19} W/cm^2, 3.509×10^{22} W/cm^2, 3.509×10^{24} W/cm^2; $\omega_0 = \pi$, $\omega_1 = 2\pi$.

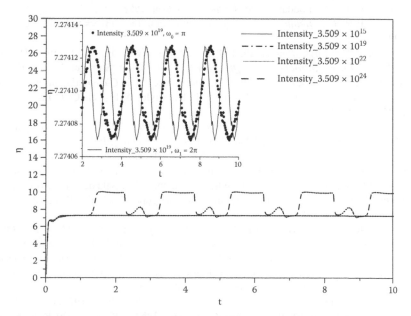

FIGURE 14.8 Time evolution of hardness (η, au) when a helium atom in its ground state is subjected to external electric fields. Intensities: $I = 3.509 \times 10^{15}$ W/cm^2, 3.509×10^{19} W/cm^2, 3.509×10^{22} W/cm^2, 3.509×10^{24} W/cm^2; $\omega_0 = \pi$, $\omega_1 = 2\pi$.

necessary TD density and current density are obtained through the solution of the pertinent generalized nonlinear Schrödinger equation. Dynamical variants of different electronic structure principles manifest themselves.

ACKNOWLEDGMENTS

We thank CSIR, New Delhi for financial assistance and Professors R. G. Parr, B. M. Deb, S. K. Ghosh and A. S. Sanz and Drs. S. Nath, S. Sengupta, A. Poddar, B. Maiti, and D. R. Roy for their help in various ways.

BIBLIOGRAPHY

1. Hohenberg, P.; Kohn, W. *Phys. Rev. B* **1964,** 136, 864.
2. (a) Runge, E.; Gross, E. K. U. *Phys. Rev. Lett.* **1984**, 52, 997; (b) Dhara, A. K.; Ghosh, S. K. *Phys. Rev. A* **1987**, 35, 442.
3. Kohn, W.; Sham, L. *J. Phys. Rev. A* **1965**, 140, 1133.
4. Madelung, E. Z. *Phys.* **1926**, 40, 322.
5. (a) Deb, B. M.; Chattaraj, P. K. *Phys. Rev. A* **1989**, 39, 1696; (b) Deb, B. M.; Chattaraj, P. K.; Mishra, S. *Phys. Rev. A* **1991**, 43, 1248; (c) Chattaraj, P. K. *Int. J. Quantum Chem.* **1992**, 41, 845; (d) Dey, B. K.; Deb, B. M. *Int. J. Quantum Chem.,* **1995**, 56, 707; (e) Chattaraj, P. K.; Sengupta, S.; Poddar, A. *Int. J. Quantum Chem.* **1998**, 69, 279.
6. Holland, P. R. *The Quantum Theory of Motion*; Cambridge University Press: Cambridge, 1993.
7. (a) Bohm, D.; *Phys. Rev. A* **1952**, 85, 166; (b) Wyatt, R. E. *Quantum Dynamics with Trajectories: Introduction to Quantum Hydrodynamics,* Springer, New York, 2007; (c) Sanz, A. S.; Gimenez, X.; Bofill, J. M.; Miret-Artés, S. in *Chemical Reactivity Theory: A Density Functional View*; ed. Chattaraj, P. K., Taylor & Francis/CRC Press: Florida, **2009,** pp. 105–119.
8. (a) Chattaraj, P. K.; Sengupta, S. *Phys. Lett. A* **1993**, 181, 225; (b) Sengupta, S.; Chattaraj, P. K. *Phys. Lett. A* **1996**, 215, 119; (c) Chattaraj, P. K.; Sengupta, S.; Poddar, A. eds. Daniel, M.; Tamizhmani, K. M.; Sahadevan, R. *Narosa, New Delhi,* **2000**, pp. 287–298; (d) Chattaraj, P. K.; Sengupta, S.; Maiti, B.; Sarkar, U. *Curr. Sci.* **2002**, 82, 101; (e) Chattaraj, P. K.; Sarkar, U. eds. Lakshmanan, M.; Sahadevan, R. *Narosa, New Delhi,* **2002**, pp. 373–382; (f) Chattaraj, P. K.; Sengupta, S.; Maiti, B. *Int. J. Quantum Chem.* **2004,** 100, 254.
9. (a) Parr, R. G.; Yang, W. *Density Functional Theory of Atoms and Molecules*, Oxford University Press: New York, 1989; (b) Geerlings, P.; De Proft, F.; Langenaeker, W. *Chem. Rev.,* **2003**, 103, 1793; (c) Chattaraj, P. K. ed. *Chemical Reactivity Theory: A Density Functional View*; Taylor & Francis/CRC Press: Florida, 2009; (d) Chattaraj, P. K.; Giri, S. *Ann. Rep. Prog. Chem. Sect. C: Phys. Chem.,* **2009**, 105, 13.
10. Parr, R. G.; Donnelly, R. A; Levy, M.; Palke, W. E. *J. Chem. Phys.* **1978**, 68, 3801;
11. (a) Parr, R. G; Pearson, R. G. *J. Am. Chem. Soc.* **1983**, 105, 7512. (b) Pearson, R. G. *Chemical Hardness: Application from Molecules to Solids*; Wiley-VCH; Weinheim, 1997.
12. (a) Berkowitz, M.; Ghosh, S. K.; Parr, R. G. *J. Am. Chem. Soc.* **1985**, 107, 6811;(b) Ghosh, S. K.; Berkowitz, M. *J. Chem. Phys.* **1985**, 83, 2976.

13. (a) Fukui, K. *Theory of Orientation and Stereoselection*; Springer: Berlin, 1973; p 134. (b) Fukui, K. *Science,* **1982**, 218, 747; (c) Fukui, K.; Yonezawa, Y.; Shingu, H. *J. Chem. Phys.* **1952**, 20, 722; (d) Parr, R. G.; Yang, W. *J. Am. Chem. Soc.* **1984**, 106, 4049.

14. (a) Parr, R. G.; Szentpaly, L. V.; Liu, S. *J. Am. Chem. Soc.* **1999**, 121, 1922; (b) Chattaraj, P. K.; Sarkar, U.; Roy, D. R. *Chem. Rev.,* **2006**, 106, 2065; (c) Chattaraj, P. K.; Roy, D. R. *Chem. Rev.,* **2007**, 107, PR46.

15. (a) Sanderson, R. T. *Science* **1951**, 114, 670; (b) Sanderson, R. T. *J. Chem. Educ.,* **1954**, 31, 238; (c) Sanderson, R. T. *Science,* **1955**, 121, 207; (d) Parr, R. G.; Bartolotti, L. J. *J. Am. Chem. Soc.* **1982**, 104, 3801.

16. (a) Pearson, R. G. *J. Am. Chem. Soc.* **1963**, 85, 3533. (b) Pearson, R. G.; Chattaraj, P. K. *Chemtracts-Inorg. Chem.* **2008,** 21, 1.

17. (a) Pearson, R. G. *J. Chem. Educ.* **1987**, 64, 561. (b) Parr, R. G.; Chattaraj, P. K. *J. Am. Chem. Soc.* **1991**, 113, 1854. (c) Chattaraj, P. K.; Liu, G. H.; Parr, R. G. *Chem. Phys. Lett.* **1995**, 237, 171.

18. (a) Politzer, P. *J. Chem. Phys.* **1987,** 86, 1072; (b) Ghanty, T. K.; Ghosh, S. K. *J. Phys. Chem.* **1993**, 97, 4951; (c) Fuentealba, P.; Reyes, O. *THEOCHEM* **1993**, 282, 65; (d) Simon-Manso, Y.; Fuentealba, P. *J. Phys. Chem. A* **1998**, 102, 2029. (e) Tanwar, A.; Pal, S.; Roy, D. R.; Chattaraj, P. K. *J. Chem. Phys.* **2006**, 125, 056101; (f) Chattaraj, P. K.; Arun Murthy, T. V. S.; Giri, S.; Roy, D. R. *THEOCHEM* **2007**, 813, 63.

19. (a) Chattaraj, P. K.; Sengupta, S. *J. Phys. Chem.* **1996**, 100, 16126; (b) Ghanty, T. K.; Ghosh, S. K. *J. Phys. Chem.* **1996**, 100, 12295; (c) Chattaraj, P. K.; Sengupta, S. *J. Phys. Chem. A* **1997**, 101, 7893; (d) Chattaraj, P. K.; Fuentealba, P.; Jaque, P.; Toro-Labbe', A. *J. Phys. Chem. A* **1999**, 103, 9307.

20. (a) Chattaraj, P. K.; Pe'rez, P.; Zevallos, J.; Toro-Labbe', A. *J. Phys. Chem. A* **2001**, 105, 4272; (b) Chamorro, E.; Chattaraj, P. K.; Fuentealba, P. *J. Phys. Chem. A* **2003**, 107, 7068; (c) Parthasarathi, R.; Elango, M.; Subramanian, V.; Chattaraj, P. K. *Theor. Chem. Acc.* **2005**, 113, 257 (d) Chattaraj, P. K.; Roy, D. R.; Giri, S. *Comput. Lett.* **2007**, 3, 223; (e) Noorizadeh, S. *J. Phys. Org. Chem.* **2007**, 20, 514; (f) Chattaraj, P. K.; Giri, S. *Ind. J. Phys.* **2007**, 81, 871; (f) Chattaraj, P. K.; Giri, S. *Ind. J. Phys.* **2008**, 82, 467; (g) Noorizadeh, S.; Shakerzadeh, E. *J. Mol. Struct. (Theochem)* **2008**, 868, 22.

21. (a) Chattaraj, P. K.; Maiti, B. *J. Phys. Chem. A* **2001,** 105, 169; (b) Chattaraj, P. K.; Maiti, B. *J. Am. Chem. Soc.* **2003,** 125, 2705; (c) Chattaraj, P. K.; Maiti, B.; Sarkar, U. *Proc. Indian Acad. Sci. (Ch. Sc.), invited article,* **2003,** 115, 195; (d) Chattaraj, P. K.; Sarkar, U. *Int. J. Quantum Chem.* **2003,** 91, 633.

22. (a) Chattaraj, P. K.; Nath, S. *Int. J. Quantum Chem.* **1994**, 49, 705; (b) Nath, S.; Chattaraj, P. K. *Pramana* **1995**, 45, 65.

23. Politzer, P.; Parr, R. G.; Murphy, D. R. *J. Chem. Phys.* **1983**, 79, 3859.

24. Chattaraj, P. K.; Sengupta, S. *J. Phys. Chem.* **1996**, 100, 16126

15 Bipolar Quantum Trajectory Methods

Bill Poirier

CONTENTS

15.1   INTRODUCTION

Over the last two decades, classical trajectory simulations (CTS) [1] have emerged as the method of choice, for computing the dynamics of atomic nuclei in molecular and condensed matter systems. Such methods are relatively easy to implement, scale linearly or quadratically in the number of atoms, and can be trivially parallelized across many cores of a supercomputing cluster or grid. On the other hand, quantum dynamical effects (tunneling, dispersion, interference, zero-point energy, etc.) are known or suspected to be important for many systems of current interest, and there is a demand for reliable numerical methods that treat quantum effects well. This has motivated a host of theoretical and computational improvements, e.g., mixed quantum-classical methods [2], semiclassical methods [3,4], centroid dynamics [5], and trajectory surface hopping (TSH) [6,7], all of which are designed, in principle, to handle large and complex systems of the sort routinely treated using CTS [1]. However, as these methods all treat quantum effects only approximately, and generally do not enable systematic convergence to the exact quantum result, it remains a vital but largely unanswered question to what extent, and for which applications, these methods actually capture the relevant quantum dynamical effects for the system in question [8].

The "conventional wisdom" states that more accurate methods, i.e., those that better represent quantum effects, require substantially greater computational (CPU) effort. However, this is not necessarily the case. For TSH for instance, one type of quantum effect—i.e., nonadiabatic transitions (essential for processes such as electron transfer)—is treated almost exactly, for a CPU cost comparable to a CTS [6,7]. This is because TSH employs CTS to propagate a molecular system on a single potential energy surface (PES) at a time, treating nonadiabatic transitions via random "hops" from one PES to another. Thus, if other quantum effects are small—e.g., in the classical limit of large action (mass and/or energy) for which CTS is presumed valid—then TSH can provide remarkably accurate results, even for pronounced intersurface coherences [6,7]. The TSH situation above suggests that highly accurate and numerically efficient quantum dynamical methods are attainable, perhaps even for single-PES dynamics. Quantum trajectory methods (QTMs) [9–13]—i.e., CTS-like simulations involving ensembles of trajectories, based (either "purely" or loosely) on the de Broglie–Bohm formulation of quantum mechanics [14–18]—are proving to be one such strategy, having already been applied successfully to model systems with hundreds of dimensions [12,13] (provided curvilinear coordinates are used, and interference is ignored). In the pure Bohm QTM formulation, the quantum trajectories are uniquely determined by the exact quantum wavefunction, $\psi(x,t)$, and (essentially) vice versa; thus, the method is, in principle, exact. Moreover, the fact that differential probability [e.g., in one dimension (1D), $\rho(x,t)\,dx$, as opposed to the probability density, $\rho(x,t) = |\psi(x,t)|^2$] is conserved along quantum trajectories implies that these go to precisely where they are needed most, and is what renders large-dimensional (large-D) computations feasible. Other properties of pure Bohm quantum trajectories are discussed elsewhere in this book, and in other sources [12,17].

This chapter is concerned with an alternate to the pure Bohm version of QTM, called the "bipolar" QTM approach [12,19–32]. The bipolar approach is motivated by a very simple question: *What happens to pure Bohm quantum trajectories in the classical limit?*—i.e., the limit in which the mass, m, or the energy, E, of the problem becomes large. Intuitively, the expected answer is very clear: *in the classical limit, quantum trajectories should approach corresponding classical trajectories*, a specific manifestation of the more general "correspondence principle," often invoked in the theory of quantum mechanics. Of the various quantum dynamical effects described in the first paragraph above, all but one are effects that disappear smoothly, in the classical limit. For systems that exhibit only these quantum effects, one finds that pure Bohm quantum trajectories do indeed approach classical trajectories in the classical limit, exactly as desired. The manner in which pure Bohm quantum trajectories deviate from classical behavior as these quantum effects (tunneling, dispersion, and zero-point energy) are "turned on" is well-understood, and widely regarded to be an elegant aspect of the Bohmian approach.

On the other hand, there is one exception to the above correspondence rule, and it is a very important one: *interference*. Specifically, for systems that exhibit substantial reflective interference, such as the 1D barrier scattering system of Figure 15.1, the pure Bohm quantum trajectories *do not* approach classical trajectories in the classical limit, and in fact, tend to *deviate increasingly* from classical behavior in this limit [12,17]. This is a serious concern for real molecular scattering applications, even in many dimensions, for which, e.g., Figure 15.1 represents the effective 1D "reaction profile" for a direct chemical reaction. The interference effect can be described as follows.

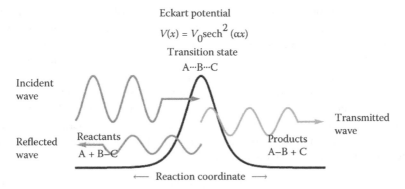

FIGURE 15.1 Schematic indicating a symmetric 1D reactive scattering barrier PES, representing a reaction profile for a typical direct chemical reaction, A + BC (reactants) → AB + C (products). The solid black curve indicates the PES, of standard Eckart form [$V(x) - V_0$ sech$^2(\alpha x)$]. The upper left sinusoidal curve indicates the left-incident plane wave, $\psi_+(x)$, moving to the right. Quantum mechanically, this is split into a transmitted wave on the far side of the barrier also moving to the right [also $\psi_+(x)$], and a reflected wave, headed to the left [$\psi_-(x)$, lower left sinusoidal curve], for energies, E, both above and below the barrier height, V_0. Classically, $E > V_0$ leads to 100% transmission, whereas $E < V_0$ leads to 100% reflection.

For classical trajectory evolution, there can be only two outcomes: if E is above the barrier, then all "reactant" trajectories incident on the barrier propagate smoothly to the far "product" side; if E is below the barrier, then all trajectories are reflected by the barrier, and then head back toward the reactant side. Quantum scattering systems, in contrast, exhibit *partial* transmission *and* reflection of the incident wave, both above *and* below the barrier. As the incident wave is scattered by the barrier, the reflected portion interferes with the incident wave headed in the opposite direction—leading to (potentially) strong oscillations in $\rho(x,t)$, which via probability conservation, must in turn lead to strong, nonclassical oscillations in the quantum trajectories. Moreover, since the quantum wavelength decreases in the classical limit, the corresponding oscillations become *more* pronounced, rather than less, in this limit.

Is it possible, therefore, to define a new kind of QTM, which has all of the same advantages as the pure Bohm QTM, yet which *also* achieves classical correspondence when there is substantial reflection interference? Such a theory would go a very long way toward justifying the validity of CTS in the classical limit, and perhaps also generating new and better approximate numerical methods. Such a theory might also allow exact QTM calculations to be performed in a more classical-like manner, thus bypassing the infamous "node problem" known to plague pure Bohm QTMs in the presence of strong interference [12]. Such a theory is the theory of the bipolar QTM.

Interference is an inherently wave-based phenomenon, and therefore cannot be directly described via a classical trajectory approach—even though it is a quantum effect that can persist in the classical limit. Nevertheless, the theory of waves, as applied to both electromagnetism (EM) and quantum scattering, does indeed offer a classical-like description of interference, through the use of the *superposition principle*. Specifically, interference is regarded as arising from the linear superposition of two or more wave components, each of which exhibits smooth, interference-free (ideally local plane-wave) behavior. Trajectory descriptions of the resultant components then exhibit classical-like behavior—e.g., superposed ray optics, in the EM case.

For many quantum reactive scattering applications, such as direct chemical reactions, two components are appropriate: one to represent the forward-reacting wave (i.e., moving from left to right in Figure 15.1), and the other describing the reverse (right-to-left) reaction, moving in the opposite direction. A "bipolar" decomposition of the wavefunction is thus indicated, i.e.,

$$\psi = \psi_+ + \psi_-, \tag{15.1}$$

where ψ_+ describes both the incident and transmitted (product) waves (as these are both positive-velocity contributions, moving to the right), and ψ_- describes the reflected (negative-velocity) wave component. The latter comes into being as a result of incident wave scattering off of the PES barrier; it therefore has vanishing probability density in the product (right) asymptote. Ideally, the two components, $\psi_\pm$, are themselves interference-free, but otherwise behave as much like a true wavefunction, ψ, as possible. Separate application of the standard Bohmian prescription to each of the $\psi_\pm$ would then yield *two* (sets of) quantum trajectories, each satisfying classical correspondence *automatically*. This is the essence of the bipolar QTM approach.

15.2 THE BIPOLAR DECOMPOSITION

The conceptual and potential numerical advantages of the bipolar QTM approach are clear at the outset. In practice however, there is an immediate difficulty to be dealt with: namely, *how does one actually define the bipolar decomposition of Equation 15.1?* The true wavefunction, ψ itself, is of course presumed to satisfy the Schrödinger equation (SE), either the time-independent (TISE) or the time-dependent (TDSE) version. The corresponding pure Bohm quantum trajectory ensemble is then uniquely specified, based on the following facts (assuming 1D, for convenience) [12,17]:

1. One quantum trajectory from the ensemble passes through each point in space, x, at every point in time, t.
2. The velocity of the quantum trajectory passing through the point (x,t) [i.e., $v(x,t)$] is obtained from the phase of $\psi(x,t)$ [i.e., $S(x,t)/h$], via

$$mv(x,t) = S'(x,t), \tag{15.2}$$

   where the prime denotes spatial differentiation.
3. The trajectory evolution can also be described via a quantum version of Newton's second law, i.e.,

$$m\ddot{x} = m\dot{v} = -V'_{\text{eff}}(x,t) = -[V'(x) + Q'(x,t)] \tag{15.3}$$

   where $V(x)$ is the PES, and $Q(x,t) = -(h^2/2m)|\psi|''/|\psi|$ the "quantum potential" [12,15–17].

Thus, the pure Bohm, or "unipolar" equations of motion are known *a priori*, and QTM development in this approach is essentially all *numerical*. In the bipolar approach, however, before any such numerical progress can be made, one must first arrive at a suitable definition for the $\psi_\pm$ components. To this end, Equation 15.1 per se is of little direct help, as it admits an infinite-dimensional range of solutions. Clearly, additional considerations must be brought into play. In this chapter, we discuss a particular bipolar strategy based on classical correspondence [19–25], although there are clearly other, potentially very different approaches that can also be taken, such as the covering function method of Wyatt and coworkers [26].

For the present, classical correspondence-based bipolar approach, we find it convenient to start with the simplest case imaginable, i.e., stationary bound state solutions of the 1D TISE, and then work in stages up to multidimensional TDSE wavepacket scattering.

15.3   STATIONARY BOUND STATES IN 1D

For stationary bound states in one (Cartesian) dimension, x, the TISE eigenstate solutions, $\psi(x)$, are L^2-integrable ($\int \rho(x)dx - 1$), and can be taken to be real-valued for all x. Together with Equation 15.1, this essentially implies that $\psi_+(x)$ must be complex conjugates of each other. Thus, $\rho_+(x) = \rho_-(x)$, and (from Equation 15.2 above) $v_+(x) = -v_-(x)$ for all x, whereas $v(x) = 0$ for $\psi(x)$ itself. One can therefore regard $\psi(x)$ as a "standing wave," constructed from a linear superposition of two equal and opposite "travelling waves," i.e., the two bipolar components, $\psi_\pm(x)$. The standing wave $\psi(x)$ can be highly oscillatory and nodal, which is certainly the case for the highly-energetically-excited eigenstates—i.e., in the classical limit of large E. However, the desired bipolar components, $\psi_\pm(x)$, should be smooth and nonoscillatory across *all* energies; moreover, the associated bipolar quantum trajectories should approach the corresponding classical trajectories, in the large-E limit. Finally, the $\psi_\pm(x)$ components should themselves resemble TISE solutions as much as possible, and in particular, must lead to bipolar densities, $\rho_\pm(x)$, that are also stationary (i.e., that exhibit no time dependence, when Equation 15.1 and the TISE are inserted into the TDSE).

In attempting to achieve all of the above requirements, a natural ansatz is to presume $\psi_\pm(x)$ components that are *themselves* mathematical TISE solutions, for the same E as $\psi(x)$ itself. Of course, these $\psi_\pm(x)$ cannot be physical TISE solutions, as they are non-L^2-integrable, and exhibit $\rho_\pm(x)$ densities that diverge in both left- and right-asymptotic limits. However, this choice is certainly stationary and node-free, and TISE-solution-like. Moreover, the set of all possible bipolar decompositions has now been reduced from an infinite-parameter space to a two-parameter space, since the TISE admits two linearly-independent solutions. The two parameters are complex-valued; however, one of these is associated with the overall scaling and phase of $\psi(x)$ itself, which as discussed above, is already predetermined. This leaves one complex-valued parameter, or alternatively, two real-valued parameters, that must still be determined.

The remaining two real-valued parameters will be chosen such that the resultant bipolar quantum trajectory ensembles, i.e., the velocity fields, $v_\pm(x)$, resemble the corresponding classical trajectory ensembles as closely as possible [19]. To this end, it is encouraging to note that the classical trajectory ensembles for bound 1D systems *are indeed* also bipolar (i.e., are characterized by double-valued velocity fields for classically allowed x values), with equal and opposite velocities given by

$$v_\pm^c = \pm\sqrt{2[E - V(x)]/m}. \tag{15.4}$$

One therefore expects $v_\pm(x)$ to be chosen, such that $|v_\pm(x)-v_\pm^c(x)|$ or $|Q_\pm(x)|$ is minimized over the classically allowed region, $V(x) < E$, and also, such that these quantities vanish in the classical limit of large E. Such a choice is possible, as discussed below. In contrast, the unipolar, pure Bohm QTM approach, *always* gets the physics wrong, because it is incapable of predicting *both* a positive- *and* negative-velocity value at a given point x—instead predicting their average value, $v(x) = 0$, which is increasingly inaccurate in the large E limit. This situation is very reminiscent of the well-known failure of Ehrenfest dynamics when applied to multisurface systems, for which a single, averaged dynamical propagation across all PESs is predicted, rather than separate propagations on each individual PES. Indeed, the latter consideration has led to the development of the much-superior, TSH approach for nonadiabatic dynamics [6,7]; a similar TSH scheme for bipolar propagation on a *single*-PES will be discussed later in this chapter (Section 15.7).

To characterize the family of available TISE solutions, the two real parameters are taken to be [19]:

1. The absolute value of the bipolar flux, $F = |\rho_\pm(x)v_\pm(x)|$, an x-independent constant.
2. The median of the enclosed action, x_0 defined such that $\int^{x0} mv_\pm(x)dx = \int_{x0} mv_\pm(x)dx = \pm(\pi h/2)(n + 1)$,

where n denotes the excitation number of the particular energy eigenstate, starting with $n = 0$ as the ground state. It is helpful to consider the phase space portraits of the relevant trajectory orbits, as is done in Figure 15.2 for the simple harmonic oscillator example, $V(x) = (k/2)x^2$, with $k = m = 1$. Classically, the v_+^c and v_-^c orbits for a given n form a closed loop, encircling a half-integer-quantized phase space volume (action) equal to $2\pi h(n + 1/2)$ (the circular curves in Figure 15.2). The associated quantum trajectories extend across all x, but always enclose a finite, and integer-quantized phase space volume of $2\pi h(n + 1)$, regardless of the particular choice of F and x_0 parameters. Moreover, by choosing F and x_0 to equal the corresponding classical values, one obtains bipolar quantum trajectories (the eye-shaped curves in Figure 15.2) that are *nearly indistinguishable* from their classical counterparts in the classically allowed region of space, and in the classical limit. As indicated in Figure 15.3, it is also true that $Q_\pm(x) \to 0$ in this region; however, in order to avoid turning points, $Q_\pm(x)$ becomes large and negative in the classically forbidden asymptotes, thus ensuring that $V_{\text{eff}}(x) < E$ and $v_+(x) > 0$ for all x, where

$$v_\pm = \pm\sqrt{2[E - V_{\text{eff}}(x)]/m}. \tag{15.5}$$

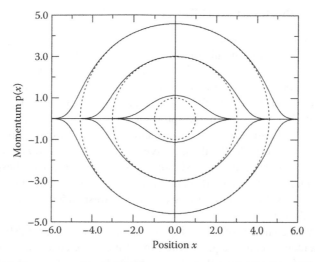

FIGURE 15.2 Phase space orbits for bipolar quantum trajectories for three different harmonic oscillator eigenstates; moving concentrically outward from the origin, these are $n = 0$, $n = 4$, and $n = 10$, respectively. Solid curves indicate bipolar quantum trajectories; dotted curves indicate corresponding classical trajectories. The correspondence principle is clearly satisfied with increasing n. (Reprinted with permission from Poirier, B., *Journal of Chemical Physics*, 121, 4501–4515, 2004.)

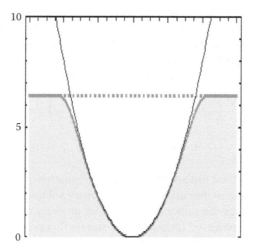

FIGURE 15.3 Bipolar effective potential, $V_{eff}(x)$ (thick solid curve), and true potential, $V(x)$ (thin solid curve), for the sixth excited harmonic oscillator eigenstate, $n = 6$. Note that $V_{eff}(x) < E$ everywhere, where the eigenenergy, E, is indicated by the thick dashed horizontal line. In the classically allowed region, $V_{eff}(x) \approx V(x)$, but $V_{eff}(x) \approx E$ in the classically forbidden region. (Modified from Poirier, B., *Journal of Chemical Physics*, 121, 4501–4515, 2004. With permission.)

15.4   STATIONARY SCATTERING STATES IN 1D

For stationary scattering states in 1D, i.e., the continuum TISE eigenstate solutions, the situation is substantially more complicated. First, there are *no* L^2-integrable solutions, and so for any given E, the two linearly-independent mathematical solutions (and arbitrary superpositions) are both physically allowed. Without loss of generality, we restrict consideration to the left-incident solution $\psi(x)$, only. Assuming for simplicity that $V(x) \to 0$ in both asymptotes, then in order to avoid interference and match classical behavior asymptotically, clearly $\psi_+(x) \propto \exp[ikx]$ in both asymptotes, and $\psi_-(x) \propto \exp[-ikx]$ in the left asymptote, where $(\hbar^2 k^2/2m) = E$. However, it can be shown that there are *no* TISE solutions $\psi_\pm(x)$ that satisfy these asymptotic conditions, which is the primary difficulty for the stationary scattering generalization of the bipolar theory. In order to achieve classical correspondence, therefore, we must consider *non*-TISE $\psi_\pm(x)$'s.

It is a well-known fact of scattering theory that for a given TISE solution, the flux of the incident stream is equal to the sum of the reflected and transmitted stream flux absolute values. Thus, the flux, F_+, associated with say, $\psi_+(x)$, far from being an x-independent constant, must decrease from the left asymptote [where $\psi_+(x)$ represents the incident wave] to the right [where it represents the transmitted wave]. The decrease in F_+ is matched by an increase in F_- for the reflected wave, $\psi_-(x)$, and is associated with dynamical *coupling* between the two components—a manifestation of scattering induced by the PES barrier. This dynamical coupling occurs only in the PES barrier region of x, i.e., where $V(x)$ is nonzero, and serves to induce transitions (or probability flow) between the two bipolar components. Borrowing from semiclassical theory [3], this leads to a unique Equation 15.1 decomposition for a given effective potential, $V_{eff}(x)$, where the resultant $\psi_\pm(x)$ are guaranteed to exhibit the correct asymptotic behavior [20–22]. However, the determination of $V_{eff}(x)$ in this formalism is unspecified, unlike in Section 15.3. The simplest possible choice is $V_{eff}(x) = 0$, leading to the following dynamical equations [20–22]:

$$\frac{d\psi_\pm}{dt} = \left(\frac{i}{\hbar}\right)(E - V)\psi_\pm - \left(\frac{i}{\hbar}\right)V\psi_\mp \tag{15.6a}$$

$$\frac{\partial\rho_\pm}{\partial t} = -F'_\pm \pm \frac{2V}{\hbar}\text{Im}[\psi_+^*\psi_-] \tag{15.6b}$$

with (d/dt) denoting total time derivative. These equations are time-dependent, but guaranteed to converge to the correct TISE stationary solution in the large-time limit. Equation 15.6b implies the important, nontrivial property that total probability for both components is conserved (related to the stream flux property described above), which is also true for other choices of $V_{eff}(x)$.

Figure 15.4 indicates the bipolar and unipolar solution densities, for a standard symmetric 1D benchmark PES (the Eckart barrier of Figure 15.1), for the choice $V_{eff}(x) = 0$. Unlike $\rho(x)$, which necessarily exhibits reflective interference in the left asymptote, both of the bipolar densities, $\rho_\pm(x)$, are *constant* in both asymptotes, as expected. For this particular system, $\rho_\pm(x)$ are also smooth and nonoscillatory in

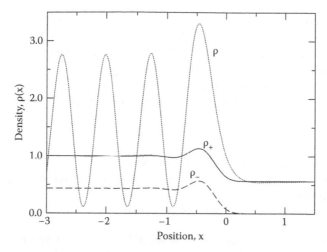

FIGURE 15.4 Computed unipolar and bipolar probability densities, for TISE energy eigenstate of the Eckart barrier system of Figure 15.1, with $E = V_0$. Solid curve: bipolar $\rho_+(x)$; dashed curve: bipolar $\rho_-(x)$; dotted curve: unipolar $\rho(x)$. The latter exhibits pronounced reflection interference in the left asymptote, but the bipolar curves are flat in both asymptotes. The transmission and reflection probabilities, i.e., right and left asymptotic values of $\rho_+(x)$ and $\rho_-(x)$, respectively, are roughly equal. (Modified from Poirier, B., *Journal of Chemical Physics*, 124, 034116, 2006. With permission.)

the barrier region, as desired. However, it can be shown that in the classical limit as $m \to \infty$, and/or E (and V_0, the barrier height) $\to \infty$, then the choice $V_{\text{eff}}(x) = 0$ leads to barrier region oscillations in $\rho_\pm(x)$, particularly when the reaction profile has a PES well (reaction intermediate, i.e., nondirect reaction). Intuitively, the reason for this undesirable behavior is clear: based on the discussion in Section 15.3, the appropriate effective potential should have $V_{\text{eff}}(x) \approx V(x)$ in the classically allowed region, and—for below barrier energies—$V_{\text{eff}}(x) \approx E$ in the barrier tunneling region. Indeed, ad hoc construction of such $V_{\text{eff}}(x)$ effective potentials *always* leads to improved (i.e., nonoscillatory) $\rho_\pm(x)$ behavior, which is very useful for practical applications. At the present stage of development, however, there does not exist a means to single out a *unique* $V_{\text{eff}}(x)$ *a priori*, as is the case for the stationary bound states, though this is an active area of ongoing investigation. This issue will resurface again in various contexts.

15.5 WAVEPACKET SCATTERING DYNAMICS IN 1D

The next generalization of the bipolar decomposition to be derived is for 1D TDSE wavepacket scattering dynamics. This is an essential prerequisite for generalization to large-D QTM calculations, for which the favorable numerical scaling requires wavefunctions that are well-localized in space, for all relevant times. The TISE scattering states, in contrast, are *delocalized*, across a very large region of space. Thus, if these states were computed directly for a large-D system, it would require prohibitive computational effort, regardless of the method used. In any event, we seek a bipolar version of the TDSE wavepacket propagation equations, presumably offering all of the advantages discussed previously in the context of TISE solutions.

In addition to previous considerations, the bipolar wavepacket evolution equations must satisfy:

- localized ψ implies localized $\psi_\pm$;
- at initial time, $\psi = \psi_+$ is the incident wavepacket;
- at final time, ψ_+ is the transmitted wavepacket, and ψ_- the reflected wavepacket;
- at all times, ψ evolution satisfies the TDSE.

For 1D systems, the above can be achieved via coherent superposition of stationary state solutions at different energies, applied separately to ψ, ψ_+, and ψ_-. Specifically, let the TISE solution for energy E be henceforth referred to as $\phi^E(x) = \phi_+^E(x) + \phi_-^E(x)$. Then,

$$\psi(x,t) = \int a(E)\varphi^E(x)\exp[-iEt/\hbar]dE, \qquad (15.7a)$$

where

$$a(E) = \langle \varphi^E | \psi(t=0) \rangle. \qquad (15.7b)$$

Applying the same expansion to the bipolar TISE components yields the corresponding $\psi_\pm(x,t)$:

$$\psi_\pm(x,t) = \int a(E)\varphi_\pm^E(x)\exp[-iEt/\hbar]dE. \qquad (15.8)$$

Substitution into Equation 15.6a then yields time evolution equations which—when recast so as to avoid all explicit dependence on the underlying TISE solutions—result in the following [24]:

$$\frac{\partial\psi_\pm}{\partial t} = -\left(\frac{i}{\hbar}\right)\left[\hat{H}\psi_\pm \pm \left(\frac{V'}{2}\right)(\Psi_\pm - \Psi_-)\right] \qquad (15.9a)$$

where

$$\Psi_\pm(x) = \int_{-\infty}^{x}\psi_\pm(x')dx'. \qquad (15.9b)$$

The resultant bipolar wavepacket components, $\psi_\pm(x,t)$, behave exactly as indicated above, i.e., for an initially left-incident wavepacket, localized far to the left of the scattering barrier, $\psi(t=0) = \psi_+(t=0)$. Over time, as ψ_+ reaches the barrier, a reflected ψ_- component comes into being in the PES barrier region. The ψ_+ component then continues to head to the right, turning into the transmitted branch of the final wavepacket, whereas ψ_- heads to the left and becomes the reflected branch.

Note from Equation 15.9a that the intercomponent coupling is proportional to $V'(x)$; thus, coupling arises only in the PES barrier region, so that in the asymptotic regions, both components satisfy free particle propagation.

For Equation 15.9 to be useful, $\psi_{\pm}$ should be smooth and interference-free for all times. In general, it is extremely difficult to concoct a bipolar theory that satisfies this property, owing to the square-root relationship between ψ and ρ, which tends to magnify the effects of interference. Nevertheless, Equation 15.9 achieves this goal very well, for a wide range of applications [24]. Figure 15.5 depicts the time evolution of the bipolar densities, $\rho_{\pm}(x,t)$, for a wavepacket scattering application (the symmetric 1D Eckart barrier discussed previously). The bipolar components are smooth and well behaved, even though $\rho(x,t)$ itself exhibits substantial reflection interference—accommodated in the bipolar treatment through the "spur" that develops on the back end of ψ_{+}, and ultimately vanishes in the large-time limit (a process known as "node healing" [12]).

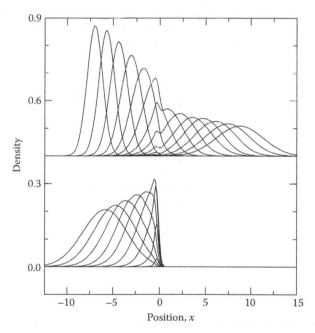

FIGURE 15.5 Bipolar component wavepacket densities, $\rho_{\pm}(x,t)$ as a function of position, x, and for a variety of times, t, for the symmetric Eckart barrier system of Figures 15.1 and 15.4. The upper family of curves represent $\rho_{+}(x,t)$ at different times, whereas the lower family of curves represent $\rho_{-}(x,t)$ (magnified by a factor of $4x$). The motion of the former over time is left-to-right, whereas that of the latter is right-to-left. At intermediate t, a stationary $\rho_{+}(x,t)$ spur forms and then dissipates in the barrier region, corresponding with the simultaneous "birth" and stationary growth of $\rho_{-}(x,t)$. After $\rho_{-}(x,t)$ has grown sufficiently large, it starts to move to the left. (Reprinted with permission from Poirier, B., *Journal of Chemical Physics*, 128, 164115, 2008.)

Although not visible on the scale of the Figure 15.5 plots, the $\rho_\pm(x,t)$ plots actually do exhibit very slight nodal behavior, in the asymptotic regions where the density values are extremely small. Thus, whereas the interference situation is *greatly* improved in the bipolar case, it has not been completely eradicated. This may be due to the fact that Equation 15.9, like Equation 15.6a, is based on the $V_{\text{eff}}(x) = 0$ choice of effective potential. As discussed in Section 15.4, a more classical-like choice for $V_{\text{eff}}(x)$ is probably more appropriate. At present, however, it is not known how to extend the derivation of Equation 15.9 for arbitrary $V_{\text{eff}}(x)$, to obtain standalone bipolar wavepacket evolution equations that make no explicit reference to the $\phi_\pm^E$.

15.6 WAVEPACKET SCATTERING DYNAMICS IN MANY DIMENSIONS

Equation 15.9 can be easily generalized for multidimensional systems with arbitrary (straight or curvilinear) reaction paths and coordinates. All that is needed is a specification of the reaction coordinate direction, $\hat{s}$, at every point in space, $\vec{x}$. The spatial derivative in Equation 15.9a then gets replaced with $\hat{s} \cdot \vec{\nabla} V$, and Equation 15.9b with a line integral along $\hat{s}$. With $\psi_\Delta = (\psi_+ - \psi_+)$, the result is [25]:

$$\frac{\partial \psi_\pm}{\partial t} = -\left(\frac{i}{\hbar}\right)\left[\hat{H}\psi_\pm \pm \left(\frac{\hat{s} \cdot \vec{\nabla} V}{2}\right)\Psi_\Delta\right] \tag{15.10a}$$

where

$$\Psi_\Delta(\vec{x}) = \int_{-\infty}^{0} \psi_\Delta(\vec{x} + s\hat{s})ds. \tag{15.10b}$$

The many-D bipolar wavepacket propagation equations above have been applied to a number of test applications [25], including the collinear $H + H_2$ chemical reaction system in Jacobi coordinates, indicated in Figure 15.6. In all cases considered, the resultant bipolar densities are smooth and well-behaved at all times, as well as satisfying all of the other desired properties as discussed above.

15.7 NUMERICAL IMPLEMENTATIONS

For quantum scattering problems for which the bipolar component propagation equations are solved using conventional means, the resultant computed $\psi_\pm(x,t)$'s can then be used to obtain bipolar quantum trajectories. Such an "analytical" approach is often applied in the pure Bohm QTM context to provide physical insight [33], and the same has been done in the bipolar case for simple 1D model systems [27]. However, the main interest here is for *synthetic* calculations, i.e., to propagate the bipolar quantum trajectories *without* knowing $\psi_\pm(x,t)$ *a priori*. To this end, there are two basic types of numerical implementations that have been considered: (1) exact numerical propagation of Equations 15.6a, 15.9a, or 15.10a, via communicating (interdependent) quantum trajectory ensembles; and (2) approximate numerical propagation, using

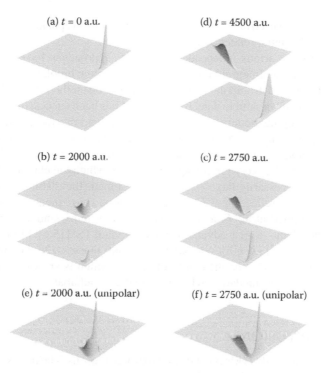

(a) $t = 0$ a.u. (d) $t = 4500$ a.u.

(b) $t = 2000$ a.u. (c) $t = 2750$ a.u.

(e) $t = 2000$ a.u. (unipolar) (f) $t = 2750$ a.u. (unipolar)

FIGURE 15.6 Bipolar wavepacket dynamics for the collinear H + H$_2$ system. Each plot represents a "snapshot" for a specific time, t, as listed. The initial wavepacket is right-incident. The reaction coordinate extends over the right-front edge of the plots, turns the bottom corner, and then heads back into the 3D plot frame. The top four plots, a–d, indicate bipolar component wavepacket densities, chronologically, in counter-clockwise order. For comparison, the bottom two plots, (e) and (f), represent the corresponding unipolar density, at the two intermediate times. Interference nodes are markedly evident in the trailing (right) edge of plot (e), but do not occur in the corresponding bipolar plot (b). (Reprinted with permission from Poirier, B., *Journal of Chemical Physics*, 129, 084103, 2008.)

independent (noncommunicating) trajectory ensembles, along the lines of CTS and TSH. While the latter is more readily applicable to extremely large molecular systems (hundreds to thousands of dimensions), the former constitutes an exact quantum dynamics method that defeats exponential scaling with system dimensionality [25]. Both methods are radical departures from the current state-of-the-art that may greatly extend the applicability of accurate quantum dynamics calculations.

Regarding (1), synthetic bipolar QTM calculations have been performed [27,28], for which two chief difficulties were encountered: (a) ψ_- is initially zero; (b) in the classical limit, bipolar quantum trajectories can get extremely close to one another, essentially because they are "trying" to cross through a caustic. Successful remedies have been found to resolve these issues [27,28], leading to the first-ever successful exact (worst relative error $\sim 10^{-4}$) synthetic QTM wavepacket calculation for a system with substantial reflection interference.

Regarding (2), various independent-trajectory approaches have been explored, including bipolar Path Integral Monte Carlo [29], quantum bipolar TSH [29], classical bipolar TSH [30], and the bipolar derivative propagation method (DPM) [31], with the last two the most successful to date. Bipolar DPM has been successfully applied to an extremely challenging model system, involving an Eckart barrier reaction coordinate, strongly coupled to up to 300 perpendicular bath modes. Accurate transmission and reflection probabilities were achieved across the entire energetic range [31]. Classical bipolar TSH is another very promising approach, in which the classical limit of large *mass*—but *not* necessarily large *energy*—is presumed (the distinction is important, because there is no reflection as $E \to \infty$) [30]. As a consequence, classical bipolar TSH can be applied at all E, and for single-PES systems, and with no forbidden energy hop problem—a great advance beyond standard TSH [5,6]. Moreover, it seems to perform quite well even for comparatively small masses. On the other hand, the results depend somewhat on the choice of $V_{\mathrm{eff}}(x)$, which is analogous to the choice of electronic states in standard TSH (e.g., diabatic vs. adiabatic).

15.8   CONCLUSIONS

In general, the study of QTMs is conducted for three reasons. The first is *foundational*: quantum trajectories provide insight into the nature of quantum mechanics itself, and particularly, classical correspondence. The second is *analytical*: for a given molecular system, quantum trajectories provide classical-like insight into the dynamics and physics of the relevant processes. The third is *synthetic*: QTMs comprise an efficient means of actually *solving* the TDSE for quantum dynamical systems. The bipolar approach offers advantages in all three areas. Foundationally, bipolar (or more generally, *multipolar*) QTMs provide true classical correspondence, even when there is substantial interference—a contingency for which the standard pure Bohm unipolar QTM is ill-equipped to succeed. Analytically, bipolar quantum trajectories offer a more classical-like, physically correct description of processes involving interference. Synthetically, bipolar QTMs greatly ameliorate—if not eradicate—the infamous "node problem" plaguing unipolar treatments, thus enabling accurate, stable synthetic QTM calculations to be performed for the first time for systems with substantial reflection interference. The bipolar approach also greatly extends the applicability, and perhaps performance, of standard TSH methodologies.

Of course, there is still more work to be done in future. Better time integrators and fitting/spatial differentiation routines are needed, in order to simplify and streamline the current synthetic bipolar QTM codes. Bipolar generalizations in broader contexts such as multisurface dynamics [23] and complex-valued QTMs [32] will be further developed. However, the most important future development may well prove to be theoretical—specifically, the derivation of a unique $V_{\mathrm{eff}}(x)$ potential function for scattering systems, as already exists for 1D stationary bound states, and discussed in Section 15.3. This will likely enable advances in a variety of different bipolar contexts.

BIBLIOGRAPHY

1. D. Frenkel and B. Smit, *Understanding Molecular Simulations* (Academic, New York, 2002).
2. G. D. Billing, Classical Path Method in Inelastic and Reactive Scattering, *Int. Rev. Phys. Chem.* **13**, 309 (1994).
3. N. Fröman and P.O. Fröman, *JWKB Approximation* (North-Holland, Amsterdam, 1965).
4. W. H. Miller, The Semiclassical Initial Value Representation: A Potentially Practical Way for Adding Quantum Effects to Classical Molecular Dynamics Simulations, *J. Phys. Chem. A*, **105**, 2942 (2001).
5. G. A. Voth, Path-Integral Centroid Methods in Quantum Statistical Mechanics and Dynamics, in I. Prigogine (ed.), *Advances in Chemical Physics* (Wiley, New York, 1996), pp. 135–218.
6. J. C. Tully, Molecular Dynamics with Electronic Transitions, *J. Chem. Phys.* **93**, 1061 (1990).
7. J. C. Tully, Nonadiabatic Dynamics, in D. L. Thompson (ed.), *Modern Methods for Multidimensional Dynamics Computations in Chemistry* (World Scientific, Singapore, 1998).
8. *Beyond the Molecular Frontier: Challenges for Chemistry and Chemical Engineering.* Organizing Committee for the Workshop on Information and Communications, Committee on Challenges for the Chemical Sciences in the 21st Century, National Research Council, 2003.
9. B. K. Day, A. Askar, and H. Rabitz, Multidimensional Wave Packet Dynamics Within the Fluid Dynamical Formulation of the Schrödinger Equation, *J. Chem. Phys.* **109**, 8770 (1998).
10. C. L. Lopreore and R. E. Wyatt, Quantum Wave Packet Dynamics with Trajectories, *Phys. Rev. Lett.* **82**, 5190 (1999).
11. E. R. Bittner, Quantum Tunneling Dynamics using Hydrodynamic Trajectories, *J. Chem. Phys.* **112**, 9703 (2000).
12. R. E. Wyatt, *Quantum Dynamics with Trajectories: Introduction to Quantum Hydrodynamics* (Springer, New York, 2005).
13. D. Babyuk and R. E. Wyatt, Multidimensional Reactive Scattering with Quantum Trajectories: Dynamics with 50–200 Vibrational Modes, *J. Chem. Phys.* **124**, 214109 (2006).
14. V. E. Madelung, Quantentheorie in hydrodynamischer form, *Z. Physik* **40**, 322 (1926)
15. D. Bohm, A Suggested Interpretation of the Quantum Theory in Terms of "Hidden" Variables. I, *Phys. Rev.* **85**, 166 (1952).
16. D. Bohm, A Suggested Interpretation of the Quantum Theory in Terms of "Hidden" Variables. II, *Phys. Rev.* **85**, 180 (1952).
17. P. R. Holland, *The Quantum Theory of Motion* (Cambridge University Press, Cambridge, 1993).
18. K. Berndl, M. Daumer, D. Dürr, S. Goldstein, and N. Zanghi, A Survey of Bohmian Mechanics, *Il Nuovo Cimento* **110B**, 735 (1995).
19. B. Poirier, Reconciling Semiclassical and Bohmian Mechanics: I. Stationary States, *J. Chem. Phys.*, **121**, 4501–4515 (2004).
20. C. Trahan and B. Poirier, Reconciling Semiclassical and Bohmian Mechanics. II. Scattering States for Discontinuous Potentials, *J. Chem. Phys.*, **124**, 034115 (2006).
21. C. Trahan and B. Poirier, Reconciling Semiclassical and Bohmian Mechanics. III. Scattering States for Continuous Potentials, *J. Chem. Phys.*, **124**, 034116 (2006).

22. B. Poirier, Development and Numerical Analysis of "Black-box" Counter-propagating Wave Algorithm for Exact Quantum Scattering Calculations, *J. Theoret. Comput. Chem.*, **6**, 99–125 (2007).
23. B. Poirier and G. Parlant, Reconciling Semiclassical and Bohmian Mechanics. IV. Multi-surface Dynamics, invited contribution, special issue to honor Bob Wyatt, *J. Phys. Chem A.* **111**, 10400–10408 (2007).
24. B. Poirier, Reconciling Semiclassical and Bohmian Mechanics. V. Wavepacket Dynamics, *J. Chem. Phys.*, **128**, 164115 (2008).
25. B. Poirier, Reconciling Semiclassical and Bohmian Mechanics. VI. Multidimensional Dynamics, *J. Chem. Phys.*, **129**, 084103 (2008).
26. D. Babyuk and R. E. Wyatt, Coping with the Node Problem in Quantum Hydrodynamics: The Covering Function Method, *J. Chem. Phys.* **121**, 9230 (2004).
27. K. Park, B. Poirier, and G. Parlant, Quantum Trajectory Calculations for Bipolar Wavepacket Dynamics in One Dimension, *J. Chem. Phys.* **129**, 194112 (2008).
28. K. Park and B. Poirier, Quantum Trajectory Calculations for Bipolar Wavepacket Dynamics Calculations in One Dimension: Synthetic Single-Wavepacket Propagation, *J. Theoret. Comput. Chem.* in press.
29. J. Maddox and B. Poirier, Bipolar Quantum Trajectory Simulations: Trajectory Surface Hopping vs. Path Integral Monte Carlo, *Multidimensional Quantum Mechanics with Trajectories*, ed. D. V. Shalashilin and M. P. de Miranda (CCP6, Daresbury Laboratory, 2009).
30. J. Maddox and B. Poirier, Bipolar Quantum Trajectory Simulations: Classical Bipolar Trajectory Surface Hopping Method, *J. Chem. Phys.*, in preparation.
31. J. Maddox and B. Poirier, The Bipolar Derivative Propagation Method for Calculating Stationary States for High-dimensional Reactive Scattering Systems, *Multidimensional Quantum Mechanics with Trajectories*, ed. D. V. Shalashilin and M. P. de Miranda (CCP6, Daresbury Laboratory, 2009).
32. T. Djama and B. Poirier, Analytical Solution of the Bipolar Decomposition for the Eckart Barrier System: Ramifications for Complex-Valued Quantum Trajectories, *J. Theoret. Comput. Chem.*, in preparation.
33. A. S. Sanz, F. Borondo, and S. Miret-Artés, Causal Trajectories Description of Atom Diffraction by Surfaces, *Phys. Rev. B* **61**, 7743–7751 (2000).

16 Nondifferentiable Bohmian Trajectories

Gebhard Grübl and Markus Penz

CONTENTS

16.1   INTRODUCTION

Quantum mechanics is often praised as a theory which unifies classical mechanics and classical wave theory. Quanta are said to behave either as particles or waves, depending on the type of experiment they are subjected to. But where in the standard formalism can a pointlike particle structure actually be found? Perhaps only to some degree in the reduction postulate applied to position measurements. In reaction to this unsatisfactory state of affairs, Bohmian mechanics introduces a mathematically precise particle concept into quantum mechanical theory. The fuzzy wave functions are supplemented by sharp particle world-lines. Through this additional structure some quantum phenomena like the double slit experiment have lost their mystery.

Clearly the additional structure of particle world-lines brings along its own mathematical problems. Ordinary differential equations are generated from solutions of partial differential equations. A mathematically convincing general treatment so far has been given for a certain type of wave functions which do not exhaust all possible quantum mechanical situations. Exactly this fact has led some workers to doubt that a Bohmian mechanics exists for all initial states Ψ_0 of a Schrödinger evolution $t \mapsto e^{-iht}\Psi_0$. We shall show for one specific case of a counterexample Ψ_0 how the problem might be resolved in general. We approximate the state Ψ_0, for which the Bohmian velocity field does not exist, by states which do have one. Their integral curves turn out to converge to limit curves which can be taken to constitute the Bohmian mechanics of the state unamenable to Bohmian mechanics at first sight.

16.2   BOHMIAN EVOLUTION FOR $\psi \in C^2$

Let $\psi : \mathbb{R} \times \mathbb{R}^s \to \mathbb{C}$ be twice continuously differentiable, i.e., $\psi \in C^2 (\mathbb{R} \times \mathbb{R}^s)$, and let ψ obey Schrödinger's partial differential equation

$$i\hbar\partial_t \psi (t,x) = -\frac{\hbar^2}{2m}\Delta \psi (t,x) + V (x) \psi (t,x) \qquad (16.1)$$

with $V : \mathbb{R}^s \to \mathbb{R}$ being smooth, i.e., $V \in C^\infty (\mathbb{R}^s)$. From ψ, which is called a classical solution of Schrödinger's equation, a deterministic time evolution $x \mapsto \gamma_x (t)$ of certain points $x \in \mathbb{R}^s$ can be derived: If there exists a unique maximal solution $\gamma_x : I_x \to \mathbb{R}^s$ to the implicit first order system of ordinary differential equations

$$\rho_\psi (t,\gamma (t)) \dot{\gamma} (t) = j_\psi (t, \gamma (t)) \qquad (16.2)$$

with the initial condition $\gamma (0) = x$, one takes γ_x as the evolution of x. Here $\rho_\psi : \mathbb{R} \times \mathbb{R}^s \to \mathbb{R}$ and $j_\psi : \mathbb{R} \times \mathbb{R}^s \to \mathbb{R}^s$ with

$$\rho_\psi (t,x) = |\psi (t,x)|^2 \text{ and } j_\psi (t,x) = \frac{\hbar}{m}\Im\left[\overline{\psi (t,x)}\nabla_x \psi (t,x)\right] \qquad (16.3)$$

obey the continuity equation $\partial_t \rho_\psi (t,x) = -\text{div}\, j_\psi (t,x)$ for all $(t,x) \in \mathbb{R} \times \mathbb{R}^s$. From now on we shall drop the index ψ from ρ_ψ and j_ψ.

For certain solutions* ψ the curves γ_x can be shown to exist on a maximal domain $I_x = \mathbb{R}$ for all $x \in \mathbb{R}^s$: If ψ has no zeros, then the velocity field $v = j/\rho$ is a C^1-vector field. v then obeys a local Lipschitz condition such that the maximal solutions are unique. If in addition there exist continuous nonnegative real functions α, β with $|v (t,x)| \leq \alpha (t) |x| + \beta (t)$ then all maximal solutions of Equation 16.2 are defined on $\mathbb{R}$ and the general solution

$$\Phi : \bigcup_{x \in \mathbb{R}^s} I_x \times \{x\} \to \mathbb{R}^s \text{ with } \Phi (t,x) = \gamma_x (t)$$

extends to all of $\mathbb{R} \times \mathbb{R}^s$ (Thm 2.5.6, Ref. [1]). Due to the uniqueness of maximal solutions the map $\Phi (t, \cdot) : \mathbb{R}^s \to \mathbb{R}^s$ is a bijection for all $t \in \mathbb{R}$. It obeys[†]

$$\int_{\Phi(t,\Omega)} \rho (t,x) \, d^s x = \int_\Omega \rho (0,x) \, d^s x \qquad (16.4)$$

for all $t \in \mathbb{R}$ and for all open subsets $\Omega \subset \mathbb{R}^s$ with sufficiently smooth boundary such that the integral theorem of Gauss can be applied to the space-time vector field (ρ, j) on the domain $\bigcup_{t' \in (0,t)} \Phi (t', \Omega)$ [2].

These undisputed mathematical facts have instigated Bohm's amendment of Equation 16.1 in order to explain the fact that *macroscopic bodies usually are localized much more strictly than their wave functions suggest.*

---

* The simplest explicitly solvable example is provided by the plane wave solution $\psi (t,x) = e^{-i|k|^2 t + ik \cdot x}$. Its Bohmian evolution Φ obeys $\Phi (t,x) = 2tk$. Another explicitly solvable case is given by a Gaussian free wave packet.

[†] Here $\Phi (t,\Omega) = \{\Phi (t,x) \,|\, x \in \Omega\}$.

In Bohm's completion of nonrelativistic quantum mechanics it is assumed that any closed system has at any time, in addition to its wave function, a position in its configuration space and that this position evolves according to the general solution Φ induced by the wave function. One says that the position is guided by ψ since Φ is completely determined by ψ (and no other forces than the ones induced by ψ are allowed to act on the position). More specifically, γ_x is assumed to give the position evolution for an isolated particle with wave function $\psi(0, \cdot)$ and position x—both at time $t = 0$.

As is common in standard quantum mechanics, $\psi(0, \cdot)$ is supposed to obey

$$\int_{\mathbb{R}^s} |\psi(0, x)|^2 \, d^s x = 1.$$

The nonnegative density $\rho(0, \cdot)$ is interpreted as the probability density of the position which the particle has at time $t = 0$. Since an initial position x is assumed to evolve into $\gamma_x(t)$, the position probability density at time t is then, due to Equation 16.4, given by $\rho(t, \cdot)$. In particular, Bohm's completion gives the position probabilities, among all the other spectral probability measures, a fundamental status, since the empirical meaning of the other ones, as for instance momentum probabilities, are all deduced from position probabilities.

There are classical solutions of Schrödinger's equation, whose general solution Φ *does not extend* to all of $\mathbb{R} \times \mathbb{R}^s$. An obstruction to do so can be posed by the zeros of ψ. In the neighborhood of such a zero the velocity field $v = j/\rho$ may be unbounded and v then lacks a continuous extension into the zero. As an example consider a time 0 wave function $\psi(0, \cdot) : \mathbb{R}^2 \to \mathbb{C}$, for which $\psi(0, x, y) = x^2 + iy^2$ within a neighborhood U of its zero $(x, y) = (0, 0)$. Within $U(0, 0)$ for the velocity field it follows that

$$\frac{m}{\hbar} v^1(0, x, y) = \Im \frac{\partial_x \psi(0, x, y)}{\psi(0, x, y)} = -\frac{2xy^2}{x^4 + y^4}.$$

Hence for $0 < |\phi| < \pi/2$ we have $v^1(r \cos\phi, r \sin\phi) \to -\infty$ for $r \to 0$ with ϕ fixed. Thus the implicit Bohmian evolution equation 16.2 is singular in a zero of the wave function whenever the velocity field does not have a continuous extension into it. As a consequence the evolution γ_x of such a zero x is not defined by Equation 16.2.

As a related phenomenon there are solutions to Equation 16.2 which begin or end at a finite time because they terminate at a zero of ψ. A nice example [3] for this to happen is provided by the zeros of the harmonic oscillator wave function $\psi : \mathbb{R}^2 \to \mathbb{C}$ with

$$\psi(t, x) = e^{-\frac{x^2}{2}} \left(1 + e^{-2it} \left(1 - 2x^2\right)\right).$$

For example, the points $|x| = 1$ are zeros of $\psi(t, \cdot)$ at the times $t \in \pi\mathbb{Z}$. They are singularities of v since

$$\lim_{t \to 0} t \Im \frac{\partial_x \psi(t, \pm 1)}{\psi(t, \pm 1)} = \pm 2.$$

Note however that

$$\Im \frac{\partial_x \psi (0, x)}{\psi (0, x)} = 0$$

for $x \neq \pm 1$.

There are more challenges to Bohmian mechanics. The notion of distributional solutions to a partial differential equation like Equation 16.1 raises the question of whether these solutions support a kind of Bohmian particle motion like the classical solutions do. After all quantum mechanics employs such distributional solutions.

16.3 BOHMIAN EVOLUTION FOR $\Psi_t \in C_h^\infty$

In standard quantum mechanics the classical solutions, i.e., the C^2-solutions of Equation 16.1, do not represent all physically possible situations. Rather, a more general quantum mechanical evolution is abstracted from Equation 16.1. It is given by the so-called weak solutions

$$\Psi_0 \mapsto \Psi_t = e^{-iht} \Psi_0 \text{ for all } \Psi_0 \in L^2 \left(\mathbb{R}^s \right)$$

with $\hbar h$ being a self-adjoint, usually unbounded Hamiltonian corresponding to Equation 16.1. The domain D_h of h does not comprise all of $L^2(\mathbb{R}^s)$, yet it is dense in $L^2 (\mathbb{R}^s)$. Since h is self-adjoint, the exponential e^{-iht} has a unique continuous extension to $L^2 (\mathbb{R}^s)$. This unitary evolution operator e^{-iht} stabilizes the domain of h as a dense subspace of $L^2 (\mathbb{R}^s)$. Thus if and only if an initial vector Ψ_0 belongs to D_h, Equation 16.1 generalizes to

$$\lim_{\varepsilon \to 0} \left\| i \frac{\Psi_{t+\varepsilon} - \Psi_t}{\varepsilon} - h\Psi_t \right\| = 0 \qquad (16.5)$$

for all $t \in \mathbb{R}$. For $\Psi_0 \notin D_h$ Equation 16.5 does not hold for any time.

Yet the construction of Bohmian trajectories needs much more than the evolution $\Psi_0 \mapsto \Psi_t$ within $L^2 (\mathbb{R}^s)$, since the elements of $L^2 (\mathbb{R}^s)$ are equivalence classes $[f]$ of functions $f \in \mathcal{L}^2 (\mathbb{R}^s)$. It rather needs a trajectory of functions instead of a trajectory of equivalence classes of square-integrable functions. If there exists a function $\psi \in C^1 (\mathbb{R} \times \mathbb{R}^s)$ such that $\Psi (t, \cdot) = [\psi (t, \cdot)]$ holds for all $t \in \mathbb{R}$, then ψ is unique and the Bohmian equation of motion Equation 16.2 can be derived from the evolution $\Psi_0 \mapsto \Psi_t$ through ψ. When does there exist such ψ?

Due to Kato's theorem (e.g., Thm X.15 of Ref. [4]) the Schrödinger Hamiltonians h, corresponding to potentials V from a much wider class than just $C^\infty(\mathbb{R}^s : \mathbb{R})$, have the same domain as the free Hamiltonian $-\Delta$, namely the Sobolev space $W^2 (\mathbb{R}^s)$. This is the space of all those $\Psi \in L^2 (\mathbb{R}^s)$ which have all of their distributional partial derivatives up to second order being regular distributions belonging to $L^2 (\mathbb{R}^s)$. Since D_h is stabilized by the evolution e^{-iht}, for any $\Psi_0 \in D_h$ there exists, for any $t \in \mathbb{R}$, a function $\psi (t, \cdot) \in \mathcal{W}^2 (\mathbb{R}^s)$ such that

$$e^{-iht} \Psi_0 = [\psi (t, \cdot)]. \qquad (16.6)$$

However, for this family of time parametrized functions $\psi\,(t,\cdot)$ the Bohmian equation of motion in general does not make sense since $\psi\,(t,\cdot)$ need not be differentiable in the classical sense.*

Therefore some stronger restriction of initial data than $\Psi_0 \in D_h$ is needed in order to supply the state evolution $\Psi_0 \mapsto e^{-iht}\Psi_0$ with Bohm's amendment. For a restricted set of initial states (x, Ψ_0) and for a fairly large class of static potentials a Bohmian evolution has indeed been constructed in Refs [5] and [6]. There it is shown that for any $\Psi_0 \in \bigcap_{n \in \mathbb{N}} D_{h^n} =: C_h^\infty$ there exists

- a (time independent) subset $\Omega \subset \mathbb{R}^s$;
- for any t a square-integrable function $\psi\,(t,\cdot)$

such that the restriction of $\psi\,(t,\cdot)$ to Ω belongs to $C^\infty\,(\Omega)$ and Equation 16.6 holds. The set Ω is obtained by removing from $\mathbb{R}^s$ first the points where the potential function V is not C^∞, second the zeros of $\psi\,(0,\cdot)$, and third those points x for which the maximal solution γ_x does not have all of $\mathbb{R}$ as its domain. Surprisingly, Ω is still sufficiently large, since

$$\int_\Omega |\psi\,(0,x)|^2\,d^s x = 1.$$

On this reduced set Ω of initial conditions a Bohmian evolution $\Phi : \mathbb{R} \times \Omega \rightarrow \mathbb{R}^s$ can be constructed. Thus if $\Psi_0 \in C_h^\infty$ and if the initial position x is distributed within $\mathbb{R}^s$ with probability density $|\psi\,(0,\cdot)|^2$ then the global Bohmian evolution γ_x of x exists with probability 1.

16.4   BOHMIAN EVOLUTION FOR $\Psi_t \in L^2 \setminus C_h^\infty$

How about initial conditions $\Psi_0 \in L^2\,(\mathbb{R}^s) \setminus C_h^\infty$? Can the equation of motion Equation 16.2 still be associated with Ψ_0? Hall has devised a specific counterexample $\Psi_0 \notin D_h$ which leads to a wave function ψ which at certain times is nowhere differentiable with respect to x and thus renders impossible the formation of the velocity field v. Therefore it has been suggested that the Bohmian amendment of standard quantum mechanics is "formally incomplete" and it has been claimed that the problem is unlikely to be resolved [7].

A promising way to tackle the problem is to successively approximate the initial condition $\Psi_0 \notin D_h$ by a strongly convergent sequence of vectors $\left(\Psi_0^n\right) \in C_h^\infty$. For each of the vectors Ψ_0^n a Bohmian evolution Φ_n exists. We do not know whether it has actually been either disproved or proved that the sequence of evolutions does converge to a limit Φ and that the limit depends on the chosen sequence $\Psi_0^n \rightarrow \Psi_0$.

Here we shall explore this question within the simplified setting of a spatially one-dimensional example. We will make use of an equation for γ_x which has already

---

* Only for $s = 1$ does Sobolev's lemma (Thm IX.24 in Vol 2 of Ref. [4]) say that $[\psi\,(t,\cdot)]$ has a C^1 representative within $\mathcal{L}^2\,(\mathbb{R})$. From such a C^1 representative $\psi\,(t,\cdot)$ the current j follows as a continuous vector field and a continuous velocity field v can be derived outside the zeros of ψ. However, v does not need to obey the local Lipschitz condition implying the local uniqueness of its integral curves.

been pointed out in Ref. [5] and which does not rely on the differentiability of j. In this case Equation 16.4 can be generalized in order to determine a nondifferentiable Bohmian trajectory γ_x by choosing $\Omega = (-\infty, x)$ in Equation 16.4 as follows.

Consider first the case of a C^2-solution of Equation 16.1 generating a general solution $\Phi : \mathbb{R} \times \mathbb{R} \to \mathbb{R}$ of the Bohmian equation of motion Equation 16.2. Since because of their uniqueness the maximal solutions do not intersect, we have $\Phi(t, (-\infty, x)) = (-\infty, \Phi(t, x)) = (-\infty, \gamma_x(t))$. From this it follows by means of Equation 16.4 that

$$\int_{-\infty}^{\gamma_x(t)} \rho(t, y)\, dy = \int_{-\infty}^{x} \rho(0, y)\, dy. \tag{16.7}$$

As a check we may take the derivative of Equation 16.7 with respect to t. This yields

$$\rho(t, \gamma_x(t))\, \dot{\gamma}_x(t) + \int_{-\infty}^{\gamma_x(t)} \partial_t \rho(t, y)\, dy = 0.$$

Making use of local probability conservation $\partial_t \rho = -\partial_x j$ we recover, by partial integration, Equation 16.2.

Now observe that Equation 16.7 for $\gamma_x(t)$ is meaningful not only when ψ is a square integrable C^2-solution of Equation 16.1 but also if $\psi(t, \cdot)$ is an arbitrary representative of Ψ_t with arbitrary $\Psi_0 \in L^2(\mathbb{R})$. In order to make this explicit let $E_x : L^2(\mathbb{R}) \to L^2(\mathbb{R})$ with $x \in \mathbb{R}$ denote the spectral family of the position operator. For the orthogonal projection E_x we have

$$(E_x \varphi)(y) = \begin{cases} \varphi(y) & \text{for } y < x \\ 0 & \text{otherwise.} \end{cases}$$

The expectation value $\langle \Psi, E_x \Psi \rangle$ of E_x with unit vector $\Psi \in L^2(\mathbb{R})$ thus yields the cumulative distribution function of the position probability given by Ψ. If we define $F : \mathbb{R}^2 \to [0, 1]$ through $F(t, x) = \langle \Psi_t, E_x \Psi_t \rangle$, then Equation 16.7 is equivalent to

$$F(t, \gamma_x(t)) = F(0, x). \tag{16.8}$$

Thus, the graph $\{(t, \gamma_x(t)) \mid t \in \mathbb{R}\}$ of a trajectory is a subset of the level set of F which contains the point $(0, x)$. If Ψ_n is a sequence in $L^2(\mathbb{R}^s)$ which converges to Ψ then

$$\lim_{n \to \infty} F_n(t, x) = \lim_{n \to \infty} \langle \Psi_n, e^{iht} E_x e^{-iht} \Psi_n \rangle = \lim_{n \to \infty} \left\| E_x e^{-iht} \Psi_n \right\|^2$$
$$= \left\| E_x e^{-iht} \Psi \right\|^2 = F(t, x)$$

because e^{-iht}, E_x, and $\|\cdot\|^2$ are continuous mappings.

Note that for any $t \in \mathbb{R}$ the function $F(t, \cdot) : \mathbb{R} \to [0, 1]$ is continuous and monotonically increasing. Furthermore $\lim_{x \to -\infty} F(t, x) = 0$ and $\lim_{x \to \infty} F(t, x) = 1$. The monotonicity is strict if $\psi(t, \cdot)$ does not vanish on any interval. Thus for any $(t, x) \in \mathbb{R}^2$ there exists at least one $\gamma_x(t) \in \mathbb{R}$ such that Equation Equation 16.8 holds. (For those values t for which $F(t, \cdot)$ is strictly increasing, there exists exactly one $\gamma_x(t) \in \mathbb{R}$ such that Equation 16.8 holds.) The function F cannot be constant in an open neighborhood of some point (t, x) if the Hamiltonian is bounded from below. Thus for any $x \in \mathbb{R}$, for which there does not exist a neighborhood on which $F(0, \cdot)$ is constant, we now *define* $\gamma_x : \mathbb{R} \to \mathbb{R}$ to be the unique *continuous* mapping for which

$$F(t, \gamma_x(t)) = F(0, x).$$

Note that $\gamma_x : \mathbb{R} \to \mathbb{R}$ is continuous, yet need not be differentiable.

16.5 HALL'S COUNTEREXAMPLE

Let us now illustrate this construction of not necessarily differentiable Bohmian trajectories by means of a solution $t \mapsto \Psi_t \notin D_h$ of the Schrödinger Equation 16.5 describing a particle confined to a finite interval on which the potential V vanishes. This solution has been used by Hall [7] as a counterexample to Bohmian mechanics. Similar ones have been used in order to illustrate an "irregular" decay law $t \mapsto |\langle \Psi_0, \Psi_t \rangle|^2$ [8]. Both works have made extensive use of Berry's earlier results concerning this type of wave function [9].

The (reduced) classical Schrödinger equation corresponding to the quantum dynamics is

$$i \partial_t \psi(t, x) = -\partial_x^2 \psi(t, x) \tag{16.9}$$

for all $(t, x) \in \mathbb{R} \times [0, \pi]$ together with the homogeneous Dirichlet boundary condition $\psi(t, 0) = \psi(t, \pi) = 0$ for all $t \in \mathbb{R}$. The corresponding Hamiltonian's domain D_h is the set of all those $\Psi \in L^2(0, \pi)$ which have an absolutely continuous representative ψ vanishing at 0 and π and whose distributional derivatives up to second order belong to $L^2(0, \pi)$. As an initial condition we choose the equivalence class of the function

$$\psi(0, x) = 1/\sqrt{\pi} \quad \text{for all } x \in [0, \pi].$$

Since within the class $\Psi_0 = [\psi(0, \cdot)]$ there does not exist an absolutely continuous function vanishing at 0 and π the equivalence class Ψ_0 does not belong to D_h. As a consequence for any t the vector $\Psi_t = e^{-iht} \Psi_0$ does not belong to D_h. This in turn implies that Ψ_t does not have a representative within the class of C^2-functions on $[0, \pi]$ with vanishing boundary values.

The Hamiltonian h is self-adjoint. An orthonormal basis formed by eigenvectors of h is represented by the functions u_k with

$$u_k(x) = \sqrt{\frac{2}{\pi}} \sin(kx) \quad \text{for } 0 \leq x \leq \pi \text{ and } k \in \mathbb{N}.$$

For $n \in \mathbb{N}$ the C^∞-function $\psi_n : \mathbb{R}^2 \to \mathbb{C}$ with

$$\psi_n(t, x) = \frac{4}{\pi\sqrt{\pi}} \sum_{k=0}^{n} \frac{e^{-i(2k+1)^2 t}}{2k+1} \sin[(2k+1)x]$$

is a classical solution to the Schrödinger equation 16.9 on $\mathbb{R}^2$ and fulfills homogeneous Dirichlet boundary conditions at $x = 0$ and $x = \pi$. Furthermore ψ_n is periodic not only in x but also in t with period 2π. More precisely $\psi(t, \cdot)$ is an odd trigonometric polynomial of degree $2n + 1$ for any $t \in \mathbb{R}$. In addition $\psi_n(t, \cdot)$ is also even with respect to reflection at $\pi/2$, i.e., it we have

$$\psi_n\left(t, \frac{\pi}{2} - x\right) = \psi_n\left(t, \frac{\pi}{2} + x\right)$$

for all $x \in \mathbb{R}$. The functions $\psi_n(\cdot, x)$ are trigonometric polynomials of degree $(2n + 1)^2$.

As is well known, the sequence $(\psi_n(0, \cdot))_{n \in \mathbb{N}}$ converges pointwise on $\mathbb{R}$. Its limit is the odd, piecewise constant 2π-periodic function $\sigma(0, \cdot)$ with

$$\lim_{n \to \infty} \sqrt{\pi}\psi_n(0, x) = \sqrt{\pi}\sigma(0, x) = \begin{cases} 1 & \text{for } 0 < x < \pi \\ 0 & \text{for } x \in \{0, \pi\}. \end{cases}$$

$\sigma(0, \cdot)$ is discontinuous at $x \in \pi \cdot \mathbb{Z}$. For any $t \in \mathbb{R}$ the sequence $(\psi_n(t, \cdot))_{n \in \mathbb{N}}$ converges pointwise on $\mathbb{R}$ to a function $\psi(t, \cdot)$. For rational t/π this function is piecewise constant [9]. However for irrational t/π the real and imaginary parts of $\psi(t, \cdot)$ restricted to any open real interval have a graph with noninteger dimension [9]. Thus for irrational t/π the function $\psi(t, \cdot)$ is nondifferentiable on any real interval. As an illustration we give in Figure 16.1 the graph of

$$x \mapsto \Re\sqrt{\pi}\psi_{500}\left(\frac{\pi}{\sqrt{12}}, \pi x\right)$$

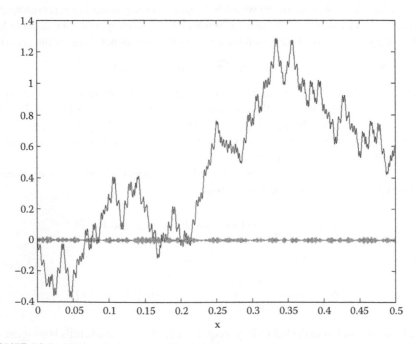

FIGURE 16.1   Real part of ψ_{500} at a fixed time.

for $0 < x < 1/2$ together with the partial sum over $k \in \{501, \ldots, 750\}$ visible as the small noisy signal along the abscissa

$$x \mapsto \Re\sqrt{\pi}\left(\psi_{750}\left(\pi/\sqrt{12}, \pi x\right) - \psi_{500}\left(\pi/\sqrt{12}, \pi x\right)\right).$$

Similarly, for given $x \in (0, \pi)$ the mapping $t \mapsto \psi(t, x)$ does not belong to the set of piecewise C^1-functions on $[0, 2\pi]$. This can be seen as follows. First note that for given x the 2π-periodic function $\psi(\cdot, x)$ has the Fourier expansion

$$\psi(t, x) = \sum_{k=1}^{\infty} c_n e^{-int} \text{ where}$$

$$c_n = \begin{cases} \dfrac{4}{\pi\sqrt{\pi}} \dfrac{1}{2k+1} \sin\left[(2k+1)x\right] & \text{for } n = (2k+1)^2 \text{ with } k \in \mathbb{N} \\ 0 & \text{otherwise.} \end{cases}$$

Assume now that $\psi(\cdot, x)$ is piecewise C^1. Then, according to a well known property of Fourier coefficients, there exists a positive real constant C such that $n\,|c_n| < C$ for all $n \in \mathbb{N}$. This implies that

$$(2k+1)\,|\sin\left[(2k+1)x\right]| < C' \quad \text{for all } k \in \mathbb{N} \tag{16.10}$$

with the positive constant $C' = \pi\sqrt{\pi}C/4$. However, for $x \notin \pi \cdot \mathbb{Z}$ there exists some real constant $\varepsilon > 0$ such that the set $\{k \subset \mathbb{N} : |\sin\left[(2k+1)x\right]| > \varepsilon\}$ contains infinitely many elements. Thus for $x \notin \pi \cdot \mathbb{Z}$ the estimate Equation 16.10 cannot hold and therefore the function $t \mapsto \psi(t, x)$ cannot be piecewise C^1 on $[0, \pi]$.

In Figure 16.2 we plot the time dependence

$$t \mapsto \Re\sqrt{\pi}\psi_{15}(\pi t, \pi/2) = \frac{4}{\pi} \sum_{k=0}^{15} \frac{(-1)^k}{2k+1} \cos\left((2k+1)^2 \pi t\right)$$

for $0 < t < 1/2$ together with the partial sum

$$t \mapsto \Re\sqrt{\pi}\left(\psi_{20}(\pi t, \pi/2) - \psi_{15}(\pi t, \pi/2)\right)$$

(noisy signal along the abscissa).

The restriction of the limit $\sigma(0, \cdot)$ to $[0, \pi]$ represents the same L^2 element as $\psi(0, \cdot)$ does. Thus $\lim_{n\to\infty}\left\|\left[\widetilde{\psi}_n(0, \cdot)\right] - \Psi_0\right\| = 0$, when $\widetilde{\psi}_n$ denotes the restriction of ψ_n to $\mathbb{R} \times [0, \pi]$. Correspondingly $\rho_n(0, \cdot) = |\psi_n(0, \cdot)|^2$ converges toward the density of the equipartition on $[0, \pi]$. Additionally, due to the continuity of the evolution operator e^{-iht}, the sequence of equivalence classes $\left[\widetilde{\psi}_n(t, \cdot)\right] \in C_h^\infty$ approximates the L^2 vector $\Psi_t = e^{-iht}\Psi_0$, i.e., for all t we have

$$\lim_{n \to \infty} \left\|\left[\widetilde{\psi}_n(t, \cdot)\right] - \Psi_t\right\| = 0.$$

Since also $E_x : L^2(\mathbb{R}) \to L^2(\mathbb{R})$ is continuous, the time dependent cumulative position distribution function $F : \mathbb{R} \times [0, \pi] \to [0, 1]$ with $F(t, x) = \langle \Psi_t, E_x \Psi_t \rangle$ obeys

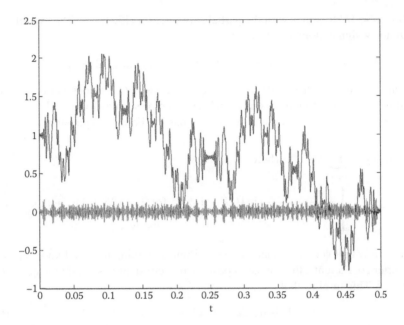

FIGURE 16.2 Real part of ψ_{15} at $x = \pi/2$.

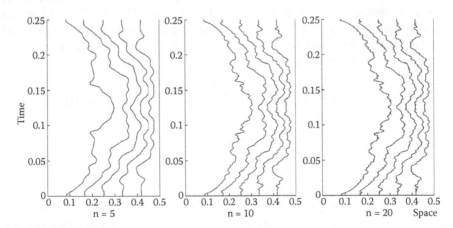

FIGURE 16.3 Level lines of F_n for $n = 5, 10, 20$.

$$F(t,x) = \lim_{n \to \infty} \left\langle \left[\widetilde{\psi_n}(t, \cdot) \right], E_x \left[\widetilde{\psi_n}(t, \cdot) \right] \right\rangle = \lim_{n \to \infty} \int_0^x |\psi_n(t, y)|^2 \, dy.$$

The level lines of the functions $F_n : \mathbb{R} \times [0, \pi] \to [0, 1]$ with $F_n(t, x) := \int_0^x |\psi_n(t, y)|^2 \, dy$ thus converge to the continuous level lines of F.

Figure 16.3 shows some level lines of F_n for $n = 5, 10, 20$ starting off at equal positions at $t = 0$. The level lines inherit the period $\pi/4$ of $F(\cdot, x)$, which has this period since the frequencies appearing in the even function $|\psi_n(\cdot, x)|^2$ are $0, 8, 16, \ldots$

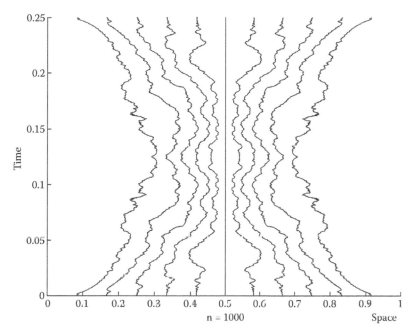

FIGURE 16.4   Level lines of F_{1000}.

Figure 16.4 shows the case $n = 1000$. Increasing n from 20 to 1000 hardly alters the level lines.

BIBLIOGRAPHY

1. B. Aulbach, *Gewöhnliche Differenzialgleichungen*, Elsevier, Amsterdam, 2004.
2. D. Dürr and S. Teufel, *Bohmian Mechanics*, Springer, Berlin, 2009.
3. K. Berndl, Global existence and uniqueness of Bohmian trajectories, arXiv:quant-ph/9509009, 1995. Published in: *Bohmian Mechanics and Quantum Theory: An Appraisal*, Eds. J.T Cushing, et al., Kluwer, Dordrecht, 1996.
4. M. Reed and B. Simon, *Methods of Modern Mathematical Physics II*, Academic press, New York, 1975.
5. K. Berndl, et al., On the global existence of Bohmian mechanics, *Comm. Math. Phys.* **173** (1995) 647–73.
6. S. Teufel and R. Tumulka, Simple proof of global existence of Bohmian trajectories, *Comm. Math. Phys.* **258** (2004) 349–65.
7. M.J.W. Hall, Incompleteness of trajectory-based interpretation of quantum mechanics, *J. Phys.* **A37** (2004) 9549–56.
8. P. Exner and M. Fraas, The decay law can have an irregular character, *J. Phys.* **A40** (2007) 1333–40.
9. M.V. Berry, Quantum fractals in boxes, *J. Phys.* **A29** (1996) 6617–29.

17 Nonadiabatic Dynamics with Quantum Trajectories

Gérard Parlant

CONTENTS

17.1 INTRODUCTION

Chemical reaction dynamics [1] can often be described in the framework of the Born–Oppenheimer approximation as the evolution from reactants to products on a single adiabatic potential energy surface (PES). However, this approximation may become invalid if the relevant PES happens to be energetically close to another one, for example near a conical intersection [2], in which case nonadiabatic transitions between these electronic states become likely [3]. Reaction dynamics on coupled PESs is governed by two quantum effects: (i) population transfer from one electronic state to another, and (ii) phase coherence between the nuclear wave packets evolving on each state, with these two effects reciprocally influencing each other. Therefore, nonadiabatic dynamics is a highly nonclassical process, which must be treated quantum mechanically. Various methods have been proposed to describe the concerted motion of nuclei on coupled electronic states and we refer the reader to Burghardt and Cederbaum [4] for a recent bibliography. In this chapter, we introduce and discuss a formally exact description of nonadiabatic dynamics using quantum trajectories.

Based on the de Broglie–Bohm formulation of quantum mechanics [5–10], the—formally exact—Quantum Trajectory Method (QTM), originally developed by Wyatt

[10,11] (a related approach has been reported by Rabitz et al. [12]), has seen a growing interest in the chemical dynamics community over the past 10 years, and now appears as a possible alternative to standard wave packet methods [13] for solving the time-dependent Schrödinger equation (TDSE). QTM is a computational implementation of the hydrodynamical approach based on the Madelung [8] ansatz, $\psi = A \exp(iS/\hbar)$, in which the probability density is partitioned into a finite number of probability *fluid elements*. Each of these elements, or *particles*, evolves along a quantum trajectory whose momentum is the gradient of the action $\vec{\nabla} S$. Contrary to classical trajectories, quantum trajectories are all coupled with one another through the nonlocal *quantum potential*, which brings in all quantum effects. In QTM, coupled equations of motion for the densities and action functions of all Bohmian particles are propagated, and from these the wave function can be recovered at each instant.

Traditionally, *analytic* quantum trajectories were extracted from conventional wave packets to provide physical insight into dynamical processes [14]. Beyond this intuitive aspect, in QTM *synthetic* quantum trajectories are propagated from scratch, to actually solve the TDSE. Since Bohmian quantum trajectories "follow" the flux of probability density, computer time is spent almost exclusively in regions of high-dynamics activity. For this reason, QTM is expected to outperform standard wave packet methods in terms of computing effort, at least in certain situations. In fact, model systems of up to hundreds of degrees of freedom have been successfully studied with QTM [10, 15, 16].

Unfortunately, a simple implementation of QTM equations of motion generally leads to numerical difficulties that limit the time during which quantum trajectories can be propagated. A computational drawback of QTM lies in the difficult evaluation of space derivatives that appear in the expression of the quantum potential (see Section 17.2.1), since these derivatives must be computed on the unstructured "grid" formed by Bohmian particles, which is moving and changing shape at each instant. An even more serious problem occurs in case of interferences—when a wave packet is reflected on a potential barrier, for example. In this case the quantum potential undergoes rapid variations (and even becomes singular at probability density nodes), thus giving rise to very irregular, numerically unstable, trajectories [17]. This is referred to in the literature as the *node problem* [10].

Various exact or approximate treatments have been proposed to deal with the node problem, and more generally with numerical difficulties encountered in the propagation of quantum trajectories. In the *bipolar decomposition* theory of Poirier et al. [18–24]—one of the most promising strategies, in our opinion—the wave function is expressed as the sum of two counter-propagating waves whose probability densities are much less oscillatory, and whose associated trajectories are much more regular, than that of the total wave function. Other approaches include the linearized quantum force [25], artificial viscosity [26], the covering function method [27], and the mixed wave function representation [28]. Of particular interest are the *adaptive grid* techniques developed by Wyatt et al. [10, 29–32], in which each particle, instead of being driven by the Bohmian flow, is assigned an arbitrary well-defined path controlled by the user. Such an adaptive grid constitutes an *arbitrary Lagrangian Eulerian*

(ALE) reference frame—as opposed to the Lagrangian "go with the flow" reference frame [10]. We will use the ALE concept later on in this chapter.

QTM has been extended to the dynamics of electronic nonadiabatic collisions by Wyatt et al. [10, 33, 34]—we will call this method the Multi-Surface Quantum Trajectory Method (MS-QTM)—and, in the context of mixed quantum states, by Burghardt and Cederbaum [4]. The features of the single surface QTM (physical insight, favorable scaling with system dimension) will naturally carry over into MS-QTM. Moreover, MS-QTM is expected to exhibit additional trajectory effects arising from the interstate coupling. In particular, quantum trajectories evolving on each surface will be affected by interstate transitions, and vice versa. For example, interstate *transfer forces* (see Section 17.2.1) can noticeably modify the course of trajectories, and eventually affect the result of the calculation.

Unfortunately, MS-QTM will also inherit the numerical drawbacks of its single-state counterpart, mentioned earlier in this Introduction. In addition, extra propagation difficulties, related to the interstate electronic coupling per se, may be anticipated, especially in case of large interstate coupling. Approximate *mixed representation* approaches that aim at decoupling single-state trajectory propagation from interstate transitions have been developed [35]* in particular in the group of Garashchuk and Rassolov [37–40] (GR).

Dynamics with quantum trajectories has been discussed in detail by Bob Wyatt in his book [10]. In this chapter we focus on the multisurface aspects of the theory. The formalism of nonadiabatic dynamics with quantum trajectories is presented in Section 17.2.1, while numerical aspects are discussed in Section 17.2.2, and an application to a simple model system follows in Section 17.3.1. Mixed representations are described in Section 17.3.2. In Section 17.3.3, the bipolar wave decomposition of Poirier [18–24], which is formally equivalent to a two electronic state problem, is briefly introduced and illustrated by an example. Finally concluding remarks are provided in Section 17.4.

17.2 THEORY

17.2.1 QUANTUM TRAJECTORY EQUATIONS FOR NONADIABATIC DYNAMICS

In this section, we set up the exact equations of motion for quantum trajectories moving on two *diabatic* electronic states $\varphi_1(e; x)$ and $\varphi_2(e; x)$ (real valued and orthonormal), where e denotes the set of electronic coordinates and x is the one-dimensional reaction coordinate. In the diabatic representation (see the Appendix), the electronic coupling matrix elements are $V_{jk}(x) = \langle \varphi_j | H_{el} | \varphi_k \rangle$, where H_{el} is the electronic Hamiltonian. The two diagonal elements $V_{11}(x)$ and $V_{22}(x)$ are the potential energy curves, while the off-diagonal elements $V_{12}(x) = V_{21}(x)$ represent the electronic coupling between states 1 and 2. The space derivative matrix elements $\langle \varphi_j | \partial/\partial x | \varphi_k \rangle$ vanish since φ_1 and φ_2 are assumed to be perfectly diabatic. Furthermore, notice that the following

---

* presented at ITAMP (Institute for Theoretical Atomic, Molecular and Optical Physics) Workshop, Harvard University, May 9–11, 2002 [36].

treatment would still be valid if the off-diagonal matrix elements V_{jk} were to depend on time (in case of a femtosecond laser pulse coupling states 1 and 2, for instance).

At each instant t, the total wave function can be expanded as:

$$\Psi(x, e, t) = \psi_1(x, t)\,\varphi_1(e; x) + \psi_2(x, t)\,\varphi_2(e; x). \tag{17.1}$$

and the coupled TDSEs for the nuclear wave functions $\psi_1(x, t)$ and $\psi_2(x, t)$ can easily be obtained, in matrix form, as [10]

$$-\frac{\hbar^2}{2m}\nabla^2 \begin{pmatrix} \psi_1 \\ \psi_2 \end{pmatrix} + \begin{pmatrix} V_{11} & V_{12} \\ V_{21} & V_{22} \end{pmatrix} \begin{pmatrix} \psi_1 \\ \psi_2 \end{pmatrix} = i\hbar\frac{\partial}{\partial t} \begin{pmatrix} \psi_1 \\ \psi_2 \end{pmatrix}, \tag{17.2}$$

where m is the reduced mass, and where ∇ stands for the partial space derivative $\partial/\partial x$. The hydrodynamic equations for the nuclear motion can be derived [10,33] by substituting into Equation 17.2, the polar form

$$\psi_j(x, t) = A_j(x, t)\exp\left[iS_j(x, t)/\hbar\right], \tag{17.3}$$

where for each state j, A_j is the amplitude, S_j is the action function, both real; A_j is positive. $P_j(x, t) = \nabla S_j(x, t)$ defines the momentum associated with the flow velocity $v_j(x, t) = P_j(x, t)/m$ of the probability fluid. In the following, we will assume an observer "going with the flow" of probability, so that the hydrodynamic equations will be written in the Lagrangian reference frame [10], which translates mathematically into the appearance of the *total* time derivative d/dt. Furthermore, in order to make the equations more readable, we will drop the x and t variables, although it must be kept in mind that the position and time dependencies of Equation 17.3 do transfer to all derived quantities.

Like their single-state counterparts, the continuity equations for the densities $\rho_j = A_j^2$,

$$d\rho_1/dt = -\rho_1\nabla v_1 - \lambda_{12},$$
$$d\rho_2/dt = -\rho_2\nabla v_2 - \lambda_{21}, \tag{17.4}$$

express the conservation of the probability fluid on each individual surface; in addition, the extra source/sink terms,

$$\lambda_{12} = -\lambda_{21} = (2V_{12}/\hbar)\,(\rho_1\rho_2)^{1/2}\sin(\Delta), \tag{17.5}$$

take into account the transfer of probability density from one state to the other one, with $\Delta = (S_1 - S_2)/\hbar$. Notice that the source/sink symmetry relation, i.e., Equation 17.5, ensures that no probability density is "lost" through state transitions in Equation 17.4. The trajectory equations are

$$dx_1/dt = p_1/m,$$
$$dx_2/dt = p_2/m. \tag{17.6}$$

In the Newtonian equations,

$$dp_1/dt = -\nabla(V_{11} + Q_{11} + Q_{12}),$$
$$dp_2/dt = -\nabla(V_{22} + Q_{22} + Q_{21}), \tag{17.7}$$

one can see that any given particle evolving on state j is subject to three force components: (i) the classical force, (ii) the quantum force, that derives from the quantum potential $Q_{jj} = (-\hbar^2/2m)\nabla^2 A_j/A_j$, which depends on the curvature of the amplitude $A_j = \sqrt{\rho_j}$; this force reflects the influence of all other particles on the same surface, and (iii) an extra coupling force, that derives from the off-diagonal quantum potential Q_{jk},

$$Q_{12} = V_{12} (\rho_2/\rho_1)^{1/2} \cos(\Delta), \tag{17.8a}$$

$$Q_{21} = V_{21} (\rho_1/\rho_2)^{1/2} \cos(\Delta), \tag{17.8b}$$

which satisfy the symmetry relation $\rho_1 Q_{12} = \rho_2 Q_{21}$. Finally, the rate of change of the action functions is given by:

$$dS_1/dt = (1/2)mv_1^2 - (V_{11} + Q_{11} + Q_{12}),$$

$$dS_2/dt = (1/2)mv_2^2 - (V_{22} + Q_{22} + Q_{21}). \tag{17.9}$$

One may notice that the momenta can be obtained from the action functions of Equation 17.9 through the relation $P_j(x,t) = \nabla S_j(x,t)$, as an alternative to the propagation of the Newtonian equations 17.7.

The exact MS-QTM described above by Equations 17.4 through 17.9 may remind the reader of the approximate Trajectory Surface Hopping (TSH) method originally designed by Tully [41,42]. However, MS-QTM and TSH have important differences. In the basic TSH algorithm, independent classical trajectories evolve along a given PES, and for each trajectory a decision is made whether to "hop" to another surface, at a particular position in the coupling region, based upon a (approximate) transition probability. In contrast, MS-QTM does not involve any trajectory hopping. Rather, the number of trajectories evolving on each electronic surface is conserved, and transfer of density and phase information between states takes place in a continuous manner.

17.2.2 ALGORITHM AND NUMERICAL DETAILS

The practical implementation of the exact MS-QTM formalism for two electronic states is done as follows. N trajectories are defined on each PES and each trajectory, identified by index i, with $i = 1, \ldots, 2 \times N$, is characterized by the functions $x_i(t)$, $p_i(t)$, $\rho_i(t)$, and $S_i(t)$. These functions of time are initialized at $t = 0$, depending on the problem under study. The ensemble of trajectories is governed by the equations of motion, i.e., Equations 17.4, 17.6, 17.7, and 17.9, which result in a set of $8 \times N$ coupled ordinary differential equations (ODEs). The system of ODEs can be propagated in time with any standard ODE algorithm [43]. In the author's work presented in Section 17.3 a Runge–Kutta algorithm with adaptive time-stepsize is utilized [43].

The propagation of the ODE system requires, at each time step, the computation of several functions of the position x and, for some of them, of their derivatives relative to x. As already mentioned, the accurate evaluation of spatial derivatives over a highly nonuniform distribution of trajectories is a challenging task. A possible strategy is to use a least squares method to carry out a *local* fit of the function to be differentiated around a given trajectory position, supported by a limited number of neighbor trajectories. This is called the Moving Least Squares (MLS) [44] method

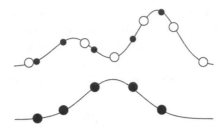

FIGURE 17.1 Schematic interpolation between two nonuniform trajectory distributions: upper state values (open circles) are interpolated at lower state trajectory positions (filled circles).

and this has to be repeated for each trajectory position. It should be noted that MLS tends to smooth out the function to be fitted, more or less so depending on the number of neighbor points taken into account. The Moving Weighted Least Squares (MWLS) introduces weights in the MLS procedure to enhance locality and compensate to some extent for the smoothing effect. We refer the reader to Chapter 5 of Bob Wyatt's book [10] where function and derivative approximations on unstructured grids are discussed in detail, and we will now focus on propagation problems related specifically to the multisurface feature of MS-QTM.

Trajectories evolving on a given state exchange information with trajectories evolving on the other state through the source/sink terms λ_{12} and λ_{21} (Equation 17.4), and through the off-diagonal quantum potentials Q_{12} and Q_{21} (Equations 17.7 and 17.9). Although trajectory distributions are generally chosen initially as uniform and identical on both states, as time evolves they become nonuniform and not coincident, so that functions pertaining to a given state must be interpolated (and possibly extrapolated) at the other state trajectory positions, as illustrated schematically in Figure 17.1. Interpolation can be carried out by means of the MLS method mentioned above, or by means of spline interpolation [43]. If wave packets on both states travel at very different velocities, one will lag behind the other, so that extrapolation may become necessary. However, extrapolation is quite uncertain and should be avoided whenever possible. For instance, in the example just mentioned the initial wave packet positions may be arranged so that they arrive at about the same time in the coupling region.

In a typical two-state collision process a wave packet is launched on, say, state 1, and one is generally interested in how much density has been transferred to state 2 when the process is over. However, since the off-diagonal quantum potential Q_{21} involves a division by $\rho_2^{(1/2)}$, one cannot initiate the process with zero density in state 2. In practice, the algorithm does tolerate starting with a very small nonzero density ϵ in state 2 and $1 - \epsilon$ in state 1 (see Section 17.3.1). A better solution is to run two different calculations sequentially, one with a "$\psi_1 + \psi_2$" coherent combination of nonzero wave packets and the other one with a "$\psi_1 - \psi_2$" such combination. The evolved sum and difference wave packets thus obtained can be combined at all times to recover the result that would be obtained when a wave packet is launched in a single channel [33].

17.3   DISCUSSION

17.3.1   Application of the Exact Multistate Quantum Trajectory Method

In order to illustrate the main features of MS-QTM, in this section the exact formalism is applied to the one-dimensional two-state model system originally used by Wyatt et al. [33], i.e., a downhill ramp and a flat curve, coupled by a Gaussian function. Due to the potential curve shapes, a wave packet traveling in the downhill direction will directly traverse the coupling region and is unlikely to be reflected, so that the interference-related numerical instabilities mentioned in the Introduction are not expected in this case. The potentials are given by $V_{11}(x) = -(c_{11}/2) \times (1+\tanh(x-a))$, $V_{22}(x) = c_{22}$, and $V_{12}(x) = V_{21}(x) = c_{12}\exp(-\beta_{12}(x - x_0)^2)$, with the following parameters: $c_{11} = 2000$ cm^{-1}, $a = -1$ a.u., $c_{22} = 10$ cm^{-1}, $c_{12} = 500$ cm^{-1}, $\beta_{12} = 3$ a.u., $x_0 = -1$ a.u.

Initially, a Gaussian wave packet (GWP) centered at $x_{10} = -4.0$ a.u. is launched on state 1 toward positive x values (i.e., in the downhill direction):

$$\psi_1\,(x, t = 0) = (2\gamma/\pi)^{(1/4)} \exp\left(-\gamma(x - x_{10})^2\right) \exp\left(ik_1 x\right), \qquad (17.10)$$

with $\gamma = 2$ a.u. and $k_1 \approx 7.4$ a.u., which corresponds to a total energy of 3000 cm^{-1} with mass $m = 2000$ a.u. In order to avoid divergence of the off-diagonal quantum potential Q_{21} (Equation 17.8b), a GWP identical to $\psi_1(x, t = 0)$ [although with a wave number k_2 corrected for the slight energy offset between $V_{11}(x_{10})$ and $V_{22}(x_{10})$] is launched on state 2 with a very small nonzero density $\epsilon = 1 \times 10^{-8}$, so that the total initial wave function is

$$\psi\,(x, t = 0) = \left(1 - \sqrt{\epsilon}\right) \times \psi_1\,(x, t = 0) + \sqrt{\epsilon} \times \psi_2\,(x, t = 0). \qquad (17.11)$$

On each state we use 60 evenly distributed trajectories, with the leftmost trajectory corresponding to a density of 1×10^{-8}. Space derivatives and function interpolations are obtained from a MWLS quadratic fitting procedure that uses the 12 nearest neighbor trajectory positions for each trajectory. The system of ODEs is propagated in time by means of a Runge–Kutta algorithm with adaptive time-stepsize [43]; a relative accuracy of 1×10^{-6} is required on all variables. Finally, for the sake of comparison, the TDSE is also solved numerically for this system by means of the Cayley version of the implicit Crank–Nicholson method [43], an exact wave packet method [13].

Figure 17.2 presents an "animation plot" of the probability densities $\rho_1(x, t)$ and $\rho_2(x, t)$ as a function of x, at times $t = 0$–50 fs, with an increment of 5 fs. Only 30 trajectories out of the 60 trajectories used in the calculation are drawn on each curve to increase readability. State 2 density clearly starts to build up at the expense of state 1 around $t = 15$ fs, when state 1 trajectories reach the coupling potential $V_{12}(x)$ centered at $x = -1$. As a consequence, state 1 density gets (moderately) deformed around $t = 20$ fs, and seems to recover its Gaussian shape once it has left the coupling region. Moreover, it is also evident from Figure 17.2 that trajectories travel faster on the downhill ramp potential than on the flat potential, as expected. Overall, the agreement between the MS-QTM result and the benchmark wave packet calculation is excellent. The slight disagreement visible mostly on state 2 densities

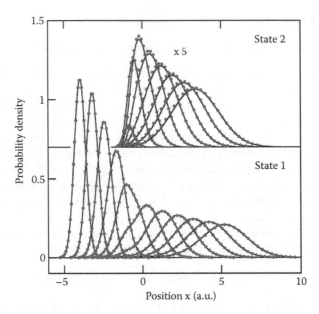

FIGURE 17.2 Probability densities $\rho_j(x,t)$ as a function of coordinate x, moving from left to right between times $t = 0$ and 50 fs, with an increment of 5 fs, for state 1 (lower panel) and state 2 (upper panel, density multiplied by 5) for Wyatt model potentials [33]; solid line: full quantum calculation; dot symbols: quantum trajectory result obtained with 60 trajectories. For clarity, one out of two trajectory symbols is omitted and symbols associated with density values smaller than 10^{-3} are omitted as well.

has not been investigated, although we may surmise that it is related to the very small density on state 2 at $t = 0$ that involves large initial values of Q_{21}. Some over-smoothing effect in the least squares fitting which uses 12 neighbors (i.e., six dot symbols in Figure 17.2) for each trajectory may play a role too.

Figure 17.3 shows a snapshot at time $t = 30$ fs of the real part of the wave packet evolving on state 1. The open circle symbols obtained from $\sqrt{\rho_i} \times \cos(i S_i/\hbar)$ for each trajectory i fit nicely on the continuous line of the exact wave packet calculation. Moreover, it is possible to synthesize the wave function at any position x by carrying out interpolations through the 60 values of ρ_i and S_i and then deriving the real and imaginary parts of the interpolated functions of x. The continuous curve (not displayed in Figure 17.3) obtained through this procedure is undistinguishable from the exact result. Thus, it turns out that quantum trajectories allow us to represent accurately this wave function (Figure 17.3) with only three points per oscillation!

In conclusion, the exact MS-QTM performs very well on the present model system [33] composed of a flat curve and a downhill ramp. Analysis of the various forces (classical, quantum, transfer forces) involved in this system can be found in Ref. [34]. Furthermore, an 11-dimensional two-state system composed of the present flat curve/downhill ramp potential together with 10 harmonic oscillators has been studied by Wyatt [10]. A total of 110, 11-dimensional trajectories were propagated

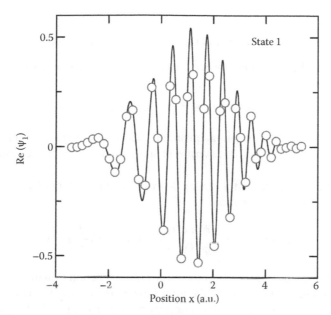

FIGURE 17.3 Real part of the wave function $Re[\psi_1(x,t)]$ at time $t = 30$ fs as a function of coordinate x, for state 1 of Wyatt model potentials [33]; solid line: full quantum calculation; open circles: quantum trajectory result obtained with 60 trajectories (some trajectories are located outside of the plotting region). The wave function synthesized from the trajectory density and amplitude (see text) is undistinguishable from the exact result.

and the wave function was synthesized, showing the gradual density transfer between states.

17.3.2 Mixed Representations

In the preceding section, MS-QTM gives excellent results on a model system that involves neither strong reflection interferences nor very large interstate coupling values, therefore an "easy" system. In the general case, however, as mentioned in the Introduction, trajectories may get numerically unstable for two reasons, which are closely interconnected (see the equations of Section 17.2.1): (i) large quantum forces due to interferences formed on each individual PES, and (ii) large transfer quantum forces due to interstate coupling. The goal of *mixed representations* is to formally separate interstate transitions from single-state nuclear motions in order to improve the propagation stability.

17.3.2.1 Mixed Coordinate-Space/Polar Representation

The mixed coordinate-space/polar representation has been introduced by Garashchuk and Rassolov [28, 37, 38]. It is based on the following decomposition of the wave function

$$\psi_j(x,t) = \phi_j(x,t) \times \chi_j(x,t), \quad j = 1,2, \tag{17.12}$$

where ϕ_j takes the usual polar form $\phi_j = a_j \exp(is_j/\hbar)$ and satisfies the single-state TDSE

$$-\frac{\hbar^2}{2m}\nabla^2\phi_j(x,t) + V_{jj}\phi_j(x,t) = i\hbar\frac{\partial}{\partial t}\phi_j(x,t), \tag{17.13}$$

which is solved by propagating quantum trajectories, with the diagonal quantum potential $(-\hbar^2/2m)\nabla^2 a_j(x,t)/a_j(x,t)$ approximated in a semiclassical manner [45–47].

The coordinate-space part $\chi_j(x,t)$ describes complex "population" amplitudes and obeys the following equation

$$K_j\chi_j(x,t) + V_{jk}\frac{\phi_k(x,t)}{\phi_j(x,t)}\chi_k(x,t) = i\hbar\frac{d}{dt}\chi_j(x,t), \tag{17.14}$$

which is obtained upon insertion of the mixed wave function Equation 17.12 into the TDSE, i.e., Equation 17.2. Notice the *total* time derivative on the right-hand side of Equation 17.14, which expresses the time dependence of $\chi_j(x,t)$ in the moving frame of reference. The kinetic energy operator K_j includes a term coupling the nonclassical momentum [46], defined as $r_j(x,t) = 2\nabla a_j(x,t)/a_j(x,t)$, with the gradient operator acting on $\chi_j(x,t)$:

$$K_j = -\frac{\hbar^2}{2m}\left(\nabla^2 + r_j(x,t)\nabla\right). \tag{17.15}$$

So far, the above formulation is exact. The left-hand side of Equation 17.14 is then evaluated by expanding $\chi_1(x,t)$ and $\chi_2(x,t)$ in a small basis set and using a linearized approximation of $r_j(x,t)$, in accordance with the semiclassical hypothesis for the quantum potential.

GR have later modified their method so that it could accommodate strong extended diabatic couplings. In this improved version of the method [38], the polar part of the wave function ϕ_j evolves on PESs whose diabatic or adiabatic character is *dynamically* determined, while the coordinate part χ_j, which describes transitions between surfaces, is generalized to a matrix form.

GR have use the mixed coordinate-space/polar representation to study the NaFH van der Waals complex [40] treated in two dimensions and the $O(^3P, ^1D) + H_2$ system [39], including four coupled electronic states for total angular momentum $J = 0$.

17.3.2.2 Decoupled Representation

In the spirit of Garashchuk and Rassolov's mixed coordinate-space/polar representation described above we have introduced [35]* the *decoupled representation*, which

---

* presented at ITAMP (Institute for Theoretical Atomic, Molecular and Optical Physics) Workshop, Harvard University, May 9–11, 2002 [36].

formally rewrites Bohmian equations of motion in order to separate interstate transitions from single-state nuclear motions. We write the nuclear wave function as the *split polar form*

$$\psi_j = \phi_j \times \chi_j \equiv a_j \exp\left(\frac{i}{\hbar}s_j\right) \times \alpha_j \exp\left(\frac{i}{\hbar}\sigma_j\right), \quad j = 1, 2, \tag{17.16}$$

(analogous to Equation 17.12 of the GR representation) where the Roman typeset amplitude a_j and action function s_j refer to single-state motion while their Greek analogs α_j and σ_j refer to interstate coupling, and where position and time dependencies have been dropped for clarity. Amplitudes and action functions in Equation 17.16 are real (amplitudes are positive) and are related to the total amplitude and action function of Equation 17.3 by $A_j = a_j \times \alpha_j$ and $S_j = s_j + \sigma_j$. Using the same Roman/Greek convention, one can further define the corresponding densities $f_j = a_j^2$ and $\omega_j = \alpha_j^2$ and momenta $p_j = \nabla s_j$ and $\pi_j = \nabla \sigma_j$, respectively, which can be completed by $\rho_j = f_j \times \omega_j$ and $P_j = p_j + \pi_j$.

Like in the GR approach described in Section 17.3.2.1, the wave function $\phi_j(x, t)$ satisfies the single-state TDSE given by Equation 17.13, from which we derive the single-state Bohmian equations of motion, expressed in the Lagrangian reference frame,

$$\frac{d}{dt}f_j = -f_j \frac{\nabla p_j}{m}, \tag{17.17a}$$

$$\frac{d}{dt}s_j = \frac{p_j^2}{2m} - V_{jj} - Q_{jj}^{\text{single}}, \tag{17.17b}$$

where the single-state quantum potential is given by $Q_{jj}^{\text{single}} = -\hbar^2/(2m)\left(\nabla^2 a_j/a_j\right)$.

The "coupled" part of the wave function, $\chi_j(x, t)$, satisfies a modified TDSE, analogous to Equation 17.14,

$$-\frac{\hbar^2}{2m}\nabla^2\chi_j - \frac{\hbar^2}{m}\frac{\nabla\phi_j}{\phi_j}\nabla\chi_j = i\hbar\frac{\partial}{\partial t}\chi_j - V_{jk}\frac{\phi_k}{\phi_j}\chi_k. \tag{17.18}$$

Our approach differs from GR's by the fact that we look for the solution $\chi_j(x, t)$ to Equation 17.18 as a polar wave function. The corresponding Bohmian equations of motion will be expressed in the same reference frame as ϕ_j in Equation 17.17. Importantly, this Lagrangian reference frame for ϕ_j constitutes an ALE frame for χ_j, which entails specific space derivative terms in the equations of motion [10,30]. Here we derive directly the Bohmian equations for the total density ρ_j and action S_j:

$$\frac{d}{dt}\rho_j = -\frac{\rho_j}{m}\nabla(p_j + \pi_j) - \frac{\pi_j}{m}\nabla\rho_j - \lambda_{jk}, \tag{17.19a}$$

$$\frac{d}{dt}S_j = \frac{p_j^2}{2m} - V_{jj} - Q_{jj} - \frac{\pi_j^2}{2m} - Q_{jj}^{\text{cpl}} - Q_{jj}^{\text{cross}} - Q_{jk}, \tag{17.19b}$$

where the source/sink terms λ_{jk} and the off-diagonal quantum potentials Q_{jk} are given by Equations 17.5 and 17.8, respectively. Bohmian equations for $\chi_j(x, t)$ can be recovered by simply "subtracting" Equations 17.17 from Equations 17.19.

As compared with the regular multisurface QTM Equations 17.4 and 17.9, the decoupled representation QTM equations 17.19 exhibit extra terms that depend on the *slip velocity* [10,30] $(p_j - P_j)/m$ (that links the Gaussian ALE frame with the true Lagrangian frame) as well as extra quantum potentials $Q_{jj}^{\text{cpl}} = -\hbar^2/(2m)\left(\nabla^2\alpha_j/\alpha_j\right)$ and $Q_{jj}^{\text{cross}} = -\hbar^2/m\left(\nabla a_j/a_j\right)\left(\nabla\alpha_j/\alpha_j\right)$.

The "total" equations of motion Equation 17.19, together with the single-state Equations 17.17, are formally exact and, in principle, may be propagated as such. However, in the example given below—two electronic states coupled by an ultrashort laser pulse—the radiative coupling is so big that the source/sink terms λ_{jk} and the off-diagonal quantum potentials Q_{jk} dominate the behavior of Equation 17.19. In this case, the Bohmian equations of motion for the "coupled" function $\chi_j(x,t)$ can be simplified [35] to give

$$d\omega_j/dt \approx -\lambda_{jk}/f_j,$$ (17.20a)

$$d\sigma_j/dt \approx -Q_{jk}.$$ (17.20b)

We now apply our decoupled-representation QTM approach described above to a model problem [48,49] of two *adiabatic* potential energy curves coupled through a laser pulse with a single frequency Ω. Assuming the rotating wave approximation [50], the laser shot on state 1 results in (i) a shift $\hbar\Omega$ bringing states 1 and 2 energetically closer to each other, and (ii) a time-dependent radiative dipole coupling between states 1 and 2. Following Grossmann [48], our matrix TDSE Equation 17.2 translates for the present model system into the dimensionless TDSEs:

$$i\frac{\partial}{\partial\tau}\psi_1(\xi,\tau) = \left[-\frac{\partial^2}{\partial\xi^2} + C^2\xi^2\right]\psi_1(\xi,\tau) + D/\cosh\left[(\tau-\tau_0)/T_c\right]\psi_2(\xi,\tau),$$

(17.21a)

$$i\frac{\partial}{\partial\tau}\psi_2(\xi,\tau) = \left[-\frac{\partial^2}{\partial\xi^2} - A\xi + B\right]\psi_2(\xi,\tau) + D/\cosh\left[(\tau-\tau_0)/T_c\right]\psi_1(\xi,\tau),$$

(17.21b)

where the potential surfaces for states 1 and 2 are a harmonic oscillator and a straight line, respectively, and where the off-diagonal coupling is independent of the dimensionless coordinate ξ. Since both potential curves $V_{jj}(\xi)$ are quadratic functions, the "decoupled" wave functions ϕ_j can be chosen as GWPs [51], whose Bohmian equivalent is easily propagated by means of Equations 17.17.

We study Grossmann's [48] model I which corresponds to the following choice of parameters: $A = 1$, $B = -12$, $C = 1/\sqrt{2}$, $D = 100$, $\tau_0 = 0.25$, and $T_c = 0.025$. This model is characterized by large radiative coupling values that allow us, as already mentioned above, to derive the simplified equations of motion Equation 17.20. Moreover, considering the very short interaction time, roughly 2.5/100 of the vibration period of state 1, we will assume that the "decoupled" GWPs $\phi_j(\xi,\tau)$ do not move or disperse during the laser pulse. As a consequence, in this particular case the associated ALE frames are reduced to fixed grids, which can be chosen as identical, so that grid interpolation from one state onto the other is not needed.

Initially the system is in the ground vibrational state of the harmonic oscillator state 1, $\phi_1(\xi, \tau = 0) = (2\gamma/\pi)^{(1/4)} \exp(-\gamma\xi^2)$, with $\gamma = 1/2\sqrt{2}$, and with an identical GWP for state 2. In practice, a very small non-zero weight is assigned to state 2 in order to avoid divergence of the off-diagonal quantum potential in Equation 17.8a: $\omega_2(\xi, \tau = 0) = 1 \times 10^{-12}$ and $\omega_1(\xi, \tau = 0) = 1 - 1 \times 10^{-12}$. The coupled differential Equations 17.17 and 17.20, where $j = 1, 2$ and $k = 1, 2$ ($j \neq k$), are propagated in time by means of a Runge–Kutta algorithm in the same conditions as in Section 17.3.1.

Let us consider the state populations, defined as $\Pi_j(\tau) = \int_{-\infty}^{+\infty} \rho_j(\xi, \tau) d\xi$. The population of the harmonic oscillator, plotted as a function of time in Figure 17.4, exhibits about two and a half oscillations (Rabi flops [50]) during the short laser pulse. Despite the very strong radiative interaction, responsible for nearly complete population swaps between states 1 and 2, one can observe an excellent agreement between exact quantum calculations and our quantum trajectory approach using only 20 trajectories. It is remarkable that such a good result is obtained with such a small number of trajectories. Moreover, the total population $\Pi_1(\tau) + \Pi_2(\tau)$ (not shown) is conserved over time within an error range smaller than 10^{-6}. Unitarity is in fact expected from Equations 17.4 and 17.5 of the exact QTM formulation which ensure

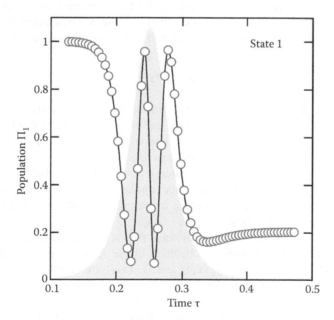

FIGURE 17.4 Population $\Pi_1(\tau)$ as a function of dimensionless time τ for Grossmann model I of two potential surfaces coupled by ultrashort laser pulses [48]; solid line: full quantum calculation; open circles: quantum trajectory result obtained with 20 trajectories; sum of probabilities $\Pi_1(\tau) + \Pi_2(\tau)$ (not shown) conserved to better than 10^{-6}. The laser pulse profile is indicated by the gray area.

the combined flux continuity. It appears to be also satisfied for the present model system by the approximate Equation 17.20a.

In conclusion, our decoupled-representation QTM was used to compute the dynamics of an ultrashort laser pulse excitation model system [48, 49], and proved to be very fast and accurate for this task. Despite very large interstate coupling values, only 20 quantum trajectories were needed to obtain accurate populations, a number that can be contrasted with the $\sim 10^4$ trajectories used in Ref. [48] to obtain a reasonable accuracy on the same system with a semiclassical method. Moreover, our approach ensures the conservation of total probability to an excellent level. Similar results have been obtained [35] on Grossmann's [48] model II, which involves even larger radiative coupling values.

17.3.3   BIPOLAR WAVE PACKETS

In this section, we briefly introduce the *bipolar decomposition* theory developed by Poirier et al. [18–24] to circumvent the *node problem*. The bipolar QTM equations of motion are formally equivalent to those of a two electronic state problem given in Section 17.2.1. As we mentioned in the Introduction, the scattering of a wave packet by a potential barrier that generates interferences between the incident wave and the reflected wave is a difficult problem for QTM because of the huge quantum forces occurring about the nodes or quasi-nodes of the wave function. In the bipolar decomposition, the wave function is expressed as the sum of two counter-propagating waves,

$$\psi(x,t) = \psi_+(x,t) + \psi_-(x,t), \tag{17.22}$$

where ψ_+ and ψ_- represent the incident/transmitted wave and the reflected wave, respectively. The main advantage of this approach is that the probability densities $\rho_+ = |\psi_+|$ and $\rho_- = |\psi_-|$ are much less oscillatory than the total probability density, and consequently the quantum trajectories associated with ψ_+ and ψ_- will be much more regular than those associated with the total wave function ψ. Besides this—very important—practical aspect, the bipolar decomposition is actually related to a more theoretical aspect of quantum trajectories, i.e., the fact that they do not satisfy classical correspondence [7, 21].

Qualitatively, when the left-incident wave packet $\psi_+(x,t)$ encounters the potential barrier, it starts to transfer probability to the $\psi_-(x,t)$ component (initially $\psi_-(x,t = 0) = 0$), because $\psi_+(x,t)$ and $\psi_-(x,t)$ are coupled together through the interaction potential (Equation 17.23). The $\psi_-(x,t)$ wave first grows in place until it reaches an appropriate size and leaves the interaction region, heading to the left, while the $\psi_+(x,t)$ wave moves straight through the interaction region. A detailed description of this process is given in Ref. [19].

Formally, $\psi_+(x,t)$ and $\psi_-(x,t)$ are solution to the following TDSE (see Ref. [19])

$$i\hbar \frac{\partial \psi_\pm}{\partial t} = \hat{H}\psi_\pm \pm \frac{V'}{2}\Psi_\Delta, \tag{17.23}$$

where

$$\hat{H} = -\frac{\hbar^2}{2m}\frac{\partial^2}{\partial x^2} + V(x),$$ (17.24)

is the Hamiltonian operator, and where the unusual off-diagonal coupling term involves the wave function integral

$$\Psi_\Delta(x) = \int_{-\infty}^{x} \left[\psi_+\left(x'\right) - \psi_-\left(x'\right)\right] dx',$$ (17.25)

as well as the space derivative of the potential V'. Here and in the following primes denote spatial differentiation, except in Equation 17.25.

From Equations 17.23, together with $\psi_\pm = R_\pm \exp[i S_\pm/\hbar]$ and $v_\pm = S'_\pm/m$, it is straightforward to derive Lagrangian QTM equations of motion [18] similar to those obtained for a two electronic state problem (Section 17.2.1):

$$\frac{dR_\pm}{dt} = -\frac{1}{2}R_\pm v'_\pm \pm \left(\frac{V'}{2\hbar}\right) \mathrm{Im}\left[\Psi_\Delta \exp(-i S_\pm/\hbar)\right],$$ (17.26)

$$\frac{dS_\pm}{dt} = \frac{1}{2}mv_\pm^2 - V - Q_\pm \mp Q_{\Delta\pm},$$ (17.27)

$$\frac{dv_\pm}{dt} = -\frac{1}{m}(V + Q_\perp \pm Q_{\Delta\pm})',$$ (17.28)

where

$$Q_\perp(x,t) = -\frac{\hbar^2}{2m}\left[\frac{R''_\pm(x,t)}{R_\pm(x,t)}\right],$$ (17.29)

and

$$Q_{\Delta\pm}(x,t) = \left[\frac{V'(x)}{2R_\pm(x,t)}\right] \mathrm{Re}\left[\Psi_\Delta(x,t)\exp(-i S_\pm(x,t)/\hbar)\right].$$ (17.30)

The coupling term in Equation 17.26, referred to as $\pm\lambda_\pm(x,t)$, describes the rate of transfer of probability amplitude between the $(\pm)$ components and is comparable to $\pm\lambda_{jk}(x,t)$ in Equation 17.4 which transfers probability density between two different electronic states.*

From these equations, Poirier et al. [18] have been able to perform numerically exact synthetic QTM calculations for a scattering system that exhibits realistic reflection and interferences.

In order to illustrate the bipolar decomposition approach, analytic bipolar quantum trajectories $x_\pm^i(t)$ from a wave packet scattered by an Eckart potential barrier are presented in Figure 17.5, where the local probability density is represented by a color, continuously varying along each trajectory i. Panels a and b correspond to trajectories x_+^i and x_-^i, respectively, with a superposition of both given in panel c.

---

* Notice, however, that λ_+ is not equal to λ_-, implying that the combined flux continuity relation is *not* satisfied here [19], contrary to the case of two different electronic states.

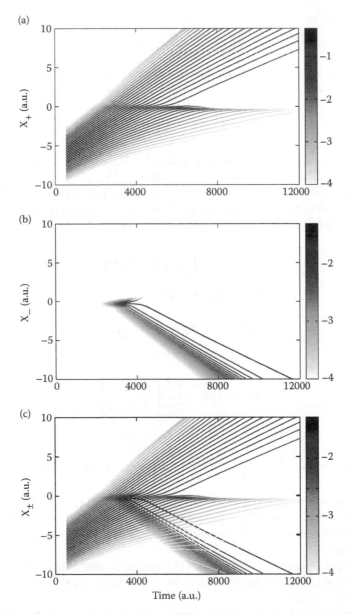

FIGURE 17.5 Color representation of analytical bipolar quantum trajectories (a) x_+ and (b) x_- for an Eckart barrier centered at $x = 0$. In panel (c), the x_+ and x_- trajectories are superimposed. The color palette shown on the right represents the base-10 logarithm of the probability density value; values smaller than 10^{-4} are invisible. On grayscale reproductions, dark regions represent high-probability density. (From Park, K., Poirier, B., and Parlant, G., *J. Chem. Phys.*, 129, 194112–16, 2008. With permission.)

The x_+^i trajectories divide into two groups. The first group is made of very regular trajectories which continue to propagate past the interaction region (centered at $x = 0$) into the product asymptote, thus forming the transmitted wave. The probability along each trajectory is conserved (except for a slight decrease due to dispersion), which translates in the figure by the fact that color does not change along a given trajectory. In the second group, trajectories tend to merge along the $x = 0$ axis. At the same time they change color and become gradually invisible. Thus, they do not conserve probability but rather, lose all of their probability which is transferred to the ψ_- component.

As for the x_-^i trajectories, they arise, as expected, in the interaction region. Initially, they stay in place for a while as they grow in probability, but eventually start heading toward negative x values to form the reflected ψ_- wave. A peculiar "fanout" effect is observed by which the x_-^i trajectories tend to spread out exponentially quickly from each other in the interaction region. Once out of the interaction region, the x_-^i trajectories exhibit a very regular shape, like the x_+^i trajectories of the first group.

In conclusion, it is clear from Figure 17.5 that the bipolar quantum trajectories that form the transmitted and reflected wave packets are very well-behaved—unlike their unipolar counterparts (not displayed), which for this system exhibit very substantial interference. The x_+^i trajectories of the second group, which merge and eventually "disappear," as well the x_-^i trajectories that "fanout" in the interaction region, are discussed in Ref. [18].

17.4 SUMMARY AND CONCLUSION

In this chapter, the extension of the QTM to the dynamics of electronic nonadiabatic transitions has been described. Wave packets representing nuclear motion on each of the coupled PESs are discretized into Bohmian particles, and the exact equations of motion for these particles reflect the classical and quantum forces on each single PES, as well as transfer of probability density between states and interstate transfer forces. On a simple two state model system, the method accurately predicts the oscillatory structure of the wave functions.

For more complex systems, mixed representations used to formally separate interstate transitions from single-state nuclear motions have been introduced. An application to ultrashort laser pulse dynamics shows that mixed representations can be very fast and very accurate.

Interestingly, the bipolar decomposition of the wave function of a single electronic state, developed by Poirier to circumvent the node problem, leads to a formalism similar to the QTM formalism for two-electronic states. Bipolar quantum trajectories have been described to illustrate the bipolar equations of motion, leading to very regular trajectories as compared to the regular monopolar trajectories for the same system.

So far, the multistate QTM has seen a relatively small development compared to the single-state QTM [10]. We believe that considerable improvement of the method can be achieved, in particular in the framework of mixed representations.

APPENDIX: DIABATIC AND ADIABATIC ELECTRONIC REPRESENTATIONS

In Section 17.2.1, we set up the QTM equations in a *diabatic*—as opposed to the well-known *adiabatic*—representation. Adiabatic electronic states are coupled through the kinetic energy operator and may change drastically [52] over a short internuclear distance range, for example in the vicinity of an avoided crossing, thus making the propagation of wave packets—and quantum trajectories as well—very difficult. However, for a counterexample, see Ref. [53]. Diabatic electronic states [54–56], as far as they are concerned, are built in such a way as to vary smoothly with internuclear distances, and are thus decoupled (as much as is possible) from the nuclear motion, but are coupled through the electronic Hamiltonian instead.

Although QTM formalism tends to be simpler in a diabatic basis, it has also been described in the adiabatic representation (see Ref. [57] and Appendix A of Ref. [4]).

BIBLIOGRAPHY

1. J.C. Polanyi, Some concepts in reaction dynamics, *Science*, **236**, 680–690 (1987).
2. W. Domcke, D.R. Yarkony and H. Köppel (eds.), *Conical Intersections*, Advances in Physical Chemistry, Vol. 15, World Scientific, London (2004).
3. M.S. Child, *Atom–Molecule Collision Theory: A Guide for the Experimentalist*, Plenum Press, New York (1979), p. 427.
4. I. Burghardt and L.S. Cederbaum, Hydrodynamic equations for mixed quantum states. II. Coupled electronic states, *J. Chem. Phys.*, **115**, 10312–10322 (2001).
5. D. Bohm, A suggested interpretation of the quantum theory in terms of "hidden" variables. I, *Phys. Rev.*, **85**, 166 (1952).
6. D. Bohm, A suggested interpretation of the quantum theory in terms of "hidden" variables. II, *Phys. Rev.*, **85**, 180 (1952).
7. P.R. Holland, *The Quantum Theory of Motion*, Cambridge University Press, Cambridge (1993).
8. V.E. Madelung, Quantentheorie in hydrodynamischer form. *Z. Phys.*, **40**, 322 (1926).
9. T. Takabayasi, The formulation of quantum mechanics in terms of ensemble in phase space, *Prog. Theoret. Phys.*, **11**, 341–373 (1954); doi:10.1143/PTP.11.341.
10. R.E. Wyatt, *Quantum Dynamics with Trajectories: Introduction to Quantum Hydrodynamics*, Springer, New York (2005).
11. C.L. Lopreore and R.E. Wyatt, Quantum wave packet dynamics with trajectories, *Phys. Rev. Lett.*, **82**, 5190–5193 (1999); doi:10.1103/PhysRevLett.82.5190.
12. F.S. Mayor, A. Askar and H.A. Rabitz, Quantum fluid dynamics in the Lagrangian representation and applications to photodissociation problems, *J. Chem. Phys.*, **111**, 2423–2435 (1999).
13. C. Leforestier, R.H. Bisseling, C. Cerjan, M.D. Feit, R. Friesner, A. Guldberg, A. Hammerich, G. Jolicard, W. Karrlein, H.D. Meyer, N. Lipkin, O. Roncero and R. Kosloff, A comparison of different propagation schemes for the time dependent Schrödinger equation, *J. Comput. Phys.*, **94**, 59–80 (1991).
14. A.S. Sanz, F. Borondo and S. Miret-Artes, Particle diffraction studied using quantum trajectories, *J. Phys.: Condens. Matt.*, **14**, 6109–6145 (2002).

15. R.E. Wyatt and K. Na, Quantum trajectory analysis of multimode subsystem-bath dynamics, *Phys. Rev. E*, **65**, 016702 (2001); doi:10.1103/PhysRevE.65.016702.

16. D. Babyuk and R.E. Wyatt, Multidimensional reactive scattering with quantum trajectories: Dynamics with 50–200 vibrational modes, *J. Chem. Phys.*, **124**, 214109–214107 (2006).

17. Y. Zhao and N. Makri, Bohmian versus semiclassical description of interference phenomena, *J. Chem. Phys.*, **119**, 60–67 (2003); doi:10.1063/1.1574805.

18. K. Park, B. Poirier and G. Parlant, Quantum trajectory calculations for bipolar wavepacket dynamics in one dimension, *J. Chem. Phys.*, **129**, 194112–194116 (2008).

19. B. Poirier, Reconciling semiclassical and Bohmian mechanics. V. Wavepacket dynamics, *J. Chem. Phys.*, **128**, 164115–15 (2008).

20. B. Poirier and G. Parlant, Reconciling Semiclassical and Bohmian mechanics: IV. Multisurface dynamics, *J. Phys. Chem. A*, **111**, 10400–10408 (2007).

21. B. Poirier, Reconciling semiclassical and Bohmian mechanics. I. Stationary states, *J. Chem. Phys.*, **121**, 4501–4515 (2004).

22. C. Trahan and B. Poirier, Reconciling semiclassical and Bohmian mechanics. II. Scattering states for discontinuous potentials, *J. Chem. Phys.*, **124**, 034115–18 (2006).

23. C. Trahan and B. Poirier, Reconciling semiclassical and Bohmian mechanics. III. Scattering states for continuous potentials, *J. Chem. Phys.*, **124**, 034116–14 (2006).

24. B. Poirier, Reconciling semiclassical and Bohmian mechanics. VI. Multidimensional dynamics, *J. Chem. Phys.*, **129**, 084103–18 (2008).

25. S. Garashchuk and V.A. Rassolov, Energy conserving approximations to the quantum potential: Dynamics with linearized quantum force, *J. Chem. Phys.*, **120**, 1181–1190 (2004).

26. D.K. Pauler and B.K. Kendrick, A new method for solving the quantum hydrodynamic equations of motion: Application to two-dimensional reactive scattering, *J. Chem. Phys.*, **120**, 603–611 (2004).

27. D. Babyuk and R.E. Wyatt, Coping with the node problem in quantum hydrodynamics: The covering function method, *J. Chem. Phys.*, **121**, 9230–9238 (2004).

28. S. Garashchuk and V.A. Rassolov, Modified quantum trajectory dynamics using a mixed wave function representation, *J. Chem. Phys.*, **121**, 8711–8715 (2004).

29. K.H. Hughes and R.E. Wyatt, Wavepacket dynamics on dynamically adapting grids: Application of the equidistribution principle, *Chem. Phys. Lett.*, **366**, 336–342 (2002).

30. C.J. Trahan and R.E. Wyatt, An arbitrary Lagrangian–Eulerian approach to solving the quantum hydrodynamic equations of motion: Equidistribution with "smart" springs, *J. Chem. Phys.*, **118**, 4784–4790 (2003).

31. L.R. Pettey and R.E. Wyatt, Wave packet dynamics with adaptive grids: The moving boundary truncation method, *Chem. Phys. Lett.*, **424**, 443–448 (2006).

32. L.R. Pettey and R.E. Wyatt, Application of the moving boundary truncation method to reactive scattering: H + H$_2$, O + H$_2$, O + HD, *J. Phys. Chem. A*, **112**, 13335–13342 (2008).

33. R.E. Wyatt, C.L. Lopreore and G. Parlant, Electronic transitions with quantum trajectories, *J. Chem. Phys.*, **114**, 5113–5116 (2001); doi:10.1063/1.1357203.

34. C.L. Lopreore and R.E. Wyatt, Electronic transitions with quantum trajectories. II, *J. Chem. Phys.*, **116**, 1228–1238 (2002).

35. J. Julien and G. Parlant, Quantum trajectory computation of ultrashort laser pulse excitation dynamics. Submitted (2010).

36. G. Parlant, *Semiclassical Nonadiabatic Dynamics with Quantum Trajectories*.

37. V.A. Rassolov and S. Garashchuk, Semiclassical nonadiabatic dynamics with quantum trajectories, *Phys. Rev. A*, **71**, 032511 (2005); doi:10.1103/PhysRevA.71.032511.

38. S. Garashchuk, V.A. Rassolov and G.C. Schatz, Semiclassical nonadiabatic dynamics using a mixed wave-function representation, *J. Chem. Phys.*, **123**, 174108–10 (2005).

39. S. Garashchuk, V.A. Rassolov and G.C. Schatz, Semiclassical nonadiabatic dynamics based on quantum trajectories for the $O(^3P,^1D) + H_2$ system, *J. Chem. Phys.*, **124**, 244307–8 (2006).

40. S. Garashchuk and V.A. Rassolov, Semiclassical nonadiabatic dynamics of NaFH with quantum trajectories, *Chem. Phys. Lett.*, **446**, 395–400 (2007).

41. J.C. Tully and R.K. Preston, Trajectory surface hopping approach to nonadiabatic molecular collisions: The reaction of H^+ with D_2, *J. Chem. Phys.*, **55**, 562–572 (1971).

42. J.C. Tully, Molecular dynamics with electronic transitions, *J. Chem. Phys.*, **93**, 1061–1071 (1990).

43. W.H. Press, B.P. Flannery, S.A. Teukolsky and W.T. Vetterling, *Numerical Recipes*, Cambridge University Press, Cambridge (1992).

44. E.R. Bittner and R.E. Wyatt, Integrating the quantum Hamilton–Jacobi equations by wavefront expansion and phase space analysis, *J. Chem. Phys.*, **113**, 8888–8897 (2000); doi:10.1063/1.1319987.

45. S. Garashchuk and V.A. Rassolov, Semiclassical dynamics based on quantum trajectories, *Chem. Phys. Lett.*, **364**, 562–567 (2002); doi:10.1016/S0009-2614(02)01378-7.

46. S. Garashchuk and V.A. Rassolov, Quantum dynamics with bohmian trajectories: energy conserving approximation to the quantum potential, *Chem. Phys. Lett.*, **376**, 358–363 (2003); doi:10.1016/S0009-2614(03)01008-X.

47. S. Garashchuk and V.A. Rassolov, Semiclassical dynamics with quantum trajectories: Formulation and comparison with the semiclassical initial value representation propagator, *J. Chem. Phys.*, **118**, 2482–2490 (2003); doi:10.1063/1.1535421.

48. F. Grossmann, Semiclassical wave-packet propagation on potential surfaces coupled by ultrashort laser pulses, *Phys. Rev. A*, **60**, 1791 (1999).

49. K.-A. Suominen, B. M. Garraway and S. Stenholm, Wave-packet model for excitation by ultrashort pulses, *Phys. Rev. A*, **45**, 3060–3070 (1992); doi:10.1103/PhysRevA.45.3060.

50. D.J. Tannor. *Introduction to Quantum Mechanics: A Time-Dependent Perspective*, University Science Books, Sausalito (2007).

51. E.J. Heller, Time-dependent approach to semiclassical dynamics, *J. Chem. Phys.*, **62**, 1544–1555 (1975).

52. V. Sidis, *Diabatic Potential Energy Surfaces for Charge Transfer Processes*, John Wiley, New York (1992), p. 73.

53. G. Parlant and D.R. Yarkony, An adiabatic state approach to electronically nonadiabatic wave packet dynamics, *Int. J. Quantum Chemi.*, **S26**, 736–739 (1992).

54. C.A. Mead and D.G. Truhlar, Conditions for the definition of a strictly diabatic electronic basis for molecular systems, *J. Chem. Phys.*, **77**, 6090–6098 (1982).

55. T.F. O'Malley, Diabatic states of molecules-quasistationary electronic states, *Adv. Atom. Molec. Phys.*, **7**, 223 (1971).

56. F.T. Smith, Diabatic and adiabatic representations for atomic collision problems, *Phys. Rev.*, **179**, 111–122 (1969).

57. J.C. Burant and J.C. Tully, Nonadiabatic dynamics via the classical limit Schrödinger equation, *J. Chem. Phys.*, **112**, 6097–6103(2000).

18 Recent Analytical Studies of Complex Quantum Trajectories

Chia-Chun Chou and Robert E. Wyatt

CONTENTS

18.1 INTRODUCTION

Bohmian mechanics, developed by Bohm in 1952, provides an alternative interpretation to nonrelativistic quantum mechanics [1–3]. In the hydrodynamic formulation of quantum mechanics, the continuity equation and the quantum Hamilton–Jacobi equation (QHJE) are obtained by substituting the wave function expressed in terms of the *real* amplitude and the *real* action function into the time-dependent Schrödinger equation. Bohm's analytical approach has been used to compute and interpret *real-valued* quantum trajectories from a precomputed wave function for a diverse range of physical processes [3–8]. In the synthetic approach, the quantum trajectory method has been developed as a computational tool to generate the wave function by evolving ensembles of real-valued quantum trajectories through the integration of the hydrodynamic equations *on the fly* [9]. Remarkable progress has been made in the use of real-valued quantum trajectories for solving a wide range of quantum mechanical problems [10].

Quantum trajectories in *complex* space in the framework of the quantum Hamilton–Jacobi formalism, developed by Leacock and Padgett in 1983 [11, 12], have recently attracted significant interest. This variant of the Bohmian approach is based on substitution of the wave function expressed by the *complex* action function into the time-dependent Schrödinger equation to obtain the *complex-valued* QHJE (this version is not the same as that in Bohm's formalism). This complex quantum hydrodynamic representation provides conceptual novelty, and also leads to new trajectory-based pictures of quantum mechanics that prove useful in computational applications. For stationary states, an accurate computational method has been proposed for the complex-valued QHJE to obtain the wave function and the reflection and transmission coefficients for one-dimensional problems [13–15]. For nonstationary states, the derivative propagation method (DPM) [16] developed in Bohmian mechanics has also been utilized to obtain approximate complex quantum trajectories and the wave function for one-dimensional barrier scattering [17–19]. In addition, this approach has been employed to describe the interference effects and node formation in the wave function [20, 21], to determine energy eigenvalues [22], and to improve the complex time-dependent Wentzel–Kramers–Brillouin method [23, 24]. Furthermore, the complex trajectory method has also been employed to analyze complex quantum trajectories and the complex quantum potential and to calculate tunneling probabilities for one-dimensional and multi-dimensional wave-packet scattering problems [25–31].

In the analytical approach, complex quantum trajectories determined from the *known* analytical form of the wave function have been analyzed for several stationary and nonstationary problems, including the free particle, the potential step, the potential barrier, the harmonic potential, and the hydrogen atom [32–39]. A unified description for complex quantum trajectories for one-dimensional problems has been presented [40]. Common features of complex quantum trajectories for one-dimensional stationary scattering problems have been analyzed for the Eckart and the hyperbolic tangent barriers [40,41]. In addition, quantum vortices form around a node in the wave function in complex space, and the discontinuity in the real part of the complex action leads to the quantized circulation integral [42]. Quantum streamlines near singularities of the quantum momentum function (QMF) and its Pólya vector field (PVF) have been thoroughly analyzed [43]. Moreover, quantum interference demonstrated by the head-on collision of two Gaussian wave packets has been explored in the complex plane [44–46]. On the other hand, several studies have been dedicated to issues related to the probability density and flux continuity in the complex plane and probability conservation along complex quantum trajectories [47–50].

In the current study, the equations of motion for complex quantum trajectories for one-dimensional time-dependent and time-independent problems will be reviewed in Section 18.2. Several recent analytical studies employing the complex quantum trajectory representation will be described in Section 18.3. Finally, we conclude with some comments and various promising topics for future investigations in Section 18.4.

18.2 QUANTUM TRAJECTORIES IN COMPLEX SPACE

18.2.1 TIME-DEPENDENT PROBLEMS

Based on the quantum Hamilton–Jacobi formalism developed by Leacock and Padgett [11, 12], the complex-valued QHJE is readily obtained by substituting the polar form of the complex-valued wave function,

$$\Psi(x,t) = \exp\left[\frac{i}{\hbar}S(x,t)\right], \tag{18.1}$$

into the time-dependent Schrödinger equation to obtain

$$-\frac{\partial S}{\partial t} = \frac{1}{2m}\left(\frac{\partial S}{\partial x}\right)^2 + V + \frac{\hbar}{2mi}\frac{\partial^2 S}{\partial x^2}, \tag{18.2}$$

where $S(x,t)$ is the complex action. As in Bohmian mechanics, the QMF is given by the guidance equation $p(x,t) = \partial S(x,t)/\partial x$. To find a quantum trajectory, we may rearrange this equation as

$$\frac{dx}{dt} = \frac{1}{m}\frac{\partial S(x,t)}{\partial x}. \tag{18.3}$$

However, since the action $S(x,t)$ is complex-valued and time remains real-valued, the trajectory requires a complex-valued coordinate. Therefore, the QMF $p(x,t)$ and the complex action $S(x,t)$ are extended to the complex space by replacing x with z. Thus, a complex quantum trajectory is defined by

$$\frac{dz}{dt} = \frac{p(z,t)}{m}. \tag{18.4}$$

In addition, the terms on the right side of Equation 18.2 correspond to the kinetic energy, the classical potential, and the quantum potential in complex space, respectively. Moreover, the QMF can be expressed in terms of the wave function through use of Equation 18.1

$$p(z,t) = \frac{\hbar}{i}\frac{1}{\Psi(z,t)}\frac{\partial \Psi(z,t)}{\partial z}. \tag{18.5}$$

Through Equation 18.2, we can obtain the equations of motion for $z(t)$, $p(z,t)$, and $S(z,t)$ for quantum trajectories in the complex space

$$\frac{dz}{dt} = \frac{p}{m}, \tag{18.6}$$

$$\frac{dp}{dt} = \frac{\partial p}{\partial z}\frac{dz}{dt} + \frac{\partial p}{\partial t} = -\frac{dV}{dz} - \frac{\hbar}{2mi}\frac{\partial^2 p}{\partial z^2}, \tag{18.7}$$

$$\frac{dS}{dt} = \frac{\partial S}{\partial z}\frac{dz}{dt} + \frac{\partial S}{\partial t} = \frac{p^2}{2m} - V - \frac{\hbar}{2mi}\frac{\partial p}{\partial z}. \tag{18.8}$$

The two terms on the right side of Equation 18.7 correspond to the classical force $f_c = -dV(z)/dz$ and the quantum force $f_q = -(\hbar/2mi)\partial^2 p/\partial z^2$, respectively. Therefore, we can determine the quantum trajectories of particles in complex space by integrating these equations and then synthesize the wave function through Equation 18.1.

For time-dependent problems, the initial state $\Psi(z, 0)$ is used to determine the initial condition $(z_0, p(z_0, 0), S(z_0, 0))$, where z_0 is the starting point of the trajectory. It is noted that the integration of Equations 18.6 through 18.8 involves the spatial derivative for the QMF. The equations of motion are not closed and general numerical methods for systems of ordinary differential equations cannot be applied to the integration of the equations of motion for quantum trajectories. However, the DPM has been developed to overcome a similar difficulty in Bohmian mechanics by solving a truncated system of equations for amplitude, phase and their spatial derivatives [16]. A computational approach through use of the DPM for solving the equations of motion in complex space using the iteration of the spatial partial derivatives of the QMF has been applied to one-dimensional and multi-dimensional scattering problems. Tannor and collaborators have applied this method, which they call Bohmian mechanics with complex action (BOMCA), to one-dimensional scattering of an initial Gaussian wave packet from a thick Eckart barrier [17–19]. In addition, they have employed this approach to describe the interference effects and node formation in the wave function [20, 21], to determine energy eigenvalues [22], and to improve the complex time-dependent Wentzel–Kramers–Brillouin method [23, 24]. Furthermore, Wyatt and collaborators have utilized this approach to investigate one-dimensional and multi-dimensional wave-packet scattering problems [25–31].

Complex quantum trajectories can be obtained for arbitrary initial positions, and the initial conditions for the quantum trajectories are determined by the initial state. Namely, a particle can start its motion at any position, and information such as the complex action is transported along its complex quantum trajectory. At a later time, when the particle crosses the real axis, we can record the information. Thus, for those particles which cross the real axis simultaneously, the wave function on the whole real axis can be synthesized using the information transported by these particles. Because the correct wave function at a specific time on the real axis must be determined by the information transported by particles arriving simultaneously at the real axis, a curve for the special initial positions of these particles in the complex space is defined as an *isochrone* [40]. The concept of isochrones has been analytically demonstrated by the free Gaussian wave packet and the coherent state in the harmonic potential in the complex plane [40].

18.2.2 TIME-INDEPENDENT PROBLEMS

For stationary states with eigenenergy E, the complex action can be expressed by $S(z, t) = W(z) - Et$ and the QMF becomes

$$p(z, t) = \frac{\partial S}{\partial z} = \frac{dW(z)}{dz} = p(z), \qquad (18.9)$$

where $W(z)$ is called the quantum characteristic function and $p(z)$ is the stationary state QMF which depends only on z. Moreover, the stationary state QMF is related to the stationary state wave function by

$$p(z) = \frac{\hbar}{i} \frac{1}{\psi(z)} \frac{d\psi(z)}{dz}. \tag{18.10}$$

Then, using the expression $S(z, t) = W(z) - Et$ and rewriting the QHJE in Equation 18.2 in terms of the stationary state QMF yield the stationary-state QHJE

$$\frac{1}{2m} p(z)^2 + V(z) + \frac{\hbar}{2mi} \frac{dp(z)}{dz} = E. \tag{18.11}$$

Similarly, we can obtain the equations of motion for stationary states from Equations 18.6 through 18.8

$$\frac{dz}{dt} = \frac{p}{m}, \tag{18.12}$$

$$\frac{dp}{dt} = \frac{2i}{\hbar} \left[E - V - \frac{p^2}{2m} \right] p, \tag{18.13}$$

$$\frac{dW}{dt} = \frac{p^2}{m}, \tag{18.14}$$

where the stationary-state QHJE in Equation 18.11 has been used.

When solving equations of motion for quantum trajectories, we encounter different difficulties for time-dependent and time-independent problems. For time-dependent problems, the system of equations of motion given in Equations 18.6 through 18.8 is not closed because the QMF is coupled to its spatial derivative. Therefore, general numerical methods for differential equations cannot be applied directly. For time-independent problems, although the system of equations of motion given in Equations 18.12 through 18.14 is closed, it is evident from Equation 18.10 that only solving the equations of motion with the *correct* initial quantum momentum $p(z_0)$ can yield the *correct* quantum trajectories belonging to the corresponding stationary states. Moreover, it is found from Equations 18.12 through 18.14 that the quantum characteristic function $W(z)$ is not coupled to z and p. Therefore, we only need to use Equations 18.12 and 18.13 to solve for quantum trajectories. In summary, the difficulty in solving time-dependent problems arises from the integration of the equations of motion, while the difficulty in solving time-independent problems arises from the specification of the initial conditions.

18.3 ANALYTICAL STUDIES OF THE COMPLEX QUANTUM TRAJECTORY FORMALISM

18.3.1 Quantum Trajectories for One-Dimensional Stationary Scattering Problems

One-dimensional stationary scattering problems including the Eckart and the hyperbolic tangent barriers have been investigated in the framework of the complex quantum Hamilton–Jacobi formalism [40, 41]. Exact complex quantum trajectories were determined by numerically integrating the equations of motion in Equations 18.12 and 18.13 using either forward or backward integrations. In the forward integration, we used the exact initial condition in the reflection region to solve the equations of motion. In the backward integration, the asymptotic QMF in the transmission region was used as the initial condition to numerically solve the equations of motion backwards. In addition, the total potential can be calculated from the exact analytical scattering wave function.

The state-dependent total potential (the sum of the classical and the quantum potentials) presents unusual and complicated structure in the complex plane. The total potential displays periodicity along the direction of the imaginary axis and reveals a complicated channel structure in the reflection region. For example, Figure 18.1a displays the real part of total potential for the one-dimensional scattering problem with the potential barrier

$$V(x) = V_0 \left[1 - \left(\frac{1 - \exp(x/a)}{1 + c \exp(x/a)} \right)^2 \right]. \tag{18.15}$$

In addition, the quantum potential of one-dimensional scattering problems shows a second-order pole or four-lobed quadrupole structure in the reflection region, and this structure originates from the asymptotic behavior of the wave function. Although the classical potentials extended analytically to complex space may show different pole structures for each problem, the quantum potentials present the same second-order pole structure in the reflection region.

In these studies [40, 41], some particles starting in the reflection region pass the barrier into the transmission region, and some rebound from the barrier back to the reflection region. In addition, localized closed trajectories form around stagnation points of the QMF in the reflection region, and they are imbedded in the walls of the channel structure. Trajectories may spiral into *attractors* or out of *repellers* near the barrier region. As shown in Figure 18.1b, for the forward integration, some particles move toward the transmission region, and some move toward the left. For the backward integration, some trajectories link the reflection and transmission regions. Some trajectories are traced from the transmission region to the repellers in the barrier region. However, some trajectories obtained by the forward integration spiral into attractors in the barrier region. Furthermore, we also derived the first-order and the second-order equations for local approximate quantum trajectories near stagnation points of the QMF, and used the first derivative of the QMF to describe the formation of attractors and repellers. Quantum trajectories present similar structures for one-dimensional stationary scattering problems.

As another example, Figure 18.2 displays quantum trajectories and the quantum momentum field for the hyperbolic tangent potential barrier given by

$$V(x) = \frac{1}{2} V_0 \left(1 + \tanh \left(\frac{x}{2a} \right) \right). \tag{18.16}$$

For the energy lower than the barrier height, particles starting from the reflection region may penetrate into the nonclassical region, and then turn back to the reflection

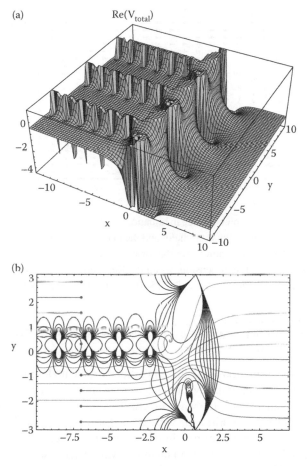

FIGURE 18.1 Scattering for the potential barrier in Equation 18.15 with $E = 1.2$, $V_0 = 1$, and $c = 0.5$: (a) Real part of the total potential; (b) quantum trajectories obtained by forward or backward integrations, and the contour map of the real part of the total potential. The initial positions for the forward integration are shown as dots. Relevant physical quantities are used in dimensionless units. (Reproduced with permission from Chou, C. C. and Wyatt, R. E., *J. Chem. Phys.*, 128, 154106, 2008. Copyright AIP 2008.)

region. The classical turning curve is defined by the curve where the scattering energy is equal to the real part of the total potential. From the stationary-state QHJE in Equation 18.11, the sum of the complex-valued kinetic energy, classical potential, and quantum potential along a complex quantum trajectory is equal to the real-valued scattering energy. The variations of these energies along complex quantum trajectories were also analyzed. Therefore, these studies not only analyzed common features of complex quantum trajectories and total potentials for these examples but also demonstrated general properties and similar structures of the complex quantum trajectories and the quantum potentials for one-dimensional time-independent scattering problems [40, 41].

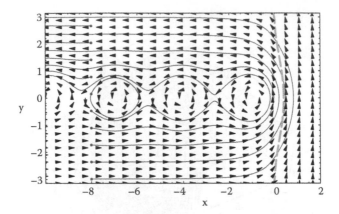

FIGURE 18.2 For the soft potential step in Equation 18.16 with $E = 0.6$: quantum trajectories (solid curves), the quantum momentum field, and the classical turning curve (dashed curve) in the principal zone. The initial positions for the forward integration are shown as dots. Relevant physical quantities are used in dimensionless units. (Reproduced with permission from Chou, C. C. and Wyatt, R. E., *J. Chem. Phys.*, 128, 154106, 2008. Copyright AIP 2008.)

18.3.2 QUANTUM VORTICES AND STREAMLINES WITHIN THE COMPLEX QUANTUM HAMILTON–JACOBI FORMALISM

Dirac predicted the formation of quantized vortices around nodes in the wave function in 1931 [51]. In quantum hydrodynamics in two or more real coordinates, quantum vortices form around nodes in the wave function and they carry quantized circulation [52–56]. Streamlines surrounding the vortex core form approximately circular loops. Such vortices play an important role in a diverse range of physical process such as the appearance of quantum vortices in quantum wave-packet studies of the collinear $H + H_2 \rightarrow H_2 + H$ exchange reaction [57–59].

Quantum vortices and streamlines in complex space were recently explored within the quantum Hamilton–Jacobi formalism [42, 43]. A quantum vortex forms around a node in the wave function in complex space, and the quantized circulation integral along a simple curve enclosing a node originates from the discontinuity in the real part of the complex action. The PVF for a complex function was introduced to interpret the circulation integral [60–62]. Pólya introduced a simple interpretation for complex contour integrals by associating a complex function $f(z)$ with a vector field $\overline{f(z)}$. If $f = u + iv$, then the associated PVF is given by $F = u - iv$, and $F = (u, -v)$ is the conjugate vector field to $f = (u, v)$ in the complex plane. The complex integral of the QMF can be interpreted in terms of the work and flux of its PVF along the contour. The real valuedness of the quantized circulation integral implies that only the work term contributes to the quantized circulation integral around a node

$$\oint_C p(z)dz = \oint_C \vec{P} \cdot d\vec{\ell} + i \oint_C \vec{P} \cdot d\vec{n} = nh, \qquad (18.17)$$

where C denotes a simple closed curve in the complex plane, $\vec{P} = (p_x, -p_y)$ is the PVF, $d\vec{\ell} = (dx, dy)$ is the tangent vector in the direction of the path C, and $d\vec{n} = (dy, -dx)$ is the normal vector pointing to the right as we travel along C.

A nonstationary state constructed from a linear combination of the ground, first, and second-excited states of the harmonic oscillator was used to illustrate the formation of a transient excited state quantum vortex.

$$\Psi(z, t) = \frac{3}{\sqrt{2}} \psi_0(z)e^{-iE_0t/\hbar} - 2\psi_1(z)e^{-iE_1t/\hbar} + \psi_2(z)e^{-iE_2t/\hbar}, \qquad (18.18)$$

where this wave function was extended to the complex space. The excited state vortices form around a higher-order node. In this example, two long-lived ground state vortices collide to form a transient first-excited state vortex, and then the first-excited state vortex immediately splits into two ground state vortices! Figure 18.3 shows the trajectories for the two nodes of the nonstationary state given in Equation 18.18, where the dimensionless units $z_d = \sqrt{\alpha}z$ and $t_d = \omega t$ have been used. At $t = 0$, the wave function has a second-order node at $z = 1$, and then the second-order node splits into two first-order nodes. At $t = \pi$, these two first-order nodes collide again to form a second order node at $z = -1$. They separate for $\pi < t < 2\pi$ and finally return to the initial position at $z = 1$.

The coupled harmonic oscillator was used as an example to present the quantized circulation integral in the multidimensional complex space. The two-particle Hamiltonian of coupled harmonic oscillators is given by

$$H = \frac{p_1^2}{2m_1} + \frac{p_2^2}{2m_2} + \frac{1}{2}m_1\omega^2x_1^2 + \frac{1}{2}m_2\omega^2x_2^2 + \frac{1}{2}k\,(x_1 - x_2)^2. \qquad (18.19)$$

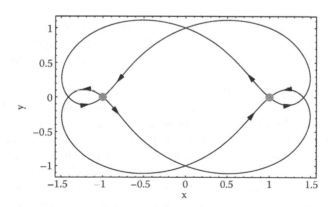

FIGURE 18.3 The trajectories for the two nodes of the nonstationary state given in Equation 18.18. At $t = 0$, these two first-order nodes coincide to form a second-order node (the right dot). At $t = \pi$, these two first-order nodes coincide again to form a second-order node (the left dot). When $t = 2\pi$, these two nodes return to the initial point. (Reproduced with permission from Chou, C. C. and Wyatt, R. E., *J. Chem. Phys.*, 128, 234106, 2008. Copyright AIP 2008.)

We consider the exact stationary state

$$\Psi_{01}(x_1, x_2) = \left(\frac{\alpha_1}{\pi}\right)^{1/4} e^{-\frac{\alpha_1}{2}X^2} \times \left(\frac{\alpha_2}{4\pi}\right)^{1/4} \left(2\alpha_2^{1/2}x\right) e^{-\frac{\alpha_2}{2}x^2}, \tag{18.20}$$

where $M = m_1 + m_2$, $\mu = m_1 m_2/(m_1 + m_2)$, $\alpha_1 = M\omega/\hbar$, $\alpha_2 = \mu\bar{\omega}/\hbar$, $\bar{\omega} = (\omega^2 + k/\mu)^{1/2}$, $X = (m_1 x_1 + m_2 x_2)/M$, and $x = x_1 - x_2$. Then, we extend the wave function in Equation 18.20 into the complex space by replacing x_1 and x_2 with z_1 and z_2. By setting $\Psi_{01}(z_1, z_2) = 0$, we obtain the corresponding *nodal plane* $z_1 - z_2 = 0$ in the four-dimensional complex space. Similarly, the circulation integral along a loop c for this case is given by

$$\Gamma = \oint_c \vec{p}(z_1, z_2) \cdot d\vec{z} = \oint_c p_1(z_1, z_2)dz_1 + \oint_c p_2(z_1, z_2)dz_2, \tag{18.21}$$

where the QMF for multidimensional problems is defined by $\vec{p} = \vec{\nabla}S = (\hbar/i)(\vec{\nabla}\Psi/\Psi)$.

In order to calculate the circulation integral, we need to choose a closed curve to "*enclose*" the two-dimensional nodal plane in the four-dimensional complex space. The two-dimensional nodal plane is given by $z_1 - z_2 = 0$, and we can find a closed curve in the plane $z_1 + z_2 = 0$ to enclose the nodal plane. The intersection of these two planes is only the point $(z_1, z_2) = (0, 0)$. Thus, we use a circle of arbitrary radius R centered on this point in the plane $z_1 + z_2 = 0$ as the closed loop to enclose the nodal plane $z_1 - z_2 = 0$. Accordingly, we let $z_1 = Re^{i\theta}$, and then $z_2 = -z_1 = -Re^{i\theta}$. Thus, we evaluate the circulation integral by integrating from $\theta = 0$ to $\theta = 2\pi$

$$\Gamma = \oint_c p_1(z_1, z_2)dz_1 + \oint_c p_2(z_1, z_2)dz_2 = \pi\hbar + \pi\hbar = h. \tag{18.22}$$

The circulation integral is quantized and equal to h. In addition, we note that each term in Equation 18.22 contributes $h/2$ to the circulation integral. Therefore, the circulation integral along a closed curve enclosing the nodal plane in complex space is quantized.

Local structures of the QMF and its PVF near stagnation points and poles were thoroughly analyzed, and the equations for approximate quantum streamlines around these points were also derived [43]. Streamlines near a stagnation point of the QMF may spiral into or away from it, or they may become circles centered on this point or straight lines passing through this point. In contrast to Bohmian mechanics in real space, streamlines near a pole display East-West and North-South opening hyperbolic structure. On the other hand, streamlines near a stagnation point of the PVF display general rectangular hyperbolic structure, and streamlines near a pole become circles enclosing the pole. Although the QMF displays *hyperbolic flow* near a node, its PVF displays *circular flow* around a quantum vortex. Furthermore, the local structures of the QMF and its PVF around a stagnation point are related to the first derivative of the QMF; however, the magnitude of the local structures for these two fields near a pole depends only on the order of the node in the wave function. These studies

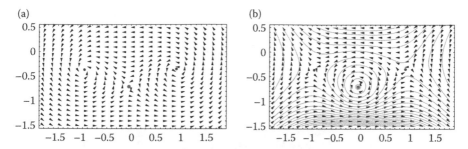

FIGURE 18.4 For $t = 1.5$: (a) the quantum momentum field; (b) the Pólya vector field and contours of the probability density (solid lines) with the stagnation points (small dots) and the pole (large dot). (Reproduced with permission from Chou, C. C. and Wyatt, R. E., *J. Chem. Phys.*, 129, 124113, 2008. Copyright AIP 2008.)

displayed the rich dynamics of streamlines in complex space for one-dimensional time-dependent and time-independent problems.

Figure 18.4 presents the QMF and the PVF around stagnation points and poles for a nonstationary state constructed from a linear combination of the ground and first-excited states of the harmonic oscillator

$$\Psi(z, t) = \psi_0(z)e^{-\frac{iE_0 t}{\hbar}} + \psi_1(z)e^{-\frac{iE_1 t}{\hbar}}. \tag{18.23}$$

The following dimensionless units are used: $z_d = z/\sqrt{\hbar/m\omega}$, $p_d = p/\sqrt{\hbar m\omega}$, $t_d = t/(1/\omega)$, and $E_d = E/\hbar\omega$. In Figure 18.4a, the positive imaginary part of the first derivatives of the QMF at these two stagnation points contributes to counterclockwise flow of the QMF, and the positive or negative real parts of the first derivatives of the QMF lead to divergence or convergence of the streamlines to the stagnation points, respectively. In addition, the QMF displays hyperbolic flow around the pole. In Figure 18.4b, the PVF displays hyperbolic flow around the two stagnation points and circular flow around the pole. In addition, the PVF is parallel to contours of the complex-extended Born probability density [42,48].

18.3.3 QUANTUM CAVES AND WAVE-PACKET INTERFERENCE

Quantum interference was investigated for the first time within the complex quantum Hamilton–Jacobi formalism [44–46]. The Gaussian wave-packet head-on collision was used as an example to demonstrate quantum interference in complex space. This process can be described by the total wave function, $\Psi(x, t) = \psi_L(x, t) + \psi_R(x, t)$ (L/R denotes left/right). Each partial wave is represented by a free Gaussian wave packet,

$$\psi(x, t) = A_t e^{-(x-x_t)^2/4\sigma_t\sigma_0 + ip(x-x_t)/\hbar + iEt/\hbar}, \tag{18.24}$$

where, for each component, $A_t = (2\pi\sigma_t^2)^{-1/4}$ and the complex time-dependent spreading is given by $\sigma_t = \sigma_0(1 + i\hbar t/2m\sigma_0^2)$ with the initial spreading σ_0. The total wave

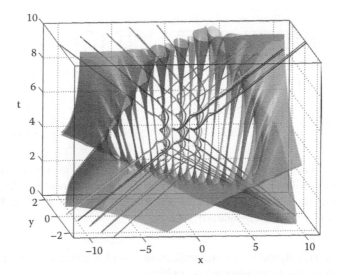

FIGURE 18.5 Quantum caves for the head-on collision of two Gaussian wave packets with the initial conditions: $x_{0L} = -10 = -x_{0R}$, $v_L = 2 = -v_R$ and $\sigma_0 = \sqrt{2}$, and maximal interference occurs at $t = 5$ in real space. These caves are formed with the isosurfaces $|\Psi(z, t)| = 0.053$ (lighter gray sheets) and $|\partial \Psi(z, t)/\partial z| = 0.106$ (darker gray sheets). The complex quantum trajectories launched from two branches of the isochrone reach the real axis at $t = 5$. All quantities are given in atomic units. (Reproduced with permission from Chou, C. C., Sanz, A. S., Miret-Artés, S., and Wyatt, R. E., *Phys. Rev. Lett.*, 102, 250401, 2009. Copyright APS 2009.)

function describing the head-on collision of two Gaussian wave packets was analytically extended to the complex plane.

As systematically analyzed for local structures of the QMF and its PVF around stagnation points and poles [42, 43], complex quantum trajectories display *helical wrapping* around stagnation tubes and *hyperbolic deflection* near vortical tubes, as shown in Figure 18.5. These tubes are prominent features of *quantum caves* in space-time Argand plots. Stagnation and vortical tubes alternate with each other, and the centers of the tubes are stagnation and vortical curves. In addition, the equation for approximate quantum trajectories around stagnation points was also derived. In contrast, Pólya trajectories display hyperbolic deflection around stagnation tubes and helical wrapping around vortical tubes. The intriguing topological structure consisting of these tubes develops quantum caves in space-time Argand plots, and quantum interference leads to the formation of quantum caves and produces this topological structure. Furthermore, the divergence and vorticity of the QMF characterizes the turbulent flow of trajectories; however, the PVF describes an incompressible and irrotational flow because of its vanishing divergence and vorticity except at poles in the complex plane.

For the case where the relative propagation velocity is larger than the spreading rate of the wave packets, trajectories keep circulating around stagnation tubes for finite times during interference and then escape as time progresses. This counterclockwise circulation of trajectories launched from different positions around the same stagnation tubes can be viewed as a resonance process. The whole process shows a type of long-range correlation among trajectories arising from different starting points. In addition, the wrapping time for an individual trajectory can be determined according to the divergence and vorticity of the QMF. The variation in the wrapping times between trajectories is connected to the distribution of nodes in the wave function and stagnation points of the QMF, and the average wrapping time over trajectories from the isochrone arriving at the real axis at maximal interference was calculated as one of the definitions for the lifetime of interference.

The rotational dynamics of the nodal line arising from the interference of these two wave packets was thoroughly analyzed in the complex plane. Since the interference features are transiently observed on the real axis only when the nodal line is near the real axis, we can define the lifetime for the interference process corresponding to the time interval for the nodal line to rotate from $\theta = -10°$ to $\theta = +10°$. As shown in Figure 18.6, the nodal line rotates counterclockwise from the initial angle to a limiting value with an angular displacement $\Delta\theta = \pi/2$, and its intersections with nodal trajectories determine the positions of nodes. In addition, the distance between nodes increases with time. For the case shown in Figure 18.6, interference features are observed on the real axis only when the nodal line is near the real axis, and this occurs between about $\theta(3.52) = -10°$ and $\theta(7.32) = 10°$. Thus, the lifetime for the interference features is about $\Delta t = 3.8$.

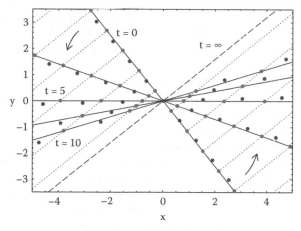

FIGURE 18.6 For the total wave function with the initial conditions: $x_{0L} = -10 = -x_{0R}$, $v_L = 2 = -v_R$ and $\sigma_0 = \sqrt{2}$, the evolution of the nodal line at $t = 0, 2.5, 5, 7.5, 10$ (black solid lines) and $t = \infty$ (black dashed line) is shown, and the arrows indicate the rotation direction. Nodes and stagnation points are denoted by dots; nodal trajectories are shown as dotted lines passing through the nodal points. All quantities are given in atomic units. (Reproduced with permission from Chou, C. C., Sanz, A. S., Miret-Artés, S., and Wyatt, R. E., *Phys. Rev. Lett.*, 102, 250401, 2009. Copyright APS 2009.)

For the case where the relative propagation velocity is approximately equal to or smaller than the spreading rate of the wave packets, the rapid spreading of the wave packets leads to the distortion of quantum caves. Because of the rapid spreading rate of the wave packets, interference features develop very rapidly and remain visible asymptotically in time. Therefore, the wrapping time becomes infinity, and this implies the infinite survival of such interference features. However, the rotational dynamics of the nodal line clearly explains the persistent interference features observed on the real axis.

In contrast with conventional quantum mechanics, two counter-propagating wave packets *always* interfere with each other in the complex plane, and this leads to a persistent pattern of nodes and stagnation points. Thus, the rotational dynamics of the nodal line in the complex plane clearly explains the transient or persistent appearance of interference features observed on the real axis, and it offers a unified and elegant method to define the lifetime for interference features. Therefore, these results show that the complex trajectory method provides a novel and insightful understanding of quantum interference at a fundamental level.

18.3.4 COMPLEX-EXTENDED BORN PROBABILITY DENSITY

Several studies have been devoted to issues concerning the probability density and flux continuity in the complex plane and probability conservation along complex quantum trajectories [47–50]. First, Poirier proposed a complexified *analytic* probability density $\rho_P(z)$ in the complex plane defined by the product of the wave function and its *generalized complex conjugate*, $\rho_P(z,t) = \Psi(z,t)\Psi^*(z^*,t)$. This probability density is consistent with the analytic continuation of the traditional probability density $\rho(x,t)$ defined along the real axis. Although this probability density is complex valued off the real axis, it satisfies a complexified flux continuity equation in the complex plane. However, it has been pointed out that this probability density produces *additional nodes* because it satisfies the Schwarz reflection principle [48]. These additional nodes are located in the mirror images with respect to the real axis of the original nodes of the wave function, and they are not a feature of the original complex-extended wave function $\Psi(z,t)$. Thus, the probability density proposed by Poirier does not match the node structure of the wave function. On the other hand, John proposed an approach to determine a conserved probability density in the complex plane by solving an appropriate equation. However, it was found that conserved probability densities may not exist in some regions in the complex plane [49].

The complex-extended Born probability density defined by replacing x with z in the Born probability density of quantum mechanics, $\rho_B(z,t) = \Psi(z,t)\Psi^*(z,t)$ where time remains real valued, was recently studied in the complex plane [48,50]. This probability density is real valued and nonnegative in complex space. It is important to note that this probability density is continuous but *not* analytic in the complex plane. Although we cannot obtain a complex-version continuity equation for the Born probability density because of its nonanalyticity, we can derive an arbitrary Lagrangian–Eulerian (ALE) rate equation for this probability density in the complex plane.

This equation gives the rate of change in the density along an *arbitrary path*. If the grid point velocity is equal to zero, the Eulerian rate equation for the density is recovered and the imaginary part of the complex total energy density determines the local rate of change in the density. If the grid point velocity is equal to the velocity of a complex quantum trajectory, we obtain the Lagrangian rate equation describing the rate of change in the density along the complex quantum trajectory and the complex quantum Lagrangian density determines the rate of change along this trajectory. Furthermore, when the grid point velocity is equal to the velocity of a Pólya trajectory, the rate of change in the probability density originates completely from the local contribution. In particular, only the complex total potential energy density contributes to the rate of change in the density along a complex path with the velocity equal to the QMF divided by $2m$ where m is the mass of the particle. The effect of this specific velocity is to cancel the contribution from the kinetic energy to the rate of change in the density. For stationary states, the rate of change in the probability density along a Pólya trajectory is equal to zero, and this result reflects the fact that the PVF is parallel to contours of the Born probability density [42,43].

The rate of change in the Born probability density along an arbitrary path can be described by the ALE rate equation; moreover, this density is closely related to the trajectory dynamics in the complex plane. The trajectory dynamics for the wave function is determined through the QMF by the dynamical equation, $dz(t)/dt = p(z,t)/m$. The PVF of the QMF contains exactly the same information as the QMF itself; hence, it also correctly reflects the information of stagnation points and poles. Furthermore, the PVF associated with the QMF is shown to be parallel to contours of the Born density. Therefore, the complex-extended Born probability density correctly reflects information on trajectory dynamics in complex space.

18.4 CONCLUDING REMARKS AND FUTURE DIRECTIONS

Analytical studies of the complex quantum trajectory formalism are based on a precomputed wave function or the exact analytical form of the wave function. Within the framework of the complex quantum Hamilton–Jacobi formalism, this brief review describes one-dimensional stationary scattering problems, quantum vortices, quantum streamlines for the QMF and its PVF around stagnation points and poles, wave-packet interference and quantum caves, and the complex-extended Born probability density. These relevant analytical studies indicate that the complex quantum trajectory method can provide a novel and insightful understanding of quantum phenomena.

Most of the previously mentioned analytical studies concentrate on one-dimensional problems. Thus, multidimensional processes in chemical physics deserve further investigation. Moreover, some fertile areas (including multimode system-bath dynamics, mixed quantum-classical dynamics, and density matrix evolution in dissipative systems) could be further explored within the complex quantum trajectory formalism.

ACKNOWLEDGMENTS

We thank the Robert Welch Foundation (grant no. F–0362) for their financial support of this research.

BIBLIOGRAPHY

1. D. Bohm, *Phys. Rev.* **85**, 166 (1952).
2. D. Bohm, *Phys. Rev.* **85**, 180 (1952).
3. P. R. Holland, *The Quantum Theory of Motion: An Account of the de Broglie–Bohm Causal Interpretation of Quantum Mechanics* (Cambridge University Press, New York, 1993).
4. A. S. Sanz, F. Borondo, and S. Miret-Artés, *Phys. Rev. B* **61**, 7743 (2000).
5. A. S. Sanz, F. Borondo, and S. Miret-Artés, *Phys. Rev. B* **69**, 115413 (2004).
6. R. Guantes, A. S. Sanz, J. Margalef-Roig, and S. Miret-Artés, *Surf. Sci. Rep.* **53**, 199 (2004).
7. A. S. Sanz and S. Miret-Artés, *J. Chem. Phys.* **126**, 234106 (2007).
8. A. S. Sanz and S. Miret-Artés, *J. Phys. A: Math. Theor.* **41**, 435303 (2008).
9. C. L. Lopreore and R. E. Wyatt, *Phys. Rev. Lett.* **82**, 5190 (1999).
10. R. E. Wyatt, *Quantum Dynamics with Trajectories: Introduction to Quantum Hydrodynamics* (Springer, New York, 2005).
11. R. A. Leacock and M. J. Padgett, *Phys. Rev. Lett.* **50**, 3 (1983).
12. R. A. Leacock and M. J. Padgett, *Phys. Rev. D* **28**, 2491 (1983).
13. C. C. Chou and R. E. Wyatt, *J. Chem. Phys.* **125**, 174103 (2006).
14. C. C. Chou and R. E. Wyatt, *Phys. Rev. E* **74**, 066702 (2006).
15. C. C. Chou and R. E. Wyatt, *Int. J. Quantum Chem.* **108**, 238 (2008).
16. C. J. Trahan, K. H. Hughes, and R. E. Wyatt, *J. Chem. Phys.* **118**, 9911 (2003).
17. Y. Goldfarb, I. Degani, and D. J. Tannor, *J. Chem. Phys.* **125**, 231103 (2006).
18. A. S. Sanz and S. Miret-Artés, *J. Chem. Phys.* **127**, 197101 (2007).
19. Y. Goldfarb, I. Degani, and D. J. Tannor, *J. Chem. Phys.* **127**, 197102 (2007).
20. Y. Goldfarb, J. Schiff, and D. J. Tannor, *J. Phys. Chem. A* **111**, 10416 (2007).
21. Y. Goldfarb and D. J. Tannor, *J. Chem. Phys.* **127**, 161101 (2007).
22. Y. Goldfarb, I. Degani, and D. J. Tannor, *Chem. Phys.* **338**, 106 (2007).
23. M. Boiron and M. Lombardi, *J. Chem. Phys.* **108**, 3431 (1998).
24. Y. Goldfarb, J. Schiff, and D. J. Tannor, *J. Chem. Phys.* **128**, 164114 (2008).
25. B. A. Rowland and R. E. Wyatt, *J. Phys. Chem. A* **111**, 10234 (2007).
26. R. E. Wyatt and B. A. Rowland, *J. Chem. Phys.* **127**, 044103 (2007).
27. B. A. Rowland and R. E. Wyatt, *J. Chem. Phys.* **127**, 164104 (2007).
28. J. K. David and R. E. Wyatt, *J. Chem. Phys.* **128**, 094102 (2008).
29. B. A. Rowland and R. E. Wyatt, *Chem. Phys. Lett.* **461**, 155 (2008).
30. R. E. Wyatt and B. A. Rowland, *J. Chem. Theory Comput.* **5**, 443 (2009).
31. R. E. Wyatt and B. A. Rowland, *J. Chem. Theory Comput.* **5**, 452 (2009).
32. M. V. John, *Found. Phys. Lett.* **15**, 329 (2002).
33. C. D. Yang, *Ann. Phys. (N.Y.)* **319**, 399 (2005).
34. C. D. Yang, *Ann. Phys. (N.Y.)* **319**, 444 (2005).
35. C. D. Yang, *Int. J. Quantum Chem.* **106**, 1620 (2006).
36. C. D. Yang, *Chaos Soliton Fract.* **30**, 342 (2006).
37. C. D. Yang, *Ann. Phys. (N.Y.)* **321**, 2876 (2006).
38. C. D. Yang, *Chaos Soliton Fract.* **32**, 312 (2007).
39. C. D. Yang, *Chaos Soliton Fract.* **33**, 1073 (2007).
40. C. C. Chou and R. E. Wyatt, *Phys. Rev. A* **76**, 012115 (2007).
41. C. C. Chou and R. E. Wyatt, *J. Chem. Phys.* **128**, 154106 (2008).
42. C. C. Chou and R. E. Wyatt, *J. Chem. Phys.* **128**, 234106 (2008).
43. C. C. Chou and R. E. Wyatt, *J. Chem. Phys.* **129**, 124113 (2008).
44. A. S. Sanz and S. Miret-Artés, *Chem. Phys. Lett.* **458**, 239 (2008).

45. C. C. Chou, A. S. Sanz, S. Miret-Artés, and R. E. Wyatt, *Phys. Rev. Lett.* **102**, 250401 (2009).
46. C. C. Chou, Ángel S. Sanz, S. Miret-Artés, and R. E. Wyatt, *Annals of Physics*, 2010, in press.
47. B. Poirier, *Phys. Rev. A* **77**, 022114 (2008).
48. C. C. Chou and R. E. Wyatt, *Phys. Rev. A* **78**, 044101 (2008).
49. M. V. John, *Ann. Phys. (N.Y.)* **324**, 220 (2009).
50. C. C. Chou and R. E. Wyatt, *Phys. Lett. A* **373**, 1811 (2009).
51. P. A. Dirac, *Proc. R. Soc. A* **133**, 60 (1931).
52. J. O. Hirschfelder, A. C. Christoph, and W. E. Palke, *J. Chem. Phys.* **61**, 5435 (1974).
53. J. O. Hirschfelder, C. J. Goebel, and L. W. Bruch, *J. Chem. Phys.* **61**, 5456 (1974).
54. J. O. Hirschfelder and K. T. Tang, *J. Chem. Phys.* **64**, 760 (1976).
55. J. O. Hirschfelder and K. T. Tang, *J. Chem. Phys.* **65**, 470 (1976).
56. J. O. Hirschfelder, *J. Chem. Phys.* **67**, 5477 (1977).
57. E. A. McCullough, Jr. and R. E. Wyatt, *J. Chem. Phys.* **51**, 1253 (1969).
58. E. A. McCullough, Jr. and R. E. Wyatt, *J. Chem. Phys.* **54**, 3578 (1971).
59. A. Kuppermann, J. T. Adams, and D. G. Truhlar, VII International Conference of Physics of Electronics and Atomic Collisions, Belgrade, Yugoslavia, p. 149 of abstracts of papers.
60. G. Pólya and G. Latta, *Complex Variables* (John Wiley & Sons, New York, 1974).
61. B. Braden, *Math. Mag.* **60**, 321 (1987).
62. T. Needham, *Visual Complex Analysis* (Oxford University Press, Oxford, 1998).

19 Modified de Broglian Mechanics and the Dynamical Origin of Quantum Probability

Moncy V. John

CONTENTS

19.1 HISTORICAL INTRODUCTION

A well-known formulation of classical mechanics of system of particles is based on solving the Hamilton–Jacobi equation

$$\frac{\partial S}{\partial t} + H\left(q_i, \frac{\partial S}{\partial q_i}, t\right) = 0, \tag{19.1}$$

where $S(q_i, \alpha_i, t)$ is Hamilton's principal function, H is the Hamiltonian, q_i are the configuration space variables, α_i are constants of integration, and t denotes time. If the Hamiltonian is independent of t and is a constant of motion equal to the total energy E, then S and Hamilton's characteristic function W are related by

$$S = W - Et. \tag{19.2}$$

Since W is independent of time, the surfaces with $W = constant$ have fixed locations in configuration space. However, the $S = constant$ surfaces move in time and may be considered as wavefronts propagating in this space [1]. For a single particle one can show that the wave velocity at any point is given by

$$u = \frac{E}{|\nabla W|} = \frac{E}{p} = \frac{E}{mv}. \tag{19.3}$$

This shows that the velocity of a point on this surface is inversely proportional to the spatial velocity of the particle. Also one can show that the trajectories of the particle must always be normal to the surfaces of constant S. The momentum of the particle is obtained as [1]

$$\mathbf{p} = \nabla W. \tag{19.4}$$

It is not clear whether this wave picture was known to classical physicists. However, in hindsight, we may now observe that there exists a correspondence between classical mechanics and geometrical optics. That is, the particle trajectories orthogonal to surfaces of constant S in classical mechanics, as we have discussed above, are similar to light rays traveling orthogonal to Huygens' wavefronts.

In modern physics, the discovery of the wave nature of matter is credited to L. de Broglie who famously postulated this phenomenon in 1923 just by reflecting upon nature's love for symmetry. On its basis, he made the astonishing prediction in the history of physics that a beam of electrons would exhibit wave-like phenomena of diffraction and interference upon passing through sufficiently narrow slits, for which he was awarded the Nobel Prize in 1929.

In spite of this, the fact that Broglie's 1924 Ph.D. thesis [2] contained the seed of a new mechanics, which can replace classical mechanics, did not get due attention for a long time. In it, and later in a paper published in 1927 [3], he proposed that Newton's first law of motion be abandoned, and replaced by a new postulate, according to which a freely moving body follows a trajectory that is orthogonal to the surfaces of equal phase of an associated guiding wave. He presented his new theory in the 1927 Solvay Conference too, but it was not well-received at that time [4].

The Schrödinger equation, which is the cornerstone of quantum mechanics, can be recast in a form reminiscent of the Hamilton–Jacobi equation by substituting $\Psi = R \exp(i S/\hbar)$ in it. The guidance relation for particle trajectories, as proposed by de Broglie, was

$$m_i \dot{\mathbf{x}}_i = \nabla_i S. \tag{19.5}$$

Solving this, we get a particle trajectory for each initial position. In 1952, D. Bohm rediscovered this formalism [5], but in a slightly different way. He noted that if we take the time derivative of the above equation, Newton's law of motion may be obtained in the form

$$m\ddot{\mathbf{x}}_i = -\nabla_i(V + Q). \tag{19.6}$$

Here

$$Q = -\Sigma \frac{\hbar^2}{2m_i} \frac{\nabla_i^2 |\Psi|}{|\Psi|} \tag{19.7}$$

is called the "quantum potential." This results in the same mechanics as that of de Broglie, but in the language of Newton's equation with an unnatural quantum potential. We shall note that the first order equation of motion Equation 19.5 suggested by de Broglie represented a unification of the principles of mechanics and optics and this should be distinguished from Bohm's second order dynamics based on Equations 19.6 and 19.7 [4]. It may also be mentioned that Bohm's revival of de Broglie's theory in pseudo-Newtonian form has led to a mistaken notion that de Broglie–Bohm (dBB) theory constituted a return to classical mechanics. In fact, de Broglie's theory was a new formulation of dynamics in terms of wave–particle duality.

It is well-known that neither the particle picture nor the trajectories have any role in the standard Copenhagen interpretation of quantum mechanics and that several physicists, including Albert Einstein, were not happy with this interpretation for various reasons. Einstein was not very happy with the de Broglian mechanics either [4]. The reason for this was speculated to be the problem it faced with real wave functions, where the phase gradient $\nabla S = 0$. For the wide class of bound state problems, the time-independent part of the wave function is real and hence while applying the equation of motion 19.5, the velocity of the particle turns out to be zero everywhere. This feature, that the particles in bound eigenstates are at rest irrespective of their position and energy, is counter-intuitive and is not a satisfactory one.

A deterministic approach to quantum mechanics, which claims the absence of this problem is the Floyd–Faraggi–Matone (FFM) trajectory representation [6–8]. In one dimension, they use a generalized Hamilton–Jacobi equation which is equivalent to that used by dBB. But FFM differs from dBB in that the equation of motion in the domain $[x, t]$ is rendered by Jacobi's theorem. For stationarity, the equation of motion for the trajectory time t, relative to its constant coordinate τ, is given as a function of x by

$$t - \tau = \frac{\partial W}{\partial E},$$ (19.8)

where τ specifies the epoch. Thus the dBB and FFM trajectory representations differ significantly in the use of the equation of motion, though they are based on equivalent generalized Hamilton–Jacobi equations.

19.2 COMPLEX QUANTUM TRAJECTORIES

By modifying the de Broglian mechanics, complex quantum trajectories were first obtained and drawn for the case of some fundamental problems in quantum mechanics, such as the harmonic oscillator, potential step, wave packets etc., in 2001 [9]. This modified dBB approach also attempts to unify the principles of mechanics and optics, but in a different way. An immediate application of the formalism was in solving the above problem of stationarity of quantum particles in bound states, in a natural way. For obtaining this representation, we shall substitute $\Psi = e^{i\hat{S}/\hbar}$ in the Schrödinger equation to obtain the quantum Hamilton–Jacobi equation (QHJE) [1]

$$\frac{\partial \hat{S}}{\partial t} + \left[\frac{1}{2m} \left(\frac{\partial \hat{S}}{\partial x} \right)^2 + V \right] = \frac{i\hbar}{2m} \frac{\partial^2 \hat{S}}{\partial x^2},$$ (19.9)

and then postulate an equation of motion similar to that of de Broglie:

$$m\dot{x} \equiv \frac{\partial \hat{S}}{\partial x} = \frac{\hbar}{i} \frac{1}{\Psi} \frac{\partial \Psi}{\partial x},$$ (19.10)

for the particle. The trajectories $x(t)$ are obtained by integrating this equation with respect to time and they will lie in a complex x-plane. It was observed that the above identification $\Psi = e^{i\hat{S}/\hbar}$ helps to utilize all the information contained in Ψ while obtaining the trajectory. (The dBB approach, which uses $\Psi = R e^{iS/\hbar}$ does not have this advantage.)

In view of the above, from here onwards, we consider x as a complex variable. Also we restrict ourselves to single particles in one dimension for simplicity. The formal procedure adopted is as follows: Identifying the standard solution Ψ as equivalent to $e^{i\hat{S}/\hbar}$, Equation 19.10 helps us to obtain the expression for the velocity field $\dot{x}$ in the new scheme. This, in turn, is integrated with respect to time to obtain the complex trajectory $x(t)$, with an initial position $x(0)$. This complex solution is obviously quite different from the corresponding dBB solution. The connection with the real physical world is established by postulating that the physical coordinate or trajectory of the particle is the real part $x_r(t)$ of its complex trajectory $x(t)$. It can be seen that also in this general case, the modified dBB trajectory $x_r(t)$ is different from the dBB trajectory.

We shall note that this new scheme differs from the dBB also in one important aspect. Here, when the particle is at some point x in the complex plane, its physical coordinate is x_r and the physical velocity is $\dot{x}_r$, evaluated at x. Thus for the same physical coordinate x_r, the particle can have different physical velocities, depending on the point x through which its complex path passes. This in principle rules out the possibility of ascribing simultaneously well-defined physical position and velocity variables for the particle [9]. The situation here is very different from that of the dBB in which for a given physical position of the particle, also the velocity is known, as obtainable from Equation 19.5. Recently, the prospects of validating the uncertainty principle using the complex trajectory representation was investigated in Ref. [10].

When $\hat{S} = \hat{W} - Et$, where E and t are assumed to be real, the Schrödinger equation gives us an expression for the energy of the particle

$$E = \frac{1}{2}m\dot{x}^2 + V(x) + \frac{\hbar}{2i}\frac{\partial \dot{x}}{\partial x}. \tag{19.11}$$

The last term resembles the quantum potential in the dBB theory. However, the concept of quantum potential is not an integral part of this formalism since the equation of motion 19.10 adopted here is not based on it.

The complex eigentrajectories in the free particle, harmonic oscillator and potential step problems, and complex trajectories for a wave packet solution were obtained in Ref. [9]. As an example, complex trajectories in the $n = 1$ harmonic oscillator are shown in Figure 19.1.

It is well-known that the QHJE, as given in Equation 19.9, was used by many physicists such as Wentzel, Pauli, and Dirac, even during the time of inception of quantum mechanics [11]. In a commendable work in 1982, Leacock and Padgett [12] have used the QHJE to obtain eigenvalues in many bound state problems, without actually having to solve the corresponding Schrödinger equation. This method was further developed by Bhalla, Kapoor, and Panigrahi [13]. However, there were no trajectories in their works; it is incorrect to ascribe complex quantum trajectories

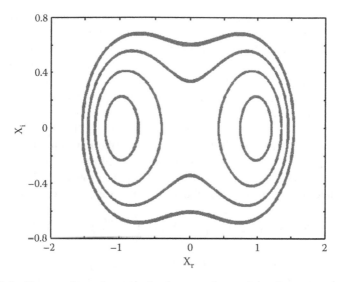

FIGURE 19.1   The complex trajectories for the $n = 1$ harmonic oscillator. (Reproduced with permission from John, M. V., *Found. Phys. Lett.*, 15, 329, 2002.)

to [12], as is done by some authors. It was the work in [9] which, by modifying the de Broglian mechanics, provided the complex trajectory representation [14, 15].

This formalism was extended to three-dimensional problems, such as the hydrogen atom, by Yang [16] and was used to investigate one-dimensional scattering problems and bound state problems by Chou and Wyatt [17,18]. Later, a complex trajectory approach for solving the QHJE was developed by Tannor and co-workers [19]. Sanz and Miret-Artés found the complex trajectory representation useful in better understanding the nonlocality in quantum mechanics [20].

19.3   PROBABILITY FROM THE VELOCITY FIELD

Instead of computing the complex trajectories $x(t)$, the complex paths $x_i(x_r)$ in the present scheme can directly be found by integrating the equation

$$\frac{dx_i}{dx_r} = \frac{\dot{x}_i}{\dot{x}_r}. \tag{19.12}$$

For doing this integration, Equation 19.10 shall be used. We may note that even for an eigenstate, the particle can be in any one of its infinitely many possible quantum trajectories, depending on its initial position in the complex plane. Therefore, the expectation values of dynamical variables are to be evaluated over an ensemble of particles in all possible trajectories. It was postulated that the average of a dynamical variable O can be obtained using the measure $\Psi^\star\Psi$ as

$$\langle O \rangle = \int_{-\infty}^{\infty} O\Psi^\star\Psi \, dx, \tag{19.13}$$

where the integral is to be taken along the real axis [9]. Also it was noted that in this form, there is no need to make the conventional operator replacements. The above postulate is equivalent to Born's probability axiom for observables such as position, momentum, energy, etc., and one can show that $\langle O \rangle$ coincides with the corresponding quantum mechanical expectation values. This makes the new scheme equivalent to standard quantum mechanics when averages of dynamical variables are computed.

However, one of the challenges before this complex quantum trajectory representation is to explain the quantum probability axiom. In the dBB approach, there were several attempts to obtain the $\Psi^*\Psi$ probability distribution from more fundamental assumptions [21]. In a recent publication which explores the connection between probability and complex quantum trajectories [22], the probability density to find the particle around some point on the real axis $x = x_{r0}$ was proposed to be

$$\Psi^*\Psi(x_{r0}, 0) \equiv P(x_{r0}) = \mathcal{N} \exp\left(-\frac{2m}{\hbar} \int^{x_{r0}} \dot{x}_i dx_r\right), \qquad (19.14)$$

where the integral is taken along the real line. This expression can be verified by direct differentiation of $|\Psi|^2$. Since it is defined and used only along the real axis, the conservation equation for probability in standard quantum mechanics is valid here also, without any modifications. This possibility of regaining the quantum probability distribution from the velocity field is a unique feature of the complex trajectory formulation. For instance, in the dBB approach, the velocity fields for all bound eigenstates are zero everywhere and it is not possible to obtain a relation between velocity and probability.

At the same time, since we have complex paths, it would be natural to consider the probability for the particle to be in a particular path. In addition, we may consider the probability to find the particle around different points in the same path, which can also be different. Thus it is desirable to extend the probability axiom to the $x_r x_i$-plane and look for the probability of a particle to be in an area $dx_r dx_i$ around some point (x_r, x_i) in the complex plane. Let this quantity be denoted as $\rho(x_r, x_i)dx_r dx_i$.

It is natural to expect that such an extended probability density agrees with Born's rule along the real line. We impose such a boundary condition for $\rho(x_r, x_i)$. Also, it is necessary to see whether probability conservation holds everywhere in the $x_r x_i$-plane. It is ideal if we have an expression for $\rho(x_r, x_i)$, which satisfies these two conditions.

Such an extended probability density was proposed in Ref. [22] and showed, with the aid of the Schrödinger equation, that the conservation equation follows from it. First, it was postulated that if ρ_0, the extended probability density at some point (x_{r0}, x_{i0}), is given, then $\rho(x_r, x_i)$ at another point that lies on the trajectory which passes through (x_{r0}, x_{i0}), is

$$\rho(x_r, x_i) = \rho_0 \exp\left[\frac{-4}{\hbar} \int_{t_0}^{t} \mathrm{Im}\left(\frac{1}{2}m\dot{x}^2 + V(x)\right) dt'\right]. \qquad (19.15)$$

Here, the integral is taken along the trajectory $[x_r(t'), x_i(t')]$. The continuity equation was shown to follow from this axiom by using the extended version of the Schrödinger equation, which gives

$$\text{Im}(E) = \text{Im}\left(\frac{1}{2}m\dot{x}^2 + V(x) + \frac{\hbar}{2i}\frac{\partial \dot{x}}{\partial x}\right) = 0, \tag{19.16}$$

since energy and time are assumed real [9,22]. This helps to write the above definition Equation 19.15 as

$$\rho(x_r, x_i) = \rho_0 \exp\left(-2\int_{t_0}^{t}\frac{\partial \dot{x}_r}{\partial x_r}dt'\right), \tag{19.17}$$

which in turn gives

$$\frac{d\rho}{dt} = -2\frac{\partial \dot{x}_r}{\partial x_r}\rho = -\left(\frac{\partial \dot{x}_r}{\partial x_r} + \frac{\partial \dot{x}_i}{\partial x_i}\right)\rho. \tag{19.18}$$

The last step follows from the analyticity of $\dot{x}$. This leads to the desired continuity equation for the particle, as it moves along: i.e.,

$$\frac{\partial \rho}{\partial t} + \frac{\partial(\rho\dot{x}_r)}{\partial x_r} + \frac{\partial(\rho\dot{x}_i)}{\partial x_i} = 0. \tag{19.19}$$

While evaluating ρ with the help of Equation 19.15 above, one needs to know ρ_0 at (x_{r0}, x_{i0}) and if we choose this point as $(x_{r0}, 0)$, the point of crossing of the trajectory on the real line, then ρ_0 may take the value $P(x_{r0})$ and may be found using Equation 19.14.

To summarize, the proposal in Ref. [22] was that the conserved, extended probability density is

$$\rho(x_r, x_i) \propto \exp\left(-\frac{2m}{\hbar}\int^{x_{r0}}\dot{x}_i dx_r\right)\exp\left[\frac{-4}{\hbar}\int_{t_0}^{t}\text{Im}\left(\frac{1}{2}m\dot{x}^2 + V(x)\right)dt'\right], \tag{19.20}$$

with the integral in the first factor evaluated over the real line and that in the second over the trajectory of the particle.

On the other hand, if we solve the conservation Equation 19.19 for time-independent problems, i.e.,

$$\frac{\partial(\rho\dot{x}_r)}{\partial x_r} + \frac{\partial(\rho\dot{x}_i)}{\partial x_i} = 0, \tag{19.21}$$

with the given boundary condition, it is possible to show that the two methods give identical results. To solve this equation, write $\rho(x_r, x_i) = h(x_r, x_i)f(p)$, where $h(x_r, x_i)$ is some solution of Equation 19.21 and p is some combination of x_r and x_i, whose value remains a constant along its characteristic curves [23]. Substituting this form of ρ into Equation 19.21, the characteristic curves are obtained by integrating the equation

$$\frac{dx_r}{\dot{x}_r} = \frac{dx_i}{\dot{x}_i} \quad \text{or} \quad \frac{dx_i}{dx_r} = \frac{\dot{x}_i}{\dot{x}_r}, \tag{19.22}$$

which is found to be the same as Equation 19.12. This demonstrates the important property that the characteristic curves for the above conservation equation are identical to the complex paths of particles in the present quantum trajectory representation.

We may now find the exact form of $f(p)$ by requiring that $\rho(x_r, 0)$ agrees with the probability $P(x_r)$, which is the boundary condition in this case. Let the integration constant in the above Equation 19.22 be the (real) coordinate of any one point of crossing of the trajectory on the real axis, denoted as x_{r0}. Since the characteristic curves are identical to the complex paths, one can take x_{r0} as the constant p along the characteristic curve and let it be expressed in terms of x_r and x_i. The assumed form for the extended probability distribution ρ may then be written as

$$\rho(x_r, x_i) = h(x_r, x_i)f(x_{r0}).$$ (19.23)

Now we can choose $f(x_{r0})$ subject to the boundary condition. At the point $x = x_{r0}$ at which the curve C crosses the real line, we demand (the boundary condition)

$$\rho(x_{r0}, 0) = h(x_{r0}, 0)f(x_{r0}) = P(x_{r0})$$ (19.24)

and obtain $f(x_{r0})$. Expressing x_{r0} in terms of x_r and x_i in $f(x_{r0})$, Equation 19.23 gives $\rho(x_r, x_i)$.

A word of caution is appropriate here. As we shall see below, there may be instances when the boundary condition overdetermines the problem and we are unable to find a solution. It is observed that this happens in those regions of the complex space where the complex trajectories do not enclose the poles of x, described as "subnets" in Ref. [9].

As an example of the proposed scheme, we consider the $n = 1$ eigenstate of the harmonic oscillator case, in which one uses Equation 19.14 to regain $P(x_r) = \mathcal{N}_1 x_r^2 \exp(-\alpha^2 x_r^2)$. In the next step, $h(x_r, x_i) = (x_r^2 + x_i^2)$ is found to be a solution of the conservation equation 19.21. $f(x_{r0})$ can now be found as

$$f(x_{r0}) \propto e^{-\alpha^2 x_{r0}^2}.$$ (19.25)

The complex paths in this case are given by $(\alpha^2 x_r^2 - \alpha^2 x_i^2 - 1)^2 + 4\alpha^4 x_r^2 x_i^2 = A^2$, a constant. Equating this constant to $(\alpha^2 x_{r0}^2 - 1)^2$, one can obtain $f(x_{r0})$ and also the extended probability density ρ, in terms of x_r and x_i. But while taking square roots, one needs to be careful. It should be noted that for the region containing the subnets in the harmonic oscillator (with $A < 1$ in the present $n = 1$ case), the boundary condition overdetermines the problem, resulting in there being no solution. But for the region outside it, one can write

$$f(x_{r0}) \propto \exp(-\sqrt{(\alpha^2 x_r^2 - \alpha^2 x_i^2 - 1)^2 + 4\alpha^4 x_r^2 x_i^2}),$$ (19.26)

and therefore,

$$\rho(x_r, x_i) \propto (x_r^2 + x_i^2)\exp(-\sqrt{(\alpha^2 x_r^2 - \alpha^2 x_i^2 - 1)^2 + 4\alpha^4 x_r^2 x_i^2})$$ (19.27)

which is plotted in Figure 19.2. It can be seen that the alternative form of $\rho(x_r, x_i)$, in terms of the trajectory integral Equation 19.20, agrees with the solution of this equation everywhere in the complex plane.

It is interesting to note that for constant potentials and for the harmonic oscillator, ρ varies as $|\dot{x}|^{-2}$, as the particle moves along a particular trajectory. This result is not contradictory to the WKB result that $\Psi^*\Psi \propto 1/v_{\text{classical}}$ for constant potentials [24], since $|\dot{x}|$ along a trajectory in our case is not the same as $v_{\text{classical}}$.

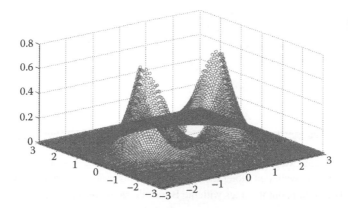

FIGURE 19.2 The extended probability density $\rho(x_r, x_i)$ for the $n = 1$ harmonic oscillator. Here, however, there does not exist a conserved probability for the subnests with $A < 1$ in the interior region. (Reproduced with permission from John, M. V., *Ann. Phys.*, 324, 220, 2009.)

19.4 PROBABILITY DISTRIBUTION FOR THE SUBNESTS

As mentioned above, Equation 19.15 gives a conserved probability along any trajectory in the $x_r x_i$-plane. But we have required that the extended probability agrees with the $\Psi^\star\Psi$ probability along the real line. It is seen that such an agreement is not possible for those trajectories which do not enclose the poles of $\dot{x}$. In the context of solving the conservation equation, it is easy to see that this is due to the boundary condition overdetermining the problem. In the trajectory integral approach, one can explain it as a disagreement of the values of ρ at two consecutive points of crossing of the trajectory on the real axis, with that prescribed by Born's rule; i.e., as a particle trajectory is traversed, if the probability ρ at one point of crossing x_{r0} on the real axis agrees with $P(x_{r0})$, then at the other point, say the point x_{r1} where the trajectory again crosses the real line, the probability calculated according to Equation 19.15 will be different from that of $P(x_{r1})$.

Given this situation, one can ask whether it is possible to find a trajectory integral definition for ρ that can agree with Born's rule (on the real line) in this region, even if it is not conserved in the extended region. Such a definition may be found similar to that in Equation 19.20:

$$\rho(x_r, x_i) \propto \exp\left(-\frac{2m}{\hbar}\int^{x_{r0}} \dot{x}_i \, dx_r\right) \exp\left[\frac{-4}{\hbar}\int_{t_0}^{t} \text{Im}\left(\frac{1}{2}m\dot{x}^2\right) dt'\right]. \quad (19.28)$$

Again the integral in the first factor is evaluated over the real line and that in the second, over the trajectory of the particle. The difference with the definition Equation 19.20 is the absence of the potential term $V(x)$ in the integrand in the second factor. However, for particles moving in zero potential regions, the two definitions coincide. We plot the extended probability density for the $n = 1$ harmonic oscillator in this region in Figure 19.3.

The probability in this region is substantially high, when compared to that in the $A > 1$ region. If this is a general feature for all subnests for large n too, it may

FIGURE 19.3   The extended probability density $\rho(x_r, x_i)$ for the $A < 1$ region of the $n = 1$ harmonic oscillator, evaluated along the trajectories. This probability density does not obey the continuity equation, but it agrees with the Born rule along the real line.

explain how the particles are confined close to the real line in the classical limit. It may also be noted that since the extended probability for $A > 1$ decreases rapidly for all $|x| \to \infty$, there is no difficulty in normalizing it for the extended plane.

19.5   DISCUSSION

In classical wave optics, the square of the amplitude is taken as proportional to the intensity of radiation. While explaining the photoelectric effect, Einstein has ascribed a certain particle nature to photons. Born remarked thus on the origin of his $|\Psi|^2$— probability axiom for material particles, in his Nobel lecture: "... an idea of Einstein gave me the lead. He had tried to make the duality of particles—light quanta or photons—and waves comprehensible by interpreting the square of the optical wave amplitudes as a probability density for the occurrence of photons. This concept could at once be carried over to the ψ-function: $|\psi|^2$ ought to represent the probability density for electrons (or other particles). It was easy to assert this, but how could it be proved?" [25]. The proof Born thought of was phenomenological, based on atomic collision processes. Ever since, Born's probability remained miraculously successful in all microscopic phenomena, though as an axiom.

In this chapter we have attempted to show that the modified de Broglian mechanics, which leads to complex trajectories, is capable of explaining the quantum probability as originating from dynamics itself. It is seen that both the Born probability along the real line and the extended probability in the $x_r x_i$-plane are obtainable in terms of certain line integrals. In the extended case, a conserved probability, which agrees with the boundary condition (Born's rule) along the real axis, is found to exist in most regions. The trajectory integral in this case is over the imaginary part of $(1/2)m\dot{x}^2 + V(x)$. For other regions (such as the subnests in the $n = 1$ harmonic oscillator case), an alternative definition for the probability satisfies the boundary condition on the real axis, though it is not a conserved one. The integrand here is simply $(1/2)m\dot{x}^2$. Since the latter is defined only for the subnests which do not enclose the poles of $\dot{x}$, there is no difficulty in normalizing the combined probability for the entire extended plane.

There were some other proposals for computing probability in the extended complex plane. One such attempt [26] is to define $\rho(x) = \bar{\Psi}(x)\Psi(x)$ where $\bar{\Psi}(x) \equiv \Psi^*(x^*)$, with x complex. The complexified flux is chosen as $j(x,t) = v(x,t)\rho(x,t) = -(i\hbar/m)\Psi^*(x^*)\Psi'$ where $v(x,t) \equiv \dot{x}$ is given by Equation 19.10 and the prime denotes spatial differentiation. With the help of the time-dependent Schrödinger equation, the author shows that, in general, $\partial\rho/\partial t \neq j'(x,t)$. This arguably leads to nonconservation of probability along trajectories. But we should remember that this negative result is based on the choices made in [26] for the probability density and flux. Moreover, this definition leads to a complex probability off the real axis, which is undesirable. Another approach is to define the probability as $\Psi^*(x)\Psi(x)$ itself [27,28]. Though this has the advantage of being real everywhere, it is not shown to obey any continuity equation anywhere. Nor is it generally a normalizable probability in the extended plane.

Notwithstanding a host of other issues to be solved for the new mechanics, the fact that it can account for Born's rule in terms of the velocity field raises our hopes for a deeper understanding of quantum probability.

BIBLIOGRAPHY

1. H. Goldstein, *Classical Mechanics*, Addison-Wesley, Reading, MA, 1980.
2. L. de Broglie, Ph.D. Thesis, University of Paris, 1924.
3. L. de Broglie, *J. Phys. Rad., 6ᵉ serie*, t. 8, (1927) 225.
4. G. Bacciagaluppi, A. Valentini, *Quantum Theory at the Crossroads*, Cambridge University Press, Cambridge, 2009; quant-ph/0609184v1 (2006).
5. D. Bohm, *Phys. Rev.* 85 (2), 166 (1952) 180.
6. E.R. Floyd, *Phys. Rev. D* 26 (1982) 1339.
7. A. Faraggi, M. Matone, *Phys. Lett. B* 450 (1999) 34.
8. R. Carroll, *J. Can. Phys.* 77 (1999) 319.
9. M.V. John, *Found. Phys. Lett.* 15 (2002) 329; quant-ph/0102087 (2001).
10. C.D. Yang, *Phys. Lett. A* 372 (2008) 6240.
11. G. Wentzel, *Z. Phys.* 38 (1926) 518; W. Pauli, in: H. Geiger, K. Scheel (Ed.), *Handbuch der Physik*, 2nd ed., Vol. 24, part 1, Springer-Verlag, Berlin, 1933, pp. 83–272; P.A.M. Dirac, *The Principles of Quantum Mechanics*, Oxford University Press, London, 1958.
12. R.A. Leacock, M.J. Padgett, *Phys. Rev. Lett.* 50 (1983) 3; R.A. Leacock, M.J. Padgett, *Phys. Rev. D* 28, (1983) 2491.
13. R.S. Bhalla, A.K. Kapoor, P.K. Panigrahi, *Am. J. Phys.* 65 (1997) 1187; R.S. Bhalla, A.K. Kapoor, P.K. Panigrahi, *Mod. Phys. Lett. A* 12 (1997) 295.
14. A.S. Sanz, S. Miret-Artés, *J. Chem. Phys.* 127 (2007) 197101.
15. Y. Goldfarb, I. Degani, D.J. Tannor, *J. Chem. Phys.* 127 (2007) 197102.
16. C.-D. Yang, *Ann. Phys. (N.Y.)* 319 (2005) 339; C.-D. Yang, *Int. J. Quantum Chem.* 106 (2006) 1620; C.-D. Yang, *Ann. Phys. (N.Y.)* 319 (2005) 444; C.-D. Yang, *Chaos, Solitons Fractals* 30 (2006) 342.
17. C.-C. Chou, R.E. Wyatt, *Phys. Rev. E* 74 (2006) 066702; C.-C. Chou, R.E. Wyatt, *J. Chem. Phys.* 125 (2007) 174103.
18. R.E. Wyatt, *Quantum Dynamics with Trajectories: Introduction to Quantum Hydrodynamics*, Springer, New York, 2005.
19. Y. Goldfarb, I. Degani, D.J. Tannor, *J. Chem. Phys.* 125 (2006) 231103.

20. A.S. Sanz, S. Miret-Artés, *Chem. Phys. Lett.* 445 (2007) 350.
21. D. Bohm, B.J. Hiley, *The Undivided Universe*, Routledge, London and New York, 1993;
    P. Holland, *The Quantum Theory of Motion*, Cambridge University Press, Cambridge,
    1993.
22. M.V. John, *Ann. Phys.* 324 (2009) 220; arXiv:0809.5101 (2008).
23. K.F. Riley, M.P. Hobson, S.J. Bence, *Mathematical Methods for Physics and Engineering*,
    Cambridge University Press, Cambridge, 1997, p. 526.
24. J.J. Sakurai, *Modern Quantum Mechanics*, Addison Wesley Longman, Singapore, 1994
    p. 104.
25. M. Born, Nobel lecture, 1954.
26. B. Poirier, *Phys. Rev. A* 77 (2008) 022114.
27. C.C. Chou, R.E. Wyatt, *Phys. Lett. A* 373 (2009) 1811.
28. C. D. Yang, *Chaos, Solitons Fractals* 42 (2009) 453.

20 Types of Trajectory Guided Grids of Coherent States for Quantum Propagation

Dmitrii V. Shalashilin

CONTENTS

20.1 INTRODUCTION

Several methods have been suggested in the literature, which employ various trajectory guided grids of frozen Gaussian wavepackets as a basis set for quantum propagation. They exploit the same idea illustrated in Figure 20.1a, that a grid can follow the wavefunction. Therefore a basis does not have to cover the whole Hilbert space of the system, as the one shown in Figure 20.1b. Among the methods considered in this chapter are the method of variational Multiconfigurational Gaussians (vMCG) and the Gaussian Multiconfigurational time-dependent Hartree (G-MCTDH) technique [1], the method of Coupled Coherent States (CCS) [2] and the recently developed Multiconfigurational Ehrenfest approach (MCE) [3], which all differ in the exact way their grids are guided. The goal of this chapter is to present all four techniques from the same perspective and compare their mathematical structure.

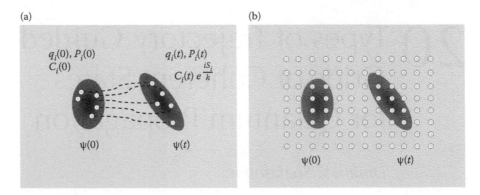

(a)

$q_i(0), P_i(0)$
$C_i(0)$

$q_i(t), P_i(t)$
$C_i(t)\, e^{\frac{iS_i}{\hbar}}$

$\psi(0)$ $\psi(t)$

(b)

$\psi(0)$ $\psi(t)$

FIGURE 20.1 Random trajectory guided grid (a) versus regular basis covering the whole dynamically important space (b).

20.2 THEORY

20.2.1 VARIATIONAL PRINCIPLE

First we consider a general derivation of the quantum equations of motion. It is well known that the time dependence of a wavefunction $\Psi(\alpha_1, \alpha_2, \ldots, \alpha_n)$ is simply that of its parameters, and the equations for the "trajectories" $\alpha_j(t)$ can be worked out from the variational principle

$$\delta S = 0 \tag{20.1}$$

by minimizing the action $S = \int L\, dt$ of the Lagrangian

$$L(\dot{\alpha}_1^*, \ldots, \dot{\alpha}_n^*, \alpha_1^*, \ldots, \alpha_n^*, \dot{\alpha}_1, \ldots, \dot{\alpha}_n, \alpha_1, \ldots, \alpha_n)$$

$$= \left\langle \Psi(\alpha_1^*, \alpha_2^*, \ldots, \alpha_n^*) \left| i\frac{\hat{\partial}}{\partial t} - \hat{H} \right| \Psi(\alpha_1, \alpha_2, \ldots, \alpha_n) \right\rangle. \tag{20.2}$$

The variational principle Equation 20.1 straightforwardly leads to the Lagrange equations of motion

$$\frac{\partial L}{\partial \alpha} - \frac{d}{dt}\frac{\partial L}{\partial \dot{\alpha}} = 0, \tag{20.3}$$

and an adjoint equation for the complex conjugate, where α is the complex vector of wavefunction parameters $\alpha = \{\alpha_1, \alpha_2, \ldots, \alpha_n\}$.

After introduction of the momentum $\mathbf{p} = \{p_1, p_2, \ldots, p_n\}$

$$p_l(\alpha_1^*, \ldots, \alpha_n^*, \alpha_1, \ldots, \alpha_n) = 2\frac{\partial L}{\partial \dot{\alpha}} = i\left\langle \Psi\left(\alpha_1^*, \ldots, \alpha_n^*\right) \left| \frac{\partial}{\partial \alpha_l} \Psi(\alpha_1, \ldots, \alpha_n)\right.\right\rangle \tag{20.4}$$

the Lagrange Equation 20.3 can be written in the Hamilton form

$$\dot{\alpha}_l = \frac{\partial\langle H\rangle}{\partial p_{\alpha_l}} = \sum_j \frac{\partial\langle H\rangle}{\partial \alpha_j^*}\frac{\partial \alpha_j^*}{\partial p_{\alpha_l}} \tag{20.5}$$

where the Hamiltonian

$$\langle \hat{H} \rangle = \langle \Psi(\alpha_1^*, \alpha_2^*, \ldots, \alpha_n^*) | \hat{H} | \Psi(\alpha_1, \alpha_2, \ldots, \alpha_n) \rangle \tag{20.6}$$

is not an explicit function of the momenta (Equation 20.6). This is the reason why the derivative of the Hamiltonian with respect to the momentum is expressed through the derivatives with respect to α and the unknown matrix $\partial \alpha_j^* / \partial p_{\alpha_l} = (D^{-1})_{lj}$ which must be obtained by inverting

$$D_{jl} = \frac{\partial p_{\alpha_l}}{\partial \alpha_j^*} = i \left\langle \frac{\partial}{\partial \alpha_j^*} \Psi(\alpha_1^*, \alpha_2^*, \ldots, \alpha_n^*) \middle| \frac{\partial}{\partial \alpha_l} \Psi(\alpha_1, \alpha_2, \ldots, \alpha_n) \right\rangle. \tag{20.7}$$

In other words Hamilton's equations can be written as a system of linear equations for the derivatives of the wavefunction parameters $\dot{\alpha}$

$$\mathbf{D}\dot{\alpha} = \frac{\partial \langle H \rangle}{\partial \alpha}. \tag{20.8}$$

Equations 20.3 and 20.8 are the main result of the work of Kramer and Saraceno [4]. Although variational principle is well known and its applications to the grids of Gaussian wavepackets can be traced back to the work of Metiu and coworkers [5], the approach of [4] is perhaps the most elegant way to derive the equations of quantum mechanics.

In the simplest case of the wavefunction presented as a superposition of orthonormal basis functions

$$|\Psi\rangle = \sum_{i=1,N} a_i |\varphi_i\rangle, \tag{20.9}$$

so that $\alpha = \mathbf{a} = \{a_1, \ldots, a_N\}$, the matrix $\mathbf{D}$ is simply a unit matrix, and the Kramer–Saraceno Equations 20.3, 20.8 yield the standard Schrödinger equation in the form of coupled equations for the amplitudes

$$i\dot{\mathbf{a}} = \mathbf{Ha} \tag{20.10}$$

where $\mathbf{H}$ is the matrix of the Hamiltonian. However, in many cases, for example in the case of Gaussian coherent state basis sets, the $\mathbf{D}$-matrix is not diagonal and the system of linear Equation 20.8 for the derivatives has to be resolved at each step of the propagation.

20.2.2 Variational Multiconfigurational Gaussians (vMCG)

Although the equations of time-dependent quantum mechanics for the wavefunction Equation 20.9, which uses a simple orthonormal basis, are very simple, Equation 20.10 can be used for systems with only a few degrees of freedom (DOF) because the size of the basis set N grows exponentially with the number of DOF. For multidimensional systems trajectory guided random basis sets can be more efficient. The main idea of the vMCG technique, which is an all-Gaussian version of the more general G-MCTDH method [1], is to apply the variational principle to the wavefunction which is represented by a superposition of frozen Gaussian wavepackets also called coherent states (CS)

$$|\Psi\rangle = \sum_{l=1,N} a_l(t)|\mathbf{z}_l(t)\rangle. \tag{20.11}$$

The coherent state represents a phase space point "dressed" with the finite width of the wavepacket determined by the parameter $\gamma = m\omega/\hbar$. For notational simplicity everywhere in this chapter it is assumed that the units are such that the Planck constant $\hbar$ and the width parameter γ are equal to one. A multidimensional CS includes coordinates and momenta of all M DOF $|\mathbf{z}_l\rangle = |z_l^{(1)}\rangle \cdots |z_l^{(M)}\rangle$. In the coordinate representation coherent states are Gaussian wavepackets

$$\langle x|z\rangle = \left(\frac{\gamma}{\pi}\right)^{\frac{1}{4}} \exp\left(-\frac{\gamma}{2}(x-q)^2 + \frac{i}{\hbar}p(x-q) + \frac{ipq}{2\hbar}\right) \tag{20.12}$$

where q is the position of the wavepacket, its classical momentum p is given by

$$z = \frac{\gamma^{\frac{1}{2}}q + i\hbar^{-1}\gamma^{-\frac{1}{2}}p}{\sqrt{2}} \quad z^* = \frac{\gamma^{\frac{1}{2}}q - i\hbar^{-1}\gamma^{-\frac{1}{2}}p}{\sqrt{2}} \tag{20.13}$$

and Equation 20.12, written for a one-dimensional CS, can easily be generalized to many dimensions.

An important property of the wavefunction Equation 20.11 is that the basis CSs are not orthogonal. The overlap is given as

$$\langle \mathbf{z}_l|\mathbf{z}_j\rangle = \Omega_{lj} = \exp\left(\mathbf{z}_l^*\mathbf{z}_j - \frac{\mathbf{z}_l^*\mathbf{z}_l}{2} - \frac{\mathbf{z}_j^*\mathbf{z}_j}{2}\right). \tag{20.14}$$

If $|\mathbf{z}\rangle$ is a multidimensional CS then the parameters of the M-dimensional wavefunction Equation 20.11 are the N amplitudes $a_{i=1,\dots,N}$ and the $N \times M$ complex parameters $z_{l=1\cdots N}^{(m=1\cdots M)}$ of the CS basis functions $|\mathbf{z}_l\rangle = |z_l^{(1)}\rangle \cdots |z_l^{(M)}\rangle$, i.e., $\alpha = \{a_1, \dots, a_N, z_1^{(1)}, \dots z_N^{(1)}, \dots, z_1^{(M)}, \dots z_N^{(M)}\}$. As has been shown in Ref. [6] the equations of vMCG are those of Kramer and Saraceno theory applied to the wavefunction Equation 20.11. Due to the apparent complexity and length of the vMCG equations they are not given in the main body of this chapter but are presented in the Appendix instead. However, the vMCG theory simply relies on the Kramer–Saraceno system of linear equations for the derivatives of the wavefunction parameters. Numerical solution of the vMCG equations is quite demanding. It requires finding the derivatives of the wavefunction parameters by resolving the linear system of size $N \times M + N$ at each time step.

20.2.3 VARIATIONAL SINGLE CONFIGURATIONAL GAUSSIAN (vSCG)

There is one case, however, when the solution of the vMCG equations can easily be obtained, which is when the wavefunction Equation 20.11 contains only a single configuration ($N = 1$)

$$|\Psi\rangle = a|\mathbf{z}\rangle. \tag{20.15}$$

The equations for the derivatives of z and a become

$$\dot{\mathbf{z}} = -i \frac{\partial \langle \mathbf{z}|\hat{H}|\mathbf{z}\rangle}{\partial \mathbf{z}^*} \tag{20.16}$$

and

$$-i\dot{a} = \left[i \frac{\dot{\mathbf{z}}\mathbf{z}^* - \dot{\mathbf{z}}^*\mathbf{z}}{2} - \langle \mathbf{z}|\hat{H}|\mathbf{z}\rangle \right] a. \tag{20.17}$$

The solution of Equation 20.16 yields a trajectory $\mathbf{z}(t)$ in the Hamiltonian which is quantum averaged over the wavepacket $\mathbf{z}$. The solution of Equation 20.17 is very simple. It is made up of a constant prefactor multiplied by the exponent of the classical action

$$a = a(0) \exp(iS) \tag{20.18}$$

where

$$S = \int \left[i \frac{\dot{\mathbf{z}}\mathbf{z}^* - \dot{\mathbf{z}}^*\mathbf{z}}{2} - \langle \mathbf{z}|\hat{H}|\mathbf{z}\rangle \right] dt. \tag{20.19}$$

The fact that the evolution of a single frozen Gaussian CS is given by the trajectory (Equation 20.16) and the amplitude Equation 20.18 as a product of a constant prefactor and the exponent of the classical action along the trajectory, has been well known since the famous work of Heller on the frozen Gaussian approximation [7].

20.2.4 Coupled Coherent States (CCS)

The method of CCS [2] attempts to decrease the computational cost of vMCG. The CCS technique uses exactly the same form of the wavefunction Equation 20.11 as vMCG with the difference that the CCS method does not find the trajectories of the coherent states z from the full variational principle. Instead the trajectories are taken in the form Equation 20.16 which is "optimal" for a single CS wavefunction Equation 20.15. Only the equations for the amplitudes are obtained from the full variational principle.

In the CCS approach the equations for the amplitudes

$$i\mathbf{\Omega E\dot{d}} = \mathbf{\Delta^2 H'Ed} \tag{20.20}$$

are usually written for the preexponential factor so that in the expression for the wavefunction the oscillating exponent is factored out of the amplitude

$$|\Psi(t)\rangle = \sum_j d_j(t) e^{iS_j(t)} |\mathbf{z}_j(t)\rangle. \tag{20.21}$$

In Equation 20.20 $\mathbf{\Omega}$ is the overlap matrix and $\mathbf{E}$ is the diagonal matrix of classical action exponents $\mathbf{E}_{lj} = e^{iS_l}\delta_{lj}$ and $\mathbf{\Delta^2 H'}$ is a matrix which includes the Hamiltonian. See Ref. [2] and the Appendix for more details.

Assuming the trajectories, instead of getting them from the full variational principle, results in huge computational savings. Only the system of N equations for the derivatives of the coupled amplitudes has to be solved comparably to the $N \times M + N$ size of the linear system in the vMCG technique. As a result CCS can afford a much larger basis set. Making assumptions about the trajectories is not an approximation but simply a reasonable choice of time-dependent basis set because the variational principle is still applied to the amplitudes of the basis CSs. The price which CCS pays for its efficiency is that the trajectories of the basis CS are not as flexible as in vMCG. Nevertheless as has been shown in many instances the CCS method is capable of accurate treatment of quantum multidimensional systems.

20.2.5 THE GAUSSIAN MULTICONFIGURATIONAL TIME-DEPENDENT HARTREE METHOD

Many problems of quantum mechanics naturally require regular basis sets for a few of the most important DOF, but a grid of randomly selected trajectory guided Gaussians for the rest of the DOF. For example, a description of nonadiabatic transitions between two or more potential energy surfaces can be done with the help of the following multiconfigurational wavefunction anzats:

$$|\Psi(t)\rangle = \sum_{k=1,N} (a_{1k}(t)|\varphi_1\rangle + a_{2k}(t)|\varphi_2\rangle)|\mathbf{z}_k(t)\rangle. \tag{20.22}$$

In each configuration $|\varphi_1\rangle$ and $|\varphi_2\rangle$ represent a regular basis for the two electronic states and $|\mathbf{z}_k(t)\rangle$ is a CS describing the nuclear DOF. Another example is the so-called system-bath problems where a small subsystem is treated on a regular basis set or grid and a random grid of Gaussians is used to describe a large number of the harmonic bath modes. Applying the Kramer and Saraceno approach in its Lagrange Equation 20.3 or Hamiltonian Equation 20.8 form to the parameters of the wavefunction Equation 20.22 would give the equations of the G-MCTDH method [3] in their simplest form, which again can be written as a system of linear equations for the time-dependent derivatives of the wavefunction parameters.* Fully variational G-MCTDH equations for the vector $\boldsymbol{\alpha} = (a_{11}, \ldots, a_{1N}, a_{21}, \ldots, a_{2N}, z_{11}, \ldots, z_{1M}, z_{N1}, \ldots, z_{NM})$ of the parameters of the wavefunction Equation 20.20 are given in the Appendix.

20.2.6 THE EHRENFEST METHOD

The equations of G-MCTDH simplify when the wavefunction Equation 20.20 includes only one configuration

$$|\Psi(t)\rangle = (a_1(t)|\varphi_1\rangle + a_2(t)|\varphi_2\rangle + \cdots)|\mathbf{z}(t)\rangle. \tag{20.23}$$

As has been shown in Ref. [4] the equations for the parameters of the wavefunction Equation 20.21 become those of the Ehrenfest dynamics [8]

---

* Although G-MCTDH allows more sophisticated ways of presenting the "system" part of the wavefunction, for simplicity only the form Equation 20.22 is considered here. Generalization is not too difficult.

$$\dot{a}_1 = i \left[i \frac{\dot{z}z^* - \dot{z}^*z}{2} - H_{11} \right] a_1 - iH_{12}a_2 \qquad (20.24)$$

$$\dot{a}_2 = i \left[i \frac{\dot{z}z^* - \dot{z}^*z}{2} - H_{22} \right] a_2 - iH_{21}a_1 \qquad (20.25)$$

for the amplitudes of "system" states along the trajectory of the "bath"

$$i\dot{z} = -\frac{\partial H^{\text{Ehr}}}{\partial z^*} \qquad (20.26)$$

where

$$H^{\text{Ehr}} = \langle \Psi | \hat{H} | \Psi \rangle = \frac{H_{11}a_1^*a_1 + H_{22}a_2^*a_2 + H_{12}a_1^*a_2 + H_{21}a_2^*a_1}{a_1^*a_1 + a_2^*a_2} \qquad (20.27)$$

is the Ehrenfest Hamiltonian, which is the Hamiltonian of the "bath" averaged over the quantum wavefunction of the "system."

Similar to the CCS theory it is convenient to write the solution for the amplitudes as a product of the oscillating exponent of the classical action and a relatively smooth preexponential factor $a_{1,2} = d_{1,2} \exp(iS_{1,2})$, for which Equations 20.24 become

$$\dot{d}_1 = -iH_{12}d_2 \exp(i(S_2 - S_1)) \qquad (20.28)$$

$$\dot{d}_2 = -iH_{21}d_1 \exp(i(S_1 - S_2)). \qquad (20.29)$$

Equations 20.24 through 20.27 represent the well-known Ehrenfest approach [8] which treats the system with a set of basis functions $|\varphi_1\rangle, |\varphi_2\rangle, \ldots$ and the bath with a single trajectory $|z(t)\rangle$.

20.2.7 The Multiconfigurational Ehrenfest Method

In Ref. [4] a method called Multiconfigurational Ehrenfest dynamics has been suggested. Similar to the G-MCTDH technique it uses a formally exact anzats Equation 20.22, which converges to the exact wavefunction, but the difference is that the CS positions $z(t)$ are not considered as variational parameters. Instead the trajectories $z(t)$ are chosen from Equations 20.25, 20.26 of the Ehrenfest method. Only the linear system Equation A.3 for the time derivatives of the amplitudes is obtained from the full variational principle and should be resolved. This results in a huge gain in computational efficiency. Only the system of $2N$ equations for the derivatives of the coupled amplitudes has to be solved comparably to the $N \times M + 2N$ size of the system Equations 20.3, 20.8 in the G-MCTDH technique. Just like in the case of the CCS method, making assumptions about the trajectories is not an approximation but simply a reasonable choice of a time-dependent basis set because the variational principle is still applied to the amplitudes. Reference [4] reports simulations of the spin-boson model with up to 2000 DOF, which is among the biggest quantum systems for which the Schrödinger equation has ever been solved numerically without approximations.

20.3 CONCLUSIONS

This chapter has shown that

(1) The equations of the vMCG and G-MCTDH techniques can be obtained as those of Kramer–Saraceno theory for multiconfigurational wavefunction.
(2) Heller's frozen Gaussian and Ehrenfest dynamics also follow from the Kramer–Saraceno approach applied to a single configuration wavefunction only.
(3) The method of CCS combines the trajectories of HFG with the vMCG equations for the amplitudes.
(4) The MCE technique uses Ehrenfest trajectories with G-MCTDH coupled equations for the amplitudes
(5) The advantage of CCS and MCE is that they reduce dramatically the number of linear equations for the time derivatives of the wavefunction parameters.
(6) Using "approximate" trajectories instead of those yielded by a fully variational approach is not an approximation but simply a choice of basis set. The choice of CCS and MCE is efficient but may be less flexible than that of the vMCG and G-MCTDH methods.

APPENDIX: VARIATIONAL PRINCIPLE AND THE EQUATIONS OF G-MCTDH

The variational principle has been applied to the derivation of equations for many types of wavefunction parametrization. For example, recently we analyzed the equations of motions for the parameters of the wavefunctions represented as a superposition of Gaussian coherent states [6]. To obtain the Lagrangian Equation 20.2 one should notice that

$$\frac{\partial}{\partial t} = \frac{1}{2}\left(\overrightarrow{\frac{\partial}{\partial t}} - \overleftarrow{\frac{\partial}{\partial t}}\right)$$

where the operators

$$\overleftarrow{\frac{\partial}{\partial t}} \quad \text{and} \quad \overrightarrow{\frac{\partial}{\partial t}}$$

act on the left and on the right. Then

$$a_l^*\left\langle \mathbf{z}_l \left| i\frac{\hat{\partial}}{\partial t} \right| \mathbf{z}_j \right\rangle a_j = a_l^*\left\langle \mathbf{z}_l \left| \frac{i}{2}\left(\overrightarrow{\frac{\partial}{\partial t}} - \overleftarrow{\frac{\partial}{\partial t}}\right) \right| \mathbf{z}_j \right\rangle a_j \qquad (A.1)$$

$$= \frac{i}{2}[a_l * \dot{a}_j - a_j\dot{a}_l*]\langle \mathbf{z}_l | \mathbf{z}_j \rangle$$

$$+ \frac{i}{2}\left[\left(\mathbf{z}_l * \dot{\mathbf{z}}_j - \frac{\dot{\mathbf{z}}_j\mathbf{z}_j*}{2} - \frac{\mathbf{z}_j\dot{\mathbf{z}}_j*}{2}\right)\right.$$

$$\left. - \left(\mathbf{z}_j\dot{\mathbf{z}}_l * - \frac{\mathbf{z}_l\dot{\mathbf{z}}_l*}{2} - \frac{\dot{\mathbf{z}}_l\mathbf{z}_l*}{2}\right)\right]a_l * a_j\langle \mathbf{z}_l | \mathbf{z}_j \rangle.$$

To perform differentiation over time in Equation A.1 one must use the formula Equation 20.14 for the CS overlap. Then the Lagrangian becomes

$$
\begin{aligned}
L &= \left\langle \Psi \left| i\frac{\hat{\partial}}{\partial t} - \hat{H} \right| \Psi \right\rangle \\
&= \sum_{kj} \frac{i}{2} [a_{1k}^* \dot{a}_{1j} - \dot{a}_{1k}^* a_{1j}] \langle \mathbf{z}_k | \mathbf{z}_j \rangle \\
&+ \sum_{kj} \frac{i}{2} \left[\left(\mathbf{z}_k^* \dot{\mathbf{z}}_j - \frac{\dot{\mathbf{z}}_j \mathbf{z}_j^*}{2} - \frac{\mathbf{z}_j \dot{\mathbf{z}}_j^*}{2} \right) - \left(\mathbf{z}_j \dot{\mathbf{z}}_k^* - \frac{\mathbf{z}_k \dot{\mathbf{z}}_k^*}{2} - \frac{\dot{\mathbf{z}}_k \mathbf{z}_k^*}{2} \right) \right] a_{1k}^* a_{1j} \langle \mathbf{z}_k | \mathbf{z}_j \rangle \\
&+ \sum_{kj} \frac{i}{2} [a_{2k}^* \dot{a}_{2j} - \dot{a}_{2k} \dot{a}_{2k}^*] \langle \mathbf{z}_k | \mathbf{z}_j \rangle \\
&+ \sum_{kj} \frac{i}{2} \left[\left(\mathbf{z}_k^* \dot{\mathbf{z}}_j - \frac{\dot{\mathbf{z}}_j \mathbf{z}_j^*}{2} - \frac{\mathbf{z}_j \dot{\mathbf{z}}_j^*}{2} \right) - \left(\mathbf{z}_j \dot{\mathbf{z}}_k^* - \frac{\mathbf{z}_k \dot{\mathbf{z}}_k^*}{2} - \frac{\dot{\mathbf{z}}_k \mathbf{z}_k^*}{2} \right) \right] a_{2k}^* a_{2j} \langle \mathbf{z}_k | \mathbf{z}_j \rangle \\
&+ \sum_{kj} \langle \mathbf{z}_k | \hat{H}_{11} | \mathbf{z}_j \rangle a_{1k}^* a_{1j} - \sum_{kj} \langle \mathbf{z}_k | \hat{H}_{22} | \mathbf{z}_j \rangle a_{2k}^* a_{2j} \\
&- \sum_{kj} \langle \mathbf{z}_k | \hat{H}_{12} | \mathbf{z}_j \rangle a_{1k}^* a_{2j} - \sum_{kj} \langle \mathbf{z}_k | \hat{H}_{21} | \mathbf{z}_j \rangle a_{2k}^* a_{2j}.
\end{aligned}
\tag{A.2}
$$

The Lagrange equations for the amplitudes are

$$
\sum_j i\dot{a}_{1j} \langle \mathbf{z}_k | \mathbf{z}_j \rangle - \langle \mathbf{z}_k | \hat{H}_{11} | \mathbf{z}_j \rangle a_{1j} - \langle \mathbf{z}_k | \hat{H}_{12} | \mathbf{z}_j \rangle a_{2j}
$$

$$
+ i \left[(\mathbf{z}_k^* - \mathbf{z}_j^*) \dot{\mathbf{z}}_j + \frac{\dot{\mathbf{z}}_j \mathbf{z}_j^*}{2} - \frac{\mathbf{z}_j \dot{\mathbf{z}}_j^*}{2} \right] \langle \mathbf{z}_k | \mathbf{z}_j \rangle a_{1j} = 0
\tag{A.3}
$$

and similarly for a_2. It is convenient to modify Equation A.2 by writing the amplitudes as a product of the oscillating exponent of the classical action and a smooth preexponential factor $a_{1j}(t) = d_{1j}(t)e^{iS_{1j}(t)}$ and $a_{2j}(t) = d_{2j}(t)e^{iS_{12}(t)}$ and using the equations for d instead of Equation A.2, which are

$$
\sum_j i\dot{d}_{1j} e^{iS_{1j}} \langle \mathbf{z}_k | \mathbf{z}_j \rangle = \sum_j \Delta^2 H_{11}'(\mathbf{z}_i^*, \mathbf{z}_j) d_{1j} e^{iS_{1j}} + \sum_j \langle \mathbf{z}_k | \hat{H}_{12} | \mathbf{z}_j \rangle d_{2j} e^{iS_{2j}} \tag{A.4}
$$

and similarly for d_{2j}. In Equation A.4

$$
\Delta^2 H_{11}' = [\langle \mathbf{z}_k | \hat{H}_{11} | \mathbf{z}_j \rangle - \langle \mathbf{z}_k | \mathbf{z}_j \rangle \langle \mathbf{z}_j | \hat{H}_{11} | \mathbf{z}_j \rangle - i \langle \mathbf{z}_k | \mathbf{z}_j \rangle (\mathbf{z}_k^* - \mathbf{z}_j^*) \dot{\mathbf{z}}_j] \tag{A.5}
$$

are the elements of the matrix $\mathbf{\Delta^2 H}'$ in Equation 20.20. The variation of z parameters yields

$$a_{1l}^* \sum_j \left\{ a_{1j} \left[i\dot{\mathbf{z}}_j - \frac{\partial H_{11}(\mathbf{z}_l^*, \mathbf{z}_j)}{\partial \mathbf{z}_l^*} \right] + a_{2j} \left[-\frac{\partial H_{12}(\mathbf{z}_l^*, \mathbf{z}_j)}{\partial \mathbf{z}_l^*} \right] \right\} \langle \mathbf{z}_l | \mathbf{z}_j \rangle$$

$$+ a_{1l}^* \sum_j i\dot{a}_{1j} (\mathbf{z}_j - \mathbf{z}_l) \langle \mathbf{z}_l | \mathbf{z}_j \rangle$$

$$+ a_{1l}^* \sum_j a_{1j} (\mathbf{z}_j - \mathbf{z}_l) \langle \mathbf{z}_l | \mathbf{z}_j \rangle \left\{ i \left((\mathbf{z}_l^* - \mathbf{z}_j^*) \dot{\mathbf{z}}_j + \frac{\dot{\mathbf{z}}_j \mathbf{z}_j^*}{2} - \frac{\mathbf{z}_j \dot{\mathbf{z}}_j^*}{2} \right) - H_{11}(\mathbf{z}_l^*, \mathbf{z}_j) \right\}$$

$$+ a_{2l}^* \sum_j \left\{ a_{2j} \left[i\dot{\mathbf{z}}_j - \frac{\partial H_{22}(\mathbf{z}_l^*, \mathbf{z}_j)}{\partial \mathbf{z}_l^*} \right] + a_{1j} \left[-\frac{\partial H_{21}(\mathbf{z}_l^*, \mathbf{z}_j)}{\partial \mathbf{z}_l^*} \right] \right\} \langle \mathbf{z}_l | \mathbf{z}_j \rangle$$

$$+ a_{2l}^* \sum_j i\dot{a}_{2j} (\mathbf{z}_j - \mathbf{z}_l) \langle \mathbf{z}_l | \mathbf{z}_j \rangle$$

$$+ a_{2l}^* \sum_j a_{2j} (\mathbf{z}_j - \mathbf{z}_l) \langle \mathbf{z}_l | \mathbf{z}_j \rangle$$

$$\left\{ i \left((\mathbf{z}_l^* - \mathbf{z}_j^*) \dot{\mathbf{z}}_j + \frac{\dot{\mathbf{z}}_j \mathbf{z}_j^*}{2} - \frac{\mathbf{z}_j \dot{\mathbf{z}}_j^*}{2} \right) - H_{22}(\mathbf{z}_l^*, \mathbf{z}_j) \right\} = 0. \tag{A.6}$$

Although they look complicated, Equations A.6 are simply a system of linear equations for the derivatives of $\mathbf{z}$. Together with Equations A.3 for the derivatives of the amplitudes they constitute the system of Kramer–Saraceno equations $\mathbf{D}\dot{\alpha} = \partial \langle H \rangle / \partial \alpha$ for the full variation of parameters. They are equivalent to those of the G-MCTDH method also based on the full variational principle. Solving the full system Equations A.3, A.6 for $2N$ amplitudes together with the $N \times M$ parameters z is an expensive computational task. The central idea of the MCE approach is to use predetermined trajectories which are close to the solution of Equation A.6. If the coherent states $|\mathbf{z}_l\rangle$ and $|\mathbf{z}_j\rangle$ are far away from each other then their overlap is small and all nondiagonal terms in Equation A.6 can be neglected. Then Equation A.6 becomes that of the individual CS Equation 20.23.

MCE uses trajectories other than those yielded by the full variational principle, but it is still a fully quantum approach, which is formally exact and can converge to accurate quantum result. This proves that the choice of Ehrenfest trajectories to guide the grid is a good one.

The vMCG equations are those of G-MCTDH Equation A.3, Equation A.6 but for a single function $|\varphi_1\rangle$ and therefore can be obtained from Equation A.3, Equation A.6 simply by dropping all terms containing the index "2". CCS is obtained from MCE in a similar fashion.

BIBLIOGRAPHY

1. I. Burghardt, H.D. Meyer, and L.S. Cederbaum, *J. Chem. Phys.*, **111**, 2927 (1999); G.A. Worth and I. Burghardt, *Chem. Phys. Lett.*, **368**, 502 (2003); I. Burghardt, M. Nest, and G.A. Worth, *J. Chem. Phys.*, **119**, 5364 (2003); I. Burghardt, K. Giri, and G.A. Worth, *J. Chem. Phys.*, 129 174101 (2008).

2. D.V. Shalashilin and M.S. Child, *J. Chem. Phys.*, **113**, 10028 (2000); D.V. Shalashilin and M.S. Child, *J. Chem. Phys.*, **114**, 9296 (2001); D.V. Shalashilin and M.S. Child, *J. Chem. Phys.*, **115**, 5367 (2001); D.V. Shalashilin and M.S. Child, *J. Chem. Phys.*, **121**, 3563 (2004); P.A.J. Sherratt, D.V. Shalashilin, and M.S. Child, *Chem. Phys.*, **322**, 127 (2006); D.V. Shalashilin and M.S. Child, *Chem. Phys.*, **304**, 103 (2004).
3. D.V. Shalashilin, *J. Chem. Phys.*, **130** 244101 (2009).
4. P. Kramer and M. Saraceno, *Geometry of the Time-Dependent Variational Principle in Quantum Mechanics* (Springer, New York, 1981).
5. S.I. Sawada, R. Heather, B. Jackson, and H. Metiu, *J. Chem. Phys.*, **83** 3009 (1985); S.I. Sawada, H. Metiu, *J. Chem. Phys.*, **84** 227 (1986).
6. D.V. Shalashilin and I. Burghardt, *J. Chem. Phys.*, **129**, 084104 (2008).
7. E.J. Heller, *J. Chem. Phys.*, **75**, 2923 (1981).
8. G.D. Billing, *Chem. Phys. Lett.*, **100**, 535 (1983); H.-D. Meyer and W.H. Miller, *J. Chem. Phys.*, **70**, 314 (1979).

21 Direct Numerical Solution of the Quantum Hydrodynamic Equations of Motion

Brian K. Kendrick

CONTENTS

21.1　INTRODUCTION

The quantum hydrodynamic formulation of quantum mechanics is intuitively appealing [1–7]. The quantum potential (Q) and its associated force $f_q = -\nabla Q$ appear on an equal footing with the classical potential and force in the equations of motion. Thus, this approach provides a unified computational method for including both classical and quantum mechanical effects in a molecular dynamics (trajectory based) calculation. However, despite their deceptively simple form, the quantum hydrodynamic equations of motion are very challenging to solve numerically. The quantum potential is non-local (i.e., the trajectories are coupled) and it is a non-linear function of the density (i.e., it depends upon the solution itself). Thus, unlike a typical classical potential, the quantum potential is manifestly time dependent and evolves in "lock-step" fashion with the dynamics. Furthermore, the quantum potential can

325

often become singular when "nodes" occur in the quantum mechanical wave function. The problematic nodes are often associated with quantum interference effects due to the reflected components of the wave function scattering off a potential barrier. Despite these formidable challenges, significant progress has been made in recent years addressing each of the issues mentioned above.

The direct numerical solution methods can be split into three categories based on the underlying reference frame: (1) Lagrangian, (2) Eulerian, and (3) Arbitrary Lagrangian–Eulerian (ALE). The first successful direct numerical solution method for solving the quantum hydrodynamic equations, called the Quantum Trajectory Method (QTM), was based on the Lagrangian frame of reference [8]. One advantage of the Lagrangian frame is the simplified equations of motion due to the computational grid moving with the fluid flow. Another advantage is the grid being optimal. That is, the grid points move with and follow the time evolution of the density so that there are few wasted grid points located in regions where the density is small or zero. However, there is a significant disadvantage associated with using a Lagrangian frame. As time evolves, the grid becomes highly non-uniform which makes an accurate evaluation of the numerical derivatives difficult. Unfortunately, the grid often becomes sparse in precisely the regions where more grid points are needed for accurate calculations (i.e., near impending node formation) [9]. The Eulerian grid has the advantage of using a fixed grid that does not change with time. Thus, the grid remains uniform as time evolves and the sparsity of grid points near nodes is avoided. Unfortunately, the number of grid points is typically much larger since the entire computational domain must be gridded and included in the calculations from the beginning. However, in practice, the Eulerian grid size problem can be mitigated by "deactivating" grid points with small density and then "activating" them as the density increases above some user specified threshold during the calculation (some care is required in order to ensure continuity in the solutions when initializing all of the field values at the newly activated grid points). As the name implies, the ALE frame combines the best properties of both the Lagrangian and Eulerian frames of reference [10]. An ALE frame can be constructed which follows the flow of the fluid while preserving a uniform grid spacing. By implementing automatic grid refinement, a nearly constant user specified grid spacing can also be maintained [11].

Within each of the three frames of reference discussed above, a given numerical approach can be further subdivided into its method for evaluating derivatives and propagating in time. One of the most successful methods for evaluating derivatives within the quantum hydrodynamic approach is the meshless Moving Least Squares (MLS) method [8, 11]. This approach has been most successful due largely to its "noise filtering" properties. In the MLS approach, the derivatives of a field are simply the coefficients of a local polynomial fit to the field values within some local radius of the desired evaluation point. The local least squares fitting tends to average out any noise or numerical errors which may be accumulating in the solution which helps to stabilize the calculations. On the other hand, the filtering also reduces the resolution or fidelity of the calculations. Unfortunately, the accuracy, stability, and convergence properties of the MLS approach are not well understood which can make it difficult to use in practice. Furthermore, the derivatives are not strictly continuous due to the different local neighborhoods used at each evaluation point. Also, the computational overhead associated with performing repeated local least squares fits often becomes significant. Mesh based approaches can also be used for evaluating derivatives such as finite differences and finite elements. The advantages of finite differences include ease of implementation, computational efficiency, higher resolution/fidelity, and continuous derivatives (to some specified order). The accuracy, stability and convergence properties of finite difference approximations are much better understood relative to the meshless MLS approach. Until recently, finite difference based approaches for solving the quantum hydrodynamic equations of motion have been plagued with instabilities. These instabilities have recently been shown to originate from the negative diffusion terms associated with the truncation error of a finite difference approximation to the quantum hydrodynamic equations [12]. By deriving the explicit analytic form of these negative diffusion terms, a stabilizing positive diffusion term can be introduced (called "artificial viscosity") which is of the same order as the errors associated with the finite difference approximation [13]. Thus, the stability of the method can be maintained as the resolution of the calculation is increased.

In any computational method, there are always inherent numerical approximations such as: grid spacing, time step, polynomial basis set size, etc. We refer to these as "numerical approximations." Other kinds of approximations which simplify or reduce the computational complexity of the original dynamical equations can also be made such as: incompressible flow, decoupling approximations (i.e., ignoring certain coupling terms), etc. We refer to these as "dynamical approximations." The difference between the numerical and dynamical approximations is that the errors associated with the numerical approximations can be reduced (via grid refinement for example) to any desired level, whereas the errors associated with a dynamical approximation are always present (since the original equations have been modified). In practice, a combination of both numerical and dynamical approximations is usually required in order to obtain a computationally feasible method (especially for systems with many degrees of freedom). In this work the focus will be primarily on the numerical approximations but a few dynamical approximations will be discussed as well.

The relevant hydrodynamic equations of motion are introduced in Section 21.2. A meshless numerical solution method based on an ALE frame of reference and MLS fitting will be reviewed in Section 21.3. This method employs automatic grid refinement, field averaging, and artificial viscosity in order to obtain accurate and stable propagation for long times. The results of example calculations for a one- and two-dimensional (2D) Gaussian wave packet scattering off an Eckart potential barrier will be presented in Section 21.4.1. A generalized artificial viscosity algorithm will be reviewed in Section 21.4.2. This method is based on "dynamic" artificial viscosity coefficients which increase and decrease automatically depending upon the local strength of the quantum force (i.e., near impending nodes). The results of a challenging example calculation using this approach to treat a quantum resonance will be presented. A dynamical approximation (the Vibrational Decoupling Scheme [VDS]) will be reviewed in Section 21.4.3. In this approach, the trajectories associated with bound vibrational degrees of freedom are decoupled which allows for linear scaling of the computational cost. Scattering results will be presented for an N-dimensional Gaussian wave packet tunneling through an Eckart barrier along a reaction path coupled to $N - 1$ harmonic oscillator (bath) degrees of freedom (for values of N up to

$N = 100$). The recently developed Iterative Finite Difference Method (IFDM) will be reviewed in Section 21.5. This grid based approach is a promising alternative to using a meshless MLS method for evaluating derivatives. The numerical errors and stability properties of the IFDM are better understood and controllable. One-dimensional scattering results based on the IFDM will be presented. Concluding remarks and future research directions will be summarized in Section 21.6.

21.2   THE QUANTUM HYDRODYNAMIC EQUATIONS OF MOTION

In the de Broglie–Bohm approach, the wave function is written in polar form $\psi(\mathbf{r}, t) = R(\mathbf{r}, t) \exp(i\, S(\mathbf{r}, t)/\hbar)$ where R and S are the real-valued amplitude and action function. Substituting this expression for ψ into the time dependent Schrödinger equation and separating into real and imaginary parts gives rise to the following two equations [1–7]

$$\frac{\partial \rho(\mathbf{r}, t)}{\partial t} + \nabla \cdot \left(\rho \frac{1}{m} \nabla S \right) = 0, \tag{21.1}$$

$$-\frac{\partial S(\mathbf{r}, t)}{\partial t} = \frac{1}{2m} |\nabla S|^2 + V(\mathbf{r}) + Q(\mathbf{r}, t), \tag{21.2}$$

where the probability density $\rho(\mathbf{r}, t) \equiv R(\mathbf{r}, t)^2$, m is the mass, and V is the interaction potential. The velocity and flux are defined as $\mathbf{v}(\mathbf{r}, t) = \nabla S/m$ and $\mathbf{j} = \rho \mathbf{v}$, respectively. Equation 21.1 is recognized as the continuity equation and Equation 21.2 is the *quantum* Hamilton–Jacobi equation. Equation 21.2 is identical to the classical Hamilton–Jacobi equation except for the last term involving the quantum potential Q which is defined as

$$Q(\mathbf{r}, t) = -\frac{\hbar^2}{2m} \frac{1}{R} \nabla^2 R = -\frac{\hbar^2}{2m} \rho^{-1/2} \nabla^2 \rho^{1/2}. \tag{21.3}$$

Equation 21.3 shows that Q depends upon the shape or curvature of ψ, and that it can become singular whenever $R \to 0$. The quantum potential gives rise to all quantum effects (i.e., tunneling, zero-point energy, etc.). Numerically it is more efficient to work with the C-amplitude given by $R = e^C$. In terms of C, Equation 21.3 becomes

$$Q = -\frac{\hbar^2}{2m} (\nabla^2 C + |\nabla C|^2). \tag{21.4}$$

The total time derivative in an ALE frame of reference is $d/dt = \partial/\partial t + \dot{\mathbf{r}} \cdot \nabla$ where the grid point velocity $\dot{\mathbf{r}}$ is *not* necessarily equal to the flow velocity $\mathbf{v}$. In an ALE frame of reference and using $\rho = e^{2C}$, Equations 21.1 and 21.2 become

$$\frac{dC(\mathbf{r}, t)}{dt} = -\frac{1}{2} \nabla \cdot \mathbf{v} + (\dot{\mathbf{r}} - \mathbf{v}) \cdot \nabla C, \tag{21.5}$$

$$\frac{dS(\mathbf{r}, t)}{dt} = L_Q + (\dot{\mathbf{r}} - \mathbf{v}) \cdot (m\mathbf{v}), \tag{21.6}$$

where L_Q is the quantum Lagrangian defined as

$$L_Q = \frac{1}{2}m|\mathbf{v}|^2 - (V(\mathbf{r}) + Q(\mathbf{r},t)).\qquad(21.7)$$

Taking the gradient of Equation 21.2 leads to the familiar equation of motion which in an ALE frame is given by

$$m\frac{d\mathbf{v}}{dt} = -\nabla(V + Q) + (\dot{\mathbf{r}} - \mathbf{v}) \cdot (m\nabla\mathbf{v}).\qquad(21.8)$$

The Lagrangian and Eulerian frames correspond to the special cases $\dot{\mathbf{r}} = \mathbf{v}$ and $\dot{\mathbf{r}} = 0$, respectively. Equations 21.5 through 21.8 constitute a set of non-linear coupled hydrodynamic (fluid like) equations which describe the "flow" of the quantum probability density. The flow lines of this "probability fluid" correspond to the quantum trajectories which are obtained by solving the equation $\dot{\mathbf{r}} = \mathbf{v}$. The quantum potential is a non-local interaction which couples these trajectories. If the quantum potential is zero, then these trajectories correspond to an uncoupled ensemble of classical trajectories. A solution to the set of Equations 21.5 through 21.8 is completely equivalent to solving the time dependent Schrödinger equation. However, despite their relatively simple appearance, these equations are notoriously difficult to solve numerically. Any small numerical error in computing the spacial derivatives can become quickly amplified as these equations are propagated in time. Most of the numerical problems can be traced to the non-linear dependence of Q on C given by Equation 21.4.

21.3   THE MLS/ALE METHOD

In this section we review the MLS/ALE method for solving the quantum hydrodynamic equations of motion [11]. This approach uses the MLS algorithm for computing derivatives and an ALE reference frame [10]. The probability fluid is discretized into n computational particles or fluid elements so that the amplitude C_i, action S_i, flow velocity $\mathbf{v}_i$, and grid points $\mathbf{r}_i$ are labeled by $i = 1, \ldots, n$. Once the initial conditions at time t are specified for all of these fields at each value of i, the coupled set of quantum hydrodynamic Equations 21.5 through 21.8 are propagated to time $t + \Delta t$ using a fourth-order Runge–Kutta time integrator. All of the fields and their derivatives which appear in these equations are evaluated using a fourth-order MLS algorithm (see Ref. [11] for more details). The solutions are monitored for convergence as the grid spacing and time step are decreased. Special attention must be paid to the "radius of support" in the MLS algorithm. The radius of support defines a spherical neighborhood around a given evaluation point. All points within this local spherical neighborhood are used in the local least squares fit. As the grid spacing is decreased, the optimal radius of support also changes. A variable radius support which depends upon the probability density was found to give the best stability (especially near the edges of the grid where a larger radius is needed).

The ALE frame requires the user to specify the grid velocities (the $\dot{r}$ in Equations 21.5 through 21.8). The idea is to choose these grid velocities in order to minimize the numerical errors. A uniform grid spacing was found to be most accurate and can be constructed by choosing the $\dot{r}_i$ to satisfy (in one spacial dimension)

$$\dot{r}_i = \frac{(r_i^{t'} - r_i^t)}{\Delta t} \tag{21.9}$$

where $t' = t + \Delta t$. The $r_i^{t'}$ and r_i^t are the particle positions on the uniform grids at times t' and t, respectively. The uniform grid at the future time step t' is constructed by first performing the time propagation from time t to t' using a Lagrangian frame (i.e., $\dot{r} = v$) in Equations 21.5 through 21.8. A uniform grid is constructed between the new end points r_1 and r_n by uniformly distributing the $n - 2$ interior points. The propagation from time t to t' is then repeated using the ALE frame with the $\dot{r}$ given by Equation 21.9. This process is repeated for each time step so that by construction the ALE grid remains uniform as time evolves. The ALE grid also follows the flow of probability density since the end points are propagated using a Lagrangian frame (i.e., the end points move along quantum trajectories). Thus, the grid is optimal in that only grid points with non-zero density are included in the calculation. However, as the wave packet spreads, the uniform grid spacing increases. In order to maintain the resolution of the calculation, an automatic grid refinement algorithm is used which increases the number of points $n \to n' = n + \Delta n$ as the calculation progresses. The number of points added (Δn) is chosen so that the effective grid spacing $(r_n - r_1)/n'$ remains less than some user specified value. The MLS algorithm is then used to interpolate new values of the fields at the new n' points. The grid refinement is not done at each time step. A threshold is set so that the grid refinement only occurs when $\Delta n > 5$. The ALE frame with automatic grid refinement produces an optimal, uniform grid with nearly constant grid spacing as time evolves.

The unitarity of the MLS/ALE method is significantly improved by using averaged fields. In this approach, the MLS/ALE solution is propagated from time t to t' and all of the fields and their derivatives (Q, V, f_q, f_c, ∇C, $\nabla \cdot \mathbf{v}$, and $\nabla \mathbf{v}$) are computed at the future time t' using the fourth-order Runge–Kutta method. These values are then averaged with their values at the previous time step t. The MLS/ALE solution is then recomputed from time t to t' using these averaged fields. In analogy with implicit finite differencing, the use of averaged fields cancels out some of the numerical errors in the calculation which results in a more accurate solution (i.e., higher order). However, since the ALE/MLS method is already fourth-order accurate in both space and time, perhaps a better analogy for using averaged fields is that it represents the first iteration in a non-linear iterative solution method (see Section 21.5).

In the absence of nodes, the MLS/ALE method with automatic grid refinement and averaged fields is an accurate and stable method for the direct numerical solution of the quantum hydrodynamic equations. However, when nodes begin to occur (due to the quantum interference associated with the reflected components of a wave packet scattering off a potential barrier for example), the singular nature of Q and f_q near the forming nodes causes the method to break down. Associated with an impending node is a "kink" in the velocity field. As a node forms, this kink develops into a sharp discontinuity at the node's location. In classical hydrodynamics, sharp discontinuities in the velocity field are often associated with shock fronts. Thus, quantum interference can be interpreted as a source of shock fronts within the quantum hydrodynamic formulation. A standard technique for stabilizing classical hydrodynamic calculations in the presence of shocks is the introduction of artificial viscosity [11, 13]. Artificial viscosity produces a stabilizing force which prevents the negative slope associated with the kink in the velocity field from becoming too large. Thus, the discontinuity in the velocity field is avoided as well as the singularities in Q and f_q. In effect, the artificial viscosity prevents the formation of a true node which allows the numerical propagation to continue for long times. The introduction of artificial viscosity terms into the equations of motion does alter the solutions. However, these effects are by construction localized to the regions near node formation which for barrier scattering occur primarily in the incoming (reactant) channel. The scattering solution in the product channel is often devoid of nodes and the solution in this region is accurate. Thus, accurate transmission probabilities and reaction rates can be computed. The goal is to choose the artificial viscosity coefficients small enough so to minimize its effects on the solution but large enough to stabilize the calculations. Example scattering problems using this approach are presented in the next section.

21.4   EXAMPLE SCATTERING PROBLEMS USING THE MLS/ALE METHOD

The MLS/ALE method described in Section 21.3 has been successfully applied to several barrier scattering calculations: (1) a one-dimensional Eckart barrier [11], (2) a 2D model collinear chemical reaction [14], (3) a one-dimensional rounded square barrier [15], and (4) an N-dimensional model chemical reaction with $N = 1, \ldots, 100$ [16]. Each of these applications will be reviewed in the following subsections.

21.4.1   ONE TO FOUR-DIMENSIONAL SCATTERING OFF AN ECKART BARRIER

Figure 21.1 plots the initial amplitude of the one-dimensional Gaussian wave packet and the Eckart barrier given by $V(r) = V_0 \, \text{sech}^2[a \, (r - r_b)]$ where $V_0 = 8000 \text{ cm}^{-1} (0.992 \text{ eV})$ is the barrier height, $a = 0.4$ determines the width, and $r_b = 7 \text{ a}_0$ is the location of the barrier maximum. The initial Gaussian wave packet at $t = 0$ is given by

$$\psi(r, 0) = (2\beta/\pi)^{1/4} \, e^{-\beta(r-r_0)^2} \, e^{i \, k(r-r_0)}, \tag{21.10}$$

where $\beta = 4 \, a_0^{-2}$ is the width parameter, $r_0 = 2 \, a_0$ is the center of the wave packet, and k determines the initial phase $S_0 = \hbar k \, (r - r_0)$ and flow kinetic energy $E = \hbar^2 k^2 / (2m)$. The initial conditions for the C-amplitude and velocity v are given by $C_0 = \ln(2\beta/\pi)^{1/4} - \beta \, (r - r_0)^2$ and $v_0 = (1/m) \, \partial S_0 / \partial r = \hbar k/m$, respectively.

The Gaussian wave packet was propagated using the ALE/MLS method including artificial viscosity [11]. The amplitude R is plotted in Figure 21.2 for the time step $t = 95.4 \text{ fs}$ and an initial kinetic energy of $E = 0.8 \text{ eV}$. The wave packet has

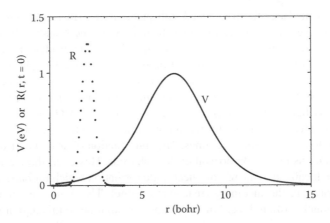

FIGURE 21.1   The amplitude R for the initial Gaussian wave packet centered at $r = 2\,a_0$ and the Eckart potential barrier centered at $r = 7\,a_0$ are plotted for the one-dimensional scattering problem.

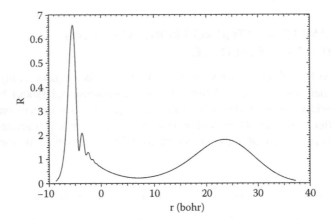

FIGURE 21.2   The amplitude R of the one-dimensional wave packet scattering from an Eckart barrier centered at $r = 7\,a_0$ is plotted at time step $t = 95.4$ fs for $E = 0.8$ eV. The wave packet tunnels through the barrier and splits into transmitted ($r > 7$) and reflected ($r < 7$) components. (Reproduced with permission from Kendrick, B. K., *J. Chem. Phys.*, 119, 5805, 2003.)

tunneled through the barrier and split into transmitted $r > 7$ and reflected $r < 7$ components. Quantum interference effects ("ripples") are clearly observed in the reflected component in the region $-4 < r < 0$. The transmitted component is smooth and Gaussian like. The transmission probability is computed by integrating the probability density ρ over the product region $r > 7$. Figure 21.3 plots several transmission probabilities as a function of time and energy. The ALE/MLS (Bohmian) results (the long-short dashed curves) are compared to the results of standard quantum mechanical calculations based on the Crank–Nicholson algorithm (the solid curves). Excellent agreement between the two sets of results is observed.

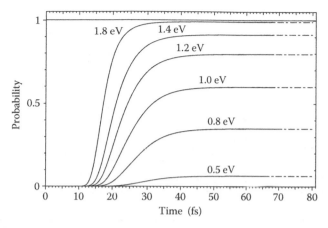

FIGURE 21.3 The transmission probability for the one-dimensional wave packet scattering from an Eckart barrier is plotted as a function of time and energy. The solid curves are computed using a Crank–Nicholson algorithm and the long-short dashed curves are computed using the MLS/ALE (Bohmian) approach. The results are essentially identical. For reference a horizontal line is plotted for unit probability. (Reproduced with permission from Kendrick, B. K., *J. Chem. Phys.*, 119, 5805, 2003.)

Figure 21.4 plots the 2D potential energy surface for a model collinear chemical reaction [14]. The potential is given by $V = V_1(s) + V_2(q; s)$ where the translational part of the potential is an Eckart function

$$V_1(s) = V_0 \operatorname{sech}^2(as), \tag{21.11}$$

and the vibrational potential is a local harmonic function

$$V_2(q; s) = \frac{1}{2}k(s)q^2, \tag{21.12}$$

where the local force constant is given by $k(s) = k_0 [1 - \sigma \exp(-\lambda s^2)]$. The initial wave packet at time $t = 0$ is also plotted in Figure 21.4. It is the product of a ground state harmonic oscillator function for the vibrational motion (q), a Gaussian wave packet for the translational motion (s), and a phase factor to give it a non-zero translational velocity. The 2D wave packet was propagated using the MLS/ALE method including artificial viscosity [11, 14]. As in the one-dimensional problem discussed above, the wave packet moves from left to right and tunnels through the barrier located at $s = 0$ splitting into transmitted ($s > 0$) and reflected ($s < 0$) components. The probability density associated with the transmitted wave packet was integrated over all s and q in the product region ($s > 0$) to obtain transmission probabilities. The transmission probabilities were in excellent agreement with those computed using the QTM method [14].

The 2D model described above was generalized to higher dimensions (N) by including additional harmonic oscillator coordinates: $q \rightarrow q^l$ where $l = 1, \ldots, N-1$ coupled to the reaction path s. The MLS/ALE method was generalized to this problem

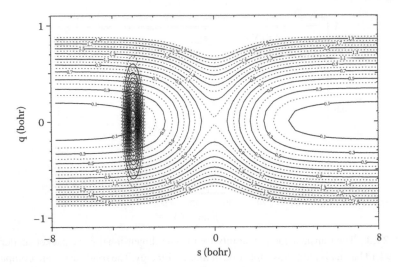

FIGURE 21.4 Contour plot of the 2D potential energy surface used for the model collinear chemical reaction. Natural collision coordinates (s, q) are used. The energy contours are indicated in eV and are measured relative to the bottom of the harmonic oscillator potential at large $|s|$. The initial wave packet centered at $s_0 = -4\,a_0$ and $q_0 = 0\,a_0$ is also plotted. (Reproduced with permission from Pauler, D. K. and Kendrick, B. K., *J. Chem. Phys.*, 120, 603, 2004.)

and a parallel computer code was developed [16]. Good performance and scalability of the parallel code was achieved which allowed for the calculation of transmission probabilities for up to $N = 4$. Unfortunately, due to the direct product structure of the grid, the computational costs exhibit the well known exponential scaling with respect to the dimensionality N. Thus, dynamical approximations are needed in order to treat higher dimensional problems (see Section 21.4.3). We refer the reader to the original papers for more details of the calculations and additional results [11, 14–16].

21.4.2 ONE-DIMENSIONAL SCATTERING OFF A ROUNDED SQUARE BARRIER

A more challenging one-dimensional scattering problem which exhibits a quantum resonance is plotted in Figure 21.5. The potential is a rounded square barrier given by [15]

$$V(r) = \frac{V_1}{2}[\tanh(a\,(r - r_1)) + 1] - \frac{V_2}{2}[\tanh(b\,(r - r_2)) + 1],\qquad(21.13)$$

where $r_1 = 0$ au, $V_1 = 50$ au, $r_2 = 1$ au, $V_2 = 20$ au. The reactant (product) channel has an asymptotic energy of $E_\alpha = 0$ au ($E_\beta = 30$ au). The parameters $a = 7$ and $b = 7$ govern the "sharpness" of the barrier's rounded edges. The incoming reactant wave packet (Ψ_{initial}) enters from the left and the transmitted part ($r > 0$) of this wave packet is correlated with the reference wave packet (Ψ_{final}) on the product side of the barrier. The initial (reactant) wave packet is given by Equation 21.10 but with

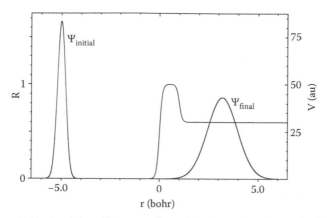

FIGURE 21.5 The amplitude of the initial (reactant) and final (product) Gaussian wave packets are plotted. Also plotted is the potential energy surface which is zero in the reactant region, 30 au in the product region and exhibits a 50 au barrier between $r = 0$ and $r = 1$ au. (Reproduced with permission from Derrickson, S. W., Bittner, E. R., and Kendrick, B. K., *J. Chem. Phys.*, 123, 54107-1-9, 2005.)

$\beta = 12\,a_0^{-2}$, $r_0 = -5\,a_0$, and $\lambda = 12$ au. The initial wave packet was propagated using the MLS/ALE method with artificial viscosity [11, 15]. However, in order to treat this more challenging problem, the artificial viscosity algorithm was generalized in two ways: (1) The artificial viscosity potential was chosen to be more localized, and (2) the artificial viscosity strength parameters were allowed to vary dynamically during the calculation [15]. The dynamical artificial viscosity coefficients were chosen to increase and decrease in magnitude proportional to the magnitude of the quantum force f_q in regions near an impending node. A rigorous derivation [12] of the artificial viscosity coefficients using finite differences confirms this choice (see Section 21.5). The dynamic coefficients allow the artificial viscosity to increase within the localized regions in parameter space where a node is beginning to form while remaining small elsewhere. Thus, the calculations can be stabilized while minimizing the global effects of the artificial viscosity on the solution.

A wave-packet correlation function approach was used to compute the scattering matrix, reaction probability, and time delay as a function of the total energy (E) [15]. In this approach, the correlation function is computed from the overlap of a time propagated wave packet with some reference wave packet in the product channel

$$C_{\beta\alpha}(t) = \int dr\, \psi_\alpha(r, t)\, \psi_\beta^*(r, 0), \qquad (21.14)$$

where $\psi_\beta(r, 0) = \Psi_{\text{final}}$ (see Figure 21.5) and $\psi_\alpha(r, t)$ is the MLS/ALE time propagated wave packet. The state-to-state scattering matrix $S_{\beta\alpha}$ is related to the Fourier transform of the correlation function. Once the scattering matrix is computed, the energy dependent reaction probability can be computed from $P(E) = |S_{\beta\alpha}|^2$ and the

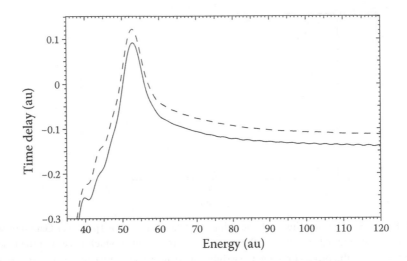

FIGURE 21.6 The time delay is plotted for both the Crank–Nicholson (solid curve) and ALE/MLS (dashed curve) calculations. Both time delays exhibit the well-known Lorentzian form near the resonance energy $E_r = 53$ au. (Reproduced with permission from Derrickson, S. W., Bittner, E. R., and Kendrick, B. K., *J. Chem. Phys.*, 123, 54107-1-9, 2005.)

energy dependent time delay can be obtained via the relation [15]

$$t_{\beta\alpha} = \hbar\,\mathrm{Im}\left(\frac{\mathrm{d}\ln S_{\beta\alpha}}{\mathrm{d}E}\right). \tag{21.15}$$

The time delay $t_{\beta\alpha}$ can be interpreted as the time for observing a wave packet that has traversed the interaction potential relative to the same wave packet which has traversed a zero potential. Figure 21.6 plots the time delays for both the MLS/ALE and Crank–Nicholson calculations as a function of E. Both sets of calculations clearly capture the reactive scattering resonance at $E = 0.53$ au. The well-known Breit–Wigner or Lorentzian form of the time delay is clearly visible in Figure 21.6 near $E = 0.53$ au. An analysis of the computed time delays which subtracts out the non-resonant (background) time delay gives the resonance lifetimes for the Crank–Nicholson and ALE/MLS calculations: $\tau = 0.230$ au and 0.235 au, respectively [15]. Argand plots of the S-matrix elements for both the Crank–Nicholson and ALE/MLS calculations exhibit the well-known "loop" in the energy trajectories of the S-matrix elements at the resonance energy. This loop structure provides additional evidence that the "bump" in the reaction probability and in the time delay is due to the presence of a quantum resonance (see Ref. [15] for more details). These calculations demonstrate that artificial viscosity can be used successfully to stabilize the calculations in the presence of severe node formation in the reactant channel while preserving the accuracy of the transmitted (product) component of the wave packet. Accurate calculations of *both* the magnitude and phase of the scattering matrix is possible which allows for the treatment of quantum resonances.

21.4.3 DYNAMICAL APPROXIMATIONS FOR TREATING QUANTUM MANY-BODY SYSTEMS

The quantum many-body problem is very difficult if not impossible to treat from first principles due to the well-known exponential scaling of the computational cost associated with standard quantum mechanical methods. The intuitive trajectory based de Broglie–Bohm formulation of quantum mechanics offers an appealing new approach to the quantum many-body problem which may result in more efficient computational methods. However, even within the quantum hydrodynamic formulation, "dynamical approximations" are needed in order make the calculations more feasible. As mentioned in the introduction (see Section 21.1), dynamical approximations involve a modification of the original (exact) equations of motion. These modifications are usually based on the idea that some coupling terms are smaller or less important than others. Thus, the goal is to identify and partition these coupling terms into those which are ignored and those which are kept. Unfortunately, due to the coupled non-linear nature of the quantum hydrodynamic equations, this goal is difficult to achieve in practice. Intuition and numerical tests are at present the modus operandi.

Several approximate methods have been developed in recent years, such as: "Linearized Quantum Force," [17, 18] the "Derivative Propagation Method," [19] and the VDS [16]. In this section, we will review the VDS. The idea behind the VDS is to take advantage of a unique property of the de Broglie–Bohm formulation of quantum mechanics. The property of interest is the well-known cancellation of the classical and quantum forces for a bound (stationary) vibrational degree of freedom (q^l). That is, the *total* force (classical plus quantum) along this degree of freedom is zero: $f_c^{q^l} + f_q^{q^l} = 0$. This condition is exact in the asymptotic reactant channel and it holds for any vibrational interaction potential (i.e., a harmonic oscillator assumption is *not* required). Setting the total force zero along the bound vibrational degrees of freedom decouples the quantum hydrodynamic equations into a set of uncoupled one-dimensional problems. Thus, the method scales linearly with respect to the number of vibrational degrees of freedom and can be easily parallelized on a computer by assigning each one-dimensional problem to a separate processor. During the course of the dynamics, some of the vibrational degrees of freedom will be coupled more strongly to the reaction path than others. For these degrees of freedom, additional couplings will need to be included or an exact treatment may be required. These couplings could be introduced in an iterative fashion or included directly during the propagation. For the vibrational degrees of freedom which are uncoupled to the reaction path (i.e., they are "spectators"), the VDS approximation is exact. A practical implementation would require the following steps:

(1) propagate the uncoupled one-dimensional solutions using the VDS everywhere;
(2) estimate the couplings between the one-dimensional solutions;
(3) re-propagate only the one-dimensional solutions that experience significant couplings and include these couplings;
(4) repeat steps (2) and (3) until the solutions converge.

The couplings can be included as additional amplitude and phase corrections for each (decoupled) one-dimensional solution [16]. Thus, the linear scaling of the computational cost is preserved while also including coupling effects.

The VDS was applied to an N-dimensional model chemical reaction using reaction path coordinates s and q^l where $l = 1, \ldots, N - 1$ [16]. In this model, the potential energy function along the reaction path was taken to be an Eckart barrier and the vibrational degrees of freedom were represented by simple harmonic oscillators (see Equations 21.11 and 21.12 and the discussion at the end of Section 21.4.1). The method was parallelized by distributing the decoupled one-dimensional problems in s for each value of the q^l among different processors. The wave packets for each decoupled one-dimensional problem were propagated using the MLS/ALE method with artificial viscosity described above in Section 21.2. Figure 21.7 plots the reaction probabilities as a function of the dimensionality N for two different energies. Figure 21.8 plots the wall-clock time as a function of dimensionality that was required to compute the reaction probabilities plotted in Figure 21.7. As expected, the computational cost scales linearly with respect to the dimensionality N. The reaction probabilities for $N = 2$ and $N = 3$ computed using the VDS were also compared to the fully coupled MLS/ALE calculations described above in Section 21.2. Excellent agreement between the VDS results and the fully coupled (exact) results was observed [16]. This indicates that the coupling terms ignored by the VDS do not affect the transmission probabilities for this model problem. These initial results are

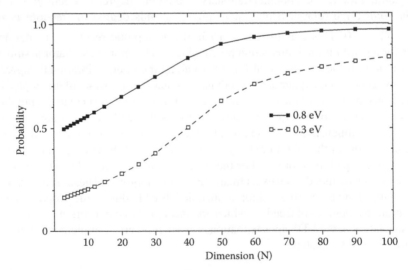

FIGURE 21.7 Converged reaction probabilities are plotted as a function of the dimension N for initial translational energies of $E = 0.3$ and 0.8 eV. The vibrational decoupling scheme was used. The data points (calculated values) are connected by line segments to help guide the eye. The reaction probability increases as the energy E is increased. (Reproduced with permission from Kendrick, B. K., *J. Chem. Phys.*, 121, 2471, 2004.)

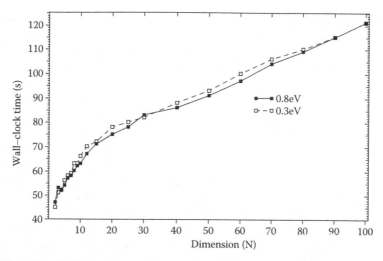

FIGURE 21.8 The wall-clock time for computing a converged reaction probability is plotted as a function of the dimension N for initial translational energies of $E = 0.8$ and 0.3 eV. The vibrational decoupling scheme was used. The data points (calculated values) are connected by line segments to help guide the eye. The wall-clock time scales linearly with respect to N and is independent of F. (Reproduced with permission from Kendrick, B. K., *J. Chem. Phys.*, 121, 2471, 2004.)

encouraging. However, for more complicated interaction potentials, additional couplings will need to be included and all four steps of the iterative procedure described above will be required.

21.5 THE ITERATIVE FINITE DIFFERENCE METHOD

The MLS/ALE method with artificial viscosity described in Section 21.3 has been successful in treating a large number of model problems including quantum resonances and fully coupled four-dimensional scattering problems. However, especially for more complicated problems, the MLS method for computing derivatives has limitations. As discussed in Section 21.1, the convergence and stability properties of the MLS method are not well understood and its derivatives are not continuous. Furthermore, the resolution or fidelity of the method and its computational overhead associated with repeated least squares fitting are often less than desirable. For these reasons, other solution methods such as those based on finite differences or finite elements are attractive. Finite difference methods are straightforward to implement, computationally efficient, and their stability and convergence properties are better understood relative to the MLS approach. However, finite difference methods have their own subtleties. The finite differencing approach is not unique and the optimal choice is often problem specific. The different approaches differ in how the time and spacial degrees of freedom are treated. In regards to time, the different approaches can be classified as "explicit," "implicit," or a combination of both. In regards to space, the different approaches are often classified as "central" or "upstream." Each approach has its advantages and disadvantages and the presence of a non-linear coupling (such as the quantum potential Q) poses additional challenges in obtaining an accurate and stable method. In particular, the presence of non-linear interaction terms requires iterative solution techniques. A straightforward application of finite differences to the quantum hydrodynamic equations usually results in an unstable solution within a few time steps. These instabilities typically originate at the edges of the grid and quickly propagate into the interior points. Until recently, the origin of these instabilities and what might be done to stabilize them has been unclear. In this section, we review a new stable finite difference approach called the IFDM for solving the quantum hydrodynamic equations of motion [12]. The method is second-order accurate in both space and time, and it converges exponentially with respect to the iteration count.

The IFDM is based on a second-order central differencing of the quantum hydrodynamic equations (Equations 21.5 and 21.8) with respect to r in an Eulerian frame of reference ($\dot{r} = 0$). The time coordinate is differenced using an average of first-order explicit and implicit differences to obtain second-order accuracy (analogous to Crank–Nicholson). For each time step, an iterative solution technique (Newton's method) is used to obtain a self-consistent solution to the resulting set of coupled non-linear finite differenced hydrodynamic equations. The number of iterations typically varies between 10 and 20 depending upon the size of the time step and the desired level of convergence. The stability of the method was investigated using the standard method of truncation error analysis. The truncation error associated with the finite difference approximation was derived and the diffusive terms proportional to Δt^2 are given by [12]

$$\gamma_v(C, v) = \alpha_1 v^2 \frac{\partial v}{\partial r} + \alpha_2 v f/m + \alpha_3 \frac{\hbar^2}{m^2} \frac{\partial v}{\partial r} \left(\frac{\partial C}{\partial r} \right)^2, \qquad (21.16)$$

$$\gamma_C(C, v) = \beta_1 v^2 \frac{\partial v}{\partial r}, \qquad (21.17)$$

where the α_i and β_1 are dimensionless constants (not necessarily all positive). The effective diffusion coefficients are given by $\Gamma_v = \Delta t^2 \gamma_v$ and $\Gamma_C = \Delta t^2 \gamma_C$. Depending upon the behavior of v, $\partial v / \partial r$ and f, the effective diffusion coefficients can be positive or negative and can change sign during the course of the calculations. Even a temporary occurrence of a negative diffusion coefficient can quickly lead to instabilities and a complete breakdown of the calculations. Equations 21.16 and 21.17 show that for a Gaussian wave packet, the effective diffusion coefficients exhibit a quadratic dependence $(r - r_0)^2$ where r_0 is the center of the Gaussian. This explains why, even for a free Gaussian wave packet, numerical instabilities begin to occur at the edges of the grid (the potentially negative diffusion coefficients are largest in magnitude there). The dependence of the effective diffusion coefficients on the total force f and $\partial v / \partial r$ also explain why calculations often become unstable in regions of large quantum force and/or large velocity gradients. These conditions are often associated with quantum interference and node formation.

A standard technique for stabilizing finite difference equations which contain potentially negative diffusion terms is called "artificial viscosity" (see also Section 21.3). In this approach, positive diffusion terms of the same functional form as those given in Equations 21.16 and 21.17 are introduced into the equations of motion. By ensuring that the added positive diffusion terms are always greater than the negative ones, the stability of the method can be maintained. For example, Figure 21.9 plots the quantum potential for a free Gaussian wave packet at the edge of the Eulerian grid where the density is extremely small ($\rho < \exp(-5,880.0)$). The Gaussian wave packet at this time step is centered at $r = -4.5$ au. The instabilities decrease as the positive artificial viscosity coefficient v_0^1 is increased (see Ref. [12] for more details). It is important to note that the artificial viscosity terms are of the *same* order as the truncation error (i.e., they are proportional to Δt^2). Thus, the stability of the calculations can be maintained even as the resolution of the calculation is improved (i.e., Δt is decreased). The artificial viscosity can also be used to stabilize the method against impending node formation (as was done in the MLS/ALE method reviewed in Section 21.3). The artificial viscosity coefficient in this case (denoted by v_0^2) is typically larger than v_0^1 and is non-zero only in regions where *both* $\partial v/\partial r < 0$ and $\rho > \rho_{min}$ (where ρ_{min} is some user specified cutoff).

The IFDM was successfully applied to the one-dimensional scattering off an Eckart barrier (see Section 21.4.1). Figure 21.10 plots the density of the IFDM solution (solid curves) at several time steps. Excellent agreement with the Crank–Nicholson calculations (dashed curves) is observed. Some differences are seen between the

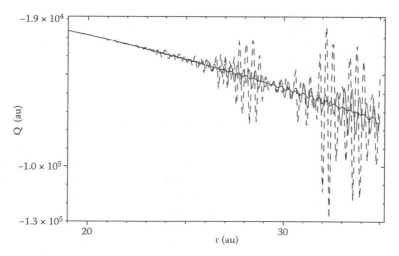

FIGURE 21.9 A magnified view of the quantum potential computed using the IFDM at the edge of the grid (for the free moving Gaussian wave packet at $t = 0.05$ au). The dashed, thin solid, and thick solid curves correspond to a $v_o^1 = 250$, 500, and 1000, respectively. The magnitude of the oscillations decreases as the artificial viscosity coefficient is increased. (Reproduced with permission from Kendrick, B. K., *J. Molec. Struct.: THEOCHEM*, 943, 158–167, 2010.)

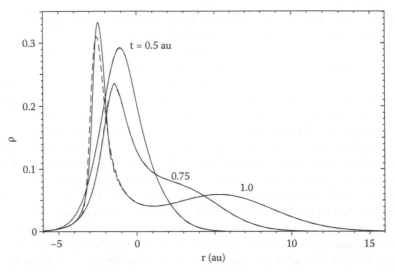

FIGURE 21.10 The probability density for the one-dimensional Gaussian wave packet scattering from an Eckart barrier is plotted at time steps $t = 0.5$, 0.75 and 1.0 au. The initial kinetic energy is $E = 50$ au. The solid curves are the IFDM results and the dashed curves are the Crank–Nicholson results (see text for discussion). (Reproduced with permission from Kendrick, B. K., *J. Molec. Struct.: THEOCHEM*, 943, 158–167, 2010.)

two solutions in the reflected part of the density at $t = 1$ au due to the effects of artificial viscosity. The transmitted part of the density ($r > 0$) is unaffected. Transmission probabilities were computed as a function of time for several energies. Excellent agreement was observed between those computed using the IFDM and those computed using the Crank–Nicholson and the MLS/ALE method [12]. Even using an Eulerian grid, the IFDM proved to be faster than the other two methods.

21.6 CONCLUDING REMARKS AND FUTURE RESEARCH

Several methods for the direct numerical solution of the quantum hydrodynamic equations of motion were reviewed. The methodologies were divided into two classes, those based on "numerical approximations" and those based on additional "dynamical approximations." Under numerical approximations, the MLS/ALE method and the IFDM methods were discussed. The MLS/ALE method is based on a meshless MLS method for evaluating derivatives, an ALE frame of reference, automatic grid refinement, averaged fields, and artificial viscosity. Collectively, these techniques result in a stable method for propagating solutions to the quantum hydrodynamic equations of motion. Applications of the MSL/ALE method include fully coupled treatments of barrier scattering up to four dimensions and a challenging one-dimensional problem which exhibits a quantum resonance. The MLS/ALE method has also been parallelized to run efficiently on large supercomputers. The newly developed IFDM method was also reviewed. This method is based on finite differencing the quantum hydrodynamic equations and the non-linear coupling (due to Q) is accurately treated using an iterative solution technique (Newton's method). The IFDM overcomes many of the limitations of the MLS approach and has been successfully applied to a one-dimensional barrier scattering problem. Under dynamical approximations, the VDS was reviewed. This method takes advantage of the unique properties of the de Broglie–Bohm formulation of quantum mechanics in regards to a bound stationary state. The vanishing of the total force for a bound stationary state was used to decouple the multi-dimensional quantum hydrodynamic equations into a set of uncoupled one-dimensional problems. The important couplings can be reintroduced into the one-dimensional solutions using an iterative solution method. The VDS scales linearly with respect to the vibrational degrees of freedom and was applied to an N-dimensional model chemical reaction with a reaction path coupled to $N - 1$ harmonic oscillator (bath) degrees of freedom. This method has also been parallelized to run efficiently on large supercomputers.

The quantum trajectory based approach is computationally attractive due to its similarities with the standard classical molecular dynamics type interface (i.e., Newton's equation of motion with forces proportional to the gradient of an interaction potential). Future work utilizing various "dynamical approximations" is needed in order to realize a quantum trajectory based molecular dynamics (i.e., QTMD) methodology, the idea being to capture the essential quantum mechanical effects such as zero point energy and tunneling at some approximate level while keeping the computational costs feasible. At the more fundamental level, the newly developed IFDM method shows great promise as an alternative to the MLS/ALE. However, more work is needed within the IFDM in order to improve the artificial viscosity algorithm in regards to node formation in the reactant region. The method is stable but the accuracy of the reflected component of the wave packet needs to be improved. Such improvements would pave the way for treating more challenging scattering and bound state problems including the calculation of long-lived quantum resonances and molecular spectra. Much progress has been made in recent years in developing computational methods for solving the quantum hydrodynamic equations. The intriguing nature and unique challenges associated with this quantum mechanical approach will no doubt continue to inspire significant progress in the years to come.

ACKNOWLEDGMENTS

This work was done under the auspices of the US Department of Energy at Los Alamos National Laboratory. Los Alamos National Laboratory is operated by Los Alamos National Security, LLC, for the National Nuclear Security Administration of the US Department of Energy under contract DE-AC52-06NA25396.

BIBLIOGRAPHY

1. E. Madelung, *Z. Phys.* **40** (1926) 322.
2. L. de Broglie, *C. R. Acad. Sci. Paris.* **183** (1926) 447; **184** (1927) 273.
3. D. Bohm, *Phys. Rev.* **85** (1952) 166; **85** (1952) 180.

4. D. Bohm, B. J. Hiley, and P. N. Kaloyerou, *Phys. Rep.* **144** (1987) 321.
5. D. Bohm and B. J. Hiley, *The Undivided Universe*, Routledge, London, 1993.
6. P. R. Holland, *The Quantum Theory of Motion*, Cambridge University Press, New York, 1993.
7. R. E. Wyatt, *Quantum Dynamics with Trajectories: Introduction to Quantum Hydrodynamics*, Springer, New York, 2005.
8. C. Lopreore and R. E. Wyatt, *Phys. Rev. Lett.* **82** (1999) 5190.
9. R. E. Wyatt and E. R. Bittner, *J. Chem. Phys.* **113** (2000) 8898.
10. K. H. Hughes and R. E. Wyatt, *Chem. Phys. Lett.* **366** (2002) 336.
11. B. K. Kendrick, *J. Chem. Phys.* **119** (2003) 5805.
12. B. K. Kendrick, *J. Molec. Struct.: THEOCHEM* **943** (2010) 158–167.
13. J. Von Neumann and R. D. Richtmyer, *J. Appl. Phys.* **21** (1950) 232.
14. D. K. Pauler and B. K. Kendrick, *J. Chem. Phys.* **120** (2004) 603.
15. S. W. Derrickson, E. R. Bittner, and B. K. Kendrick, *J. Chem. Phys.* **123** (2005) 54107-1-9.
16. B. K. Kendrick, *J. Chem. Phys.* **121** (2004) 2471.
17. V. A. Rassolov and S. Garashchuk, *J. Chem. Phys.* **120** (2004) 6815.
18. S. Garashchuk, *J. Phys. Chem. A* **113** (2009) 4451.
19. C. J. Trahan, K. Hughes, and R. E. Wyatt, *J. Chem. Phys.* **118** (2003) 9911.

22 Bohmian Grids and the Numerics of Schrödinger Evolutions

D.-A. Deckert, D. Dürr, and P. Pickl

CONTENTS

22.1 INTRODUCTION

Bohmian mechanics [1, 2] is a mechanical theory about the motion of particles. In contrast to quantum mechanics, Bohmian mechanics is a complete physical theory which leaves no interpretational freedom; in particular, it is free of paradoxes like Schrödinger's cat. This completeness is achieved by introducing a definite configuration of particles for all times which is governed by the Bohmian velocity law. The predictions of Bohmian mechanics agree with those of quantum mechanics whenever the latter are unambiguous. Besides that, it turns out that in some situations the Bohmian velocity law is a convenient tool for the numerical integration of the Schrödinger equation. There is a lot of literature which explains the physical side of Bohmian mechanics, see e.g. Ref. [2]. The role of this chapter is to explain the advantages in using Bohmian mechanics for numerics.

Let us discuss the Bohmian equations of motion in units of mass $m = \hbar = 1$. First, one has the Schrödinger equation

$$i\frac{d\psi_t(x)}{dt} = \left(-\frac{\nabla_x^2}{2} + V(x)\right)\psi_t(x). \tag{22.1}$$

Here $x = (x_1, \ldots, x_N)$ denotes the configuration in configuration space $\mathbb{R}^{3N}$. Furthermore, we use the notation $\nabla_x = (\nabla_{x_1}, \ldots, \nabla_{x_N})$ where ∇_{x_k} is the gradient with respect to $x_k \in \mathbb{R}^3$.

Let ψ_t be a solution of the N-particle Schrödinger equation with initial condition $\psi_t\big|_{t=0}(x) = \psi^0(x)$. Given the initial positions $q_1^0, \ldots, q_N^0$ of the particles, the Bohmian velocity law then determines the positions $q_1(t), \ldots, q_N(t)$ of the particles at any time t. In this way Bohmian mechanics describes for given ψ_t and any given initial configuration $q(0) := (q_k^0)_{1 \leq k \leq N}$ the motion of the configuration space trajectory $q(t) := (q_k(t))_{1 \leq k \leq N}$. Note that *one* such configuration space trajectory describes the motion of *all* N particles.

The trajectories of the N particles $q_k(q_k^0, \psi^0; t) \in \mathbb{R}^3$, $1 \leq k \leq N$, are solutions of the Bohmian velocity law

$$\frac{dq_k(t)}{dt} = \Im\left[\frac{\psi_t^* \nabla_{q_k} \psi_t}{\psi_t^* \psi_t}(q_1, \ldots, q_N)\right] \tag{22.2}$$

with $q_k(t = 0) = q_k^0$ as initial conditions and ψ^* denoting the complex conjugate of ψ, so that $\psi^* \psi = |\psi|^2$. Interpreting $\psi^* \psi$ as an inner product, the generalization to spin wavefunctions is straightforward [2]. Using configuration space language the Bohmian trajectory of an N particle system is an integral curve $q(t) \in \mathbb{R}^{3N}$ of the following velocity field on configuration space

$$v^\psi(q, t) = \Im\left[\frac{(\psi_t^* \nabla_q \psi_t)(q)}{(\psi_t^* \psi_t)(q)}\right], \tag{22.3}$$

i.e.,

$$\frac{dq(t)}{dt} = v^\psi(q, t). \tag{22.4}$$

Statistical Bohmian Mechanics. Like in quantum mechanics the exact initial configuration q^0 of the N particles is unknown. One usually starts with a probability distribution for the initial configuration given by the density $|\psi^0|^2$ (Bohmian mechanics explains why, see Ref. [2, 3]).

Therefore, in Bohmian mechanics one has to consider an ensemble of many configuration space trajectories which is initially $|\psi^0|^2$-distributed. Each of these configuration space trajectories describes a different possible physical time evolution of all the N particles and must not be confused with different one-particle trajectories.

Under general conditions one has global existence and uniqueness of Bohmian configuration space trajectories, i.e., the integral curves do not run into nodes of the wavefunctions and they cannot cross [4]. The uniqueness, i.e., non-crossing property, is due to the fact that the Bohmian law is given by a velocity vector field: at each configuration point there is a unique vector, prescribing the velocity of the configuration in that point. To avoid misunderstandings it is important to emphasize that the non-crossing of Bohmian configuration space trajectories has nothing to do with possible crossings of particle paths. The crossing of particle paths is in fact allowed in Bohmian mechanics. For example the situation where two particle paths cross at time t_0 is described by just one Bohmian trajectory: at time t_0 the position of both particles is equal, let's say x_0, i.e $q(t_0) = (x_0, x_0)$.

To analyze Bohmian mechanics statistically we must introduce an ensemble density $\rho(q, t)$. Such a density fulfills the continuity equation, which for the Bohmian flow reads

$$\frac{\partial \rho(x,t)}{\partial t} + \nabla_x \left(\rho(x,t) v^\Psi(x,t) \right) = 0.$$

The empirical import of Bohmian mechanics arises now from equivariance of the $|\psi|^2$-measure: one readily sees that by virtue of (Equation 22.3) the continuity equation is identically fulfilled by the density $\rho(x,t) = |\psi(x,t)|^2$, known as the quantum flux equation,

$$\frac{\partial |\psi(x,t)|^2}{\partial t} + \nabla_x \left(|\psi(x,t)|^2 v^\Psi(x,t) \right) = 0.$$

In contrast to Hamiltonian statistical mechanics where typical trajectories are described by stationary measures (like microcanonical or canoncial distributions) in Bohmian mechanics the typical trajectories are described by the equivariant measure $\rho(x,t) = |\psi(x,t)|^2$. In other words the notion of equivariance replaces the notion of stationarity This means that if one looks at an ensemble of Bohmian trajectories such that the respective initial configurations $q(0) = (q_1^0, \ldots, q_N^0)$ are distributed according to $\rho_0 = |\psi^0(x_1, \ldots, x_N)|^2$ at time $t = 0$, then for any time t the configurations $q(t) = (q_1(q_1^0, \psi^0; t), \ldots, q_N(q_N^0, \psi^0; t))$ are distributed according to $\rho_t = |\psi_t(x_1, \ldots, x_N)|^2$. Because of this, Bohmian mechanics agrees with all predictions made by orthodox quantum mechanics whenever the latter are unambiguous [3,5].

Hydrodynamic Formulation of Bohmian Mechanics. In this section we recast the Bohmian equations of motion in a form which is suitable for numerical analysis.

We first write ψ in the Euler form

$$\psi(x,t) = R(x,t)e^{iS(x,t)}$$

where R and S are real-valued functions. Equation 22.3 together with Equation 22.1 separated into their real and complex parts gives the following set of differential equations (which are very much analogous to the Hamilton–Jacobi equation of Hamiltonian mechanics)

$$\frac{dq}{dt} = \nabla_q S(q,t)$$

$$\frac{\partial R(x,t)}{\partial t} = -\frac{1}{2} \nabla_x (R(x,t)\nabla_x S(x,t))$$

$$\frac{\partial S(x,t)}{\partial t} = -\frac{1}{2} (\nabla_x S(x,t))^2 - V(r) + \frac{1}{2} \frac{\nabla_x^2 R(x,t)}{R(x,t)}.$$

We now introduce the functions $R(t)$, $S(t)$ defined by replacing the configuration point x by the configuration space trajectory $q(t)$:

$$R(t) = R(q(t),t) \quad S(t) = S(q(t),t).$$

Taking the derivatives with respect to time, for example,

$$\frac{dR(q(t),t)}{dt} = \frac{\partial R(q(t),t)}{\partial t} + \left. \frac{\partial R(x,t)}{\partial x} \right|_{x=q(t)} \frac{dq}{dt}$$

yields, using the notation $q = q(t)$,

$$\frac{dq}{dt} = \nabla_q S(q,t) \tag{22.5}$$

$$\frac{\partial R(q,t)}{\partial t} = -\frac{1}{2} R(q,t) \nabla_q^2 S(q,t) \tag{22.6}$$

$$\frac{\partial S(q,t)}{\partial t} = \frac{1}{2}\left(\frac{dq}{dt}\right)^2 - V(q) + \frac{1}{2}\frac{\nabla_q^2 R(q,t)}{R(q,t)}. \tag{22.7}$$

Numerical Integration. To solve a differential equation like the Schrödinger equation numerically one has to first discretize the problem. Instead of the full function $\psi(x,t)$ where x varies in configuration space $\mathbb{R}^{3N}$ one considers $\psi(x_k,t_l)$ where $x_1, \ldots, x_n$ and $t_1, \ldots, t_m$ are discrete "grid" points in configuration space and on the time axis.

When choosing the grid one has to take care of the following two things:

- The values $\psi(x_k,t_l)$ define an approximation $\widetilde{\psi}$ for ψ. For example one could think of $\widetilde{\psi}(x,t) = \psi(x_k,t_l)$ where for x_k one chooses the grid point closest to x and t_l closest to t (as explained later one usually employs a fitting algorithm to compute the numerical approximation $\widetilde{\psi}$ from the discrete data points $(\psi(x_k,t_l)_{1\le k\le n, 1\le l\le m})$. To reproduce from this discretization the physically relevant features of the full wavefunction one has to take care that the relevant regions in configuration space are well covered by the grid points.

  The sense of approximation needed in physics is the L^2-distance, i.e., in the above example

$$\left(\int |\widetilde{\psi}(x,t) - \psi(x,t)|^2 dx\right)^{1/2}. \tag{22.8}$$

- Of course the choice of grid points also influences the propagation of errors, as for example, the accuracy in computing derivatives depends on the grid.

In the usual approaches the grid points are fixed once and stay fixed for all times. Here one would have to densely distribute grid points in all regions of configuration space where the wavefunction may have support during the whole time interval of interest (this region is, when not confined to a potential trap, usually unknown).

This has the disadvantages that:

- Once the grid is fixed there may be regions in configuration space into which the wavefunction propagates and which are not covered by grid points.
- On the contrary there might be regions in configuration space where the wavefunction never lingers or stays only for a short time. Including such grid points in the computation for all times only slows down the numerical simulation without leading to better accuracy.

Bohmian Grids. One idea of avoiding these problems is to introduce a time dependent grid which is adapted to the time evolution of ψ_t. For Equation 22.8 the natural choice for the grid point trajectories for the Schrödinger equation are the Bohmian configuration space trajectories. To understand this recall equivariance: choosing the initial grid points $|\psi^0|^2$-distributed guarantees that they will stay $|\psi_t|^2$-distributed for all times.

This co-moving grid spreads dynamically according to the spreading of $|\psi_t|^2$ and the grid points will primarily remain in regions of space where $|\psi_t|^2$ is large while avoiding regions of nodes or tails of $|\psi_t|^2$ which are numerically problematic and physically irrelevant. The support of the wavefunction is then for all times well covered by grid points and there are only a few grid points in the physically uninteresting regions of configuration space, i.e., in regions where $|\psi_t|^2$ is small.

Such grids naturally optimize the approximation of the wavefunction in the sense of Equation 22.8. This idea goes back to Bob Wyatt [6]: instead of considering the equations on a fixed grid one regards the set of Equations 22.5 through 22.7 which can be integrated simultaneously and give the best adapted co-moving grid on the fly.

Wyatt's algorithm is particularly interesting for:

- Long-time simulations in scattering situations which usually (for methods using a fixed grid) demand a huge number of grid points and therefore a huge computational effort.
- Simulations of entangled oligo particle systems which usually either neglect entanglement and give inaccurate results or demand huge computational effort since one has to consider grid points in all $3N$ dimensions of a configuration space of N particles.

In order to perform the numerical integration of the set of differential Equations 22.5 through 22.7 we follow the straightforward method described in Ref. [6]. The only crucial part in this is computing the derivatives involved in the set of differential equations which we shall discuss below in detail.

22.2 PROBLEMS OF THE NUMERICAL IMPLEMENTATION

The quality of the numerical implementation can be measured in terms of the proper motion of the grid points which ideally should move along the Bohmian configuration space trajectories. Once they deviate from the latter equivariance will be violated, which means that the numerical approximation becomes bad in the L^2 sense (Equation 22.8).

The motion of the simulated grid points can go bad:

- when the grid point trajectories approach each other;
- when the wavefunction develops nodes;
- at the boundary of the grid.

We shall explain: Whenever two Bohmian configuration space trajectories approach each other the Bohmian time evolution will prohibit a crossing of them (because otherwise uniqueness would be violated). The mechanism which prevents crossings may however fail in numerical simulations due to numerical errors.

Therefore, it is vital for stable numerical simulations to understand this mechanism in terms of the dynamics given by Equations 22.5 through 22.7. The important term which is responsible for the mechanism to work is the quantum potential in Equation 22.7:

$$\frac{1}{2}\frac{\nabla_q^2 R(q,t)}{R(q,t)} \ . \tag{22.9}$$

Thinking of equivariance the particles are always $|\psi_t(x)|^2$-distributed, thus the approach of two particles goes hand in hand with an increase of $|\psi_t(x)|^2$ at the positions of the two particles. This leads to a bump in $R(q,t)$ on the microscopic scale defined by the grid points. At such a bump, $\nabla_q^2 R(q,t)$ is negative of course, and a negative quantum potential $\nabla_q^2 R(q,t)/R(q,t)$ in turn accelerates the two approaching grid points away from each other by virtue of Equation 22.7 and in this way prevents crossings (see Figure 22.1).

A numerical simulation which does not reproduce the derivatives on a microscopic level is not sensitive to bumps. Thus it does not prevent the crossing of grid point trajectories. Also the time step has to be chosen sufficiently small, at least a few times smaller than the distance of two approaching grid points divided by the relative Bohmian velocity. Otherwise there might not be enough iterations for the quantum potential to decelerate the approaching grid points.

Things can go wrong in particular when $R(q,t)$ is very small: a small value of $R(q,t)$ can cause a large quantum potential (Equation 22.9) and therefore large accelerations resulting in large velocities.

Typical situations where $R(q,t)$ is small are "nodes" and "tails" of the wavefunction. Nodes are local zeros of the wavefunction and tails are extended regions in

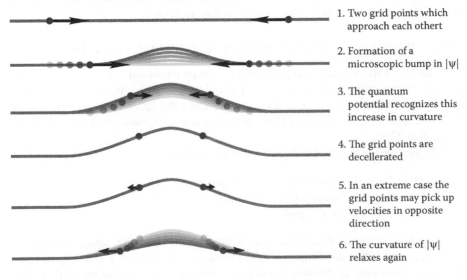

1. Two grid points which approach each othert

2. Formation of a microscopic bump in $|\psi|$

3. The quantum potential recognizes this increase in curvature

4. The grid points are decellerated

5. In an extreme case the grid points may pick up velocities in opposite direction

6. The curvature of $|\psi|$ relaxes again

FIGURE 22.1 Deceleration of approaching configuration space trajectories.

configuration space where the wavefunction is small. For example one typically has tails near the boundary of the grid. A small numerical error there can cause erroneously huge quantum potentials because the denominator in Equation 22.9 goes to zero. Note that this problem is of a conceptual kind since at the boundary there is a generic lack of knowledge of how the derivatives of R or S in Equations 22.5 through 22.7 behave; see Figure 22.2 for two example methods of fittings which will be explained later on in more detail.

We describe briefly how this conceptual problem can cause severe numerical instability: For this suppose that only the last grid point in the upper left plot of Figure 22.2 is lifted a little bit upwards by, e.g., some numerical error. Then the resulting quan-

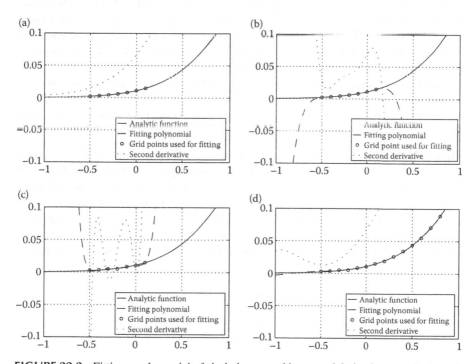

FIGURE 22.2 Fitting a polynomial of sixth degree and its second derivative near the boundary. Upper left: the function to fit smoothly decays. Both least squares and polynomial fitting give identical fitting polynomials mimicking this decay. Upper right: polynomial fitting was used while the function value at the fourth grid point was shifted upwards by 10^{-4} (not visible in the plot) to simulate a numerical error in order to see how sensitive polynomial fitting reacts to such an error. Lower left: all values at the grid points were shifted by an individual random amount of the order of 10^{-3} and polynomial fitting was used. Lower right: all values at the grid points were shifted by the same random numbers as on the lower left while the additional grid points were also shifted by individual random amounts of the order of 10^{-3} and least squares fitting was used. By comparison with upper left plot one observes the robustness of least squares fitting to such numerical errors at the boundary.

tum potential will cause the last grid point to move toward the second last one. This increases the density of grid points and thus again increases R in the next step such that this effect is self-amplifying (in fact growing super-exponentially) and will affect the whole wavefunction quickly.

22.3 FITTING ALGORITHMS: WHAT THEY CAN DO AND WHAT THEY CAN'T

In the following we shall explain strategies to circumvent the aforementioned problems of numerical implementations of Bob Wyatt's idea. As said, in order to perform the numerical integration the only crucial part in this is computing the derivatives involved in the set of differential equations. There are several techniques known and Wyatt [6] gives a comprehensive overview. One technique called *least squares fitting* is commonly used. This technique turns out to be numerically stable, however, it can be inappropriate with respect to the issues discussed in Section 22.2 and thus gives large numerical errors for general initial conditions and potentials.

On the other hand there are several other fitting techniques, among them polynomial fitting. The latter turns out to give very accurate numerical results in the first few time integration steps, however, it soon becomes unstable. Understanding the source of instabilities, however, makes it possible to stabilize polynomial fitting.

Fitting Methods. Let us quickly explain the two fitting methods used to compute the derivatives encountered in Equations 22.5 through 22.7 in the later numerical example: *least squares fitting* and *polynomial fitting*. Both provide an algorithm for finding, e.g., a polynomial* of degree, say $(m - 1)$, which is in some sense to be specified close to a given function $f : \mathbb{R} \to \mathbb{R}$ known only on a set of, say n, pairwise distinct data points, $(x_i, f(x_i))_{1 \leq i \leq n}$, as a subset of the graph of f. The derivative can then be computed from the fitting polynomial by algebraic means. For the further discussion we define

$$y: = (f(x_i))_{1 \leq i \leq n}$$
$$X: = (x_i^{(j-1)})_{1 \leq i \leq n, 1 \leq j \leq m}$$
$$a = (a_i)_{1 \leq i \leq m} \in \mathbb{R}^m$$
$$\delta = (\delta_i)_{1 \leq i \leq n} \in \mathbb{R}^n.$$

Using this notation the problem of finding a fitting polynomial to f on the basis of n pairwise distinct data points of the graph of f reduces to finding the coefficients of the vector a obeying the equation

$$y = X \cdot a + \delta$$

such that the error term δ is in some sense small. Here the dot $\cdot$ denotes matrix multiplication. The fitting polynomial is given by $p(x) = \sum_{j=1}^{m} a_j x^{j-1}$.

---

* Or more general an element of an m-dimensional vector space of functions. The coefficients of the design matrix X determine which basis functions are used.

- Polynomial Fitting: For the case $n = m$ we can choose the error term δ to be identically zero since X is an invertible square matrix and $a = X^{-1} \cdot y$ can be straightforwardly computed.
- Least Squares Fitting: For arbitrary $n > m$ there is no longer a unique solution and one needs a new criterion for finding a unique vector a. The algorithm of least squares fitting therefore uses the minimum value of the accumulated error $\Delta := \sum_{i=1}^{n} w(x - x_i)\delta_i^2$ where $w : \mathbb{R} \to \mathbb{R}$ specifies a weight dependent on the distance between the i-th data point x_i and some point x where the fitting polynomial is to be evaluated. The vector a can now be determined by minimizing Δ as a function of a by solving

$$\left(\frac{\partial \Delta(a)}{\partial a_j} \right)_{1 \le j \le m} = \left(-2 \sum_{i=1}^{n} w(x - x_i)(y - X \cdot a)_i x_i^{j-1} \right)_{1 \le j \le m} = 0 .$$

Note that for $m = n$ and any non-zero weight w the algorithm of least squares fitting produces the same a as the polynomial fitting would do since for a determined by polynomial fitting the non-negative function $\Delta(a)$ is zero and thus a naturally minimizes the error term.

The advantage of polynomial fitting is that the fitting polynomials actually go through all the data points, $(x_i, f(x_i))_{1 \le i \le n}$. However, when data points lie unfavorably the resulting fitting polynomial may have high oscillations, see e.g., Figure 22.2.

The advantage of least squares fitting is that it produces very smooth fitting polynomials. However, the fitting polynomials do not reproduce the derivatives of f on the scale of the grid points, see e.g., Figure 22.3.

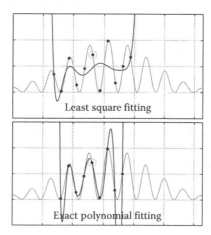

FIGURE 22.3 Comparison of least squares fitting and polynomial fitting. The stars denote the data points used for the fitting, the red function denotes the wavefunction to be fitted and the blue function denotes the fitting polynomials. Note that the least squares fitting polynomial does not even reproduce the correct sign of the derivative at the marked point.

As we shall explain below, both disadvantages lead to serious problems when implementing Wyatt's idea. However, with little more effort one can establish fitting algorithms for which both disadvantages appear only in a mild way:

- One idea is to fit a polynomial to the data points $(x_i, f(x_i))_{1 \le i \le n}$ such that it fulfills the following two conditions: First, it goes through the relevant points where we want to evaluate the derivatives of the function and its two next neighbors. Second, it gives the best fit in the least squares sense for all remaining data points. Therefore, one needs to choose a polynomial degree $(m - 1)$ with $3 < m < n$. The advantage hereby is that although one reproduces the most relevant points the resulting fitting polynomials are comparably smooth.
- Another idea is to use fitting polynomials with a degree $(m - 1)$ with $m > n$. This way the fitting polynomial is underdetermined. Among all the polynomials of degree $(m - 1)$ which go through all the n data points one chooses the one with least oscillations as fitting polynomial. The oscillations could be measured by the largest derivative at all data points.

We shall refer to such fitting methods as "hybrid fitting methods." Let us point out how the different fitting methods react to the numerical problems stated in Section 22.2.

When the Grid Point Trajectories Approach Each Other. Let us first explain why least squares fitting is inappropriate in cases of grid point trajectories which approach each other. Although the algorithm of least squares fitting is well known for its tendency to stabilize numerical simulations by averaging out numerical errors, exactly this averaging is responsible for not keeping the grid points on their Bohmian trajectories in such cases, and thus it is incapable of preventing grid points from crossing: recall the role of the quantum potential discussed above, cf. Figure 22.1, and the importance of reproducing the microscopic structure of the wavefunction during numerical fitting. It is absolutely vital for a numerical simulation to implement a fitting algorithm that reconstructs not only R, respectively S, in an accurate way but also its second derivative. The tendency of the least squares fitting algorithm is to average out those aforementioned small bumps in R, respectively S. As the simulation is then blind to recognize these bumps in the wavefunction, it is incapable of recognizing grid points moving toward each other. Grid point trajectory crossings become unavoidable, and the motion of the grid points is no longer Bohmian.

Note that since the averaging occurs on a microscopic scale, situations in which grid points move only very slowly or generically do not move toward each other (for example a free Gaussian wave packet or one approaching a potential creating only soft reflections) are often numerically doable with least squares fitting. On the other hand numerical simulations with least squares fitting in general situations like the one we shall discuss in the numerical example below are bound to break down as soon as two grid points get too close to each other.

A good numerical method must therefore take note of the small bumps and must prevent their increase. Such a numerical method is provided by polynomial fitting, since there the polynomials go through all grid points. For this reason polynomial fitting recognizes the bumps of R and S, and hence, the resulting quantum potential recognizes the approaching grid points. Now recall the physics of the Bohmian time evolution which, as we have stressed in the introduction, prevents configuration space trajectories from crossing each other. Therefore, one can expect this method to be self-correcting and hence stabilizing. For a comparison of polynomial and least squares fitting see Figure 22.3.

When the Wavefunction Develops Nodes. As described above the problem of approaching grid point trajectories appears in particular at nodes of the wavefunction. At nodes R is very small and all effects of the quantum potential Equation 22.9 are amplified. Averaging out the derivatives on the length scale of grid points may lead to crossing of grid point trajectories, i.e., the simulated grid point trajectories may behave unphysically. On the other hand, also exaggeratedly large derivatives coming from possible highly oscillating polynomials (as, for example, produced by polynomial fitting) are amplified. Our later numerical example shows that the former problem is more severe than the latter. Nevertheless, choosing another fitting method (we expect the hybrid fitting methods to work very well in general situations) will give even better results than polynomial fitting.

Note, however, that by equivariance in the vicinity of the nodes of the wavefunction one expects only a few grid points that may behave badly.

At the Boundary of the Grid. It has to be remarked that polynomial fitting creates a more severe problem at the boundary of the supporting grid than least squares fitting. This problem is of a conceptual kind since at the boundary there is a generic lack of knowledge of how the derivatives of R or S behave, see Figure 22.2.

At the boundary the good property of polynomial fitting, namely to recognize all small bumps, works against Bohmian grid techniques. On the contrary, least squares fitting simply averages these small numerical errors out—see the lower right plot in Figure 22.2. The conceptual boundary problem, however, remains and will show up eventually also with least squares fitting. Also here we expect better results from the above mentioned hybrid fitting methods.

Using Equivariance for Stabilization. As discussed in the Introduction, if the grid points are distributed according to $|\psi^0|^2$, they will remain so for all times. The advantage in this is that now the system is over-determined and the actual density of grid points must coincide with R^2 for all times. This could be used as an on the fly check whether the grid points behave like Bohmian trajectories or not. If not, the numerical integration does not give a good approximation to the solution of Equations 22.5 through 22.7. It could also be considered to stabilize the numerical simulation with a feedback mechanism balancing R^2 and the distribution of the grid points which may correct numerical errors of one or the other during a running simulation.

22.4   A NUMERICAL EXAMPLE IN MATLAB®

The simple numerical example given in this section illustrates the increased numerical stability when polynomial fitting instead of least squares fitting (least squares fitting should only be used to ease the boundary problem). This example, however, is clearly not the optimal method; it is rather meant to support our arguments given so far and can easily be improved, e.g., using hybrid fitting methods. We shall consider a numerical example in one dimension with one particle. Therefore, in what follows $N = 1$ and the n trajectories are the possible configuration space trajectories (used as a co-moving grid) of this one particle. We remark, as mentioned earlier, that Gaussian wave packets are unfit for probing the quality of the numerical simulations, because all grid point will be moving apart from each other and no crossing has to be feared. Therefore, we take as initial conditions two superposed free Gaussian wave packets with a small displacement together with a velocity field identically zero and focus our attention on the region where their tails meet, i.e., where the Bohmian trajectories will move toward each other according to the spreading of the wave packets, see Figure 22.4. The physical units given in the figures and the following discussion refer to a wave packet on a length scale of $1\text{Å} = 10^{-10}$ m and a Bohmian particle with the mass of an electron.

The numerical simulation is implemented as in Ref. [6] with minor changes. It is written in MATLAB® using IEEE Standard 754 double precision. The source code can be found in [7]. In order to ease the boundary problem we use a very strong least squares fitting at the boundary. Note that this only eases the boundary problem and is by no means a proper cure. In between the boundary, however, the number of data points for the fitting algorithm and the degree of the fitting polynomial can be chosen in each run. In this way it is possible to have either polynomial fitting or least squares fitting in between the boundary. The simulation is run twice: first with the polynomial fitting and then with the least squares fitting algorithm in between the boundaries.

We take seven basis functions for the fitting polynomial in both cases. Note that the choice should at least be greater or equal to four in order to have enough information about the third derivative of the fitting polynomial. In the first run with polynomial fitting the number of grid points used for fitting is seven and in the second run for the least squares fitting we choose nine, which induces a mild least squares behavior. For the numerical integration we have used a time step of 10^{-2} fs with a total

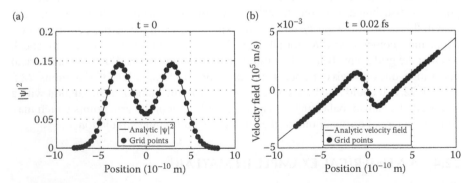

FIGURE 22.4 Left: Initial wavefunction at time zero. Right: Its velocity field after a very short time.

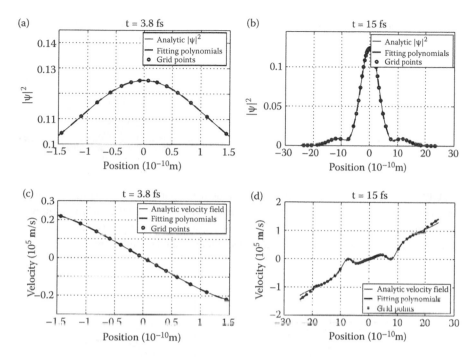

FIGURE 22.5 Simulated $|\psi|^2$ together with the velocity field at times $t = 3.8$ fs and $t = 15$ fs using polynomial fitting in between the boundary. Left: the center part of the wavefunction where grid points move toward each other. Right: the whole wavefunction at a later time. The grid points have all turned and move apart from each other. The kinks at $\pm 20 \cdot 10^{-15}$ m in the velocity field in the lower right plot are due to the transition from least squares fitting at the boundary to polynomial fitting in between the boundary (see paragraph: The Boundary Problem in Section 22.3).

number of 51 grid points supporting the initial wavefunction. The grid points have not been $|\psi^0|^2$ distributed because in order to see only the differences between least squares fitting and polynomial fitting the distribution of the grid points does not matter too much.

Results. The first run with polynomial fitting yields accurate results and does not allow for trajectory crossing way beyond 5000 integration steps, i.e., 50 fs, see Figures 22.5 and 22.6. The second run with least squares fitting reports a crossing of trajectories already after 430 integration steps, i.e., 4.3 fs, and aborts, see Figures 22.7 and 22.8. The numerical instability can already be observed earlier, see the left-hand side of Figure 22.7. An adjustment of the time step of the numerical integration does not lead to better results in the second run. The crossing of the trajectories of course occurs exactly in the region were the grid points move fastest toward each other. Referring to the discussion before, the small bumps in R created by the increase of the grid point density in this region are not seen by the least squares algorithm and

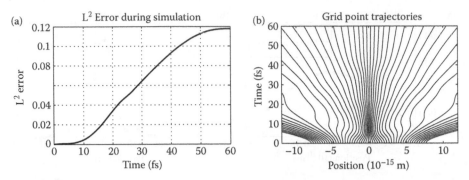

FIGURE 22.6 Both plots belong to the simulation using polynomial fitting in between the boundary. Left: L^2 distance between the simulated wavefunction and the analytic solution of the Schrödinger equation. Right: a plot of the trajectories of the grid points. Note how some trajectories initially move toward each other, decelerate, and finally move apart.

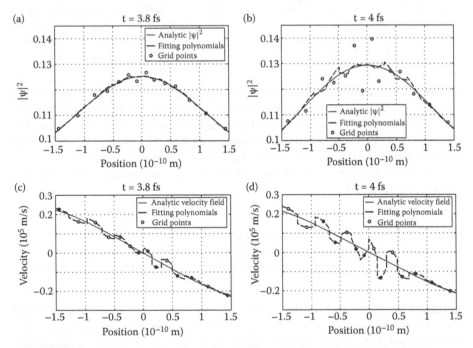

FIGURE 22.7 Left: same as left-hand side of Figure 22.5 but this time using least squares fitting throughout. Right: a short time later. Note in the upper left plot, the fitting polynomials of least squares fitting fail to recognize the relatively big ordinate change of the grid points. So the grid points moving toward each other are not decelerated by the quantum potential. Finally after $t = 4.3$ fs a crossing of the trajectory occurs.

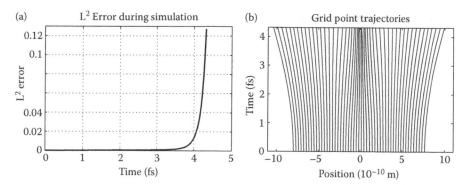

FIGURE 22.8 Both plots belong to the simulation using least squares fitting throughout. Left: L^2 distance between the simulated wavefunction and the analytic solution of the Schrödinger equation. Right: a plot of the trajectories of the grid points. Note by comparison with right-hand side of Figure 22.6 the failure of least squares fitting to recognize grid point trajectories moving toward each other until they finally cross and the numerical simulation aborts.

thus are not seen by the numerical integration of Equations 22.5 through 22.7, compare Figures 22.5 and 22.7. The approaching grid points are not decelerated by the quantum potential and finally cross each other, see Figure 22.7, while in the first run they begin to decelerate and turn to the opposite direction during the time between 3 fs and 5 fs, see Figure 22.5. To visualize how the two algorithms "see" R^2 and the velocity field during numerical integration these entities have been plotted in Figures 22.5 and 22.7 by merging all fitting polynomials in the neighborhood of every grid point together (from halfway to the left neighboring grid point to halfway to the right neighboring grid point). In Figure 22.7 one clearly observes the failure of the least squares fitting algorithm to see the small bumps in R.

Note that in some special situations in which the time step of the numerical integration is chosen to be large, polynomial fitting may lead to trajectory crossing as well. This may happen when the number of time steps in which the simulation has to decelerate two fast approaching grid points is not sufficient. This effect is entirely due to the fact that numerical integration coarse-grains the time. The choice of a smaller time step for the simulation will always cure the problem as long as other numerical errors do not accumulate too much.

The relevant measure of quality of the numerical simulation is naturally the L^2 distance between the simulated wavefunction Re^{iS} and the analytic solution of the Schrödinger equation ψ_t, i.e., $\left(\int dx \,|\psi_t(x) - R(x,t)e^{iS(x,t)}|^2\right)^{1/2}$, and is spelled out on the left-hand side of Figures 22.6 and 22.8 for both runs.

BIBLIOGRAPHY

1. S. Goldstein. Bohmian Mechanics. Contribution to *Stanford Encyclopedia of Philosophy*, ed. by E.N. Zalta, 2001. http://plato.stanford.edu/entries/qm-bohm/
2. D. Dürr and S. Teufel *Bohmian Mechanics*. Springer, Berlin, 2009.

3. D. Dürr, S. Goldstein, and N. Zanghì. Quantum Equilibrium and the Origin of Absolute Uncertainty. *J. Statist. Phys.*, 67:843–907, 1992.
4. S. Teufel and R. Tumulka. Simple Proof for Global Existence of Bohmian Trajectories. *Commun. Math. Phys.*, (258):349–365, 2005.
5. D. Dürr, S. Goldstein, and N. Zanghì. Quantum Equilibrium and the Role of Operators as Observables in Quantum Theory. *J. Statist. Phys.*, 116:959–1055, 2004.
6. R.E. Wyatt. *Quantum Dynamics with Trajectories*. Springer, New York, 2005.
7. D.-A. Deckert, D. Dürr, and P. Pickl. *Quantum Dynamics with Bohmian Trajectories*. arXiv:quant-ph/0701190, 2007.

23 Approximate Quantum Trajectory Dynamics in Imaginary and Real Time; Calculation of Reaction Rate Constants

Sophya Garashchuk

CONTENTS

23.1 INTRODUCTION

In this chapter we formulate quantum trajectory dynamics in imaginary time which seamlessly connects to Bohmian dynamics in real time. Approximate determination of the quantum potential and mixed representation of wavefunctions with nodes make this trajectory approach stable and scalable to high dimensions, while including the leading quantum effects into wavefunction evolution. Reaction probabilities, energy eigenstates and reaction rate constants can be obtained by propagating an ensemble of quantum trajectories in real and/or imaginary time.

Most of the quantum-mechanical (QM) theoretical studies of thermal reaction rate constants, $k(T)$, and cumulative reaction probabilities, $N(E)$, are based on the trace expressions [1], that involve evolution operators—the Hamiltonian operator in real time and the Boltzmann operator in inverse temperature, equivalent to imaginary time—and the symmetrized flux operator,

$$\bar{F} = \frac{\iota}{2m}[\hat{p}, \delta(x - x_0)].$$

(23.1)

The function $\delta(x - x_0)$ is the Dirac δ-function; x is the reaction coordinate, m is the reduced mass conjugate to x, and x_0 is the location of the surface dividing reactants and products. In atomic units the cumulative reaction probability is

$$N(E) = 2\pi^2 \text{Tr}[\bar{F}\delta(E - \hat{H})\bar{F}\delta(E - \hat{H})].$$

(23.2)

The thermal reaction rate constant,

$$k(T) = Q^{-1}(T) \int_0^\infty C_{ff}(t)dt,$$

(23.3)

can be found from the flux–flux correlation function $C_{ff}(t)$,

$$C_{ff}(t) = \text{Tr}[e^{-\beta\hat{H}/2}\bar{F}e^{-\beta\hat{H}/2}e^{\iota\hat{H}t}\bar{F}e^{-\iota\hat{H}t}].$$

(23.4)

In the above equations T labels temperature, $Q(T)$ is the reactant partition function, and $\beta = (k_B T)^{-1}$ is the thermal evolution variable, where k_B is the Boltzmann constant. The real time is labeled t; x_0 is set to zero for both dividing surfaces. While $N(E)$ is a more general quantity from which the reaction rate constants can be computed for any temperature,

$$2\pi k(T)Q(T) = \int_{-\infty}^\infty e^{-\beta E}N(E)dE,$$

(23.5)

Equations 23.3 and 23.4 can be better suited for numerical evaluation due to the damping effect of the Boltzmann operator, $\exp(-\beta\hat{H})$, suppressing wavefunction oscillations.

In general, to describe the dynamics of several (more than four) nuclei, approximate wavefunction evolution methods are needed, because the conventional grid- or basis-based QM methods [2] scale exponentially with the system size. The trajectory-based propagation techniques, such as those of Refs [3–8], combining intuitiveness and low cost of classical mechanics with a description of the leading quantum effects on dynamics of nuclei provide appealing alternatives. Several recent approaches [9–14] are based on the de Broglie–Bohm formulation of the Schrödinger equation, where the wavefunction is represented by the quantum trajectories propagated with approximations to its derivatives. We develop dynamics with globally determined approximations to the quantum potential: In Section 23.2 we show how to compute the cumulative reaction probability Equation 23.2 by propagating Gaussian-based eigensolutions of the flux operator [15] using the variational Approximate Quantum Potential (AQP) [9] and the mixed wavefunction representation [16, 17]; in Section 23.3 we describe the imaginary-time quantum trajectory dynamics with the momentum-dependent quantum potential (MDQP) [18] suitable for computation of low-lying energy eigenstates; in Section 23.4 we combine real and imaginary-time dynamics to compute reaction rates from Equation 23.3 to Equation 23.4. For implementations that are cheap and scalable to many dimensions, the MDQP is determined by fitting the trajectory momenta in a small basis, also used to evolve wavefunctions with nodes.

All derivations are given in one Cartesian dimension, $x - (-\infty, \infty)$, in atomic units. ∇ denotes differentiation with respect to x for compactness. Planck's constant is explicitly included in several places to emphasize the $\hbar$-dependence of the quantum potential. One-dimensional numerical illustrations are given for anharmonic wells and for the Eckart barrier modeling the transition state of H_3.

23.2   CALCULATION OF N(E) WITHIN THE FLUX OPERATOR FORMALISM WITH APPROXIMATE QUANTUM TRAJECTORIES

As first observed numerically [19], in a finite basis the singular operator $\bar{F}$ given by Equation 23.1 has just two nonzero eigenvalues, which enabled efficient QM calculations of $N(E)$ for tetratomic systems using the transition state wavepackets [20, 21]. The two eigenvalues are negatives of each other and the corresponding eigenvectors are complex conjugates,

$$\bar{F}|\phi^{\pm}\rangle = \pm\lambda|\phi^{\pm}\rangle, \quad |\phi^{+}\rangle^{*} = |\phi^{-}\rangle. \tag{23.6}$$

Substituting the spectral representation of $\bar{F}$ and the integral expression of $\delta(E - \hat{H})$,

$$2\pi\delta(E - \hat{H}) = \int_{-\infty}^{\infty} e^{-\imath\hat{H}t}e^{\imath Et}dt,$$

into Equation 23.2, $N(E)$ can be written through the wavepacket correlation functions $C^{\pm}$,

$$C^{\pm} = \langle\phi^{+}(0)|\phi^{-}(t)\rangle, \tag{23.7}$$

$$N(E) = \lambda^{2}\left(\left|\int_{-\infty}^{\infty} C^{+}e^{\imath Et}dt\right|^{2} - \left|\int_{-\infty}^{\infty} C^{-}e^{\imath Et}dt\right|^{2}\right). \tag{23.8}$$

The eigensolutions of $\bar{F}$ in a finite basis, derived analytically by Seideman and Miller [22], can be expressed in terms of the Dirac δ-function [15],

$$\phi^{\pm} = \frac{\delta(x)}{\sqrt{2\langle\delta|\delta\rangle}} \mp \frac{\imath\nabla\delta(x)}{\sqrt{2\langle\nabla\delta|\nabla\delta\rangle}}, \quad \lambda = \frac{\sqrt{\langle\delta|\delta\rangle\langle\nabla\delta|\nabla\delta\rangle}}{2m}. \tag{23.9}$$

In principle, the limit of any δ-function sequence [23] can be used in Equation 23.9, for example the sinc-function sequence [24]. The Gaussian-based δ-function sequence,

$$\delta(x) = \lim_{\gamma \to \infty} \sqrt{\frac{\gamma}{\pi}} \exp(-\gamma x^2), \qquad (23.10)$$

can be advantageous for many trajectory-based propagation methods (including the AQP approach) [3–7,9], because they are exact for Gaussian wavepackets and because the initial wavefunction localized in coordinate space can be efficiently represented with trajectories. Therefore, we approximate the eigensolution Equation 23.9 with

$$\phi^{\pm}(x) = \left(\frac{2\gamma}{\pi}\right)^{1/4} \exp(-\gamma x^2) \left(\frac{1}{\sqrt{2}} \pm \imath \sqrt{2\gamma} x\right), \quad \lambda = \frac{\gamma}{m\sqrt{8\pi}}, \qquad (23.11)$$

choosing a large value of the width parameter γ which gives $N(E)$ independent of γ. Short-time analytical propagation in the parabolic approximation to the barrier can be introduced to widen the highly localized initial wavefunction Equation 23.11 as described in Ref. [15].

23.2.1 REAL-TIME VARIATIONAL APPROXIMATE QUANTUM POTENTIAL

The AQP method [9] is based on the quantum trajectory formulation [25, 26] of the usual time-dependent Schrödinger equation (TDSE),

$$\left(-\frac{\hbar^2}{2m}\nabla^2 + V\right)\psi(x,t) = \imath\hbar\frac{\partial}{\partial t}\psi(x,t). \qquad (23.12)$$

The wavefunction expressed in terms of the real amplitude $A(x,t)$ and the real phase $S(x,t)$,

$$\psi(x,t) = A(x,t)e^{\imath S(x,t)/\hbar}, \qquad (23.13)$$

is represented by an ensemble of trajectories characterized at time t by positions x_t, momenta p_t and action functions S_t. The trajectory momentum is the gradient of the wavefunction phase,

$$p(x,t) = \nabla S(x,t). \qquad (23.14)$$

In the Lagrangian frame of reference,

$$\frac{d}{dt} = \frac{\partial}{\partial t} + \frac{p}{m}\nabla, \qquad (23.15)$$

Newton's laws of motion for the quantum trajectories follow from the TDSE:

$$\frac{dx_t}{dt} = \frac{p_t}{m}, \quad \frac{dp_t}{dt} = -\nabla(V + U), \qquad (23.16)$$

$$\frac{dS_t}{dt} = \frac{p_t^2}{2m} - (V + U). \qquad (23.17)$$

The function U denotes the quantum potential,

$$U = -\frac{\hbar^2}{2m} \frac{\nabla^2 A(x,t)}{A(x,t)},$$ (23.18)

which vanishes in the classical limit $\hbar \to 0$ for nonzero wavefunction amplitudes. The wavefunction density $A^2(x,t)$ satisfies the continuity equation which, taking into account evolution of the volume element δx_t associated with each trajectory,

$$\delta x_t = \delta x_0 \exp\left(\int_0^t \frac{\nabla p_\tau}{m} d\tau\right),$$ (23.19)

results in conservation of the probability within δx_t, i.e., of the trajectory "weight" $w(x_t)$,

$$w(x_t) = A^2(x_t)\delta x_t, \qquad \frac{dw(x_t)}{dt} = 0.$$ (23.20)

Therefore, $A(x,t)$ can be reconstructed from trajectory positions and weights.

The concept of the trajectory weights is central to the approximate implementation of Equations 23.16 and 23.17 of Garashchuk and Rassolov [9, 27]. The AQP (labeled $\tilde{U}$) is determined variationally, thus conserving the energy of the trajectory ensemble, from linearization of the nonclassical component, $r(x,t)$, of the momentum operator:

$$r(x,t) = \frac{\nabla A(x,t)}{A(x,t)}.$$ (23.21)

Minimization of $\langle(r - \tilde{r})^2\rangle$, where $\tilde{r}$ is a linear function, has a solution in terms of the moments of the trajectory distribution:

$$r(x,t) \approx \tilde{r}(x,t) = -\frac{x - \langle x\rangle_t}{2(\langle x^2\rangle_t - \langle x\rangle_t^2)}.$$ (23.22)

$\tilde{U}$ and its gradient are determined analytically,

$$\tilde{U} = -\frac{\hbar^2}{2m}\left(\tilde{r}^2(x,t) + \nabla\tilde{r}(x,t)\right).$$ (23.23)

The AQP method is exact for Gaussian wavepackets and describes wavepacket bifurcation, moderate tunneling, and zero-point energy for times dependent on the anharmonicity of V. It is numerically cheap and stable due to the "mean-field"-like procedure of determining AQP: with Equation 23.20 the expectation values in Equation 23.22 are computed as sums over trajectories labeled with index k,

$$\langle\Omega(x)\rangle_t = \int \Omega(x)A^2(x,t)dx = \int \Omega(x_t)A^2(x_t)\delta x_t = \sum_k \Omega(x_t^{(k)})w^{(k)}.$$ (23.24)

Using the mixed wavefunction representation, the correlation functions of Equation 23.7 can also be computed as single sums over trajectories, without finding $A(x,t)$.

23.2.2 WAVEPACKET EVOLUTION AND COMPUTATION OF CORRELATION FUNCTIONS IN THE MIXED COORDINATE/POLAR REPRESENTATION

To propagate the flux operator eigenfunctions Equation 23.11, we use the mixed coordinate/polar wavefunction representation introduced in Ref. [16],

$$\phi(x,t) = \psi(x,t)\chi(x,t), \tag{23.25}$$

where $\psi(x,0)$ is a normalized Gaussian and $\chi(x,0)$ is its polynomial prefactor. The coordinate part, $\chi(x,t)$, is represented in the Taylor basis $\vec{f}(x) = \{1, x, \ldots\}$ of size N_a,

$$\chi(x,t) = \vec{f}(x) \cdot \vec{a}(t). \tag{23.26}$$

The polar part, $\psi(x,t)$, is an initially nodeless function solving the TDSE Equation 23.12. Substitution of Equations 23.25 and 23.26 into the TDSE for $\phi(x,t)$, multiplication by $\psi^*(x,t)\vec{f}(x)$ and integration over x gives the matrix equation for $d\vec{a}/dt$,

$$\imath \langle \vec{f} \otimes \vec{f} \rangle \frac{d\vec{a}}{dt} = \left(\frac{\langle \nabla \vec{f} \otimes \nabla \vec{f} \rangle}{2m} - \imath \frac{\langle p \vec{f} \otimes \nabla \vec{f} \rangle}{m} \right) \vec{a}. \tag{23.27}$$

All integrals are evaluated according to Equation 23.24. For example, the matrix elements of the last term of Equation 23.27 are

$$\langle p \vec{f} \otimes \nabla \vec{f} \rangle_{ij} = \sum_k p_t^{(k)} f_i(x_t^{(k)}) \nabla f_j(x_t^{(k)}) w^{(k)}.$$

The polar part $\psi(x,t)$ is described in terms of quantum trajectories, initiated with zero initial momenta at positions sampling the Gaussian part of $\phi^{\pm}(x)$ given by Equation 23.11, propagated in the presence of AQP. The functions $\chi^{\pm}(x,0) = a_1(0) \pm a_2(0)x$ are the polynomial prefactors of $\phi^{\pm}(x)$. Restricting $\chi^{\pm}$ to a linear form, for symmetric V the coefficients are:

$$a_1(t) = a_1(0), \quad a_2(t) = a_2(0) \exp\left(-\int_0^t \frac{\langle p\,x \rangle_\tau + \imath/2}{m \langle x^2 \rangle_\tau} d\tau \right). \tag{23.28}$$

The mixed representation reduces the propagation error and cost, because it allows for stable trajectory dynamics of the nodeless (at least for short time) $\psi(x,t)$ and it simplifies calculations of the correlation functions, $C^{\pm}$, in Equation 23.7. $C^{\pm}$ can be expressed via the trajectory weights if the evolution operator is equally partitioned between bra and ket,

$$C^{\pm}(2t) = \sum_k w^{(k)} e^{2\imath S_t^{(k)}} \left(a_1(t) + a_2(t)x_t^{(k)} \right) \left(a_1(t) \mp a_2(t)x_t^{(k)} \right). \tag{23.29}$$

23.2.3 NUMERICAL ILLUSTRATION

Here the wavepacket formulation of Equations 23.7, 23.8, and 23.11 is used to compute the reaction probability, $N(E)$, for a one-dimensional model of the H_3 transition state—for the Eckart barrier scaled to have $m = 1$,

$$V = V_0 \cosh^{-2}\alpha x. \tag{23.30}$$

In scaled units the barrier height is $V_0 = 16$ and $\alpha = 1.3624\ a_0^{-1}$.

The purpose was to assess the accuracy of the AQP description of the tunneling regime. The implementation was not optimized. The trajectory positions were distributed uniformly. For linear $\chi(x,t)$, the only quantities needed to determine the quantum force and the polynomial coefficient a_2 in Equation 23.28 are $\langle x^2 \rangle$ and $\langle px \rangle$, which could be computed with relative errors below 2×10^{-5} with 1000 trajectories. We needed many more trajectories (10^6) to obtain low-tunneling probabilities with uniform sampling. The numerical cost scales linearly with the number of trajectories. In principle, the AQP parameters and coefficients $\vec{a}$ can be simply stored from a calculation with a few thousand trajectories and used to propagate a larger number of independent trajectories, possibly, with more effective trajectory sampling.

The cumulative reaction probability, $N(E)$, obtained for $\gamma = 1000\ a_0^{-2}$ is shown in Figure 23.1. The trajectories were propagated up to $t = 1.5$, which, due to the doubling of time in Equation 23.29, would be sufficient to resolve probabilities above 10^{-9} in a conventional QM wavepacket calculation. For energies below V_0 the AQP probabilities above 10^{-5} were quite accurate. The AQP probabilities below this value were too high. Nevertheless, there was a noticeable improvement over the parabolic barrier approximation: for $E - 0.52$, the lowest energy resolved, the parabolic barrier dynamics overestimated $N(E)$ by 2×10^4, while the AQP probability was 200 times smaller than that. As argued in Ref. [9], out of the two approximations made—the linear quantum force and the linear form of $\chi(x,t)$—the former is the main source of error.

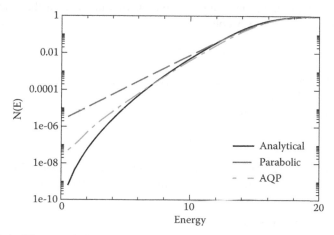

FIGURE 23.1 The cumulative reaction probability, $N(E)$, for the Eckart barrier obtained from propagation of the flux eigenvectors using the AQP trajectory dynamics and mixed wavefunction representation. A comparison is made with the analytical QM probability and with the probability derived from the analytical wavefunctions evolving in the parabolic potential in Equations 23.7 and 23.8.

23.3   TRAJECTORY DYNAMICS IN IMAGINARY TIME WITH THE MOMENTUM-DEPENDENT QUANTUM POTENTIAL

Direct computation of the reaction rate constants according to Equations 23.3 and 23.4 requires Boltzmann evolution of wavefunctions, equivalent to propagation in imaginary time. Another type of problem solvable in imaginary time is the determination of low-lying energy levels and eigenfunctions [28], in particular of the energy levels in high-dimensional systems with Monte Carlo methods [29–31]. The approach is based on the simple fact that in imaginary time,

$$\hat{H}\psi(x, \tau) = -\hbar\frac{\partial\psi(x, \tau)}{\partial\tau}, \quad \tau > 0 \tag{23.31}$$

the higher energy components of a wavefunction $\psi(x, \tau)$ decay faster than the lower energy components: a wavefunction expressed in terms of the eigenstates $\{\psi_k\}$ of $\hat{H}$ ($E_k > 0$),

$$\hat{H}\psi_k(x) = E_k\psi_k(x), \quad \psi(x, \tau) = \sum_k c_k e^{-E_k\tau}\psi_k(x), \tag{23.32}$$

decays to the lowest energy state of the same symmetry. Some recent examples are calculations performed for malonaldehyde [32] and for CH_5^+ [33].

Substitution of the time variable, $t \to -\iota\tau$ ($\tau > 0$), transforms the real-time TDSE Equation 23.12 into the diffusion Equation 23.31. In semiclassical treatments of Equation 23.31 [34, 35] transformation of $p \to -\iota p$ and $S \to -\iota S$, led to the classical equations of motion of trajectories in the "inverted" potential (the negative of the original V). The same transformation of Equations 23.16, however, leads to singular dynamics of a Gaussian wavefunction [36]. Such behavior is related to the nonunique representation of a real wavefunction in terms of the amplitude and "phase," as Equation 23.13 becomes

$$\psi(x, \tau) = A(x, \tau)e^{-S(x,\tau)/\hbar},$$

and re-partitioning of the wavefunction between A and S to avoid singularities was successfully employed.

For a real positive wavefunction we use an unambiguous exponential form,

$$\psi(x, \tau) = e^{-S(x,\tau)/\hbar}. \tag{23.33}$$

The wavefunction Equation 23.33 is the same as in the Zero Velocity Complex Action method of Ref. [37] implemented in imaginary time on a fixed grid.

23.3.1   EQUATIONS OF MOTION

Substitution of the wavefunction Equation 23.33 into Equation 23.31 and division by $\psi(x, \tau)$ gives

$$\frac{\partial S(x, \tau)}{\partial\tau} = -\frac{1}{2m}(\nabla S(x, \tau))^2 + V + \frac{\hbar}{2m}\nabla^2 S(x, \tau). \tag{23.34}$$

Using τ as a time variable and defining the trajectory momentum by Equation 23.14 in the Lagrangian frame of reference Equation 23.15, Equation 23.34 leads to the following equations of motion,

$$\frac{dx}{d\tau} = \frac{p}{m}, \quad \frac{dp}{d\tau} = \nabla(V + U). \tag{23.35}$$

Compared to the real-time equations 23.16 and 23.18, in imaginary time we have the "inverted" potential and the MDQP,

$$U(x, \tau) = \frac{\hbar}{2m} \nabla^2 S = \frac{\hbar \nabla p}{2m}. \tag{23.36}$$

Note that imaginary-time $U(x, \tau)$ is proportional to $\hbar/m$ and, thus, vanishes in the classical limit as its real-time counterpart Equation 23.18. In many dimensions Equation 23.36 generalizes to $U(\vec{x}, \tau) = \hbar \vec{\nabla} \cdot \vec{\nabla} S/(2m)$. Evolution of the imaginary-time action function is

$$\frac{dS}{d\tau} = \frac{p^2}{2m} + V + U, \tag{23.37}$$

and the energy of a trajectory is

$$\varepsilon = \frac{\partial S}{\partial \tau} = -\frac{p^2}{2m} + V + U. \tag{23.38}$$

Remarkably, for the wavefunction Equation 23.33 evolution of the volume element δx_τ given by Equation 23.19 *cancels* the contribution of MDQP to the average quantities,

$$\langle \Omega \rangle = \int_{-\infty}^{\infty} \Omega(x) e^{-2S(x)/\hbar} dx = \sum_k \Omega(x_\tau^{(k)}) e^{-2\tilde{S}_k/\hbar} \delta x_0^{(k)}. \tag{23.39}$$

$\tilde{S}$ indicates the "classical" action function computed along the *quantum* trajectory,

$$\tilde{S}_\tau = \int_0^\tau \left(\frac{p_t^2}{2m} + V(x_t) \right) dt. \tag{23.40}$$

For a Gaussian wavefunction evolving in a quadratic potential, $S(x, \tau)$ is a quadratic function of x, $p(x, \tau)$ is linear in x, U is a time-dependent constant and the quantum force, ∇U, is zero. Therefore, average x- or p-dependent quantities of classical ($U = 0$) and QM evolutions are the same. This quantum/classical similarity was one of the motivations behind the real-time Bohmian mechanics with Complex Action (BOMCA) in complex space [38].

23.3.2 APPROXIMATE IMPLEMENTATION AND EXCITED STATES

To implement Equations 23.35 and 23.37 one has to know derivatives of p determining U and ∇U. Since our goal is a cheap approximate methodology, we find derivatives from the least squares fit [39] to $p(x, \tau)$ weighted by $\psi^2(x, \tau)$. Representing $p(x, \tau)$

in a small basis $\vec{f}(x)$ of size N_b, $p(x,\tau) \approx \vec{f}(x) \cdot \vec{b}(\tau)$, the optimal coefficients $\vec{b}$ minimizing $\langle (p - \vec{f} \cdot \vec{b})^2 \rangle$ are

$$\vec{b} = \mathbf{M}^{-1}\vec{P}. \tag{23.41}$$

$\mathbf{M}$ is a time-dependent symmetric matrix of the basis function overlaps,

$$M_{ij} = \langle f_i | f_j \rangle = \sum_k f_i(x_\tau^{(k)}) f_j(x_\tau^{(k)}) \exp(-2\tilde{S}_\tau^{(k)}) \delta x_0^{(k)}. \tag{23.42}$$

The vector $\vec{P}$ includes the trajectory momenta averaged with the basis functions,

$$P_i = \langle p f_i \rangle = \sum_k p_\tau^{(k)} f_i(x_\tau^{(k)}) \exp(-2\tilde{S}_\tau^{(k)}) \delta x_0^{(k)}. \tag{23.43}$$

Placing the minimum of V at $x = 0$ we use the Taylor basis,

$$\vec{f}(x) = (1, x, x^2, \dots)^T, \tag{23.44}$$

to determine MDQP and to evolve wavefunctions with nodes in the mixed coordinate space/trajectory representation analogous to Equation 23.25.

The nodeless $\psi(x,\tau)$ solving Equation 23.31 evolves in imaginary time into the ground state wavefunction according to Equation 23.32. For $\chi(x,\tau)$ represented in a basis $\vec{f}(x)$ of size N_c,

$$\chi(x,\tau) = \vec{f}(x) \cdot \vec{c}(\tau), \tag{23.45}$$

the time-dependent coefficients $\vec{c}$ are obtained from Equation 23.31 proceeding as in Section 23.2,

$$\frac{d\vec{c}}{d\tau} = -\frac{\hbar}{2m}\mathbf{M}^{-1}\mathbf{D}\vec{c}. \tag{23.46}$$

$\mathbf{D}$ is the matrix of the basis function derivatives,

$$D_{ij} = \langle \nabla f_i | \nabla f_j \rangle = \sum_k \nabla f_i(x_\tau^{(k)}) \nabla f_j(x_\tau^{(k)}) \exp(-2\tilde{S}_\tau^{(k)}) \delta x_0^{(k)}. \tag{23.47}$$

To find the ground state of a system we start with a Gaussian $\psi(x,0)$, since its evolution is exact in a quadratic potential for the smallest meaningful basis, $N_b = 2$. Although a linear approximation to $p(x,\tau)$ gives zero quantum force, which does not affect the positions of the trajectories, MDQP is still required to define the total energy. To obtain N_s eigenstates we evolve N_s mixed representation functions $\phi_n = \chi_n \psi$ ($n = 1, \dots, N_s$). The functions χ_n are represented in the Taylor basis Equation 23.44, and their coefficients are written as a matrix $\mathbf{C} = (\vec{c}_1, \vec{c}_2, \dots, \vec{c}_{N_s})$. The number of states, N_s, is no larger than the basis size, N_c. The initial functions χ_n will have the correct number of nodes, if the initial values C_{ij} are chosen for $\phi_n(x,0)$ to be, for example, the eigenfunctions of a harmonic oscillator. At each time step wavefunctions of the lower energy are projected out:

$$\chi_j^{new} = \chi_j - \sum_{k,k<j} \langle \chi_j | \chi_k \rangle \chi_k, \tag{23.48}$$

which in terms of $\mathbf{C}$ and the overlap matrix Equation 23.42 is

$$C_{ij}^{new} = C_{ij} - \sum_{k,k<j} \langle \chi_j | \chi_k \rangle C_{ik}, \quad \langle \chi_j | \chi_k \rangle = (\mathbf{C}^T \mathbf{M} \mathbf{C})_{jk}. \tag{23.49}$$

Both the determination of MDQP and the evolution of $\mathbf{C}$ require inversion of the overlap matrix $\mathbf{M}$. For $N_c = N_b$ the calculation of the excited states requires little effort in addition to the quantum trajectory propagation. The outlined approach is approximate and is expected to work for a few low-energy eigenstates where the modest basis size is sufficiently accurate.

23.3.3 CALCULATION OF THE ENERGY LEVELS

The approximate MDQP approach is illustrated for anharmonic systems, which proved to be challenging for real-time exact and approximate quantum trajectory dynamics [40–42]. The difficulty is the inherent instability of the real-time Bohmian trajectories describing stationary eigenstates: such trajectories should not move, therefore classical and quantum forces have to cancel each other *exactly*. Otherwise, small displacements of the trajectory positions lead to rapid "decoherence" of the trajectories and, consequently, to the loss of quantum effects. In contrast, the imaginary-time trajectories are not stationary—their dynamics reflects the decay of the wavefunction. For the harmonic oscillator, the imaginary-time trajectories "roll off" a parabolic barrier, which is the inverted well centered at $x = 0$, as shown on Figure 23.2. The same is qualitatively true for anharmonic single-well potentials in one dimension.

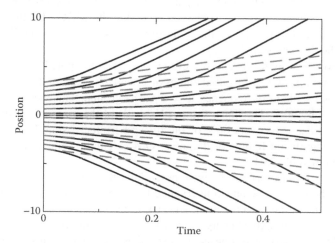

FIGURE 23.2 Imaginary-time quantum trajectories for the quadratic (dash) and quartic (solid line) oscillators. The quantum potential is found approximately using $N_b = 4$ for the latter potential.

TABLE 23.1

The Energy Levels ($n = 0, \ldots, 3$) for the Anharmonic Well of Equation 23.50 Obtained from the Imaginary-Time Quantum Trajectory Dynamics. The Bottom Line Contains the Exact QM Values from the Diagonalization of the Hamiltonian Matrix. In the Harmonic Approximation to the Potential, the Energy Levels are $E_n^{quad} = n + 1/2$

N_b	N_c	E_0	E_1	E_2	E_3
2	4	0.76007	2.3139	4.5298	7.1670
4	4	0.80358	2.7371	5.2856	8.2111
6	6	0.80375	2.7379	5.1829	7.9435
	QM	0.80377	2.7379	5.1719	7.9424

For illustration we consider a particle of mass $m = 1$ a.u. in a well with quartic anharmonicity [36],

$$V = x^4 + x^2/2. \tag{23.50}$$

The parameter $\gamma = 1 \, a_0^{-2}$ specifies a Gaussian initial wavefunction,

$$\psi(x, 0) = \left(\frac{2\gamma}{\pi}\right)^{1/4} e^{-\gamma x^2}, \tag{23.51}$$

evolved up to $\tau = 1.0$ a.u. using 500 trajectories. The trajectories corresponding to V of Equation 23.50 and its quadratic part, $V^{quad} = x^2/2$, are shown on Figure 23.2. Four basis functions, $N_b = 4$, were used to define the MDQP for the full V. The trajectories spread out faster in the anharmonic case, but the overall behavior is similar to the dynamics in V^{quad}.

Four lowest energy levels and wavefunctions were computed using four and six functions to represent χ_n. The initial values of $\mathbf{C}$ were taken as for the excited states of the harmonic oscillator of frequency $w = 2\gamma$. The ground state energy of V exceeds that of V^{quad} by 60%. The linear fitting ($N_b = 2$) of the momentum, exact for the harmonic oscillator, recovers 86% of the difference. Increasing N_b gives energies within four significant figures of the exact QM calculation for nearly all levels as summarized in Table 23.1.

The quantum trajectory results were essentially the same for $\gamma = 2$ and $0.5 \, a_0^{-2}$. In general, the decay time and convergence to the ground state will depend on the closeness of the initial wavefunction, $\psi(x, 0)$, to the ground state and on the energy levels. To obtain the highest energy eigenstate, $n = 3$, the projection procedure of Equation 23.48 necessitated time steps as small as $d\tau = 10^{-4}$. This eigenstate is the fastest to decay and, if the lower energy components are not removed from $\phi_3(x, \tau)$, which is a mixture of eigenstates, very frequently, then this wavefunction can decay to a lower energy eigenfunction.

TABLE 23.2

The Energy Levels for the Double Well for $n = 0, \ldots, 3$, Obtained with Approximate MDQP, $N_b = N_c = 6$, and by the Diagonalization of the Hamiltonian Matrix. The Initial Width Parameter is $a = 0.75$. The Normalized Energies of ϕ_n are Computed at Time τ Listed in the Second Column

m [a.u.]	τ [a.u.]	E_0	E_1	E_2	E_3
1	2.0	0.397	1.122	2.395	3.842
	QM	0.397	1.122	2.378	3.841
20	8.0	0.133	0.152	0.345	0.473
	10.0	0.131	0.152	0.348	0.479
	QM	0.133	0.152	0.332	0.465

The double well of Ref. [43], considered for $m = 1$ and $m = 20$,

$$V = (x^2 - 1)^2/4, \qquad (23.52)$$

presents a more challenging application. As m is increased the ground state function changes from a single peak to a double peak. The trajectory dynamics is more complicated, because unlike purely divergent trajectories of single wells, in the double well some trajectories stay in the barrier region. For $m = 20$, approximation of $p(x, \tau)$ within a small basis did not give fully converged eigenfunctions. Using larger bases, in principle, should give converged eigenvalues, but this was not pursued as it seemed impractical.

Table 23.2 shows the four lowest energy levels computed with $N_b = N_c = 6$. The initial width parameter is $\gamma = 0.75 \; a_0^{-2}$ for all calculations. The agreement of the wavefunction energies with the exact QM energy levels is fairly good, though the energies of ϕ_n for $m = 20$ weakly depend on τ after initial decay. The accuracy of the lower energy levels, which is better than 1.5%, resolves the $E_1 - E_0$ energy splitting of 0.02 a.u.; the accuracy for $n = 2$ and 3 is 5 and 3%, respectively. The accuracy for the $m = 1$ system is better than 1% for all levels. The ground state wavefunctions for $m = 1$ and $m = 20$ are shown in Figure 23.3. Note that the double peak for $m = 20$ has evolved from the Gaussian function (without the polynomial prefactors).

23.4 COMPUTATION OF REACTION RATES FROM THE IMAGINARY AND REAL-TIME QUANTUM TRAJECTORY DYNAMICS

Now we are ready to compute the reaction rate constants using Equation 23.4 within the approximate quantum trajectory dynamics in real and imaginary time and the mixed wavefunction representation [44]. Boltzmann evolution of the singular flux operator eigenvectors composed of all energy components—the infinite temperature situation—can be viewed as "cooling." Due to rapid decay of the high-energy components the wavefunctions become nonsingular for any finite temperature. In fact, we

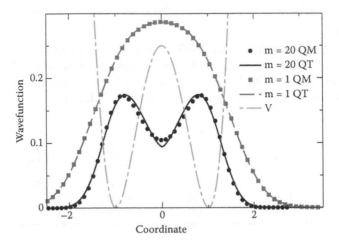

FIGURE 23.3 The ground state of the double well for $m = 20$ and $m = 1$ after propagation up to $\tau = 10$ a.u. and $\tau = 2$ a.u. respectively, computed with the bases of size $N_b = N_c = 6$. Symbols indicate QM wavefunctions obtained with the split-propagator evolution. The quantum trajectory (QT) results are shown with lines. The classical potential is represented with the dash.

can use eigensolutions of the "thermalized" flux,

$$\bar{F}_\beta = e^{-\beta \hat{H}/2} \bar{F} e^{-\beta \hat{H}/2}, \tag{23.53}$$

rather than those of $\bar{F}$. The inverse of "high temperature," parameter β_0 similar to γ of Equation 23.11, will define the initial wavepackets. The eigenfunctions of $\bar{F}_\beta$ given by Pollak [45] have the same structure as Equation 23.11 and can be propagated in imaginary time with quantum trajectories to the desired final β (low temperature) as described in Section 23.3.3, and then followed by the real-time trajectory propagation as described in Section 23.2.3. For a parabolic barrier

$$V^{par} = V_0 - \frac{mw^2 x^2}{2},$$

the eigensolutions of $\bar{F}_\beta \Phi_\beta^\pm = \pm \lambda_\beta \Phi_\beta^\pm$ are

$$\Phi_\beta^\pm = \phi^\pm(x) e^{-V_0 \beta}, \qquad \lambda_\beta = \frac{w}{\sqrt{8\pi} \sin(\beta w)} \tag{23.54}$$

where $\phi^\pm(x)$ is given by Equation 23.11 and the parameter γ defines the width of the thermal flux operator eigenvectors,

$$\gamma = \frac{mw}{2 \tan(\beta w/2)}. \tag{23.55}$$

Note that Equation 23.54 is valid only at high temperatures (small β) where $\gamma > 0$. We define the initial wavefunctions $\Phi_{\beta_0/2}^\pm$ for small values of β_0 (in the example below equivalent to $4V_0$), so that evolution in the parabolic approximation V^{par} to the anharmonic barrier V is accurate. These wavefunctions undergo Boltzmann evolution, followed by real-time propagation according to Equation 23.4,

$$k(T)Q(T) = \lambda_{\beta_0/2}^2 \int_{-\infty}^{\infty} \left(|\langle \phi_{\beta/4}^+(0)|\phi_{\beta/4}^+(t)\rangle|^2 - |\langle \phi_{\beta/4}^+(0)|\phi_{\beta/4}^-(t)\rangle|^2 \right) dt, \quad (23.56)$$

where

$$\phi_{\beta/4}^\pm(0) = \exp\left(-\frac{(\beta - \beta_a)\hat{H}}{4} \right) \Phi_{\beta_0/2}^\pm. \quad (23.57)$$

In the mixed wavefunction representation the correlation functions $\langle \phi_{\beta/4}^+(0)|\phi_{\beta/4}^\pm(t)\rangle$ are given by Equation 23.29. The Boltzmann evolution is accomplished as described in Section 23.3.3. The trajectory part of the thermal function, $\phi(x, \beta/4)$, is real, thus, the initial phase and momenta of the real-time trajectories is zero. The relation between the trajectory attributes at the start of the real time propagation on the left-hand-side of Equation 23.58 and at the end of the imaginary-time propagation on the right-hand-side of Equation 23.58 is straightforward:

$$x_k(0) = x_k(\beta/4)$$
$$w_k(0) = \psi^2(x_k)\delta x_{\beta/4}^{(k)} = \exp(-2\tilde{S}_k)\delta x_0^{(k)}. \quad (23.58)$$

Numerical results for the Eckart barrier Equation 23.30 obtained with approximate trajectories are shown in Figure 23.4 as a function of $k_B T$ (or β^{-1}). The top of the barrier is equivalent to 5000 K. Five thousand trajectories were propagated with the linear fitting of the imaginary-time p to get MDQP, the linear fitting of the nonclassical momentum, $A^{-1}\nabla A$ to obtain AQP, and linear representation of χ. Compared to the calculation of $N(E)$ in Section 23.2, we need far fewer trajectories at low energies because the real-time weights depend on the wavefunction decay in imaginary time Equation 23.58. QM results for the parabolic and Eckart barriers are obtained from Equation 23.5 using analytical expressions for the appropriate $N(E)$ and integrating from 0 to ∞ in energy. kQ shown on Figure 23.4 is the Laplace transform of $N(E)$ shown in Figure 23.1. The ranges of the axes are the same in both figures. We were able to compute kQ for a lower temperature than the corresponding energy resolved in the $N(E)$ calculation. The accuracy of kQ obtained from the approximate quantum trajectory dynamics is superior to that of $N(E)$.

23.5   SUMMARY

We have presented the imaginary-time formulation of the quantum trajectory dynamics, which, unlike the straightforward mapping of real-time Bohmian equations [36], is unique and nonsingular and, unlike BOMCA [38], unfolds in real phase space. The exponential form of the wavefunction Equation 23.33 leads to the trajectory formulation of the diffusion equation 23.31, where the dynamics of the trajectory ensemble unfolds in the presence of the MDQP, $U = \nabla p/(2m)$, added to the "inverted"

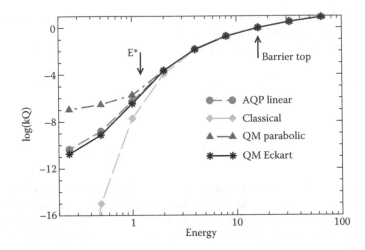

FIGURE 23.4 kQ for the Eckart barrier obtained in the mixed wavefunction formulation with imaginary and real-time AQP evolution. Comparison is made to the analytic QM expressions for the Eckart barrier, and to kQ computed for the parabolic barrier.

classical potential. The contribution of MDQP to average quantities is compensated by the change in the trajectory volume element—trajectory contributions depend on the "classical" action function computed along the quantum trajectory. For a Gaussian wavefunction, whose initial width parameter in our formulation defines the initial actions and momenta of the trajectories, MDQP is a time-dependent constant producing zero quantum force.

For a cheap global implementation—a single approximation works for the entire trajectory ensemble—MDQP is found from the least squares fit of the trajectory momenta in a small (two to six functions) polynomial basis. A nodeless wavefunction decays to the ground energy state. The same basis can be used to define polynomial prefactors or χ-functions to represent wavefunctions with nodes, which decay to the excited energy states provided that the lower energy contributions are projected out. For illustration four lowest energy eigenstates were computed for anharmonic single and double wells. The Taylor basis of six functions gave accurate (four significant figures) results for the single well. For the more challenging double-well problem, the accuracy was better than 1% for the light particle; for the heavy particle the accuracy was not as high (a few percent for $n = 2$ and 3, and under 1.5% for $n = 0$ and 1), but still adequate to resolve the $E_1 - E_0$ energy splitting, which is about 15% of the zero-point energy. Moreover, the two maxima of the ground state wavefunction emerged from the evolution of the initial Gaussian function, which is a major challenge for real-time quantum trajectories.

We have also shown how to compute reaction rate constants using the imaginary-time quantum trajectory dynamics followed by the AQP dynamics in real time. The mixed wavefunction representation is convenient for propagation of the flux operator eigenfunctions as shown by computing $N(E)$ and of the thermal flux operator eigenfunctions as shown by computing kQ for the Eckart barrier. Partitioning the wavefunction into a nodeless part represented by the quantum trajectories and a polynomial prefactor describing the nodal structure allows stable, more accurate trajectory propagation, and computation of the correlation functions in terms of trajectory weights rather than amplitudes (which would entail additional approximation and numerical cost). AQP dynamics led to overestimation of $N(E)$ below 10^{-5}, though it gave an improvement over the parabolic barrier probabilities. Computation of kQ was faster and more accurate than computation of $N(E)$, because low-tunneling probabilities were described through the decay of wavefunctions in imaginary time for the former, rather than by the long-time correlation of real-time wavepackets for the latter. Future work will include high-dimensional applications.

ACKNOWLEDGMENTS

This work was supported by the University of South Carolina.

BIBLIOGRAPHY

1. W. H. Miller, S. D. Schwartz, and J. W. Tromp. Quantum mechanical rate constants for bimolecular reactions. *J. Chem. Phys.*, 79:4889–4898, 1983.
2. J. C. Light and T. Carrington, Jr. Discrete variable representations and their utilization. *Adv. Chem. Phys.*, 114:263–310, 2000.
3. E. J. Heller. Time-dependent approach to semiclassical dynamics. *J. Chem. Phys.*, 62: 1544, 1975.
4. E. J. Heller. Frozen Gaussians: A very simple semiclassical approximation. *J. Chem. Phys.*, 75:2923, 1981.
5. R. C. Brown and E. J. Heller. Classical trajectory approach to photodissociation: The Wigner method. *J. Chem. Phys.*, 75:186–188, 1981.
6. M. F. Herman and E. Kluk. A semiclassical justification for the use of non-spreading wavepackets in dynamics calculations. *Chem. Phys.*, 91:27, 1984.
7. K. G. Kay. Integral expressions for the semiclassical time-dependent propagator. *J. Chem. Phys.*, 100:4377–4392, 1994.
8. J. V. Van Vleck. The correspondence principle in the statistical interpretation of quantum mechanics. *Proc. Nat. Acad. Sci. U.S.A.*, 14:178–188, 1928.
9. S. Garashchuk and V. A. Rassolov. Energy conserving approximations to the quantum potential: Dynamics with linearized quantum force. *J. Chem. Phys.*, 120:1181–1190, 2004.
10. S. Garashchuk and V. A. Rassolov. Semiclassical dynamics based on quantum trajectories. *Chem. Phys. Lett.*, 364:562–567, 2002.
11. A. Donoso, Y. J. Yeng, and C. C. Martens. Simulation of quantum processes using entangled trajectory molecular dynamics. *J. Chem. Phys.*, 119:5010–5020, 2003.
12. J. B. Maddox and E. R. Bittner. Estimating Bohm's quantum force using Bayesian statistics. *J. Chem. Phys.*, 119:6465–6474, 2003.
13. J. Liu and N. Makri. Monte Carlo Bohmian dynamics from trajectory stability properties. *J. Phys. Chem. A*, 108:5408–5416, 2004.
14. C. J. Trahan, K. Hughes, and R. E. Wyatt. A new method for wave packet dynamics: Derivative propagation along quantum trajectories. *J. Chem. Phys.*, 118:9911–9914, 2003.

15. S. Garashchuk and T. Vazhappilly. Wavepacket approach to the cumulative reaction probability within the flux operator formalism. *J. Chem. Phys.*, 131(16):164108, 2009.

16. S. Garashchuk and V. A. Rassolov. Modified quantum trajectory dynamics using a mixed wavefunction representation. *J. Chem. Phys.*, 121(18):8711–8715, 2004.

17. S. Garashchuk. Description of bound reactive dynamics within the approximate quantum trajectory framework. *J. Phys. Chem. A*, 113:4451–4456, 2009.

18. S. Garashchuk. Quantum trajectory dynamics in imaginary time with the momentum-dependent quantum potential. *J. Chem. Phys.*, 132, 2010.

19. T. J. Park and J. C. Light. Quantum flux operators and thermal rate-constant—collinear H+H$_2$. *J. Chem. Phys.*, 88:4897–4912, 1988.

20. D. H. Zhang and J. C. Light. Cumulative reaction probability via transition state wave packets. *J. Chem. Phys.*, 104(16):6184–6191, 1996.

21. J. C. Light and D. H. Zhang. The quantum transition state wavepacket method. *Faraday Discuss.*, 110:105–118, 1998.

22. T. Seideman and W. H. Miller. Transition-state theory, siegert eigenstates, and quantum-mechanical reaction-rates. *J. Chem. Phys.*, 95:1768–1780, 1991.

23. G. Arfken. *Mathematical Methods for Physicists*, 3rd edition. Academic Press, New York, 1985.

24. F. Matzkies and U. Manthe. Accurate reaction rate calculations including internal and rotational motion: A statistical multi-configurational time-dependent hartree approach. *J. Chem. Phys.*, 110:88–96, 1999.

25. E. Madelung. Quantum theory in hydrodynamic form. *Z. Phys.*, 40:322–326, 1926.

26. D. Bohm. A suggested interpretation of the quantum theory in terms of "hidden" variables, I and II. *Phys. Rev.*, 85:166–193, 1952.

27. S. Garashchuk and V. A. Rassolov. Quantum dynamics with Bohmian trajectories: Energy conserving approximation to the quantum potential. *Chem. Phys. Lett.*, 376:358–363, 2003.

28. S. Janecek and E. Krotscheck. A fast and simple program for solving local Schrödinger equations in two and three dimensions. *Comp. Phys. Comm.*, 178(11):835–842, 2008.

29. D. Blume, M. Lewerenz, P. Niyaz, and K. B. Whaley. Excited states by quantum Monte Carlo methods: Imaginary time evolution with projection operators. *Phys. Rev. E*, 55:3664–3675, 1997.

30. D. M. Ceperley and L. Mitas. Monte Carlo Methods in Quantum Chemistry. *Adv. Chem. Phys.*, 93, 1996.

31. W. A. Lester Jr., L. Mitas, and B. Hammond. Quantum monte carlo for atoms, molecules and solids. *Chem. Phys. Lett.*, 478(1–3):1–10, 2009.

32. A. Viel, M. D. Coutinho-Neto, and U. Manthe. The ground state tunneling splitting and the zero point energy of malonaldehyde: A quantum Monte Carlo determination. *J. Chem. Phys.*, 126:024308, 2007.

33. C. E. Hinkle and A. B. McCoy. Characterizing excited states of CH_5^+ with diffusion Monte Carlo. *J. Phys. Chem. A*, 112:2058–2064, 2008.

34. W. H. Miller. Classical path approximation for the Boltzmann density matrix. *J. Chem. Phys.*, 55:3146–3149, 1971.

35. N. Makri and W. H. Miller. Coherent state semiclassical initial value representation for the Boltzmann operator in thermal correlation functions. *J. Chem. Phys.*, 116:9207–9212, 2002.

36. J. Liu and N. Makri. Bohm's formulation in imaginary time: Estimation of energy eigenvalues. *Molec. Phys.*, 103:1083–1090, 2005.

37. Y. Goldfarb, I. Degani, and D. J. Tannor. Semiclassical approximation with zero velocity trajectories. *Chem. Phys.*, 338:106–112, 2007.
38. Y. Goldfarb, I. Degani, and D. J. Tannor. Bohmian mechanics with complex action: A new trajectory-based formulation of quantum mechanics. *J. Chem. Phys.*, 125(23):231103, 2006.
39. W. H. Press, B. P. Flannery, S. A. Teukolsky, and W. T. Vetterling. *Numerical Recipes: The Art of Scientific Computing*, 2nd edition. Cambridge University Press, Cambridge, 1992.
40. E. R. Bittner. Quantum initial value representations using approximate bohmian trajectories. *J. Chem. Phys.*, 119:1358–1364, 2003.
41. V. A. Rassolov and S. Garashchuk. Bohmian dynamics on subspaces using linearized quantum force. *J. Chem. Phys.*, 120:6815–6825, 2004.
42. S. Garashchuk and V. A. Rassolov. Stable long-time semiclassical description of zero-point energy in high-dimensional molecular systems. *J. Chem. Phys.*, 129:024109, 2008.
43. D. Blume, M. Lewerenz, and K. B. Whaley. Quantum Monte Carlo methods for rovibrational states of molecular systems. *J. Chem. Phys.*, 107:9067–9078, 1997.
44. S. Garashchuk. Calculation of reaction rate constants using approximate evolution of quantum trajectories in imaginary and real time. *Chem. Phys. Lett.*, 491:96–101, 2010.
45. E. Pollak. The symmetrized quantum thermal flux operator. *J. Chem. Phys.*, 107(1):64–69, 1997.

24 Dynamical Systems Approach to Bohmian Mechanics

F. Borondo

CONTENTS

24.1 INTRODUCTION

The predictive power of quantum mechanics (QM) plays a role of paramount importance in the understanding of matter and its fundamental laws [8]. Despite this great success, the standard formulation of QM, usually known as *the Copenhagen interpretation* [10], has always suffered from interpretational difficulties [7]. For this reason, David Bohm introduced in the 1950s an alternative version [3], which has experienced a revitalization in the past few years, supported by a new computationally oriented point of view. The so-called *Bohmian mechanics* (BM) constitutes a true theory of quantum dynamics [6, 18], and many interesting practical applications, including the analysis of the tunneling mechanism [9], scattering processes [12, 14], or the classical–quantum correspondence [13], just to name a few, have been revisited using this novel point of view.

Despite this recent development, it is surprising how little attention has been paid to put this theory on firm grounds. In particular, the nature of the trajectories on which BM is based is largely unknown, and this has led in the past to the publication of erroneous results [1]. Trajectories in BM are intrinsically chaotic [4] due to the effect of the vortices [17] induced in the associated velocity field by the nodal structure of the pilot wave [3]. Taking this into account, the aim of the present chapter is to examine BM from the point of view of *dynamical systems* [15], a field that provided, in the past, mathematical support for so-called *chaos theory*.

24.2   AN OVERVIEW OF NONLINEAR DYNAMICS

Dynamical systems (DS) theory is the branch of applied mathematics used to describe the behavior of complex DS. These dynamics can be discrete, and then given by a map: $\vec{x}_{n+1} = f(\vec{x}_n)$, or continuous in time, and then given by a flux specified by its velocity vector field: $\vec{v}(t) = (d\vec{x}/dt) = \vec{f}[\vec{x}(t)]$. Each state of the system is characterized by the *state vector*, $\vec{x}(t) = (x_1(t), x_2(t), \ldots)$, which is usually represented by a point in the so-called *phase space* (PS). The time evolution of this point gives the corresponding *trajectory*. Fluxes can be dissipative, and then the trajectories will approach asymptotically specific PS structures (fixed points [FPs], limit circles or tori, and strange attractors), or Hamiltonian, which in turn can be conservative or not. The best example to understand the difference is the one-dimensional (1D) pendulum. In a real pendulum, friction dissipates energy. This damps the oscillations, bringing the pendulum to a stop (attractor) after some time. However, in the ideal case where friction is negligible, the oscillations repeat themselves forever, and no attractors exist for its dynamics.

In what follows, for simplicity we will restrict the discussion to conservative Hamiltonian systems. In this case the dynamics follow the Hamilton equations of motion

$$\dot{x}_i = \partial H/\partial P_i; \quad \dot{P}_i = -\partial H/\partial x_i, \tag{24.1}$$

where $H(\vec{x}, \vec{P})$ is the Hamiltonian function giving the energy of the system, and $P_i = \partial L/\partial x_i$ are the conjugate momenta.

24.2.1   ONE-DIMENSIONAL HAMILTONIAN SYSTEMS

Let us start by considering the PS structure of a 1D pendulum. Although this is in principle a trivial system, from the pedagogical point of view it allows us to introduce most of the fundamental concepts of nonlinear dynamics in a very easy way. The Hamiltonian function is given by

$$H = \frac{P_x^2}{2m\ell} - mg\ell \cos x, \tag{24.2}$$

for which two types of motion are possible: (a) *libration* (or vibration), when $E < mg\ell$, and (b) *rotation* when $E > mg\ell$. In the first case P_x and x are periodic functions of t, while in the second only the momentum behaves in this way, with x increasing or decreasing continuously, according to the sign of P_x. The corresponding PS trajectories are ellipses and undulated functions, as plotted in Figure 24.1(a). The direction of the flux has also been indicated with arrows. Two equilibrium points at $x = 0$ (stable) and π (unstable) exist. In the jargon of nonlinear dynamics these points are referred to as FPs, since they do not move: *elliptical* (EFP) for the first one,

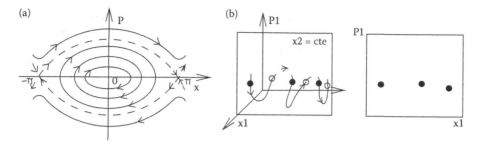

FIGURE 24.1 (a) Phase space portrait of the pendulum. (b) Construction of a Poincaré surface of section.

since in its neighborhood trajectories are ellipses, and *hyperbolic* (HFP) for the second, since the nearby motion describes hyperbolas. The "trajectory" separating these two motions ($E = mg\ell$) is called the *separatrix*, and has an infinitely long period. Moreover it has two branches, upper and lower, crossing at the HFPs, giving rise to four invariant manifolds, two of them *incoming* (according to the flux) and the others *outgoing*. Moreover, in this system each outgoing curve, when followed in time, goes into an incoming one, corresponding in the pendulum to a motion starting slightly off the upper equilibrium point, then moving away from it, and finally returning to the same point from the opposite direction.

24.2.2 Two-Dimensional Hamiltonian Systems

The PS in this kind of system is four dimensional. However, since energy is conserved the motion takes place in a 3D manifold, which is usually hard to visualize. This can be conveniently done by using the *Poincaré surface of section* (PSOS), which consists in representing the intersection of the trajectory with a chosen surface (remember that a plane has two surfaces). Figure 24.1b illustrates the method. In this, a unique (temporal) sequence of points,

$$(x_1(t_0), P_1(t_0)) \quad \xrightarrow{T} \quad (x_1(t_1), P_1(t_1))$$
$$(x_1(t_1), P_1(t_1)) \quad \xrightarrow{T} \quad (x_1(t_2), P_1(t_2)) \tag{24.3}$$
$$\cdots \qquad\quad \cdots \qquad\qquad \cdots ,$$

is obtained. The transformation T is simply a map, which for conservative systems belongs to the class of area-preserving maps. Usually, the results obtained from a large number of trajectories, all with the same energy, are represented in a single plot, called the *composite PSOS*.

By simple inspection of a PSOS, one can tell which is the dynamical regime, regular or chaotic, of the corresponding trajectory. Indeed, in the chaotic regime only energy is conserved, and then trajectories wander around all the (3D) accessible PS, crossing through (almost) all points in a finite 2D area. In this case, the (quasi)ergodic hypothesis is fulfilled, and the methods and conclusions of statistical mechanics apply.

At the other extreme, we have the case in which there are as many constant of motion as degrees of freedom. In our case, this implies that the motion takes place in a 2D manifold, which is dynamically invariant and has the topological structure of a torus. This result is not difficult to probe, but can be more easily illustrated by considering the case of two uncoupled harmonic oscillators:

$$H_0 = \frac{P_1^2}{2m_1} + \frac{P_2^2}{2m_2} + \frac{1}{2}m_1\omega_1 x_1^2 + \frac{1}{2}m_2\omega_2 x_2^2. \tag{24.4}$$

By performing a transformation to good action–angle variables [15], $(I_1, \theta_1, I_2, \theta_2)$,

$$x_i = \sqrt{\frac{2I_i}{m_i\omega_i}} \, \sin\theta_i, \quad P_i = \sqrt{2m_i\omega_i I_i} \, \cos\theta_i. \tag{24.5}$$

we find that the Hamiltonian is given by a very simple expression: $H = I_1\omega_1 + I_2\omega_2$, and so is the corresponding motion, since

$$
\begin{aligned}
\dot{\theta}_i &= \frac{\partial H}{\partial I_i} = \omega_i \quad \rightarrow \quad \theta_i = \theta_i(t_0) + \omega_i t, \\
\dot{I}_i &= -\frac{\partial H}{\partial \theta_i} = 0 \quad \rightarrow \quad I_i = \text{constant}.
\end{aligned}
\tag{24.6}
$$

These expressions indicate that each oscillator describes in its PS a circle (topologically equivalent to the ellipse) of radius I_i, with θ_i the associated polar angle. Accordingly, the total PS is a 2D torus, $\mathbb{T}^2$, the Cartesian product of the two circles, $\mathbb{T}^1 \otimes \mathbb{T}^1$.

The motion in $\mathbb{T}^2$ can be of two types depending on the relative value of the two frequencies: $\alpha = \omega_1/\omega_2$. If $\alpha = n/m \in \mathbb{Q}$, the frequencies are in resonance, and trajectories are mere 1D lines, closing themselves on $\mathbb{T}^2$, similarly to the Ouroboros [11], the mythical snake eating itself. The corresponding PSOS consists of n or m points, periodically visited. In this case the motion is regular and periodic. On the other hand, if $\alpha \notin \mathbb{Q}$ the trajectories never close, densely filling the surface of the torus. The PSOS consists of a closed line, and the motion is regular and quasiperiodic.

Generic Hamiltonian systems are neither ergodic nor completely integrable, and they show a mixed PS consisting of coexisting regular and chaotic regions, with islands of regularity embedded in the chaotic sea. This structure can be understood in terms of the celebrated KAM theorem, due to Kolmogorov, Arnold, and Moser [15], Roughly stated, it establishes that, in slightly perturbed systems, a finite measure of the original tori persists, which are those characterized by an "irrationally enough" frequency ratio. This condition is given for 2D systems of the form: $H = H_0 + \varepsilon H'$, where H_0 is integrable and H' is a generic perturbation, as

$$\left| \frac{\omega_1}{\omega_2} - \frac{n}{m} \right| > \frac{K(\varepsilon)}{m^{5/2}}, \tag{24.7}$$

where $K(\varepsilon)$ is a complex function of ε. These tori which survive according to the KAM condition (therefore called *KAM tori*) are only distorted by the perturbation.

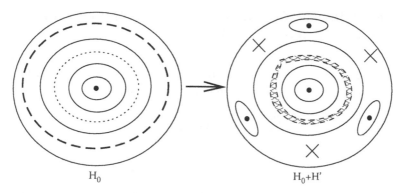

FIGURE 24.2 Schematic illustration of the tori destruction described by the KAM and Poincaré–Birkhoff theorems. Full, dashed and dotted lines represents very irrational, not so irrational and rational tori, respectively.

The remaining tori are destroyed, but through two different mechanisms. All resonant tori, i.e., $\alpha \in \mathbb{Q}$, are destroyed for any value of ε, since they do not fulfill condition Equation 24.7, and their fate is dictated by the Poincaré–Birkhoff theorem [2], which establishes that $2km$, $(k = 1, 2, \ldots)$ FPs on the tori remain, half of them are EFPs and the other half HFPs, alternating along the original invariant curve, as shown in Figure 24.2. The former are surrounded in turn by new smaller islands of stability, to which the above mechanism of tori destruction also applies *ad infinitum*. On the other hand, we have the hyperbolic points that are connected by the separatrix in H_0. This smooth curve gets very complicated when the perturbation is turned on ($\varepsilon \neq 0$) giving rise to the so-called *homoclinic tangle*, described by Poincaré in his pioneering work giving rise to chaos theory. This result is not unexpected since the separatrix represents the frontier between librations and rotations, two topologically completely different motions. Finally, tori characterized by $\alpha \notin \mathbb{Q}$ are destroyed, but give rise to fractal invariant sets called *cantori*. The nature of these objects in 2D systems can be easily understood in terms of their PSOS, which is analogous to the Cantor middle third set [5], which presents "holes" at all scales.

24.2.3 TIME-REVERSIBLE SYSTEMS

The so-called KAM scenario that we have just described in the previous subsections also applies to DS exhibiting time-reversibility. Although this fact is known in the mathematical literature [16], we believe that it has passed largely unnoticed by most physical researchers working in the field of DSs.

The time-reversible symmetry can be defined straightforwardly. A dynamical system, $\dot{\vec{x}} = X(\vec{x}, t)$, is time-reversible if there exists an involution, $\vec{x} = \Theta(s)$, that is a change of variables satisfying $\Theta^2 = \text{Id}$ and $\Theta \neq \text{Id}$, such that the new system results in $\dot{s} = D\Theta^{-1}(s)X(\Theta(s), t) = -X(s, t)$. One of the dynamical consequences of reversibility is that if $\vec{x}(t)$ is a solution, then so is $\Theta(\vec{x}(-t))$. This fact introduces symmetries in the system giving rise to relevant dynamical constraints.

24.3 BOHMIAN MECHANICS FROM THE POINT OF VIEW OF DYNAMICAL SYSTEMS

The BM formalism is based on the polar expression of the wavefunction: $\psi(\vec{x}, t) = R(\vec{x}, t) \exp[i S(\vec{x}, t)]$ ($\hbar$ is taken equal to one throughout the chapter), which after substitution in the time-dependent Schrödinger equation gives

$$\frac{\partial R^2}{\partial t} = -\nabla \cdot (R^2 \nabla S), \tag{24.8}$$

$$\frac{\partial S}{\partial t} = -\frac{(\nabla S)^2}{2} - V - \frac{1}{2} \frac{\nabla^2 R}{R}. \tag{24.9}$$

The second equation here is known as the "quantum" Hamilton–Jacobi equation, due to the extra "quantum" potential: $Q = \nabla^2 R/(2R)$. Its form allowed Bohm to derive, for spinless particles, a Newtonian equation of motion: $\ddot{\vec{x}} = -\nabla V(\vec{x}) - \nabla Q(\vec{x})$, from which "quantum trajectories" can be calculated. Alternatively, one can consider the associated velocity vector field

$$X_\psi = \nabla S = \frac{i}{2} \frac{\psi \nabla \psi^* - \psi^* \nabla \psi}{|\psi|^2}, \tag{24.10}$$

and obtain the trajectories from that.

24.3.1 MODEL

As an example, we choose to study the 2D isotropic uncoupled harmonic oscillator, whose Hamiltonian can be written, without loss of generality, as

$$\hat{H} = -\frac{1}{2} \left(\frac{\partial^2}{\partial x^2} + \frac{\partial^2}{\partial y^2} \right) + \frac{1}{2}(x^2 + y^2), \tag{24.11}$$

considering the particular wavefunction consisting of a mixture of the eigenstates $|00\rangle$, $|10\rangle$, and $|01\rangle$, with coefficients $\mathcal{A} = A + iD$, $\mathcal{B} = B + iE$, and $\mathcal{C} = C + iF$, respectively. Moreover, we assume the condition $BC \neq EF$ to ensure the existence of a single vortex in the velocity field, $(\dot{x}, \dot{y})$, at any time.

24.3.2 RESULTS

To examine the characteristics of the corresponding trajectories, we first integrate $(\dot{x}, \dot{y})$ with a 7/8-th order Runge–Kutta–Fehlberg method, and then monitor the associated dynamics by computing stroboscopic PSOSs. This is done by plotting the position of the trajectory, $(x(t), y(t))$, at times $t = 2\pi n$, for $n = 1, 2, \ldots$ and for several initial conditions. Some results are shown in Figure 24.3. (Do not pay attention at this point to the parameters defining the trajectories that will be defined later.) As can be seen, the results indicate that our system follows the behavior dictated by the KAM theorem. In particular, the left plot shows that we are in a completely integrable case, with all motions circumscribed to invariant tori. Moreover, we have three FPs,

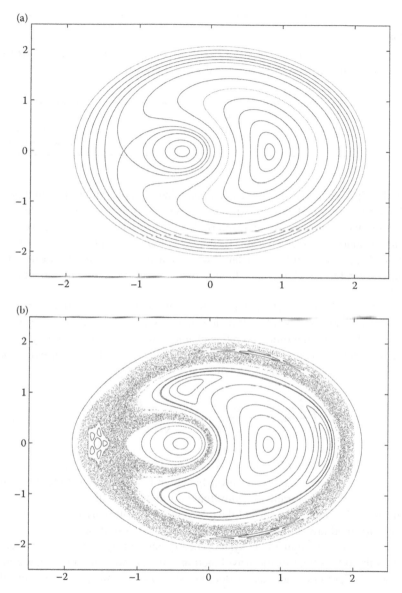

FIGURE 24.3 Poincaré surfaces of section for quantum trajectories generated from the velocity field Equation 24.12 with $a = A/\sqrt{2}B = 0.4$ and $b = A/\sqrt{2}C = 0.4$ (left) or 0.44 (right).

two stable and one unstable, clearly visible on the $y = 0$ axis, which correspond to resonant motions. Also, the four manifolds emanating from the HFP are connected with a separatrix defining three separate regions in our PS picture. The outer one is filled with circular shaped tori, the left one with elliptically shaped tori, and the third

TABLE 24.1

Dynamical Characteristics of the Quantum Trajectories Generated from the Different Possibilities in the Canonical Model Equation 24.12

	Integrable	Hamiltonian	Time Reversible	Stroboscopic Sections
$A = 0, B \neq 0, C \neq 0$	yes	no	yes	Ellipses around origin
$A \neq 0, B = C$	yes	yes	yes	Left panel in Figure 24.3
$A \neq 0, B \neq C$	no	no	yes	Right panel in Figure 24.3

one with kidney shaped tori. On the other hand, the right plot shows that in this case a band of chaotic motion has grown around the separatrix as a result of the destruction of the surrounding tori. Also as a result of this process a conspicuous 3:1 resonance, with a secondary 5:1 one, have survived, in accordance with the Poincaré–Birkhoff theorem. These are identified as islands of regularity near the border of the picture. The interaction has also broken the kidney shaped torus corresponding to another 3:1 resonance.

At first sight the KAM character exhibited by the behavior of our system is not obvious, since the corresponding flux does not look Hamiltonian or time-reversible. However, it has a hidden time-reversible character that can be unveiled by a suitable orthogonal transformation and time shift (see the details in Ref. [4]), which brings the pilot wavefunction into the following canonical form

$$\psi = \left(\frac{\hat{A}e^{-it}}{\sqrt{\pi}} + \frac{2x\hat{B}e^{-2it}}{\sqrt{2\pi}} + \frac{2yi\hat{C}e^{-2it}}{\sqrt{2\pi}} \right) e^{-\frac{1}{2}(x^2+y^2)}, \qquad (24.12)$$

where $\hat{A} = A$, $\hat{B} = B + C$, and $\hat{C} = B - C$. Now, depending on the values of these parameters, our system can be rigourously shown to behave in different dynamical ways, ranging from integrable to Hamiltonian or time-reversible.

The dynamical characteristics of the different possibilities arising from our canonical velocity field are summarized in Table 24.1, which represents a true road map to navigate across the dynamical system, i.e., quantum trajectories, that are defined based on the pilot effect of the wavefunction. Finally, note that the generic model, i.e., when E, F, or G do not vanish, does not satisfy any of the properties considered in the table.

ACKNOWLEDGMENTS

Support from MICINN–Spain under contracts No. MTM2006-15533, No. MTM2009–14621, and i-MATH CSD2006-32, Comunidad de Madrid under contract SIMUMAT S-0505/ESP-0158, and CEAL (UAM–Banco de Santander) is gratefully acknowledged.

BIBLIOGRAPHY

1. Alcantara, O. F., Florencio, J., and Barreto F.C.S. 1999. Chaotic dynamics in billiards using Bohm's quantum mechanics. *Phys. Rev. E* 58:2693–2695.
2. Berry, M.V. 1978. Regular and irregular motion. *AIP Conf. Proc.* 46:16–120.
3. Bohm, D. 1952. A suggested interpretation of the quantum theory in terms of "hidden" variables. I and II. *Phys. Rev.* 85:166–179 and 180–193.
4. Borondo, F., Luque, A., Villanueva, J., and Wisniacki, D.A. 2009. A dynamical systems approach to Bohmian trajectories in a 2D harmonic oscillator. *J. Phys. A* 42:495103(14pp).
5. Cantor Set, Wikipedia, The Free Encyclopedia, http://en.wikipedia.org/wiki/Cantor_set (as at November 2009).
6. Holland, P. 1995. *The Quantum Theory of Motion.* Cambridge: Cambridge University Press.
7. Jammer, M. 1989. *The Conceptual Development of Quantum Mechanics.* Woodbury, N.Y.: American Institute of Physics.
8. Lam, K.S. 2009 *Non-Relativistic Quantum Theory: Dynamics, Symmetry, and Geometry.* Singapore: World Scientific.
9. Lopreore, C.L. and Wyatt, R.E. 1999. Quantum wave packet dynamics with trajectories. *Phys. Rev. Lett.* 82:5190–5193.
10. Von Neuman, J. 1955. *Mathematical Foundations of Quantum Mechanics.* Princeton: Princeton University Press.
11. Ouroboros, Wikipedia, The Free Encyclopedia, http://en.wikipedia.org/wiki/Ouroboros (as at November 2009).
12. Philippidis, C., Dewdney, C., and Hiley, B.J. 1979. Quantum interference and the quantum potential. *Nuovo Cim. B* 52:1528.
13. Sanz, A.S., Borondo, F., and Miret-Artes, S. 2001. On the classical limit in atom-surface diffraction. *Europhys. Lett.* 55:303–309.
14. Sanz, A.S., Borondo, F., and Miret-Artes, S. 2002. Particle diffraction studied using quantum trajectories. *J. Phys.: Condens. Matter* 14:6109–6145.
15. Schuster, H.G. 1988. *Deterministic Chaos.* Weinheim: VCH.
16. Sevryuk, M.B. 2008. *Reversible Systems.* Berlin: Springer–Verlag.
17. Wisniacki, D.A., Pujals, E.R., and Borondo, F. 2006. Vortex interaction, chaos and quantum probabilities. *Europhys. Lett.* 73:671–676.
18. Wyatt, R.E. 2005. *Quantum Dynamics with Trajectories. Introduction to Quantum Hydrodynamics.* New York: Springer-Verlag.

Index

Printed and bound by CPI Group (UK) Ltd, Croydon, CR0 4YY

18/10/2024

01776264-0016